FUNDAMENTALS OF ELECTRIC POWER ENGINEERING

IEEE Press
445 Hoes Lane
Piscataway, NJ 08854

FUNDAMENTALS OF ELECTRIC POWER ENGINEERING

From Electromagnetics to Power Systems

MASSIMO CERAOLO

DAVIDE POLI

IEEE

IEEE PRESS

WILEY

Library of Congress Cataloging-in-Publication Data:

Ceraolo, Massimo.
Fundamentals of electric power engineering : engineering: from electromagnetics to power systems / Massimo Ceraolo, Davide Poli.
pages cm
Summary: "Covers topics such as: circuits, electrical machines and drives, power electronics, power system basics. new generation technologies"– Provided by publisher.
Includes bibliographical references and index.
ISBN 978-1-118-67969-2 (hardback)
1. Electric power systems. I. Poli, Davide. II. Title.
TK1001.C444 2014
621.31–dc23

2013044541

10 9 8 7 6 5 4 3 2 1

Everything should be made as simple as possible, but not simpler.

Albert Einstein

CONTENTS

PREFACE

Electrical engineering is a field of engineering that in general deals with the study and application of electricity, electronics, and electromagnetism. Depending on how it is intended in different areas of the world, it may cover a wide range of subfields, including electronics, digital computers, power engineering, telecommunications, control systems, and signal processing.

This broad range of fields of interest can be split into two main areas:

- what we could call *signal-oriented* electrical engineering, for which electric quantities (voltage, current, etc.) are used to carry signals—for example, inside TV sets or computers or through electromagnetic waves;
- what we could call *power-oriented* electrical engineering, for which electric quantities are used to manage and transfer power—for example, in power lines, electric machines, rectifiers, or inverters.

Indeed, from a practical point of view, these two kinds of electrical engineering are very different from each other. The approach to analysing applications is different; the physical objects that fulfil different functions tend also to be very different. For instance, while the dimensions of signal-oriented electric devices have shrunk by several orders of magnitude in the last decades (think of computers or mobile phones, for instance), the dimensions of apparatuses tend to be much more stable whenever large powers are involved (e.g., the size of a 100 kW electric motor or a 100 MW power station has not changed significantly).

Electric power engineering is a branch of industrial engineering, while signal electrical engineering is generally not considered as such.

University students of nonelectrical engineering, such as students of civil, mechanical, aerospace, chemical, or even control engineering, will be interested in power-oriented electrical engineering. Mechanical or aerospace engineers, for instance, should be able to understand in detail how an electric motor or an electric drive works. Civil engineers might need to understand how electrical installations are built and how the external power system feeding them operates. Basic information about the safety aspects of electricity might interest all of them.

This book aims to give university teachers support to teach nonelectrical engineering students all they need to know about basic electric phenomena, circuits, and electric machines and drives, as well as the basics of electric safety and an introduction to how large power systems are built and operated.

It might also be useful for professional engineers who want to have a source of updated, though concise, information of nearly everything that happens in power-oriented electrical engineering.

We have made every effort to explain each subject in the simplest way. However, in the case of more complex concepts, we have tried, rather than feigning simplicity, to illustrate them as clearly as possible. Taking our cue from Albert Einstein, we have adopted the approach of *making everything as simple as possible, but not simpler*.

This book tends to say things in a concise way. This is because we believe that this way it offers a stimulus to students: were we to describe each path of reasoning in detail, we would not stimulate the student's independent flow of reasoning.

In Chapters 3 and 4, special attention has been devoted to ensuring that a clear distinction is made between physical systems and their mathematical models. This is important, not only for the study of the topics in this book, but also as a lesson to students on how engineering proceeds: first we model reality, then we analyse the models. If the results of our analysis are not satisfactory, this might depend on the model we originally chose, and perhaps a more accurate version must be selected and used.

A few final comments:

- We have used graphic symbols, mathematical formulations, and even wording in compliance with the standards of the International Electrotechnical Commission. Details of this are in Chapter 1.
- We have tried to ensure that the exercises offer not only the chance to consolidate theoretical knowledge but also an opportunity for further learning. The reader is first led by hand through worked examples embedded in the text. At the end of Chapters 4–7 and 10–12 there are several exercises of increasing complexity and a solution is outlined for the least straightforward of these.

- At the end of some chapters we have added short biographies of the scientists who have given their names to the laws described in the chapter. Although very concise, these biographies offer some idea of the period and background of each scientist's discoveries. Whenever available, the correct pronunciation of the scientists' names is given, based on the International Phonetic Association Alphabet.

READERSHIP AND PREREQUISITES

This book is aimed at all students of engineering, with the exception of electrical engineering students (who require additional details of each topic). The book should also be very useful for nonelectrical qualified engineers, who may not have retained good support material from their student years or who may need to brush up on their knowledge of the fundamentals of electrical engineering without resorting to specialist books.

The most important prerequisite in order to take full advantage of the book is some knowledge of electromagnetics. The best results will be obtained easily by students who have already attended a university-level electromagnetics course. However, we have made every effort to make the book accessible to students who have only basic knowledge of electromagnetic phenomena from their secondary school studies. Chapter 2 will help these students to refresh their memory and to become acquainted with the symbols and approach used in the book.

LEVELS OF READING

This book caters to different levels of study. The most important example concerns electronic power conversion and Chapter 8. Electronic power conversion is widely used in electric drives and power systems. However, to fully understand it requires time. Although Chapter 8 contains a fairly lengthy description of how power electronics components are composed and operate, this chapter is not essential to an understanding of Chapters 10–12 (which deal with electric drives in addition to electric machines). Chapter 15 also deals with electronic converters in power systems, but, again, students are not required to have first studied Chapter 8. Needless to say, if there is time available in a university course to include Chapter 8, it follows that Chapters 10–12 and 15 can then be studied at a higher level of understanding.

The book also has some "more in depth" boxes. These contain extra information that is not essential for a clear understanding of the rest of the chapter to which they belong. "More in depth" boxes are included to trigger the curiosity of the reader, who can decide whether or not to read or study them.

ACKNOWLEDGMENTS

We would like to thank all our friends who have contributed to this project. In particular, we thank Luca Sani for his careful revision of the chapters on electric machines and drives.

Massimo Ceraolo
Davide Poli

ABOUT THE AUTHORS

Massimo Ceraolo (b. 1960)

He received his MSc degree, with honors, in Electrical Engineering from the University of Pisa in 1985. He is a Full Professor of Electric Power Systems since 2002. At present, he teaches Networks, Components and Electric Systems and Electric and Hybrid Vehicles at the University of Pisa.

Davide Poli (b. 1972)

He received his MSc degree, with honors, and his PhD in Electrical Engineering from the University of Pisa, respectively, in 1997 and 2001. He is an Assistant Professor of Electric Power Systems since 2001. At present, he teaches Power Quality and Power System Reliability at the University of Pisa.

PART I

PRELIMINARY MATERIAL

Fundamentals of Electric Power Engineering: From Electromagnetics to Power Systems, First Edition.
Massimo Ceraolo and Davide Poli.

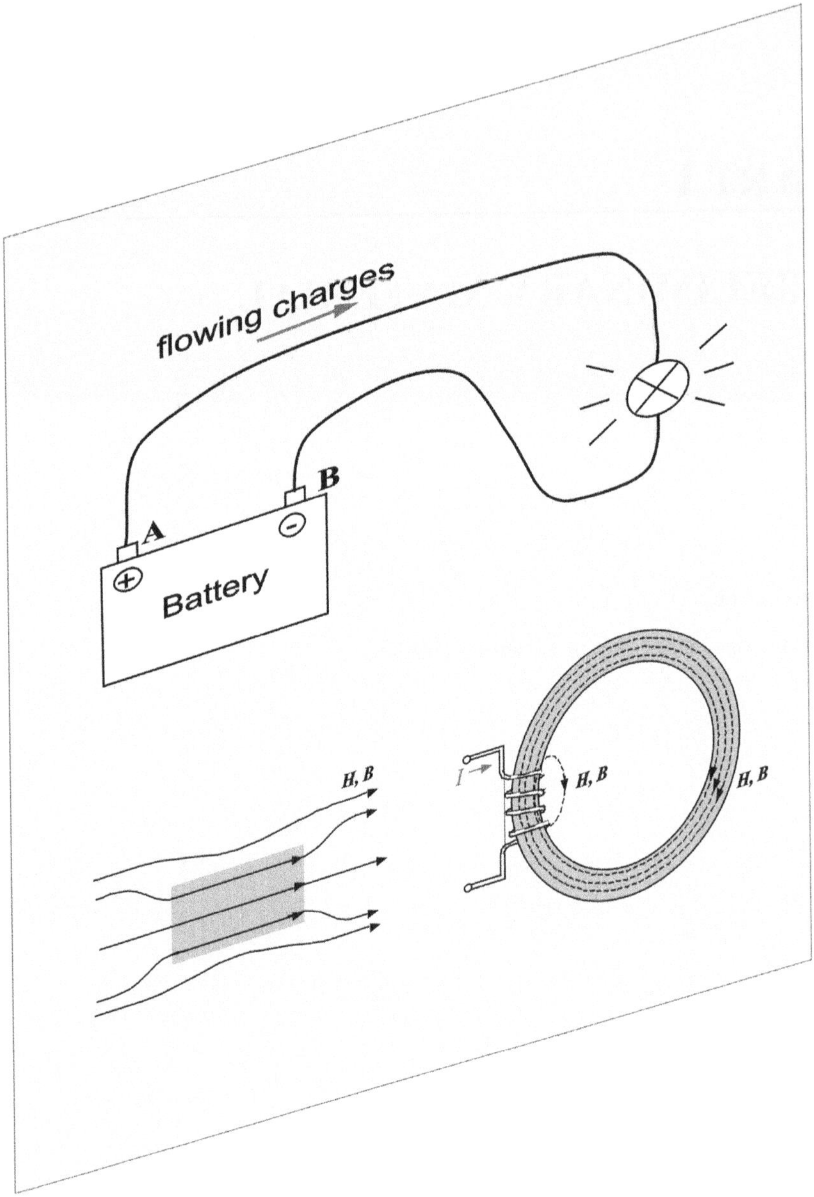
flowing charges
A
B
Battery
H, B
I
H, B
H, B

1

INTRODUCTION

1.1 THE SCOPE OF ELECTRICAL ENGINEERING

It is universally agreed that Electrical Engineering is a branch of engineering that deals with the phenomena of electricity. Apart from this vague statement, however, there is no worldwide agreement on the actual scope of an electrical engineer. In particular, there are two main approaches:

- In some cases, electrical engineering is considered as encompassing those disciplines in which electrical quantities (voltages, currents) are used to transfer *signals* (e.g., in computers, radio and TV sets, etc.) and those in which electrical quantities are used to manage and transfer *energy* and *power* (electrical machines and lines, electrical household installations, etc.). This approach is, for instance, normally followed in North America.
- In other cases, electrical engineering is considered to be involved only when electrical quantities are used to *transfer* and *convert* energy and power. This approach is usually followed in Europe. This kind of electrical engineering is often called *electric power engineering*.

This book follows this second approach, hence its title. Generally speaking, the whole scope of Electric Power Engineering comprises everything needed to manage electric energy, from its generation to its final utilization.

Fundamentals of Electric Power Engineering: From Electromagnetics to Power Systems, First Edition.
Massimo Ceraolo and Davide Poli.

The word "generation" might be a bit misleading since energy cannot, indeed, be generated: the term means production of electricity by conversion from other forms of energy. For instance, the electric alternators of large oil or gas power plants "generate" electricity by conversion from mechanical energy, in turn obtained by using other machines, like steam or gas turbines. Photovoltaic panels are another example of electricity generators: they produce electric energy through the conversion of solar radiation.

The final utilization of electricity very often involves another conversion; for example, the final energy form can be heat (in heaters or ovens) or mechanical energy (in industrial electric motors, in electric cars, etc.). There are cases, however, in which electric energy is used as such; the most significant example is for supplying computers or other electronic apparatuses.

Between generation and final utilization, electric energy can be transformed several times (for instance in power transformers, which raise the voltage while lowering the current and vice versa), and transferred for distances of up to hundreds or thousands of kilometres, by means of power lines.

Indeed, it will be seen in this book, especially in Part IV, that the more power to be transmitted, the higher the required voltage level. Therefore, the power system has low voltage (LV) parts (for instance, power in homes and offices is always LV), medium voltage (MV) parts (the alternators of large power plants generate power at MV level), and high voltage (HV) and extra high voltage (EHV) levels.

All these apparatuses, which convert or transfer energy, are therefore parts of a great system, one of the largest that mankind has ever built, that encompasses the generation, transformation, transmission, distribution, and utilization of electric energy and is called a *power system*. All this can be visualized in the diagram in Figure 1.1, which shows the main functions of a power system along with the different energy forms involved.

A typical situation includes electricity generation in power plants, transformation toward high voltage in transformers, transmission toward load centres, transformation into medium or low voltage, distribution to single loads, and conversion to final usage. In the figure, the term "Bulk Generation" refers to large-scale centralized facilities, which inject their production into the transmission grid. "Distributed Generation" (DG) is instead composed of a large number of small-scale power plants, installed

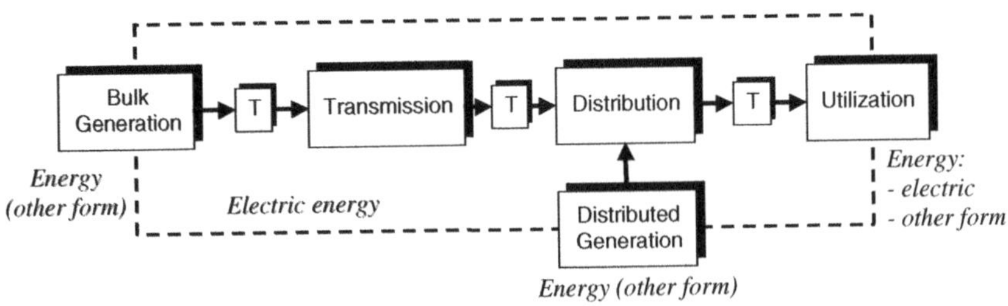

FIGURE 1.1. General structure of a (full) electric power system (T blocks indicate transformation made by power transformers).

close to the final users and directly connected to a distribution network. In the same figure, "T" represents the transformation performed by power transformers; other forms of electric-to-electric transformation are usually made inside the biggest blocks; for instance, utilization could involve a combined rectifier/inverter pair that allows maximum flexibility to the electric motor speed variation. The DG can be provided with power transformers, depending on the voltage of the generator and of the receiving network.

The outline of the electric power system shown in Figure 1.1 is much simplified, in line with the aims of this chapter. A more detailed and accurate description of the electric power system will be given in Part IV of this book.

When a *power system* is mentioned, what is usually intended is one of the very large networks that link power plants (large or small) to loads, by means of an electric grid that may be as large as a continent, such as the whole of Europe or North America. A power system, in this sense, extends from a very large power plant (e.g., having thousands of MW of generated power) right up to either the lamp that might now be lit on your table or the sockets giving electricity to loads from the nearest wall! Smaller power systems could be made of sections of a larger system. Examples are shown in Figure 1.2.

Figure 1.2a contains several components (breaker, cable, motor), which operate together and are connected to a feeding network. The subsystem represented in

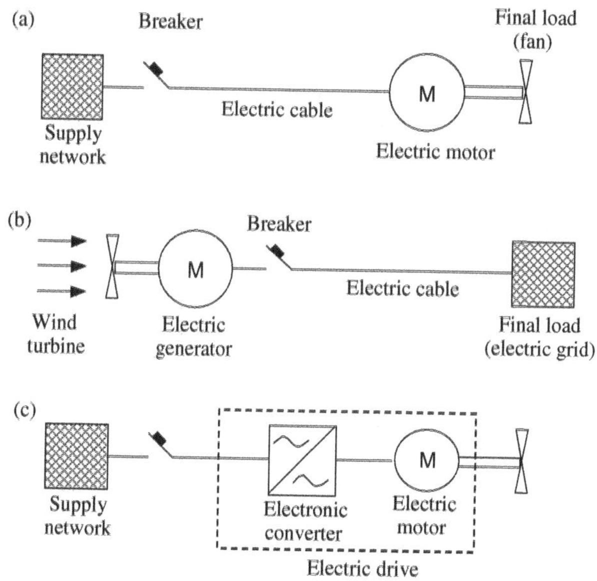

FIGURE 1.2. An electric power system fed by a supply network: a partial electric power system.

Figure 1.2a could be one of the final users of the electric energy in the utilization block shown in Figure 1.1.

Figure 1.2b contains many of the same components as Figure 1.2a, but its purpose is totally different. Instead of the fan, we have a wind turbine, which has some similarities to a large fan, but with the power flow reversed: it receives power from an air flow to produce mechanical energy, while a fan uses mechanical energy to obtain an air flow. The subsystem represented in Figure 1.2c could be one of the small power plants contained in the block of distributed generation shown in Figure 1.1.

Finally, Figure 1.2c contains a variation of the system in Figure 1.2a. The presence of the electronic converter allows much greater flexibility in the use of the electric motor and, in particular, allows variable speed operation of the fan. The electronic converter modifies electrical quantities, thus transforming *electricity into electricity*, differently from motors and generators that convert, respectively, mechanical energy into electricity and vice versa. Electronic converters tend to be increasingly present in power systems, even though they are not in evidence in the simplified diagram in Figure 1.1.

A very large number of power systems like the ones shown in Figure 1.2 operate only when connected to the mains—for example, a feeding network.

A power system such as that shown in Figure 1.1 is called a *full power system*, since its operation does not require feeding points from other electricity sources and the produced electricity is supplied to loads.

Power systems that are fed instead by an external electricity source or that produce (by conversion from other sources) electricity and convey it to a larger grid are called *partial power systems*.

There are also full power systems that are much smaller than the large power systems (such as those of Europe or North America) discussed earlier, but still, on a smaller scale, perform the basic functions of generation, distribution, and utilization of energy. An example is the small system created to feed a building yard, along with the cables and loads. Another example is the electric system on board electric cars: battery, inverter, motor, and accessory parts.

It should be stressed that a power system is basically composed of power lines and apparatuses that convert energy (*energy converters*). Power lines are relatively simple in their inner structure and do not need a great deal of explanation, especially in an introductory book.

The energy converters that are of interest to electric power engineering can be divided into two categories:

- Apparatuses for converting electricity into other forms of energy and vice versa. With reference to Figure 1.1, these are usually at the source of the system ("Bulk Generation" block), where electric energy is produced through conversion from other forms of energy and at its end ("Utilization" block), where electricity, when not used as such, is converted into other forms. Of great importance are the apparatuses that convert electricity into mechanical energy and vice versa—that is, those used for *electromechanical-energy conversion* (electromechanical converters).

- Conversion from electrical energy into electrical energy with different characteristics—that is, *electric-to-electric energy conversion*. This kind of conversion is carried out by power transformers (like those shown in Figure 1.1), but also in other situations. For instance, electricity can be converted from alternating current (AC) into direct current (DC) (using rectifiers) or from DC into AC (using inverters), and so on. This kind of conversion, not explicitly shown in the simplified diagram of Figure 1.1, is becoming increasingly frequent in power systems; each of the larger blocks in Figure 1.1 can contain electric-to-electric conversion apparatuses. For instance, an electric motor is often fed by an inverter, to form a system called *electric drive*.

1.2 THIS BOOK'S SCOPE AND ORGANIZATION

Nonelectrical engineers do not need to know the details of electric power systems; however, they need to master its basic functions in order to be able to exploit their application and to effectively collaborate with electrical engineers in more complex cases.

Since this book is intended for use in courses of one or two semesters, the authors have had to make important decisions on how deeply each topic should be dealt with. Our final decision was to focus on showing (a) how a physical system can be modelled using circuits and (b) how circuits can be analysed. Once readers have gained the ability to "solve" circuits—that is, to numerically compute currents, voltages, and power—they will have gained sufficient knowledge of the phenomena in any electric device; at that point, the way has been paved for learning more about electric machines, drives, and power systems.

To pursue its objectives, the book has been divided into the following four parts:

Part I: Preliminary Material. This part contains two very different chapters, both of which are introductory to the book's core material. Chapter 1 includes miscellaneous topics such as a discussion of the very meaning of electrical (or electric power) engineering, as well as an overview of the scope and organization of the book. Chapter 2, on the other hand, creates a bridge between the core material of this book and the student's previous knowledge. It is organized into two levels, and students can select the one most appropriate to their previous knowledge of electromagnetism.

Part II: Electric circuit Concept and Analysis. The main purpose of this part, as mentioned previously, is to show readers how to handle electric circuits. For this, we have adopted an innovative approach: readers will learn that circuits are mathematical/graphic tools to model physical systems operating with electric quantities. We will show that, because they are models, the results we obtain from mathematically solving circuits are accurate only to the extent to which they correctly model physical systems. We will also explain that they are zero-dimensional models, while actual systems are distributed-parameter, that is, three-dimensional. This explanation is useful not only for building a sound base of electrical engineering knowledge, but also as a significant example of how engineering is practiced in all of its fields.

Part III: Electric Machines and Power Converters. We saw in the previous paragraph that electric power systems contain several apparatuses to convert energy. These are the main subjects of Part II of this book.

This part combines three disciplines that are traditionally distinct: *electric machines* (machines for electromechanical conversions, plus the power transformer), *power electronics* (dealing with electric-to-electric conversion, different from that of power transformers), and *electric drives*.

The aim is to interpret the modern world, where these disciplines are strictly related to each other, and to present information in the form best suited to readers of this book, whether or not they are electric engineering students.

Part IV: Power Systems Basics. The description of power systems in Section 1.1 is very concise. Depending on how this book is used, more detailed information about the whole electric power system may be required. This is given in Part IV of the book, which contains (a) a description of the structure and operation of the system and (b) basic information about the risks of electricity for livestock and about how to prevent accidents.

Since this book is intended for courses of one or two semesters, some parts have been written in such a way that they can be omitted for shorter programmes.

Each chapter of the book starts with a "For the Instructor" box explaining the approach to be followed, along with (whenever possible) advice as to what can be safely omitted in shorter courses.

Examples of one-semester courses that can be taught using this book are:

- A course subsequent to a circuit course. In this case, Part II can be omitted and the material for the course can be drawn from Parts III and IV.
- A one-semester course on the fundamentals of electric power engineering, as the only electrical engineering course in a programme. In this case, Chapters 8, 14, and 15 can be totally omitted. If further reductions are necessary, Chapter 13 can also be omitted.

1.3 INTERNATIONAL STANDARDS AND THEIR USAGE IN THIS BOOK

1.3.1 International Standardization Bodies

Since it is written in the third millennium, this must be a *global* book. It is therefore intended for any reader from anywhere in the world. This means that the graphics and conventions for drawings and writing equations must be independent, as far as possible, of individual country preferences.

To ease the reciprocal exchange of information (and objects), common standards have been set by international organisations, in particular:

- The *Bureau International des Poids et Mesures* (BIPM), whose task is ("*to ensure world-wide uniformity of measurements and their traceability to the International System of Units (SI).*"[1] It thus provides indications on how to numerically evaluate and indicate measurements of different quantities.
- The *International Electrotechnical Commission* (IEC), which is "*the world's leading organization that prepares and publishes International Standards for all electrical, electronic and related technologies.*"[2] The IEC is one of three global sister organizations (IEC, ISO, ITU) that develop International Standards for global use.
- The *IEEE standards Association* (IEEE-SA), which "*brings together a broad range of individuals and organizations from a wide range of technical and geographic points of origin to facilitate standards development and standards related collaboration.*"[3]

Of these, the most important organization for the purposes of this book is the IEC; however, some basic information from the SI can be found in the publication [s2], which provides a good interpretation of the BIPM documents; some ISO standards, such as [s9], might also be of interest.

There are some fields of Electrical Engineering in which IEEE standards are an acknowledged important international reference; in these cases, reference is also made to IEEE standards (such as [s5] for harmonics control, quoted in Chapter 8).

A detailed presentation of standards is far beyond the scope of this book, but a systematic adoption of all the agreed standards (with some minor deviations) will help the reader to become accustomed to them and to remember them for many years to come.

1.3.2 The International System of Units (SI)

The International System of Units, SI[4] for short, defines the units of measure to be used all over the world, for measuring the different quantities used in any field of science or technology, from physics to engineering.

However, there are many situations in which people do not comply with the SI and use other units of measure. For instance, aeroplane altitudes are commonly indicated in feet, ship speeds in knots, and engine powers in HP. This is totally unjustified in the majority of cases. Dealing with thousands of different units of measure traditionally used in different countries, and with the corresponding conversion factors, only adds

[1]Text drawn from [s1].
[2]Text drawn from [s3].
[3]Text drawn from [s4].
[4]This international abbreviation comes from the French version of the name: Le Système International d'Unités.

TABLE 1.1. The Five Base Quantities and Units Considered in this Book

Quantity	Unit Name	Unit Symbol	Preferred Symbols for the Quantity
Length	metre[a]	m	l, s, r
Mass	kilogram	kg	m
Time	second	s	t
Current	ampere	A	I, i
Thermodynamic temperature	kelvin	K	T

[a]The American spelling "meter" is also acceptable. In this book, whenever there are differences, the British spelling is always chosen, as defined in the *Oxford English Dictionary*.

undue effort and uncertainty to the work of technicians as well as ordinary people. The use of non-SI units, except in very limited cases for which specific justifications may exist, is even more questionable in books addressed to the younger generations since this could cause the perpetuation of these errors, thus slowing, if not jeopardising, the whole process of universal dissemination of the SI and the benefits it can bring.

In this book, therefore, the SI units of measure are always used, with virtually no exceptions. Only units considered by the SI itself to be "non-SI units accepted for use with the SI" are used, such as minutes (min), hours (h), and days (d), because of their widespread use in everyday life.[5]

Students are strongly advised to use SI units as much as possible. Once accustomed to them, they will find it natural to use them always. This way, one day in the future, the entire population (or most of it) will use a single unit for a single quantity, which will make life easier for everyone.

The SI defines seven *base units*, which by convention are independent, as well as many other derived units, one for each quantity, expressed in relation to the base units.

In this book, only quantities in relation to a subsystem of five of the SI base units will be used; the base units of interest of this book will thus only be those shown in Table 1.1. Please note the style of writing in accordance with international standards, applied both to base and derived units:

- The initial letter of a unit is in lowercase and no accents are used (see the example of the unit for electric currents).
- The symbol of a unit *must* always be written either uppercase or lowercase as given. For example, kilogram has the symbol kg (not Kg), ampere A, and so on. Symbols may be composed of more than a single letter (such as Pa for pascal, the standard unit of pressure).[6] Finally, symbols must be written in roman

[5]For a complete list of such SI-accepted, non-SI units and notes on their usage, see [s2].

[6]The ISO standard provides a strict rule for the case of units: when the unit is represented by the initial of a scientist's name, it must be uppercase: V for volt, A for ampere, and so on; otherwise it should be lowercase: kg for kilogram, lx for lux, and so on. The only exception is the liter, for which the use of "L" instead of "l" is accepted, to avoid confusion with the number 1 (one).

(not italic) type, regardless of the type used in the surrounding text, and must not be followed by a dot, unless at the end of the sentence (e.g., "a current of 2 A is generated", *and not* "a current of 2 A. is generated").

As far as temperature is concerned, thermodynamic temperature is mentioned in Table 1.1 because in the SI it is used to define the Celsius temperature, using the very well-known equation

$$t = T - T_0$$

where t and T are the same temperature, measured in degrees Celsius and kelvin, respectively, while the reference temperature T_0 equals 273.15 K.[7]

When the numerical value of a quantity is given, its unit of measure must be shown alongside its numerical value: $I = 2$ A means that I is twice the value of the SI standard current, the ampere. What is not widely known is that the same expression can also be written $I/\mathrm{A} = 2$. As surprising as this might be, it is very rational: the numerical value of any quantity is always the ratio of the quantity to the reference value (in this case the SI base quantity).

This way of expressing units of measure of quantities is recommended also when the unit of measure refers to several numerical values, such as in tables or plots. This is visually expressed in the following tables, in which the recommended way is compared to another common way of expressing units of measure in table or plot headings—that is, within square brackets:

RECOMMENDED VERSION

Object	T/K
one	216
two	218
three	222

NONRECOMMENDED VERSION

Object	T[K]
one	216
two	218
three	222

1.3.3 Graphic Symbols for Circuit Drawings

The circuit drawings in this book are written according to the latest international standards. A summary of the symbols used is contained in the following table.

[7] Here there is a clear exception to the general rule requiring the units of measure to be lowercase: the addition of word "degree" changes the word "Celsius" from a unit name to the name of the scientist, thus requiring an initial capital.

The following symbols are a selection from International ISO/IEC/IEEE standards ([s6], [s9] and [s10]). In rare cases some small deviation from the standard is used, and the reason for this choice is to be found in the *notes* column.

Component	Symbol	Notes
Voltage source	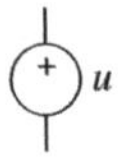	(1) the vertical line represents the ideal wire on which the source is applied. (2) the "+" sign indicates the polarity of voltage u when $u > 0$; for greater clarity, an optional "–" may be added opposite "+".
Sinusoidal voltage source		Voltage source symbol can also be used.
Current source		(1) The vertical line represents the ideal wire on which the source is applied. (2) The arrow sign indicates the direction of current i when $i > 0$.
Resistor		(1) The aspect ratio should be 3:1. (2) In this book, especially throughout Chapter 2, rectangles also model generic branches, but they will have a different aspect ratio (see next row).
Passive element (with impedance)		In AC circuits the resistor symbol is commonly used to represent a generic *passive element* (with impedance)—for example an *R–L* (resistor-inductor) or an *R–L–C* (resistor-inductor-capacitor) series.
Generic branch		Generic branch, which can be a resistor, an inductor, a source, or any other component or combination of components. Aspect ratio: 2:1 or less.
Inductor		The aspect ratio should be 4:1.
Capacitor		
Transformer—form 1	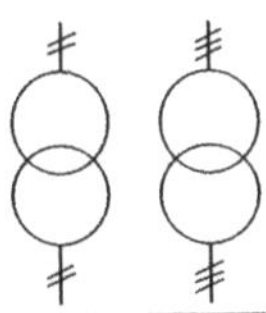	The two symbols refer to single-phase and three-phase transformers. They are used in single-line representations only.

(*Continued*)

Component	Symbol	Notes
Transformer—form 2		The IEC does not provide specific symbols for *ideal transformers* (circuit elements) and *transformers* (machines that can be modeled with varying degrees of detail). In this book the ideal transformer will be indicated using a letter *i* in the scheme.
Coil		The IEC considers this symbol to be obsolete. However, it does not provide a specific symbol for coils. In this book, coils are intended as physical objects (usually with some resistance and inductance) while inductors are ideal components with inductance only.
Three-phase synchronous machine		The IEC symbol requires the textual information to be as follows: • M for asynchronous motor • G for asynchronous generator • MS for synchronous motor • GS for synchronous generator In this book, deviation is made from this standard, since circuit elements represent objects, and the object is the synchronous or asynchronous machine; *motor* and *generator* are just operating modes of the object.
Three-phase asynchronous machine		

1.3.4 Names, Symbols, and Units

Equations relating to electric phenomena and circuits appear in this book according to the latest international standards. This is because of the *global* nature of this book and to ease communication between people of different countries or regions of the world. Readers can thus be confident that the graphic conventions used throughout the book closely match (with very few, well-motivated exceptions) those of internationally agreed standards, and they are strongly advised to become familiar with them and to use them now and in the future.

The basic rules set by these standards for writing SI units of measure have already been presented in Section. 1.3.2. Other rules, closely followed in this book, are as follows:

- Symbols that identify physical quantities are written in italics (e.g., V or v for potential, V for volume, m for mass, I or i for currents, and so on.[8] Subscripts are

[8]The only exception in this book, in line with the majority of books, regards quantities represented by uppercase Greek characters: these are not written in italics, simply because this makes them easier to distinguish. For example, when a mechanical speed is indicated using an uppercase omega, it will be written as Ω rather than $\mathit{\Omega}$.

written in italics when they refer to physical quantities (e.g., C_p for thermal capacitance at constant pressure p), but in roman (upright) in all other cases ($\mathrm{C_g}$ for gas thermal capacitance, U_{av} for average voltage).

- Vectors and matrixes are represented using bold type (e.g., $\boldsymbol{E}$ and $\boldsymbol{B}$ for electric and magnetic fields);
- Time-varying quantities, whenever possible, are expressed using lowercase letters (such as i for currents, u for voltages) while quantities that are constant over time are expressed using uppercase symbols (I and U, respectively, for current and voltage).
- Complex numbers, as stated in [s9], are indicated by underlining the related symbol; for example, $\underline{U}=\underline{Z}\,\underline{I}$ is Ohm's law for alternating circuits, expressed by means of complex numbers. Conjugates of complex numbers are referred to using an asterisk: $\underline{\mathrm{Z}}^*$ is the conjugate of $\underline{\mathrm{Z}}$.

In addition to these general rules, the following additional conventions, compliant with standards, though not mandatory, are used in this book:

- When sinusoidal voltages are given, their peak values are indicated by the peak sign "^" above the relevant symbol (e.g., the peak of a sinusoidal voltage u is $\hat{U}$).
- Symbols representing integer numbers (e.g., $i, j, k, \ldots$) are shown in italics since this makes them easier to read and is very common practice in books and articles.

The names of the quantities used in the book, their symbols, and their unit of measure are also taken from the above-mentioned international standards and are shown in the following table.

The official standard has been simplified in some cases. For instance, "electric field" is used instead of the official name "electric field strength." This is for reasons of both simplicity and rationality; other names would otherwise also need to be changed: for instance, "electric current intensity" would have to be used instead of "electric current."

Quantity	Symbol	Unit		Notes
		Name	Symbol	
Electric potential	v, V	volt	V	
Voltage, potential difference[a]	u, U	volt	V	
Electric current	i, I	ampere	A	
Electromotive force	e, E	volt	V	
Current density	J	ampere per square metre	$\mathrm{A/m^2}$	
Resistance	R	ohm	Ω	
Conductance	G	siemens	S	
Inductance (or self-inductance)	L	henry	H	
Impedance	Z	ohm	Ω	

(*Continued*)

Quantity	Symbol	Unit		Notes
		Name	Symbol	
Electric field electric field strength	$\boldsymbol{E}$	volt per metre	V/m	
Electric flux density	$\boldsymbol{D}$	coulomb per square metre	C/m^2	
Permeability	μ	henry per metre	H/m	$\boldsymbol{B} = \mu \boldsymbol{H}$
Magnetic field, magnetic field strength	$\boldsymbol{H}$	ampere per metre	A/m	
Magnetic flux density	$\boldsymbol{B}$	tesla	T	
Magnetic flux	ϕ, Φ	weber	Wb	
Linked flux	Ψ	weber	Wb	
Permittivity	ε	farad per metre	F/m	$\boldsymbol{D} = \varepsilon \boldsymbol{E}$
Phase difference	φ	—	—	
Reluctance	$\mathcal{R}$	one per henry	H^{-1}	
Resistivity	ρ	ohm metre	Ωm	
Volumic charge[b]	ρ	coulomb per cubic metre	C/m^3	

[a]The name "voltage", commonly used in the English language, is the term preferred by IEC, but is an exception to the principle that a quantity name should not refer to the name of a unit. Another term, equivalent to voltage, is "tension".
[b]Also (known as) volume density of charge.

1.3.5 Other Conventions

An important decision, and one for which no solution is suggested by international standards, regards the decimal marker. The 22nd General Conference on Weights and Measures (CGPM) decided in 2003 that "the symbol for the decimal marker shall be either the point on the line or the comma on the line."

In this book the decimal marker is shown as a point on the line.

For vector products, the two following symbols, again from International Standards, are used everywhere:

- Result $\boldsymbol{c}$ of *dot* product between $\boldsymbol{a}$ and $\boldsymbol{b}$: $\boldsymbol{c} = \boldsymbol{a} \cdot \boldsymbol{b}$
- Result $\boldsymbol{c}$ of *cross* product (or *vector* product) between $\boldsymbol{a}$ and $\boldsymbol{b}$: $\boldsymbol{c} = \boldsymbol{a} \times \boldsymbol{b}$

1.4 SPECIFIC CONVENTIONS AND SYMBOLS IN THIS BOOK

In addition to conventions stipulated by relevant international standards, steps have been taken to ensure a uniform style throughout this book. This additional standardization is in the form of simple conventions as shown in this section.

1.4.1 Boxes Around Text

For easy reference, boxes are used to emphasize very important pieces of information. The following types of boxes are used:

> **Convention:** *Name of convention*
>
> Contains adopted conventions, such as the one used to indicate voltage polarity. Normally the adopted conventions are drawn from International standards; when this does not occur, the decision is commented on and justified.

> **Definition:** *Name of definition*
>
> Contains the definition of new concepts (such as a circuit) or quantities (such as the ampere as unit of measure of a current).

> **Law:** *Name of law*
>
> States some fundamental laws of electromagnetism or circuits, such as the charge conservation law or Kirchhoff's laws.

> **Result:** *Name of result*
>
> The main results of the analyses carried out are evidenced in boxes, so that they are easily spotted at a glance.

> **Rule:** *Name of rule*
>
> Practical rules to be applied to obtain particular results are also boxed. An example is the rule that allows a circuit-like physical system to be dealt with using the abstract circuit concept.

1.4.2 Grayed Boxes

Sometimes the text is evidenced in grayed boxes. Two types are used: *more in depth* boxes and *for the instructor* boxes.

The "more in depth" boxes can be found throughout the chapters, and offer in-depth insight to the basic concepts in the general text. Although not essential for acquiring a basic knowledge of the topics, their visual appearance is such that the reader is stimulated to read (and possibly study) them.

The "for the instructor" boxes are to be found only at the beginning of a chapter, just below the table of contents, and explain the rationale behind the choices made, to help teachers plan their presentation of topics in class.

The appearance of these boxes is as follows:

More in Depth

This is a simple *more in depth* box.

For the Instructor

This is a simple *for the instructor* box.

1.4.3 Terminology

Terminology in any textbook should be free from strict standards. However, when a lot of books share the same terminology, this is useful for readers wishing to refer to several sources. Therefore, whenever possible, terms drawn from International Standards such as [s4] or [s7] are used.

As regards circuits, some deviation from standards was advisable in some cases; therefore in Chapter 2, in the section "definitions", the most important definitions relating to circuits are reviewed, and deviations from International Standards evidenced.

1.4.4 Acronyms

Minimum use is made of acronyms to facilitate reading. The only acronyms used in the book, also written in full on occasion, are those shown in the following table:

Acronym	In Full
AC	alternating current
DC	direct current
EMF	electromotive force
PM	permanent magnet
PPU	power processing unit
KCL	Kirchhoff's current law
KVL	Kirchhoff's voltage law
rpm	revolutions per minute
rms	root mean square

1.4.5 Reference Designations

A selected reference list is included at the end of this book. The reference number contains a letter indicating the nature of the reference. For instance, [s2] is an international standard (as indicated by the letter "s"), [bc1] is a circuit-related book (as indicated by the letter "c"), and [p2] is a scientific paper.

2

THE FUNDAMENTAL LAWS OF ELECTROMAGNETISM

For the Instructor

Readers of this book are expected to have at least some knowledge of the basic laws of electromagnetism—that is, the laws of Gauss, Ampère (with Maxwell's added displacement current), and Faraday, as well as Gauss's law for magnetism and the definitions of $\boldsymbol{E}$ and $\boldsymbol{B}$ by means of induced forces.

Nevertheless, in this chapter we offer a concise presentation of all these laws.

This way we provide some basic knowledge to those who have not previously encountered them; however we also recommend that you read this chapter even if you have previous knowledge of electromagnetics, so as to refresh your memory and to become acquainted with the symbols used in this book (which are basically the ones recommended by international standardization bodies).

Generally speaking, you can omit to read the *More in Depth* boxes. If, however, you have already attended an electromagnetics course at university level, these boxes will act as a bridge between your previous studies and the material in this book.

We will start with the definitions of electric and magnetic fields, which link the worlds of mechanical and electromagnetic phenomena.

Fundamentals of Electric Power Engineering: From Electromagnetics to Power Systems, First Edition.
Massimo Ceraolo and Davide Poli.

This chapter cannot substitute a good book of electromagnetism. Engineers, in particular, may want to have stronger knowledge of *applied* electromagnetics. For this purpose, a good book is reference [be2].

Note. In this chapter the general rule that lowercase symbols are used for time-varying quantities and uppercase for constant quantities is not followed. Forces, currents, potentials, and fields used in this chapter can all be time-varying.

2.1 VECTOR FIELDS

In physics and engineering, it is very frequent to evaluate actions among objects occurring at a distance. For instance, when two point masses are considered in a space which is ideally free from any other mass, the gravitational force exists between them that is expressed by the well-known formula

$$F_g = \gamma \frac{Mm}{d^2} \tag{2.1}$$

where γ is a constant, d is the distance between the two masses M and m, and the force is attractive and directed from one mass to the other.

Similarly, when two positive charges Q and q are present in a space free from any other charge, the force exerted is

$$F_C = k \frac{Qq}{d^2} \tag{2.2}$$

where k is a constant, d is the distance between the two charges, and the force is repulsive and directed from one charge to the other.

Near the surface of the earth, the gravity force is instead given by

$$F_{g0} = mg \tag{2.3}$$

and is always directed vertically toward the ground.

This formula is immediately obtained from (2.1), considering the value of the universal constant γ, the earth's mass M, and the earth's radius r.

However, instead of considering the forces as being due to two objects (two masses, two charges, etc.), it is very useful in these cases to imagine that the first of these has somewhat *altered* the space, so that when the second system is present, it feels a force applied to it. For instance, the presence of a first mass M can cause the space to be altered so that, at any point, a force per unit mass exists, whose value is obtained from the expression

$$\boldsymbol{F}_{\mathrm{gu}} = \gamma \frac{M}{r^2} \boldsymbol{r}_{\mathrm{u}}$$

where $\boldsymbol{r}_{\mathrm{u}}$ is a unit length vector having its head in the centre of the space occupied by M and its tail in the position where $\boldsymbol{F}_{\mathrm{gu}}$ is computed.

Thus we think of the space as having at each point, by effect of the presence of M, a virtual force $\boldsymbol{F}_{gu}$, per unit mass. In other words, once a mass m is positioned at a point **P**, this mass will feel an attractive force toward M, whose strength is $F_g = F_{gu}m$.

This virtual force, defined for the different points of space, is what we call *a force field*. The process is very similar for the Coulomb force (2.2). In this case, the force is considered to be due to a first charge (let it be Q) and evaluated per unit of the second charge, according to the relation

$$\boldsymbol{F}_{Cu} = k\frac{Q}{r^2}\boldsymbol{r}_u \tag{2.4}$$

In comparison to the gravitational field, here we have the difficulty that charges can be either positive or negative. $\boldsymbol{F}_{Cu}$ is the force exerted, by effect of the presence of charge Q, on a unit charge *assumed to be positive.*

This is another example of force field. In this case, once a charge q is positioned at a point **P**, this charge will feel a force toward Q, whose strength is $F = F_{Cu}q$.

Equation (2.4) is valid also with regard to signs. When the product Qq is positive, the force tends to move the two charges apart from each other; otherwise the force is attractive.

More in general, *a vector field* is a mathematical expression defining a vector in all points of a subset of space. Vector fields are very useful in analysing natural phenomena, as we will soon see later in this chapter.

An important scalar quantity that can be derived for a generic vector field $\boldsymbol{G}$ is the so-called *line integral* or *path integral*. If we consider a path l through space followed by a point (e.g., a point mass or a point charge), the corresponding line integral I is the integral of the scalar product of vector field $\boldsymbol{G}$ and the line element of the path $\mathrm{d}\boldsymbol{l}$ (an infinitesimal vector whose direction is the path tangent):

$$I = \int_l \boldsymbol{G} \bullet \mathrm{d}\boldsymbol{l}$$

Note that when the vector field is a per unit force field, the path integral is the per unit work the field produces. For instance, the quantity

$$W_{ABu} = m\int_{\mathbf{A}}^{\mathbf{B}} \boldsymbol{F}_u \bullet \mathrm{d}\boldsymbol{l}$$

is the work the field per unit mass $\boldsymbol{F}_u$ produces when the mass m moves (by whatever cause) from point **A** to point **B**, along a given path.

Some of the vector fields are *conservative*. A force field is said to be conservative if the work it does between two generic points **A** and **B** does not depend on the path chosen between **A** and **B**, or, equivalently, if the work done along any closed curve is always null.

$\boldsymbol{F}$ is conservative $\Leftrightarrow W_l = \oint_l \boldsymbol{F} \bullet \mathrm{d}\boldsymbol{l} = 0$ whatever closed curve l is chosen.[1]

[1]Line integrals along closed curves are often called cyclic integrals.

or, equivalently,

$$\boldsymbol{F} \text{ is conservative} \Leftrightarrow W_{AB} = \int_{\mathbf{A}}^{\mathbf{B}} \boldsymbol{F} \bullet \mathrm{d}\boldsymbol{l} \text{ is the same whatever the path from } \mathbf{A} \text{ to } \mathbf{B}.$$

This is true in particular for the force fields we have considered per unit charge or per unit mass:

$$\boldsymbol{F}_{\mathrm{u}} \text{ is conservative} \Leftrightarrow W_{AB\mathrm{u}} = \int_{\mathbf{A}}^{\mathbf{B}} \boldsymbol{F}_{\mathrm{u}} \bullet \mathrm{d}\boldsymbol{l} \text{ is the same whatever the path from } \mathbf{A} \text{ to } \mathbf{B}.$$

Since the work between **A** and **B** does not depend on the path followed, it is a function of points **A** and **B** only. This function is called *potential* and is normally indicated in physics textbooks using the letter V.

$$W_{AB\mathrm{u}} = \int_{\mathbf{A}}^{\mathbf{B}} \boldsymbol{F}_{\mathrm{u}} \bullet \mathrm{d}\boldsymbol{l} = V(\mathbf{A}) - V(\mathbf{B}) \tag{2.5}$$

If we consider the Coulomb field, $\boldsymbol{F}_{\mathrm{u}} = \boldsymbol{F}_{\mathrm{C}}/m$ and the potential V is called *electric potential.*

Potential is of the utmost importance in electricity. The common *voltage* of our household electrical installation is indeed an "electric potential difference"—that is, the difference assumed by the potential function V between the two terminals of the electricity supply.

2.2 DEFINITION OF *E* AND *B*; LORENTZ'S FORCE LAW

You may have learned in your previous studies of electromagnetics that a charge q in the vicinity of another charge Q is subject to a force, which can be expressed using Coulomb's law. Coulomb's law was also mentioned in Section 2.1 as a very significant example of force field.

Because of its importance, what in Section 2.1 was called $\boldsymbol{F}_{\mathrm{Cu}}$, i.e., Coulomb force per unit charge has a name of its own, the *electric field*, and the specific symbol $\boldsymbol{E}$.

Now we can express Coulomb's law in more detail than in equation (2.4), as follows:

$$\boldsymbol{E} = \frac{Q}{4\pi\varepsilon r^2}\boldsymbol{r}_{\mathrm{u}} \tag{2.6}$$

where

- $\boldsymbol{r}_{\mathrm{u}}$ is a unit length vector with its tail in the centre of the space occupied by Q and its head directed toward the position where $\boldsymbol{E}$ is computed.
- ε is a physical characteristic of the medium in the position where $\boldsymbol{E}$ is computed, called the *permittivity*. Values of permittivity of different materials are easily found in books and on the Internet.

Remember that, using the electric field concept, the force exerted on a charge q located in the point of space where $\boldsymbol{E}$ exists is subject to the force

$$\boldsymbol{F}_{\mathrm{C}} = q\boldsymbol{E} \tag{2.7}$$

Equation (2.7) defines the electric field (for stationary charges) $\boldsymbol{E}$, as the force per unit charge occurring to a charge at distance r from the field of origin (where Q resides) and with the same direction as $\boldsymbol{r}_{\mathrm{u}}$. So we can issue the following:

> **Definition:** *Electric field*
>
> The *electric field strength* or *electric field* in a point of space is the force per unit charge that acts on a stationary charge located in that point of space.

Equation (2.7) defines the electric field in a point of space without direct link to (2.6). This is intentional, since the electric field in a point of space is in general produced by a system of several charges in different zones of space, and therefore the electric field configuration is typically much more complex than the one indicated by equation (2.6).

The electric field $\boldsymbol{E}$ is conservative, and therefore it allows a potential function V so that:

$$\int_{\mathrm{A}}^{\mathrm{B}} \boldsymbol{E} \bullet \mathrm{d}\boldsymbol{l} = V(\mathbf{A}) - V(\mathbf{B}) \tag{2.8}$$

Because of its importance, the potential V has its own unit of measure, the volt, whose symbol is V. Hence it follows that the electric field $\boldsymbol{E}$ can be expressed in N/C (newton per coulomb) or V/m (volt per meter).

If the same charge q is moving in a point of space at a speed $\boldsymbol{v}$ in the vicinity of a magnet, it is also subject to a force which is orthogonal to $\boldsymbol{v}$. This can be experimented by observing a current-carrying conductor (inside which charges flow) near a permanent magnet, such as a compass needle.

In a similar manner to (2.2) the magnet can be imagined as altering the behaviour of space in which q moves by creating a field $\boldsymbol{B}$, which is related to the force produced on the moving charge by the law

$$\boldsymbol{F}_{\mathrm{L}} = q\,\boldsymbol{v} \times \boldsymbol{B} \tag{2.9}$$

This expression clarifies that the force is orthogonal to $\boldsymbol{v}$ and $\boldsymbol{B}$, proportional to q, and that it also defines its direction.

Equation (2.9) defines the *magnetic flux density* field $\boldsymbol{B}$. So we can issue the following:

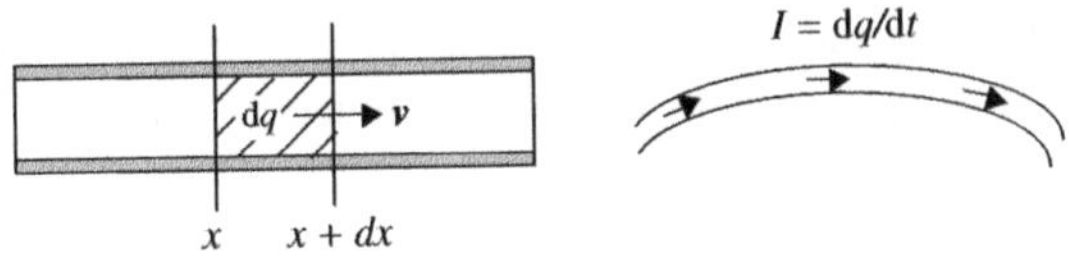

FIGURE 2.1. Charges flowing inside a conductor.

> **Definition:** *Magnetic flux density*
>
> *Magnetic flux density* in a point of space is a vector that creates a force exerted on a moving charge, in that point of space where no electric field exists, according to the relation $\boldsymbol{F}_{\mathrm{L}} = q\,\boldsymbol{v} \times \boldsymbol{B}$.

The general expression of the electric field is obtained by summing the terms $\boldsymbol{F}_{\mathrm{C}}$ and $\boldsymbol{F}_{\mathrm{L}}$:

$$\boldsymbol{F} = \boldsymbol{F}_{\mathrm{C}} + \boldsymbol{F}_{\mathrm{L}} = q(\boldsymbol{E} + \boldsymbol{v} \times \boldsymbol{B}) \tag{2.10}$$

Equation (2.10) is called the *Lorentz force law,* and $\boldsymbol{F}$ is called the Lorentz force.[2]

We will often need to deal with charges continuously flowing within wires instead of point charges. For these, a modified version of equation (2.9) can be derived. Consider the left part of Figure 2.1.

During the time $\mathrm{d}t$ the charge $\mathrm{d}q$ traverses section x of the conductor shown. The current $i(t)$ is defined (this concept will be discussed several times in this book) as the rate at which the charge traverses any conductor's cross section at a given time:

$$i(t) = \frac{\mathrm{d}q}{\mathrm{d}t}$$

Therefore it is

$$\mathrm{d}q\boldsymbol{v} = i\,\mathrm{d}t\frac{\mathrm{d}\boldsymbol{x}}{\mathrm{d}t} = i\,\mathrm{d}\boldsymbol{x}$$

The effects of an existing flux density $\boldsymbol{B}$ on the conductor current i that flows through the full conductor length (cf. the right-hand part of Figure 2.1) can be imagined as the sum of the forces generated on each conductor segment $\mathrm{d}\boldsymbol{x}$: $\boldsymbol{F}(t) = \int_l \mathrm{d}\boldsymbol{F}$. The force $\mathrm{d}\boldsymbol{F}$, in turn, can be imagined as being due to the current element $\mathrm{d}I = I\mathrm{d}x$. Therefore,

[2]Some authors call equation (2.9) the Lorentz force law, and this has suggested our subscript. However, modern texts tend to call (2.10) the Lorentz law. Just for your curiosity: one of Lorentz's greatest achievements is the so-called *Lorentz transformation*, which is the basic foundation of main pillar relativity theory. Lorentz's transformation applied to Coulomb's law gives (2.10).

an equivalent representation of formula (2.9), but referring to a current on a conductor instead of a free charge q, is

$$\mathrm{d}\boldsymbol{F} = \int_l I\, \mathrm{d}\boldsymbol{x} \times \boldsymbol{B} = \mathrm{d}\boldsymbol{I} \times \boldsymbol{B} = \boldsymbol{I} \times \boldsymbol{B}\, \mathrm{d}x \tag{2.11}$$

where $\mathrm{d}\boldsymbol{x}$ is an infinitesimal length of the conductor carrying current I, directed as I.

Equation (2.11) is normally referred to as Laplace's equation of electromagnetically induced force, or *Laplace's force law.*

The forces produced on charge flows by a magnetic flux density field $\boldsymbol{B}$ are expressed in a simple way when the speed $\boldsymbol{v}$ or the direction of the conductor in which the current l flows are orthogonal to $\boldsymbol{B}$:

$$\begin{matrix} F = v \cdot B \cdot q \\ F = I \cdot B \cdot l \end{matrix} \quad (\text{when } \boldsymbol{v} \perp \boldsymbol{B}) \tag{2.12}$$

To find the direction of the force vector, we can use the right-hand rule: we consider, in the order given, the directions of the first and second factors of (2.12). If they are oriented like the index and middle fingers of the right hand, then the corresponding forces are oriented like the thumb of the same hand. Similarly, if the thumb has the same direction as $\boldsymbol{v}$ or $\boldsymbol{I}$ while the index has the same direction as $\boldsymbol{B}$, the force will have the direction of the middle finger. Finally, the procedure can be repeated by aligning the middle finger with $\boldsymbol{v}$ or $\boldsymbol{I}$ and aligning the thumb with $\boldsymbol{B}$.

If the order of the indexes were considered in the wrong way, the direction of the third finger would give an erroneous (opposite) direction for the force.

2.3 GAUSS'S LAW

An important consequence of Coulomb's law (2.6) is Gauss's law.

Before proceeding to Gauss's law, let us first define the flux ϕ of a vector field $\boldsymbol{G}$ across a surface A. It is a scalar quantity defined as follows:

$$\phi = \iint_A \boldsymbol{G} \bullet \mathrm{d}\boldsymbol{A} \tag{2.13}$$

where $\mathrm{d}\boldsymbol{A}$ is the elemental surface—that is, a vector having as magnitude the elemental scalar surface $\mathrm{d}A$ which is orthogonal to the surface itself. Particularly important is the flux traversing a closed surface A (a closed surface is one surrounding a volume, e.g. the surface of a sphere, a cube, etc.).

Gauss's theorem states that:

Law: *Gauss's law*

The flux of the electric field through any closed surface $A(v)$, surrounding volume v of space containing the electric charge Q, measured as coming out, equals Q/ε.

Mathematically, it can be expressed as follows:

$$\iint_A E \bullet \mathrm{d}\boldsymbol{A} = Q/\varepsilon \tag{2.14}$$

A more precise formulation of Gauss's law is provided in the following *More in Depth* box.

More in Depth A: *Gauss's law*

Gauss's law is one of the fundamental laws of electromagnetism; however, it can simply be regarded as a consequence of (2.2) and thus will not often be mentioned in relation to electric power engineering.

To express Gauss's law, it is useful to first define the vector $\boldsymbol{D}$ as the *electrical flux density* proportional to the electric field, by means of the permittivity defined when discussing equation (2.6):

$$\boldsymbol{D} = \varepsilon \boldsymbol{E}$$

The flux across a surface $\boldsymbol{A}$ of the electrical flux density is called *electric flux* Ψ:

$$\Psi = \iint_A \boldsymbol{D} \bullet \mathrm{d}\boldsymbol{A}$$

Gauss's law can be stated a follows: the electrical flux through a closed surface $A(v)$ surrounding volume v of space containing the electric charge Q equals Q, that is,

$$\Psi = \oiint_{A(v)} \boldsymbol{D} \bullet \mathrm{d}\boldsymbol{A} = \oiint_{A(v)} \varepsilon \boldsymbol{E} \bullet \mathrm{d}\boldsymbol{A} = \iiint_v \rho \, \mathrm{d}v = Q \tag{2.A.1}$$

2.4 AMPÈRE'S LAW AND CHARGE CONSERVATION

In the previous section, we introduced equation (2.9), allowing us to evaluate the force exerted on a moving charge q by what has been called magnetic field $\boldsymbol{B}$. Indeed that equation was introduced thinking of the force exerted on a moving charge by a nearby magnet, and the magnetic flux density $\boldsymbol{B}$ is thought of as a modification of the space at the point in which q resides, due to the presence of the magnet.

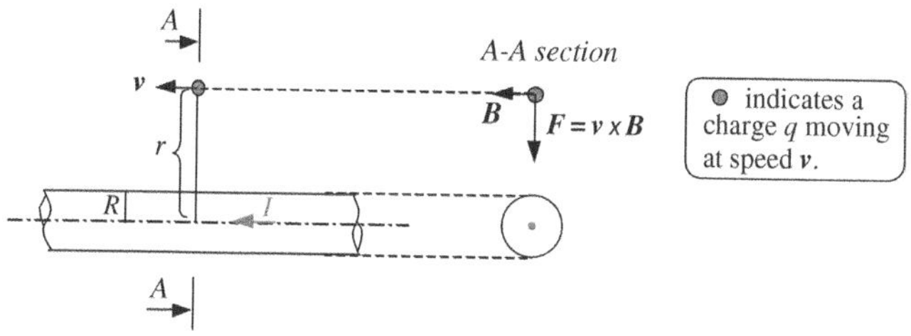

FIGURE 2.2. Force interaction between a conductor carrying current I and the charge q moving at the speed $\boldsymbol{v}$.

It can be experimentally observed that the same effect on a moving charge q can be induced by an electric current (i.e., a continuous flow of charge) in a conductor.

The situation can be visualized in Figure 2.2. The dot in the centre of the conductor's cross-sectional area in the right-hand part of the figure indicates, by convention, that the current in this conductor is moving upward from the page (i.e., it is directed toward the reader's eyes).

Therefore, current I in the conductor produces, at the point of the moving charge, a magnetic flux density $\boldsymbol{B}$.

Another field used in electromagnetics, closely related to the magnetic flux density, is the *magnetic field* $\boldsymbol{H}$. Fields $\boldsymbol{H}$ and $\boldsymbol{B}$ are related to each other by the following relation:

$$\boldsymbol{B} = \mu \boldsymbol{H} \tag{2.15}$$

in which μ, called magnetic permeability, is a constant characteristic of the matter existing in the point of space in which $\boldsymbol{H}$ and $\boldsymbol{B}$ are evaluated.

A relation which is more complex than equation (2.15) will be discussed in the latter part of Chapter 7.

The correlation between $\boldsymbol{H}$, $\boldsymbol{B}$, and I for the system shown in Figure 2.2 is

$$B = \frac{\mu I}{2\pi r} \tag{2.16}$$

where $H = |\boldsymbol{H}|$ and the direction of $\boldsymbol{H}$ is along the circumference in the plane orthogonal to the wire direction containing the point where $\boldsymbol{H}$ is evaluated.

Another significant case is when a coil with N turns is wound along a toroid such as the one in the figure below. Here the toroid length l is much greater than the cross-sectional diameter in all points of the toroid:

$$H \cong \frac{NI}{l} \tag{2.17}$$

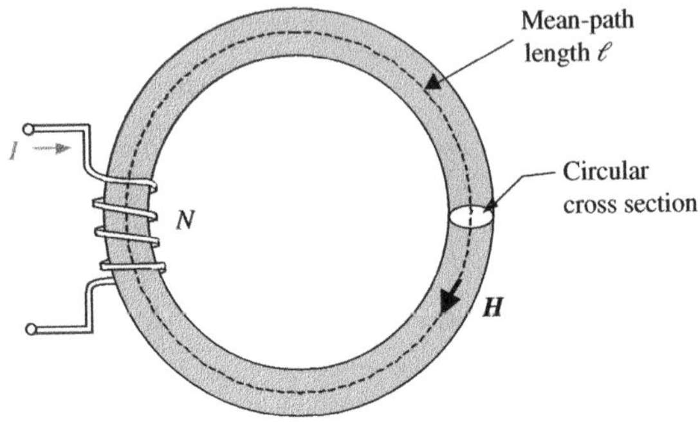

The quantity NI, interpreted as a "force" that tends to generate the magnetic field, is called "magnetomotive force" (MMF).

Magnetic field $\boldsymbol{H}$ (and therefore magnetic flux density B, related to it by the proportionality relation (2.15), is nonconservative. Therefore the work along a closed path can be nonzero.

For instance, the works of H and B along the circular path shown in the figure above are

$$W_H = \oint_l \boldsymbol{H} \bullet \mathrm{d}\boldsymbol{l} = Hl = NI \neq 0, \quad W_B = \oint_l \boldsymbol{B} \bullet \mathrm{d}\boldsymbol{l} = \mu NI \neq 0$$

Equations (2.16) and (2.17) constitute two common forms of **Ampère's law**.

More in Depth B: *Pointwise Ampère's law and Maxwell's extension*

To generalize equation (2.16), we consider current I as the flux of a vector "current density" $\boldsymbol{J}$ through the cross section of a conductor:

$$I = \iint_A \boldsymbol{J} \bullet \mathrm{d}\boldsymbol{A}$$

The current density $\boldsymbol{J}$ is due to moving charges in a conductive material (metal, electrolytic solution), so that $\mathrm{d}I = \boldsymbol{J} \bullet \mathrm{d}\boldsymbol{A}$ is the current flowing into the infinitesimal surface $\mathrm{d}A$, as indicated in the diagram below.

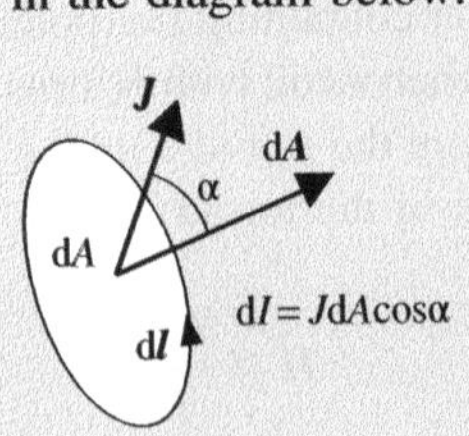

This way we can introduce a new version of Ampère's law, of which the first of (2.16) is just a particular application as follows:

$$\oint_l \boldsymbol{H} \bullet \mathrm{d}\boldsymbol{l} = \iint_{A(l)} \boldsymbol{J} \bullet \mathrm{d}\boldsymbol{A} \tag{2.B.1}$$

that is, the work of the magnetic field along a path l is equal to the current crossing any surface $A(l)$ with l as contour. From (2.B.1) we can immediately obtain (2.16): in this case the left-hand term of (2.B.1) is equal to $2\pi rH$ and the right-hand term is equal to I.

Ampère's law was modified by Maxwell, who expanded on the term "current density" to create *total current density*, that is, the sum of *conduction* and *displacement* current densities:

$$\boldsymbol{J}_\mathrm{t} = \boldsymbol{J} + \frac{\partial \boldsymbol{D}}{\partial t}$$

where $\partial \boldsymbol{D}/\partial t$ is by definition the *displacement current density* ($\boldsymbol{D}$ is defined in the More in Depth box **A**).

The Ampère–Maxwell law is therefore

$$\oint_l \boldsymbol{H} \bullet \mathrm{d}\boldsymbol{l} = \iint_{A(l)} \boldsymbol{J}_t \bullet \mathrm{d}\boldsymbol{A} = \iint_{A(l)} \left(\boldsymbol{J} + \frac{\partial \boldsymbol{D}}{\partial t}\right) \bullet \mathrm{d}\boldsymbol{A} \tag{2.B.2}$$

Remember that for the equation signs to be correct, the curve l must be traversed in a direction that an observer, looking at the head of vector $\boldsymbol{A}$, sees as counter-clockwise.

Maxwell's addition to Ampère's law has had a tremendous impact on electromagnetics and enables the explanation and mathematical description of the generation and propagation of electromagnetic waves. Although Maxwell's addition to Ampère's law is never used in this book in the form (2.B.2), its consequences are always present behind the scenes.

Another important law of electromagnetism is the so-called continuity law (or charge conservation law). In its simplest formulation it states the following:

Law: *Continuity equation (charge conservation)*

In steady state, the electric charge contained in a given region of space is constant.

However, in variable conditions, the charge contained in a given region of space can change.

A more general and complete approach to the continuity equation makes use of Ampère's, Maxwell's, and Gauss's equations and produces a result that is valid not only in steady state but also in variable conditions. It is provided in the following *More in Depth* box.

More in Depth C: Continuity equation

The most complete version of the continuity equation law can be obtained by merging (2.A.1) with (2.B.2).

Consider a volume v bound by surface A, as in the diagram below. If the surface is crossed by an arbitrary plane π, the curve l is determined and A is decomposed into its constituting parts A_1 and A_2.

Equation (2.B.2) says that $\oint_l \boldsymbol{H} \bullet \mathrm{d}\boldsymbol{l} = \iint_{A_1(l)} \boldsymbol{J}_t \bullet \mathrm{d}\boldsymbol{A}_1 = -\iint_{A_2(l)} \boldsymbol{J}_t \bullet \mathrm{d}\boldsymbol{A}_2$

Therefore, since $A = A_1 + A_2$, $\iint_A \boldsymbol{J}_t \bullet \mathrm{d}\boldsymbol{A} = 0$ whatever the volume v bound by surface A.

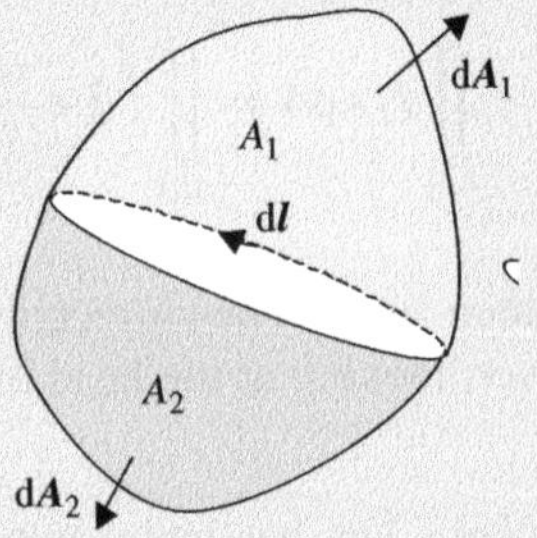

The new version of continuity equation is therefore

$$\oiint_{A(v)} \boldsymbol{J}_t \bullet \mathrm{d}\boldsymbol{A} = 0 \tag{2.C.1}$$

that is, "the total charge leaving a fixed volume v bound by surface A is always null," where the total charge flow is obviously the flow of the total current density as defined in (2.B.2).

An equivalent form is

$$\oiint_{A(v)} \boldsymbol{J} \bullet \mathrm{d}\boldsymbol{A} = -\frac{\partial Q}{\partial t} \tag{2.C.2}$$

This relation in fact states that 'the charge leaving any fixed volume v bound by surface A of space through current density $\boldsymbol{J}$ equals the variation of the charge accumulated inside v'.

The equivalence of (2.C.2) to (2.C.1) can be proved as follows:

$$\oiint_A \boldsymbol{J}_t \bullet \mathrm{d}\boldsymbol{A} = 0 \Rightarrow \oiint_A \boldsymbol{J} \bullet \mathrm{d}\boldsymbol{A} = -\oiint_A \frac{\partial \boldsymbol{D}}{\partial t} \bullet \mathrm{d}\boldsymbol{A} = -\frac{\partial}{\partial t}\oiint_A \boldsymbol{D} \bullet \mathrm{d}\boldsymbol{A} = -\frac{\partial Q}{\partial t}$$

i.e., equation(2.C.2)

In the latter passage, (2.A.1) in the form $\oiint_{A(v)} \boldsymbol{D} \bullet \mathrm{d}\boldsymbol{A} = Q$ was exploited.

2.4.1 Magnetic Field and Matter

Equation (2.15) indicates generic proportionality between magnetic field and magnetic flux density. The magnetic permeability μ of the matter containing the two fields is important.

The reference value is the permeability of vacuum $\mu_0 = 4\pi \cdot 10^{-7}$ H/m. For many materials, permeability is very close to this value (air, water, lead, copper, aluminium, etc.) Some have slightly greater permeability than a vacuum and are called *paramagnetic*, while others have slightly less and are called *diamagnetic*.

Other materials, which are very important in electric power engineering, are called *ferromagnetic* and have $\mu \gg \mu_0$. The ratio for ferromagnetic materials μ/μ_0 is usually greater than 10^5; for instance, highly purified iron (from which the name ferromagnetic derives) is around $2 \cdot 10^5$.

Indeed, in ferromagnetic materials the link between $\boldsymbol{H}$ and $\boldsymbol{B}$ is not a simple proportionality like that of (2.15); rather, a more complex relation applies. This complexity is discussed to some extent in Chapter 7 of this book.

It is demonstrated in electromagnetics that a consequence of $\mu \gg \mu_0$ for ferromagnetic materials is that whenever there is some source of magnetic field in a space containing air (or other nonferromagnetic material) and ferromagnetic material, the magnetic field concentrates in the latter, as illustrated in Figure 2.3.

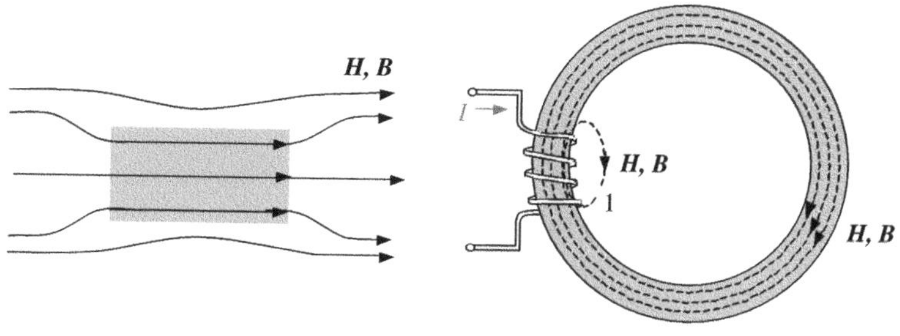

FIGURE 2.3. Field lines tend to flow inside iron: the cases of an open and a toroidal shape.

In particular, the right-hand part of the figure shows a coil wound around a ferromagnetic toroid. If it carries some current, the corresponding magnetic field flows nearly entirely within the toroid itself, as a consequence of the very high ratio μ_{Fe}/μ_{air}. This can be visualized by considering that the great majority of field lines are inside the toroid and only a small part flows, at least partly, through air; or, equivalently, by considering that the largest part of the flux in the cross section of the coil flows in the toroid and only a small percentage of it, called leakage flux, flows, at least partly, through air.

This is the basic reason why electric machines always have iron cores: they are based on magnetic phenomena, and the iron core is very important for causing magnetic flux to flow, in a concentrated manner, inside predefined flux tubes. This will be first discussed in Chapter 7, in which we introduce magnetic circuits, and then in all the other chapters in Part III of this book.

2.5 FARADAY'S LAW

Before introducing Faraday's law, consider again definition (2.13) of the flux of a vector field across a surface. If this definition is applied to magnetic flux density $\boldsymbol{B}$, it becomes

$$\phi = \iint_A \boldsymbol{B} \bullet \mathrm{d}\boldsymbol{A} \tag{2.18}$$

which defines the magnetic flux ϕ.

In addition, we need to define the concept of "electromotive force," often abbreviated as EMF. The name of this quantity suggests that it is *an entity that causes charges to flow in a wire*. If, for instance, we connect a small battery to a portable lamp, the battery exerts some force on the charges inside the conductors, and inside the lamp itself, and causes them to flow through the lamp wires (see diagram below). Because of this behaviour, this entity is called *electromotive force*.

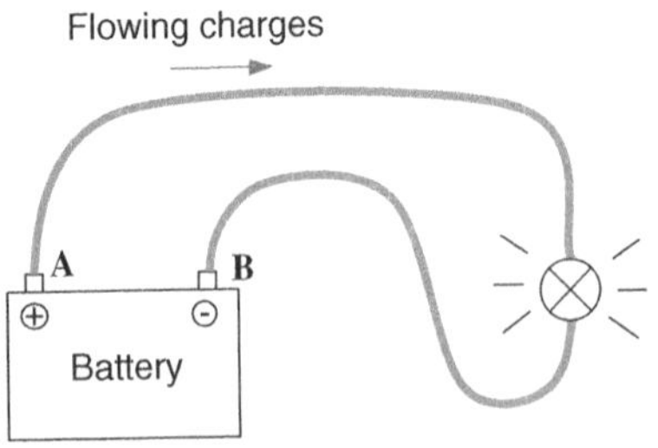

Although the entity causing charges to flow in the conductor is called electromotive force, it is not a real "force" according to the concept of force used in mechanics. In fact, more strictly speaking, an electromotive force is any cause that *creates a difference of electric potentials* in two points of an electric wire. For instance the battery of the previous figure can be imagined as creating a difference of potentials $U_{AB} = V_A - V_B$. Any such difference of potentials is called voltage. Referring again to (2.5) and (2.8), it is clear that U_{AB} is the work done by the battery per unit charge flowing. So if in a given time charge Q flows, the total energy delivered by the battery is the work done, that is,

$$E = W_{ABu}Q = (V_{\mathbf{A}} - V_{\mathbf{B}})Q = U_{\mathbf{AB}}Q \tag{2.19}$$

The battery's electromotive force is the cause that creates U_{AB}, and it is equal to U_{AB} as measured when the load (here the lamp) is disconnected from the circuit.

Equation (2.19) confirms that "electromotive forces" are not actual forces because they are conceived in mechanics but, instead, work per unit charge.

The mathematical expression of EMF will depend on the physical mechanism that produces it. The mechanism of producing electromotive forces in batteries will not be discussed here.

Naturally, batteries create electromotive forces that are constant over time (in reality they vary, but very slowly). In many cases in electric engineering we use time-varying electromotive forces that can be produced, for instance, through the application of *Faraday's law*.

Faraday's law indicates that when a *coil* (i.e., one or more loops of a wire) is traversed by time-varying magnetic flux, an electromotive force is produced that tends to circulate charges in the load to which the coil is connected, and this law expresses a mathematical formulation of that EMF. Consider the diagram below. It shows a coil (also called winding) constituted by N turns, traversed by the same flux ϕ which is a function of time. Faraday's law states that

$$E_{\mathrm{f}} = -\frac{\mathrm{d}\psi}{\mathrm{d}t} = -N\frac{\mathrm{d}\phi}{\mathrm{d}t} \tag{2.20}$$

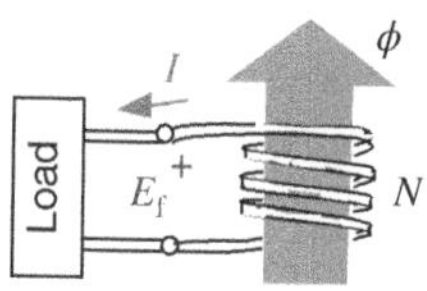

where ϕ is the *magnetic flux* traversing the coil and E_{f} is the electromotive force produced by the flux variation. The so-called *linked flux* ψ is simply equal to $N\phi$.

The minus sign of E_{f} has a special meaning. It shows that the induced E_{f} is such that, *if connected to an outside load, it tends to cause a current that generates a flux in opposition to the variation of* ϕ.

In other words, when ϕ is increasing, E_{f} tends to cause a current in a load that produces a flux opposite to ϕ (such as a positive I in the figure); when ϕ is decreasing,

E_f tends to cause a current that enforces ϕ (i.e. a current in the opposite direction to I). *The "+" marker in the previous diagram thus indicates the positive terminal when the flux is increasing.*

The fact that current I in the figure creates a flux which is opposite to ϕ can be verified using the right-hand rule.[3]

Often flux ϕ and linked flux ψ are simply due to the current circulating in the wire where E_f appears. In this case, it is useful to evaluate E_f in relation to I instead of ϕ or $\psi = N\phi$. This is easily done by relating the reference directions of I and ϕ to each other through the right-hand rule. I and ϕ in the diagram below correspond to the right-hand rule.

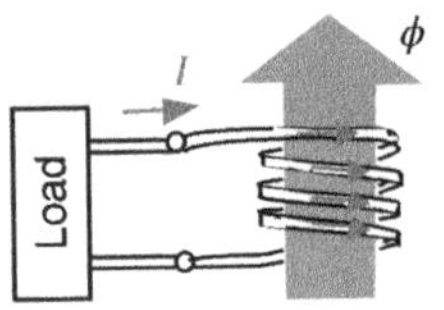

In this case Faraday's law is simply

$$E_f = L\frac{di}{dt} \quad \text{or} \quad E_f = -L\frac{di}{dt}$$

depending on the reference signs assumed for i and E_f (see diagram below).

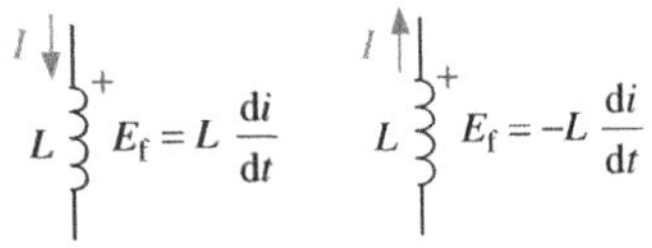

Now (as always in circuits) the "+" marker indicates the reference sign: when $E_f > 0$ the actual polarity will correspond to the marker, if $E_f < 0$ is opposite to it. The expression "electromotive force" is often abbreviated as e.m.f. or EMF.

We can summarise Faraday's law as follows:

> **Law:** *Faraday's law*
>
> When a time-varying magnetic flux traverses a coil, an electromotive force is produced that is proportional to the linked flux ψ time derivative.
>
> The sign of this electromotive force tends to cause the flow of a current which opposes the change of ψ.

[3]The rule is applied as follows. Using the right hand, curl the fingers and align them to the wire loop windings in such a way that the finger tips are oriented like arrowheads in the direction of the loop current; the thumb indicates the reference direction of flux (the arrowhead is its tip).

More in Depth D: *Pointwise Faraday's law*

Using the magnetic and electric fields of different points of space, Faraday's law can be stated as follows: "The line integral of the electric field strength $\boldsymbol{E}$ over a closed path l equals the flux of the magnetic flux density time derivative across any surface $A(l)$ bound by l."

$$\oint_l \boldsymbol{E} \bullet \mathrm{d}\boldsymbol{l} = -\iint_{A(l)} \frac{\partial \boldsymbol{B}}{\partial t} \bullet \mathrm{d}\boldsymbol{A} \qquad (2.D.1)$$

The minus sign in equation (2.D.1) is a consequence of the definition assumed for $\boldsymbol{E}$, $\boldsymbol{B}$.

Clearly, changing a sign in the definition for $\boldsymbol{E}$ or $\boldsymbol{B}$ would turn that sign into a plus. However, this is not advisable since the signs assumed here are the same as those used all over the world and are even included in International standards.

2.6 GAUSS'S LAW FOR MAGNETISM

The solenoidal nature of magnetic fields implies that the flux of $\boldsymbol{B}$ (or, equivalently, the flux of $\boldsymbol{H}$) across any closed surface is always zero, whatever the field and the surface:

$$\oiint_A \boldsymbol{B} \bullet \mathrm{d}\boldsymbol{A} = 0 \qquad (2.21)$$

Note that relation (2.21) has the same structure as Gauss's law (2.14), except that the right-hand term is zero. This is sometimes interpreted as "no magnetic charge exists in nature."[4]

The distinct similarity between (2.21) and (2.14) justifies the name given to this law, even though it is the result of several studies and has no specific link with Gauss's work.

Gauss's law for magnetism can be expressed in simpler terms as follows:

The force lines of $\boldsymbol{B}$ *and* $\boldsymbol{H} = \boldsymbol{B}/\mu$ *are always closed loops.*

[4] In fact, magnets are always constituted by magnetic dipoles: no individual "positive" or "negative" magnetic charges have ever been experimentally found.

2.7 CONSTITUTIVE EQUATIONS OF MATTER

2.7.1 General Considerations

In general, electric and magnetic phenomena occur in different ways depending on whether they occur in an empty space or in point of space where matter exists.

In fact, matter reacts to electric and magnetic phenomena, modifying them: the way in which matter interacts with fields is defined by the so-called *constitutive equations* (or *constitutive relations*).

Here we disregard the possibility of anisotropic behaviour—that is, bchaviour of matter that depends on the direction of fields applied to it—since it is well beyond the scope of this book. Thus the constitutive equations of matter are the following:

$$\boldsymbol{D} = \varepsilon \boldsymbol{E}$$
$$\boldsymbol{B} = \mu \boldsymbol{H}$$
$$\boldsymbol{J} = \sigma \boldsymbol{E}$$

where ε, μ, and σ are constants, in general depending on the point of space considered, and are called *permittivity*, *permeability*, and *conductibility*, respectively. These constants were used in the preceding sections in the description of electromagnetic laws in basic terms, since in basic electromagnetics, by tradition, there is no rigorous distinction between laws and constitutive equations.

The third of these constitutive equations—that is, $\sigma \boldsymbol{E} = \boldsymbol{J}$ or $\boldsymbol{E} = \rho \boldsymbol{J}$ (where $\rho = 1/\sigma$ is the resistivity of the substance considered)—is often called *pointwise Ohm's law*.

2.7.2 Continuous Charge Flow Across Conductors

Because of its importance, pointwise Ohm's law requires some further explanation. It can be helpful to have a look inside a current-carrying conductor.

Classic electromagnetics theory assumes that conductors contain electrons, that is, charges able to move freely within the conductor. Indeed, these charges move very rapidly as a consequence of thermal agitation, with speeds on the order of 10^6 m/s. This movement, however, causes them to collide with the conductor lattice. When an external electric field (i.e., a force per unit charge) is applied, the charges are accelerated in such a way that a drift speed, on the order of 10^{-4} m/s, is superimposed on the thermal speed. This drift speed of charges is macroscopically seen as a current (charge per unit of time) flow.

This justifies the fact that an electric field causes some current to flow and that the greater the electric field, the greater the charge flow, and therefore the current.

The force applied to the charges induces them to accelerate, so the resulting speed, if there were no collisions, would increase indefinitely. The effect of collisions is, however, similar to some damping of the charging movement; therefore, applying an external force to a charge cloud causes the speed to rise to a regime value determined by the equilibrium between force applied and force due to damping, in turn due to collisions. It can be demonstrated, however, that this regime speed is reached in times that are very small with respect to the application of the external electrical field, on the order of 10^{-14} s: this time is so small that it can be neglected in all the applications of electric power engineering. It is thus possible to state pointwise Ohm's law, in which the electric field is assumed to be proportional to the electric current density, that is, the force applied to the charges is proportional to the charge (regime) drift speed.

This result can be obtained mathematically by taking into account only the effects of electric field $\boldsymbol{E}$, and therefore force $\boldsymbol{F}_1$ of (2.10). Using this simplification, the electric charge speed can be found as the solution of the following equation:

$$m\frac{\mathrm{d}v}{\mathrm{d}t}+\frac{m}{\tau}v=eE \qquad (^\circ)$$

where m and e are the electron mass and charge, respectively, and E is the electric field; τ, the time constant of the equation, can be computed from known physical parameters: for copper, as already mentioned, it is around 10^{-14} s.

Given equation (°), if an external field E is applied to a charge q, it will first be accelerated, but will soon (3–5 times the value of τ) reach the regime velocity $v_{\text{reg}}=eE\tau/m$. The value of τ is so small that in nearly all the circuits and systems studied by electrical engineers the transient in which v varies can be neglected and electrons can be assumed as all having the same regime velocity:

$$v\cong\frac{\tau}{m}eE$$

Drift speed and current are related to each other. Consider the diagram below, showing a piece of conducting material carrying current. In the infinitesimal time interval dt the charges flow in the conductor at the drift speed $v=\mathrm{d}x/\mathrm{d}t$, and the charge contained in the volume of conductor along dx at the end of dt will have flown through the cross section of the conductor.

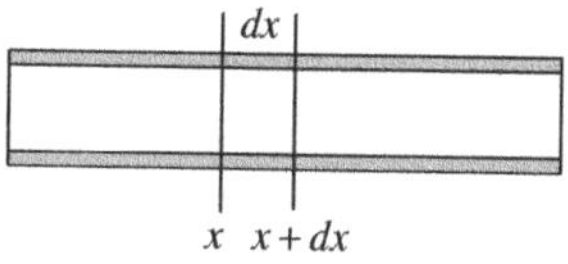

The charge flow can be correlated to the charges in the conductor. If n indicates the number of electrons per unit volume and e the electric charge of an electron it will be:

$$\mathrm{d}Q=ne\,\mathrm{d}xS=I\,\mathrm{d}t \qquad \text{and} \qquad \mathrm{d}x=v\,\mathrm{d}t=\frac{\tau}{m}eE\cdot\mathrm{d}t \quad \text{(see also footnote 5)}$$

[5]Note that this equation allows us to write a relation between electric current and charge speed on the conductor: $I=\mathrm{d}Q/\mathrm{d}t=neS\,\mathrm{d}x/\mathrm{d}t=neSv$.

Consequently,

$$ne\frac{\tau}{m}eE \cdot \mathrm{d}t \cdot S = I\,\mathrm{d}t \Rightarrow J = \frac{I}{S} = ne^2\frac{\tau}{m} \cdot E \qquad \text{or} \qquad J = \sigma E \quad \text{with } \sigma = ne^2\frac{\tau}{m}$$

which allows us, using reasonable values for n, e, m, and σ, to evaluate τ, again giving a value of about 10^{-14} s.

The formula obtained, expressing the dependence of conductivity on various parameters, is able to explain the experimental result that the conductivity of conducting materials decreases with temperature. In metals, nearly all the electrons move freely through the lattice, and therefore any increase in temperature will not increase the number of free electrons; however, the temperature increase will cause the lattice atoms to vibrate strongly, thus raising the probability of electrons hitting them during their free movement and reducing the mean free time between collisions τ.

It is also interesting to note that even in insulating materials, the proposed model for (very small) conductions works well, even though the free electrons are very limited in number. For insulating materials, however, the increase in free electrons according to temperature will be significant, and the effect of the increased atom vibration will be more so. This explains the experimental result that the resistivity of insulating materials reduces as temperature increases.

A final point on pointwise Ohm's law is that it relates the total force field acting on the charge within conductors in the direction of charge flow and not only that produced by electric charges in the space near the point considered.[6] In cases when more than a force field is present, we can write

$$\boldsymbol{E}_t = \rho \boldsymbol{J}$$

where $\boldsymbol{E}_t$ is the total force field acting on the charges, per unit of charge. For instance, in Section 3.3.1, when a microscopic view of circuits justifies the use of the abstract circuit concept in physical DC circuits, the following is used:

$$\boldsymbol{E}_t = \boldsymbol{E}_c + \boldsymbol{E}_s$$

where $\boldsymbol{E}_c$ is due to chemical reactions and $\boldsymbol{E}_s$ is due to electric charges.

2.8 MAXWELL'S EQUATIONS AND ELECTROMAGNETIC WAVES

During the eighteenth century the great scientist James Clerk Maxwell made an outstanding contribution to the study of electromagnetism. He revised all the equations previously introduced by other scientists and put them in good order.

[6]The demonstration based on equation (°) neglected the second term of equation (2.10). However, the demonstration could be repeated, with more complex mathematical passages, for the more general case.

The whole set of electromagnetic phenomena could then be studied by making use of his four equations, which were a reworking of Faraday's, Ampere's and Gauss' laws, as well as Gauss's law for magnetism. All these laws have been recalled in this chapter. Naturally the three constitutive equations of matter recalled in Section 2.7 should be added.

Maxwell also wrote these laws in differential forms, which have proven to be very useful to scientists who study complex electromagnetic phenomena. These forms are, however, beyond the scope of this book and will not be discussed here.

In addition to this activity of reordering previous laws, Maxwell gave another exceptional contribution to electromagnetics: he modified Ampere's law, adding the term $\partial \boldsymbol{D}/\partial t$ to the current density vector $\boldsymbol{J}$ to obtain the total current density $\boldsymbol{J}_{\mathrm{t}}$ (see More in Depth B). This addition dramatically changed the whole landscape of electromagnetic phenomena.

With this addition, Maxwell was able to mathematically show that a variable electric field produces a magnetic field, and vice versa. The two fields are closely concatenated to form a unique comprehensive phenomenon, called the electromagnetic field.

The electromagnetic field is ubiquitous in our lives: all wireless transmissions, from radio and TV signals to mobile phone communication and Wi-Fi networks, to name just a few, make use of the electromagnetic field as a medium. Even light is a form of electromagnetic field. Electromagnetic fields, however, are beyond the scope of this book.

Is it possible to study electric and magnetic phenomena while disregarding the fact that time-varying electric and magnetic fields are concatenated? The answer is yes, within given limits. It can indeed be said that physical systems in which electric and magnetic phenomena occur can be studied while disregarding electromagnetic field propagation *whenever their size is small in comparison to the length of the electromagnetic wave.*

It could be demonstrated, starting from Maxwell's equations written in their differential form, that the electromagnetic wave moves at the speed $c = 1/\sqrt{\varepsilon\mu}$ which, for empty space and air, equals:

$$c = \frac{1}{\sqrt{8.854 \times 10^{-12}\ \mathrm{F/m} \times 4\pi \times 10^{-7}\ \mathrm{H/m}}} \approx 300{,}000\ \mathrm{km/s}$$

If we want to study systems with a maximum size of a few metres while disregarding radiation of electromagnetic fields, we must limit the study to relatively slow variation of all the quantities (voltages, currents, etc.).

In fact for the electromagnetic wave, as for all waves, the following simple relation between frequency f, speed v, and wavelength λ applies:

$$v = \lambda f$$

As a consequence, if we want, for instance, to limit our scope to systems not larger than 10–100 metres, all analyses can be made while disregarding the radiation of electromagnetic fields if the following constraint is met:

$$\lambda \gg 10 \div 100\,\mathrm{m} \Rightarrow f \ll 300 \cdot 10^6\,\mathrm{m/s}/(10 \div 100\,\mathrm{m}) = 3 \div 30\,\mathrm{MHz} \qquad (2.22)$$

> **Rule:** *Validity of analysis of electric and magnetic phenomena while disregarding the concatenation of electric and magnetic fields*
>
> In the case of systems of up to 10–100 m in size, electric and magnetic field concatenation can be disregarded when operational frequencies up to a few megahertz are considered.

This entire book up to Chapter 15 is written on the basis that rule (2.22) is valid!

Chapter 15, indeed, deals with much larger systems, up to a few thousand kilometres. These are studied at the so-called industrial frequency, that is, 50 Hz or 60 Hz. For these frequencies, the wavelength of the electromagnetic field is 5000–6000 km. Thus, even for these large systems, when they are operated at the industrial frequency, electromagnetic radiation can be neglected.

2.9 HISTORICAL NOTES

2.9.1 Short Biography of Faraday

Michael Faraday[7] (London, England 1791–London, England 1867) was a chemist and physicist and is considered one of the best experimentalists who ever lived. He first taught himself chemistry, working in particular on electrochemistry. He founded the fundamental laws of electrolysis. Later he also worked successfully on electromagnetism.

His main achievements were the quantitative evaluation of the electric charge per mole of electrons, today called Faraday's constant, and the mathematical expression of induction law, today called Faraday's law. This law is very important in Electrical Engineering and will be quoted many times throughout this book.

He also made important contributions in other fields of physics.

His importance in electromagnetism is sealed by the adoption in the SI of the farad (symbol: F) for the unit of measure of capacitance of capacitors.

2.9.2 Short Biography of Gauss

Johann Carl Friedrich Gauss[8] (Braunschweig, Germany 1777–Hannover, Germany, 1855) is considered to be one of the greatest mathematicians ever. He was also an

[7]Pronounced (from *Oxford English Dictionary*): /ˈfærədeɪ/
[8]Pronounced (from *Oxford English Dictionary*): /gaʊs/

excellent physicist and astronomer. As a mathematician he made many contributions to various fields: the demonstration of the fundamental theorem of algebra, the graphical representation of complex numbers, and the distribution of statistical errors (today called the Gaussian curve) and theorems relating to the curvature of surfaces. He also first hypothesized the existence of non-Euclidean geometries.

His contribution to electromagnetics was also very important. His fundamental theory is today called Gauss's flux theorem.

2.9.3 Short Biography of Maxwell

James Clerk Maxwell (Edinburgh, Scotland, 1831–Cambridge, England, 1879) was one of the greatest physicists of all times, whose work is often compared to that of Newton and Einstein. His most important contributions are in gas kinetic theory and electromagnetism. As regards the latter, he unified all the electromagnetism laws, which proved to be highly effective; all electromagnetics phenomena were then summarized in his four famous equations. From these equations, he demonstrated the existence of electromagnetic waves and that electricity, magnetism and even light are all manifestations of the same phenomenon, the electromagnetic field.

His analyses included the introduction of a new term in Ampère's law [the term $\partial \boldsymbol{D}/\partial t$ in (2.B.2)], with the concept of *displacement current*, allowing him to develop his theory and demonstrate the electromagnetic wave.

At the time of his death, Maxwell's theory of electromagnetism was one of several. Its validity was established only in 1887, when it was experimentally verified by Hertz.

2.9.4 Short Biography of Ampère

André Marie Ampère (Lyon 1775, Marseille 1836) was a French physicist and mathematician, credited for laying down some of the basis of electromagnetism.

Ampère showed that two parallel wires carrying electric currents attract or repel each other, depending on whether the currents flow in the same or opposite directions, respectively. On the basis of these experiments and by applying his deep knowledge of mathematics, he formulated the famous law that bears his name (see Section 2.4).

Ampère offered courses in mathematics, theoretical and experimental physics, philosophy, and astronomy at the University of Paris, at the prestigious Collège de France, and at the new École Polytechnique.

The SI unit of measurement of electric current, the ampere, is named after him.

2.9.5 Short Biography of Lorentz

Hendrik Antoon Lorentz (Arnhem—The Netherlands 1853, Haarlem 1928) was a Dutch physicist who refined Maxwell's electromagnetic theory since his doctoral thesis, "The theory of the reflection and refraction of light," presented in 1875.

Before the existence of electrons was proved, he proposed that light waves were due to oscillations of an electric charge in the atom. Lorentz developed a rigorous mathematical theory of the electron's behaviour, for which he received the Nobel Prize in 1902.

Professor of mathematical physics at Leiden University, he also derived the transformation equations subsequently used by Albert Einstein to describe space and time in his relativity theory. Lorentz's transformation applied to Coulomb's law gives the Lorentz's force law described in Section 2.2.

PART II

ELECTRIC CIRCUIT CONCEPT AND ANALYSIS

Fundamentals of Electric Power Engineering: From Electromagnetics to Power Systems, First Edition.
Massimo Ceraolo and Davide Poli.

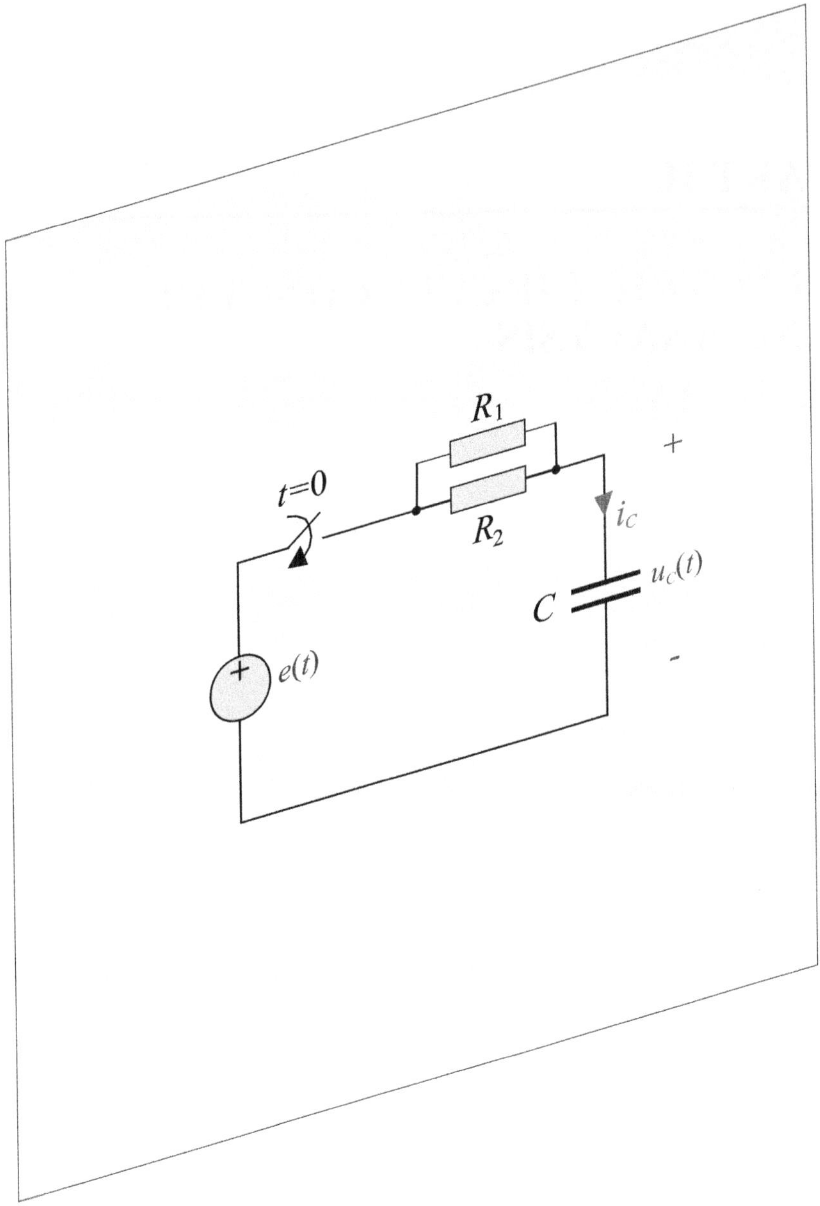

R1
t=0
R2
+
iC
uC(t)
C
e(t)
-

3

CIRCUITS AS MODELLING TOOLS

For the Instructor

This chapter differs in approach from similar textbooks, in that it makes a clear distinction between physical systems and models.

Since no uniform terminology exists in common books, the following definitions have been adopted:

- *Circuital systems*, that is, physical systems constituted by devices connected by wires, which make good candidates for modelling using lumped-component models.
- *Circuits*, that is, actual (lumped-component) models.

This approach has several advantages:

- It helps the student (i.e., the future engineer) to become accustomed to (a) the importance of *modelling* a physical system and (b) the corresponding need to evaluate the effects of what is disregarded in the process. This activity will be useful throughout the student's working life.

Fundamentals of Electric Power Engineering: From Electromagnetics to Power Systems, First Edition.
Massimo Ceraolo and Davide Poli.

- It allows a clear distinction between electromagnetism laws (i.e., Maxwell's equations written in more or less complicated forms) and Kirchhoff's laws, which are assumed to be valid, by definition, for all circuits.

The authors' decision to make this shift from the more frequent approach was based on years of teaching experience and of tackling students' daily needs and doubts. It is hoped that this approach will be appreciated by teachers and, consequently, that students will also find it useful.

3.1 INTRODUCTION

Electrical engineers spend much of their time working with electric circuits, or simply *circuits*. The word *circuits* is used to indicate graphical/mathematical tools used in descriptions of systems in which charges flow constantly (so-called direct-current circuits), or are sinusoidal (so-called alternating-current circuits), or in which they vary in a general way. They are able to model electronic boards, home and industrial electrical installations, the inner behaviour of electrical machines, and so on.

In this book, the physical systems that circuits are intended to model, whenever precision is necessary, will be called *circuital systems*.

More in Depth

Written texts (textbooks, articles, brochures) often fail to emphasize the difference between a physical system and its model. This is erroneous, because physical systems have a given behaviour that can be modelled with different degrees of precision.

For instance, the wheel of a car may simply be an ideal rigid, rolling cylinder for some purposes. But if we want to evaluate the forces generated by collisions, elasticity must be taken into account; to evaluate energy losses we must also consider rubber hysteresis, and if we want to evaluate the response to lateral forces, further characteristics must be assessed.

Differentiating between physical systems and their models is very important because in order to draw a model from a system, assumptions have to be made, and the results of the model analysis can be applied to the given system only if these assumptions are met to a sufficient degree of precision.

That is why in this book we have taken great care to distinguish between physical systems and their models—that is, circuital systems and circuits.

FIGURE 3.1. A simple circuital system.

Consider, for instance, the simple system shown in Figure 3.1. It shows an electric sinusoidal generator feeding two lamps with the interposition of a couple of wires, which are represented 'thick' because, in a physical system, they not only have length but also width and depth.

Obviously, our analysis of this system would be greatly simplified if, instead of having to simultaneously analyse all points of space using the laws of electromagnetism (see Chapter 1), it were possible to write independent equations of the individual elements involved and link them using additional congruence equations.

A qualitative analysis of Figure 3.1 shows that the generator is connected to the lamps with long wires, while there are short connections at the two horizontal sides of the circuit.

It is intuitively understood that, if the effect of space around the wires is not significant for what happens inside the system components, the system in Figure 3.1 can be studied as being constituted by connection of its main elements, as indicated in Figure 3.2. The elements are the generator, the two lamps, and the two long wires, while the short wires used for connecting the lamps to the longer wires can be ignored. All elements have *terminals*—that is, points used to connect them to the other circuital elements. In the figure, terminals are indicated as small white circles. Thin lines represent ideal wires—that is, graphical symbols indicating that what happens in the circuit is exactly the same as if the components at each end were directly connected to each other. Thin lines are therefore like ideal wires, able to transfer electric charge from one point to another of space in an ideal way—that is, in such a way that charges at one end are immediately transferred to the other end.

This step from the system in Figure 3.1 to the one in Figure 3.2 is dramatically important: it implies that we have modelled a *spatially distributed system* (for which all the fields connected to electromagnetic phenomena, such as electric field $\boldsymbol{E}$, magnetic field $\boldsymbol{H}$, current density field $\boldsymbol{J}$, etc., have a value at each point of space) as a

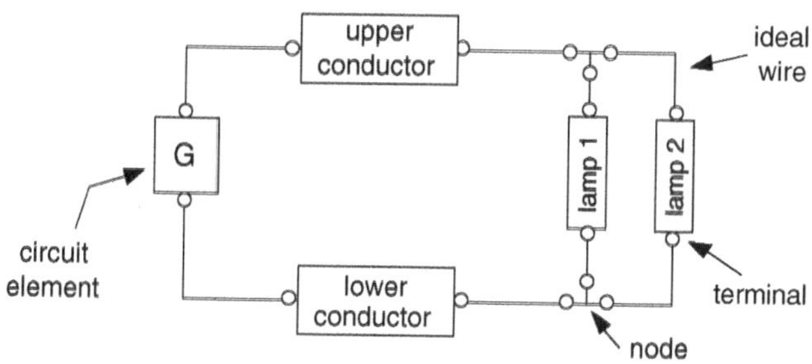

FIGURE 3.2. Circuital approximation (circuit modelling) of the system in Figure 3.1.

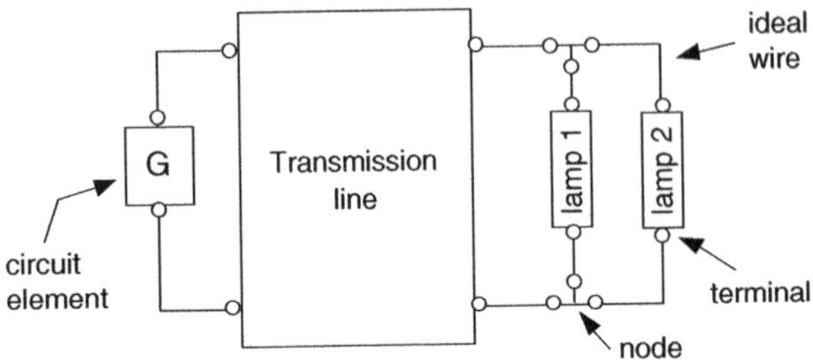

FIGURE 3.3. A circuit modelling the system of Figure 3.1, different from that of Figure 3.2.

lumped-elements system, whose behaviour is uniquely determined by the behaviour of the boxes connected by ideal wires—expressed in terms of equations we will call *constitutive equations* of the components (or elements)—and by the connections made by the wires.

More specifically, we have modelled the distributed-parameter circuital system we started out with, through a circuit which is a lumped-parameter model and much simpler to analyse.

The ease of analysing electromagnetic systems using circuits will become increasingly clear as the study of this book proceeds. But what kind of systems can be studied as circuits?

To give a very simplified answer, we can say that these are what we earlier called circuital systems—that is, systems which are physically constituted by (a) devices (such as the generator, lamps, and long wires of Figure 3.1) that perform complex actions, (b) the wires connecting them, and (c) some space around these elements whose effects on the electrical phenomena inside them can be disregarded.

These qualitative considerations are often too simplified. It must be stressed that if a circuit is not correctly modelled on a physical system, the results of its analysis (calculated currents and voltages) will differ from the actual values of the original circuit, and this could lead to significant, or even serious, errors. Therefore, in this and the next two chapters, some discussion on the basic hypotheses of allowing a circuital system to be modelled as a circuit will be presented.

The construction of a lumped-parameter model of a physical system is not unique. For instance, if the current between the upper and lower conductors of the system in Figure 3.1 cannot be ignored, the model shown in Figure 3.3 may be considered; here the transmission line is a single lumped component which can also take into account phenomena occurring between its upper and lower conductors.

3.2 DEFINITIONS

To deal with circuits effectively, we need precise definitions of the terms used. Although there is some uniformity in textbooks as regards terms and definitions, some significant differences exist. Therefore, it is useful to summarise here all the major definitions related to circuits used in this book.

The authors have carefully consulted the International Electrotechnical Vocabulary (IEV) [s4], which is the most authoritative source of standardized terminology for all electrical engineers all over the world. Therefore, the definitions used here usually comply with those in the IEV. However, our wording is original and rather simplified, given the book's scope and readership. Where there are significant differences from the IEV, brief explanatory notes are added.

The following definitions are given in alphabetical order for the reader's convenience.

Branch. A *branch* is a circuit element with two terminals.

Branch-Based Circuit. A *branch-based circuit* is an electric circuit in which all elements, except nodes, have two terminals.

The IEV does not give any definition for circuits with only two terminal components, other than nodes. A definition specific to this book is necessary because of the importance of this circuit topology.

Circuit (Electric Circuit). An *electric circuit,* or simply *circuit,* is a graphical—mathematical tool that constitutes a lumped-component model of a circuital system, consisting of lumped components connected to each other. These components are the circuit elements. Since circuits are models of real systems, circuits with different levels of accuracy can be produced for the same physical system.

Notes

1. See note on the definition of circuital systems.
2. Circuit behaviour is defined by the inner behaviour of circuit elements and by their interconnection: no influence is possible between what happens outside and what happens inside circuit wires and elements.
3. In this definition, no constraints are shown on the inner structure or behaviour of the circuit elements. However, *ideal transformers,* very special circuit elements which will be discussed in Chapter 5, play a particular role in circuits: they are interconnecting elements for forming *electric networks.*

Circuit Element. A *circuit element* is a component of a circuit that is connected to other components by means of connection points, called terminals.

Circuital System. A circuital system is a physical system containing elements connected to each other through wires in such a way as to form one or more closed loops. A circuital system is a spatially distributed (three-dimension) system.

IEV does not provide specific terminology to distinguish between physical systems and mathematical models.

Ideal Wire. An *ideal wire* is a branch with terminals of the same potential value, so that the voltage across its terminals is zero.

Node. A *node* is a point in a circuit in which three or more circuit elements are connected to each other. A node may be represented in expanded form, when connection of several elements is graphically represented as two or more connections. See the example in Figure 3.17a.

Although this definition does not conform exactly to the one in the IEV, it is very commonly used in textbooks.

Network (Electric Network). An *electric network*, or simply *network*, is a set of circuits separated from each other by ideal transformers. Two circuits separated by ideal transformers are also normally called *magnetically coupled circuits.*

Notes

1. In IEV, electric circuits and electric networks are treated as being equivalent.
2. Networks and ideal transformers will be introduced in this book in Section 5.3.3.

3.3 CHARGE CONSERVATION AND KIRCHHOFF'S CURRENT LAW

3.3.1 The Charge Conservation Law

The reader should be already acquainted with the idea of *charge conservation.*

The charge conservation principle states that "the charge in a given region of space remains constant over time." However, studies carried out mainly in the nineteenth century show that a charge which is the sum of individual charged elements (electrons, ions) is not conserved, but can accumulate in elements of space. The issue was solved by Maxwell, who defined a new form of charge, called displacement charge, which enables us to state the following:

Law: *Total charge conservation*

The total charge (sum of conduction and displacement charge) in any given region of space remains perfectly constant over time.

Conduction charge is the charge that moves through conductor wires (electrons) or conductive solutions (ions), while the displacement current is caused by the *variation of the displacement field over time*. This is detailed in the following *More in Depth* box, as well as in the next chapters.

More in Depth

This block is intended for those with some previous knowledge of the electromagnetic behaviour of a capacitor, and some basic knowledge of the behaviour of a simple R-C circuit in which a capacitor is charged.

In the circuit shown below, if initially there is no charge in the conducting plates of capacitor C, when switch sw is closed, some current flows in the circuit, and net charge enters the closed curve L.

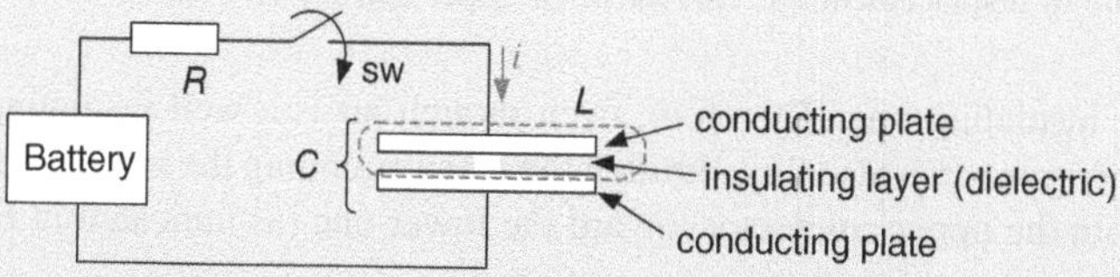

Charge is thus accumulated inside the conducting plates, and therefore the charge conservation principle does not apply. If, however, we assume that a special current, called *displacement current*, flows in the insulating layer (also called the *dielectric*), i.e. the space between the conducting plates, and that this current is exactly equal to the current i flowing in the conductors outside the capacitor, then total charge conservation applies, the total charge being due to both conduction and displacement currents. In this case the total conducting plate charge will not change, since the dielectric will absorb a *displacement* current that is equivalent to the conduction current entering the upper conducting wire.

3.3.2 Charge Conservation and Circuits

In the qualitative analysis shown in Section 3.1, it was stated that circuital systems may be correctly modelled by means of circuits when "the effects of space around the wires is not significant for what happens inside the system components."

If the space around system components must not affect what occurs inside them, no charge must flow in that space. Consider again the system in Figure 3.1, shown also in Figure 3.4.

The reader will already know that although materials are commonly classified as either conducting or insulating, a perfect conductive material does not exist,[1] nor does

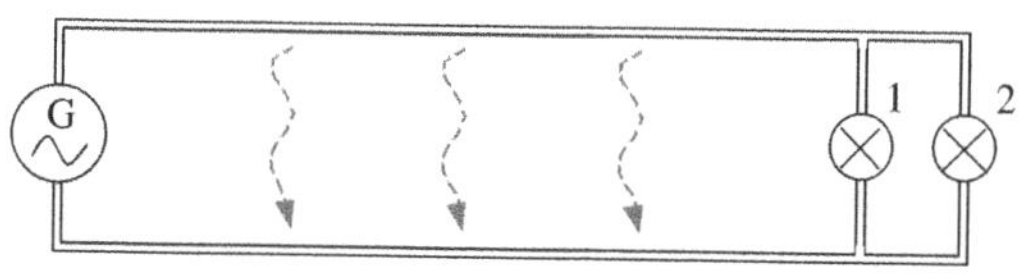

FIGURE 3.4. The system of Figure 3.1, showing stray currents.

[1]The case of "superconductivity," which occurs at very low temperatures, is a separate subject and is not discussed here.

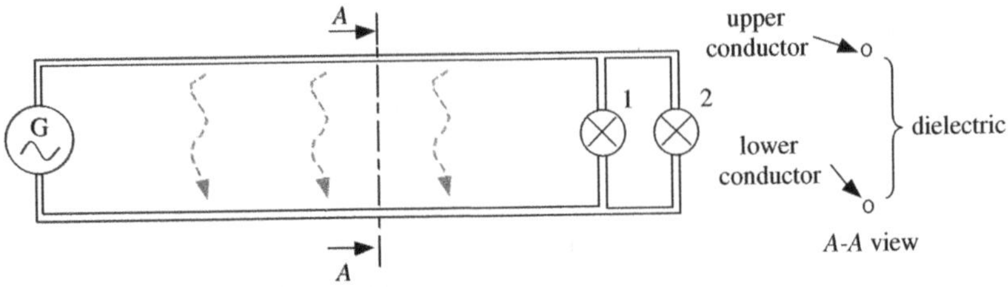

FIGURE 3.5. Stray currents through perfectly insulating means can be due to capacitive effects or due to displacement currents between upper and lower wires.

a perfectly insulating one. Therefore, even though air is a well-insulating material, some stray currents can also flow through the air surrounding the wires—for example, flowing from the upper conductor toward the lower one (as indicated in Figure 3.4), and vice versa.

If the system shown in Figure 3.1 is represented by the circuit of Figure 3.2, it should be possible to neglect these stray currents. Indeed, this is very often acceptable; although in some rare cases, stray currents are taken into account and different models are used.

It is not enough to consider that the air surrounding the conductors is perfectly insulating in the step from Figure 3.1 to Figure 3.2. It must be noted that the system of Figure 3.1 is subject to variable quantities, since the generator is a component (which we have not yet discussed in depth) able to determine a variable voltage at its ends. Therefore all the quantities of the system in Figure 3.1 (currents, voltages, fields at all of the points of space) vary with time. When quantities vary with time, displacement currents can flow between the upper and lower wires even though the air might be assumed to be perfectly insulating. Indeed, upper and lower conductors can be thought as being the two conductor plates of a capacitor whose insulating layer is the air between them (Figure 3.5). Again, in the process that leads from circuital systems to lumped-parameter circuits, neither conductive nor displacement currents must flow through the air (or space) interspersed between the ideal wires of any circuit.

Consider again one of the lumped-component models proposed for the system of Figure 3.1—that is, the circuit also shown in Figure 3.3 and in Figure 3.6. Here, for simplicity's sake, the terminals are no longer specifically shown (the reader should keep in mind that they always exist).

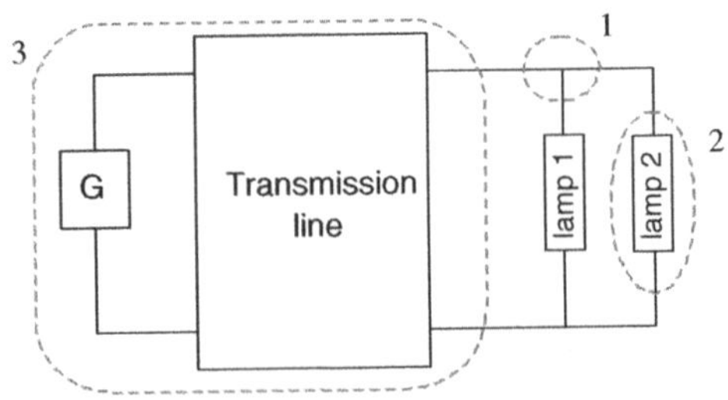

FIGURE 3.6. Possible closed curves around circuit parts.

In this figure, three different closed curves cross the circuit: a curve surrounding a circuit node (type 1), one surrounding a single lumped element (type 2), and another surrounding a group of elements and wires (type 3).

The charge conservation principle applies to these; therefore the global charge entering the curves, whether conductive or displacement, must be equal to zero. But it was also said that whenever a circuit is created, all displacement currents between wires are negligible.

This is equivalent to saying that displacement currents can occur only inside circuit elements.

Consequently, it can be concluded that the total (conductive) charge entering through some of the wires that traverse any of the curves must be exactly equal to the sum of those that flow out. This is equivalent to saying that the sum of the charge entering though some of the wires is identical to the sum of the charge exiting from the others.

This is the rationale behind Kirchhoff's law—that is, the charge conservation law for circuits. In the next sections it will be expressed in a way which is more formal (and more useful for practical computations).

3.3.3 The Electric Current

From their basic knowledge of electromagnetic phenomena, readers should already be aware that electric charges inside conductor materials (namely electrons) are not linked to atoms, but can move freely. If, therefore, a wire is made of conducting material, these charges can move freely from one end of it to another.

Also in other conductors, such as electrolytic solutions, there are charges which are free to move; while in conductor media the charge carriers are just electrons and are negatively charged, in an electrolytic solution the charges are carried by ions that can be either positive or negative.

In general, therefore, charge carriers in a conducting wire can move from left to right or from right to left, and they can be positively or negatively charged.

It has been experimentally verified that a positive charge moving from left to right is equivalent to a charge of the same size but with an opposite sign moving from right to left (Figure 3.7a)[2] and that the combined effect of a positive charge Q^+ and a negative charge Q^- is equivalent to a generic charge Q, assumed to be positive, moving in the same direction as Q^+ (Figure 3.7b).

Therefore, in the general case in which both positive charges move in either direction in a conductor medium, the analysis can be performed by computing an equivalent positive charge defined as

$$Q = Q_1 - Q_2$$

[2]For simplicity's sake, this book, like most textbooks, will often say "a charge moves" rather than "a charge carrier moves."

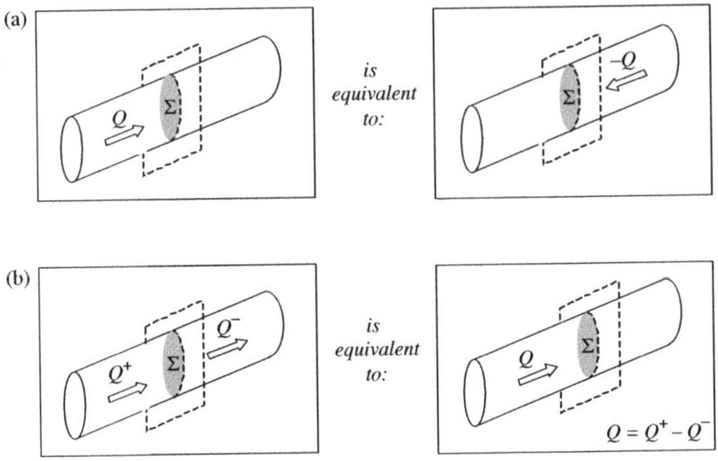

FIGURE 3.7. Charge flow equivalence: (a) A positive Q from left is equivalent to the opposite charge $-Q$ from right. (b) A comprehensive charge $Q = Q^+ - Q^-$ from left is equivalent to $Q^+ > 0$ from left and $Q^- < 0$ from right.

where:

- Q_1 is the algebraic sum of charges in the same direction as the direction assumed for Q;
- Q_2 is the algebraic sum of charges moving in the opposite direction to that assumed for Q.

If we consider this equivalent charge Q (flowing in time t through a cross section Σ of a conductor (Figure 3.7b), the current I flowing in the conductor is defined as

$$I = \frac{Q}{t} \qquad \text{for continuous flow at a constant rate}$$

If the flow of charge varies over time, an infinitesimal time interval $\mathrm{d}t$ can be considered in which the charge flowing through Σ is also infinitesimal, and it can be indicated as $\mathrm{d}q$ (remember that by convention, lowercase letters indicate quantities that vary with time; see Section 1.3.4). In this case the current $i(t)$, by definition, is

$$i(t) = \frac{\mathrm{d}q(t)}{\mathrm{d}t} \qquad \text{for any flow (constant or variable)}$$

Thus, the electric current is defined as the rate at which the charges flow. Since charges are measured in coulomb (symbol: C) in the SI, the electric current will therefore be measured in coulomb per second. Because of its importance, the electric current has a unit of measure of its own, called ampere (symbol: A): one ampere is one coulomb per second.

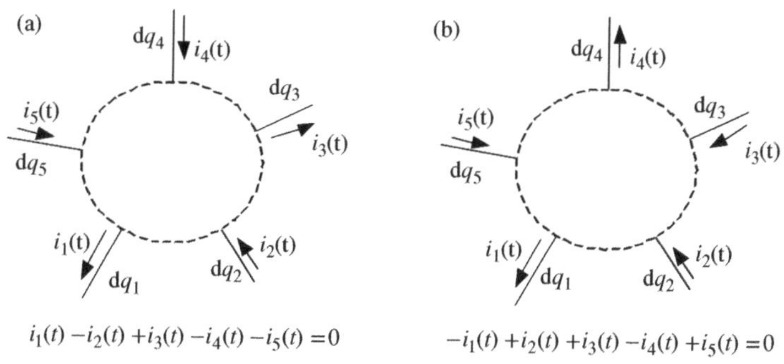

FIGURE 3.8. Examples of charge conservation law in n-terminal elements (arrows indicate actual charge flow directions).

Definition: *Ampere* (unit of measure of current)

The unit of measure of *current* is the *ampere* (symbol: A).

One ampere is the charge flow of one coulomb per second.[3] In formula: 1 A = (1 C)/(1 s)

To further clarify the definition and its sign, it can be stated that a current of 1 A moving from left to right can be equivalent to either (a) one positive coulomb charge crossing a given surface in one second from left to right or (b) one negative coulomb charge crossing a given surface in one second from right to left.

3.3.4 Formulations of Kirchhoff's Current Law

Consider a generic closed curve drawn in a circuit, in such a way that it does not cross any lumped component of the circuit, but only wires.

The situation is represented in Figure 3.8a, where just the curve and the traversing wires are shown (the parts of the circuit inside and outside the curve are omitted).

Since we are considering a *circuit*, no charge can flow outside conductors and lumped elements; only conductive charge can flow from the inner to the outer curve and vice versa, through wires.

Needless to say that while for clarity we have shown only five wires traversing the closed curve, our reasoning is valid for any number of wires.

Imagine that this charge (considered as a positive charge, as discussed earlier) flows from the outside of the loop toward the inside, as shown by the arrows near the currents in the figure.

[3]Students should know from their study of electromagnetism that this is not the SI's definition, although the SI does not contradict it; it is proposed here because it is more effective for students' understanding at this point of the book.

Let dq_k be the infinitesimal charge entering or exiting the curve in the infinitesimal time interval dt from wire k, where the directions are represented by arrows. The charge conservation for circuits (Section 3.2.2) tells us that the charge inside the dashed curve must remain constant, and therefore the charge entering must equal the one exiting:

$$dq_2 + dq_4 + dq_5 = dq_1 + dq_3 \qquad \text{(charge flows as per Figure 3.8a)}$$
$$dq_2 + dq_3 + dq_5 = dq_1 + dq_4 \qquad \text{(charge flows as per Figure 3.8b)}$$

and, dividing by the corresponding time interval dt and moving all terms on the left-hand sides, we obtain

$$-i_1(t) + i_2(t) - i_3(t) + i_4(t) + i_5(t) = 0 \qquad \text{(currents directed as per Figure 3.8a)}$$
$$-i_1(t) + i_2(t) + i_3(t) - i_4(t) + i_5(t) = 0 \qquad \text{(currents directed as per Figure 3.8b)}$$

Note that in some cases dependence on time t is explicitly shown by $i(t)$, while in other cases it is implicit. In all cases, however, the equations indicating the charge conservation law, whether expressed in terms of charges or currents, are valid at any time.

Unfortunately, we do not know in advance the actual direction of charge flows. To allow writing of equations ahead of knowing them, we consider the numbers $q(t)$ and i(t) as real numbers, which can therefore have positive or negative values.

Now we must remember what we already discussed in Section 3.2.3—that is, that charge flowing in one direction is exactly equivalent to a flow in the opposite direction of a charge having the opposite sign. Therefore it is very natural to state the following convention:

Convention: *Current sign*

The number indicating a current i is a real number indicating the charge flow per unit of time; its positive or negative sign indicates whether or not the flow direction is in agreement with the arrow near the current name i. The arrow is called reference direction of current i.

With the help of this convention, instead of dealing with actual charge flow directions, we can use the *reference* directions. Therefore, the current equations of Figure 3.8 are still valid when the arrows are not the actual current directions, but just *reference* (or, for some authors, *assumed*) current directions.

This means that we can write the equations shown in Figure 3.8, considering currents as being real numbers, whose value will be positive whenever the actual charge flow[4] is in agreement to the assumed directions and negative in the opposite case.

[4]Remember that we consider flow of *positive* charges.

Now consider a special set of assumed directions: all directions entering the zone for which we want to state charge conservation law. In this case this law can be stated simply as follows: the sum of currents at all terminals of a circuit element, assumed to be entering, must always be identically zero.

This is obviously mathematically equivalent to saying that the sum of currents of all terminals of an element, assumed to be leaving it, must always be identically zero.

In the following, whenever this does not cause ambiguity, the word "assumed" is omitted; therefore, for a branch current indicated by a symbol, the expressions "current assumed entering . . ." and "current assumed exiting . . ." will be substituted by "current entering . . ." and "current exiting. . . ."

When analysing a circuit, it often occurs that current reference directions do not all enter or leave a curve. Thus, alternative formulations are needed for generic assumed directions. They can be as follows (subscript "*in*" indicating assumed sign entering the curve, "*out*" indicating signs assumed exiting):

$$\sum_{k=1}^{n\_in} i_{k,in}(t) = \sum_{k=1}^{n\_out} i_{k,out}(t) = 0; \qquad \text{the sum of all currents entering a closed curve in a circuit identically equals the sum of all currents leaving it.} \tag{3.1}$$

$$\sum_{k=1}^{n\_in} i_{k,in}(t) - \sum_{k=1}^{n\_out} i_{k,out}(t) = 0; \qquad \text{if all currents are taken as entering, the algebraic sum of all currents entering a closed curve in a circuit is identically zero.} \tag{3.2}$$

When charge conservation is applied to a curve of types 1, 2, or 3 in Figure 3.6, it gives rise to specific versions, stating that the sum of currents assumed to be entering any node, any circuit element, or any subcircuit is always zero.

In the previous examples (for instance, those shown in Figures 3.2 and 3.3), circuit elements have two or more terminals, through which they can exchange charge with their exterior. It will be seen that, except for nodes, in the large majority of cases circuit elements will have two terminals. Given their importance, they need a name of their own; therefore they will be called *branches*.[5]

When analysing circuit branches, the charge conservation law is applied from the very first stage, simply using a unique value for the currents at their two ends. Consider Figure 3.9. It is clear that the use of a single current flowing in branch A (e.g., i_A shown in Figure 3.9b is more convenient than using different names for the currents at the two sides of element A, as indicated in Figure 3.9a); and add equations stating their algebraic equality.

Now we can state Kirchhoff's current law in three equivalent forms used in circuit analysis.

[5]The reader is advised to refer to Section 3.2 for definitions of branches and branch-based circuits.

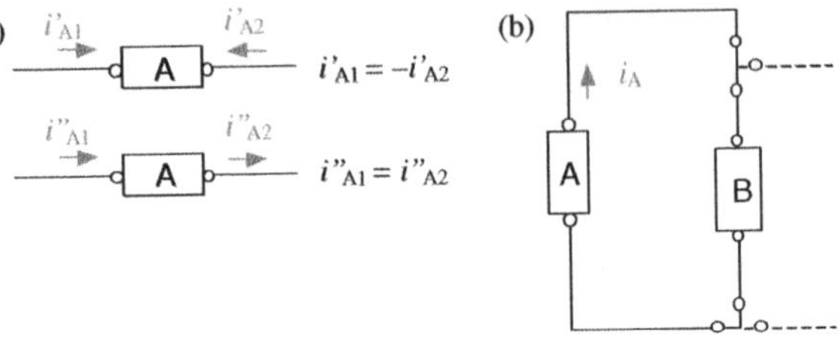

FIGURE 3.9. Charge conservation for branches: (a) Stated by equations. (b) Implicitly defined using a single branch current.

Since in the large majority of cases in everyday work situations Kirchhoff's law is applied to nodes, it is expressed here with reference to nodes. However, it should now be clear that it is still valid if we substitute the word "node" with "closed curve":

> **Law:** *Kirchhoff's Current Law* (KCL)
>
> In any circuit:
>
> **Form 1:** The algebraic sum of all of the currents entering any node in a circuit is zero.
>
> **Form 2:** The algebraic sum of all of the currents leaving any node in a circuit is zero.
>
> **Form 3:** The sum of the currents entering any node in a circuit equals the sum of the currents leaving it.
>
> KCL applies to any circuit; the three forms are still valid and the word "entering" or "leaving" is replaced with "assumed to enter" or "assumed to leave," respectively.

Example 1. Consider the circuit represented in Figure 3.10a.

The KCL equations—for instance, written in all forms 1—are shown in Table 3.1 below.

The assumed directions for currents are totally arbitrary. In Figure 3.10b, which refers to the same circuit as Figure 3.10a, some of them are different, resulting in the different systems of equations listed in Table 3.2.

Once solved, these systems will obviously give the same values for i_A, i_B, i_E, i_{FG}, and opposite values for i_C and i_D. This means that once any result is evaluated according to a given direction, it is the same, regardless of the direction assumed for writing the equations.

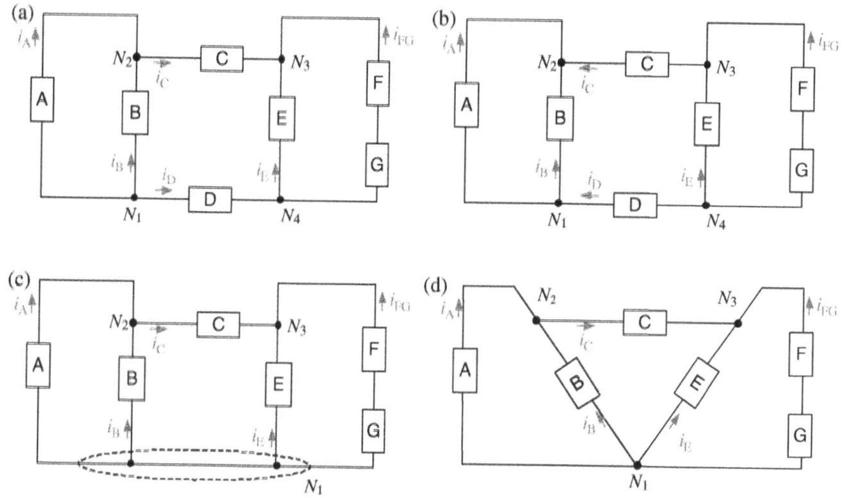

FIGURE 3.10. Example circuits showing practical application of KCL.

TABLE 3.1. KCL Equations Relating to Figure 3.10a

	Form 1	Form 2	Form 3
N_1:	$-i_A - i_B - i_D = 0$	$i_A + i_B + i_D = 0$	$0 = i_A + i_B + i_D$
N_2:	$i_A + i_B - i_C = 0$	$-i_A - i_B + i_C = 0$	$i_A + i_B = i_C$
N_3:	$i_C + i_E + i_{FG} = 0$	$-i_C - i_E - i_{FG} = 0$	$i_E + i_{FG} + i_C = 0$
N_4:	$i_D - i_E - i_{FG} = 0$	$-i_D + i_E + i_{FG} = 0$	$i_D = i_E + i_{FG}$

TABLE 3.2. KCL Equations Relating to Figure 3.10b

	Form 1	Form 2	Form 3
N_1:	$-i_A - i_B + i_D = 0$	$i_A + i_B - i_D = 0$	$i_D = i_A + i_B$
N_2:	$i_A + i_B + i_C = 0$	$-i_A - i_B - i_C = 0$	$i_A + i_B + i_C = 0$
N_3:	$-i_C + i_E + i_{FG} = 0$	$i_C - i_E - i_{FG} = 0$	$i_E + i_{FG} = i_C$
N_4:	$-i_D - i_E - i_{FG} = 0$	$i_D + i_E + i_{FG} = 0$	$0 = i_D + i_E + i_{FG}$

Consider now the circuit represented in Figure 3.10c. It is different from the previous ones, because element D has been eliminated. In this circuit, node N_1 is a unique node which is just shown in expanded form. Indeed the same circuit could be rewritten as in Figure 3.10d, perfectly equivalent to the one in Figure 3.10c, but with node N_1 unexpanded.

These two figures show that expanded representation may render the circuit more readable, for instance allowing reporting all the elements in horizontal or vertical circuit branches. The KCL equations for the circuits shown in Figures 3.10c and d are listed in Table 3.3.

The KCL equations for circuits reported in Figures 3.10c and 3.10d are as shown in Table 3.3.

TABLE 3.3. KCL Equations Related to Figures 3.10c and 3.10d

	Form 1	Form 2	Form 3
N_1:	$-i_A - i_B - i_E - i_{FG} = 0$	$i_A + i_B + i_E + i_{FG} = 0$	$0 = i_A + i_B + i_E + i_{FG}$
N_2:	$i_A + i_B - i_C = 0$	$-i_A - i_B + i_C = 0$	$i_A + i_B = i_C$
N_3:	$i_C + i_E + i_{FG} = 0$	$-i_C - i_E - i_{FG} = 0$	$i_C + i_E + i_{FG} = 0$

3.4 CIRCUIT POTENTIALS AND KIRCHHOFF'S VOLTAGE LAW

3.4.1 The Electric Field Inside Conductors

Consider the very simple circuital system shown in Figure 3.11, containing just a battery, a conductor, and an electric load in the form of a lamp.

Remember that inside the conductor, as everywhere in space, there is proportionality between current density $\boldsymbol{J}$ and the total electrical field $\boldsymbol{E}_t$ (see Section 2.7):

$$\boldsymbol{E}_t = \rho \boldsymbol{J} \qquad (\boldsymbol{E}_t = \boldsymbol{E}_b + \boldsymbol{E}_c) \tag{3.1}$$

where $\rho = 1/\sigma$ is the resistivity of the material at the point where $\boldsymbol{E}_t$ and $\boldsymbol{J}$ are evaluated.

Equation (3.1) is discussed in Section 2.7.2.

Electric field $\boldsymbol{E}_b$ is caused by the input of external power that the battery receives from chemical potential energy, while $\boldsymbol{E}_c$ is due to charges distributed along the surfaces of conductor wires; it is therefore conservative and its work along the circuit is null.

These surface charges might have the appearance of those depicted in Figure 3.11 with plus (+) and minus (−) signs. Much more on surface charges in conductors can be found in Chapter 8 of [be1].

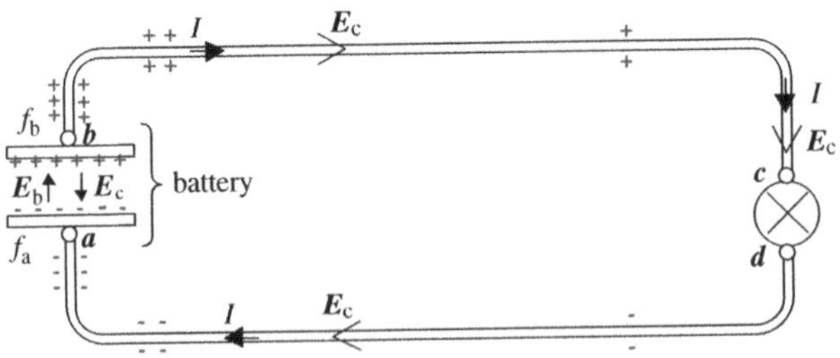

FIGURE 3.11. A simplified version of the circuital system of Figure 3.1: only one lamp is present and the voltage generated by the generator is constant over time.

Integrating (3.1) around the loop of the circuit gives

$$\oint \boldsymbol{E}_{\mathrm{t}} \cdot \mathrm{d}\boldsymbol{l} = \oint \rho \boldsymbol{J} \cdot \mathrm{d}\boldsymbol{l} = \left(\rho_{\mathrm{b}} \frac{l_{\mathrm{b}}}{S_{\mathrm{b}}} + \rho_{\mathrm{c}} \frac{l_{\mathrm{c}}}{S_{\mathrm{c}}} + \rho_{\mathrm{l}} \frac{l_{\mathrm{l}}}{S_{\mathrm{l}}}\right) I = (R_{\mathrm{b}} + R_{\mathrm{c}} + R_{\mathrm{l}}) I \tag{3.2}$$

where R_{b}, R_{c}, and R_{l} indicate, respectively, the physical characteristics of battery b, conductor c, and lamp l, called "resistances," which will be further discussed in the next chapter. In this section attention must be given to the type of subscripts. For instance "b" means "battery", while "b" indicates point $\boldsymbol{b}$ of the system in Figure 3.11.

Exploiting the conservative nature of $\boldsymbol{E}_{\mathrm{c}}$, which implies the cyclic integral of $\boldsymbol{E}_{\mathrm{c}}$ to be zero, we obtain equation (3.2):

$$\int_a^b \boldsymbol{E}_{\mathrm{b}} \cdot \mathrm{d}\boldsymbol{l} = (R_{\mathrm{b}} + R_{\mathrm{c}} + R_{\mathrm{l}}) I$$

which shows that the power generated by the battery is dissipated in the circuit elements (including battery inner resistance) as a result of their resistivity.

The example given, however, also has the very important purpose of showing that in this circuit an electric field $\boldsymbol{E}_{\mathrm{c}}$ is generated all around the loop, because of surface charges, and therefore has the nature of an electrostatic field; it is conservative and thus has a potential function V.

Between any two points of the circuital system we can define a potential difference or voltage U. For instance, between $\boldsymbol{a}$ and $\boldsymbol{b}$ or $\boldsymbol{c}$ and $\boldsymbol{d}$ it is:

$$U_{ba} = V_b - V_a, \qquad U_{cd} = V_c - V_d$$

A voltage expressed as potential difference is often called *voltage across terminals*. For instance, U_{ba} is the voltage across terminals $\boldsymbol{b}$ and $\boldsymbol{a}$.

A circuit that most naturally represents the system of Figure 3.11 is the one shown in Figure 3.12.

The ideal wires are needed to separate the four basic elements, but obviously the potentials at the ends of ideal wires must be the same (e.g., $V_a = V_{a'}$, etc.). If all the consecutive branch voltages are summed, we have

$$U_{ba} + U_{cb} + U_{dc} + U_{ad} = (V_b - V_a) + (V_c - V_b) + (V_d - V_c) + (V_a - V_d) = 0 \tag{3.3}$$

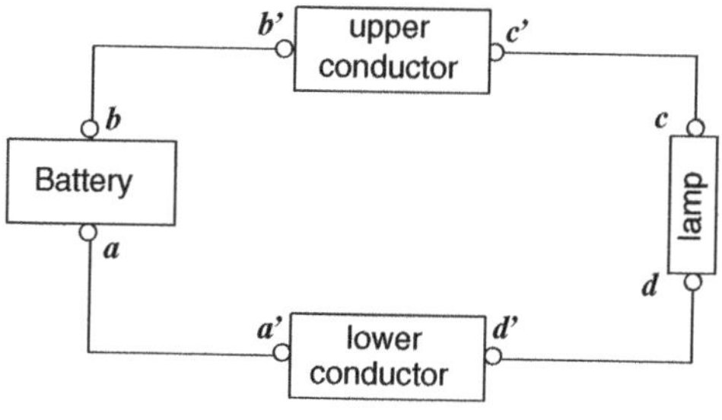

FIGURE 3.12. A circuital representation of the system in Figure 3.11.

where the equality of zero is due to the fact that all the potential values appear twice, with opposite signs.

Equation (3.3) shows that the consecutive *voltages across terminals* (i.e., potential differences) in the loop constituted by the circuit have a sum equal to zero. This is very similar to the well-known law of physics which states that the work performed by a conservative field in any closed loop is zero: the sum of the voltages is indeed the work, per unit charge, performed by the electric field $\boldsymbol{E}_c$ present in the conductor.

Equation (3.3) is interesting, but it would be much more useful if it could be extended to larger families of circuital systems and to the related circuits.

A question arises immediately: does this result depend on the fact that the electric source is a battery (i.e., a constant voltage), or is it valid also in the case of time-varying sources, such as the sinusoidal voltage source considered in Figure 3.1? Indeed, result (3.3) comes from the physical model of the system in Figure 3.11, which led us to consider the presence of the electric field $\boldsymbol{E}_c$ inside the conductor, created only by the charges appearing on the surface of the circuit conductors. It was implicitly assumed that no other fields were induced by the possible presence of other electromagnetic phenomena outside the system shown in Figure 3.11—for example, time-varying magnetic fields produced by other systems not shown, which, according to Faraday's law, would have induced an additional contribution to (that would have increased or decreased) the conductor's inner electric field.

An analysis of this topic requires knowledge that we will acquire in Chapter 5. The following is a preview of the result to be found there.

The result (3.3) will be valid also in cases in which the voltage source varies with time, given that the induction phenomena have negligible effects; that is, the voltage induced by effect of Faraday's law in the considered loop is negligible.

Or, in other words, *the consecutive voltages across terminals (i.e., potential differences) in a one-loop circuital system have a sum that is equal to zero whenever the voltage induced by effect of Faraday's law in the considered loop is negligible.*

Faraday's law is discussed in Section 2.5.

This is congruent with the initial rule stated above, for creating circuits from circuital systems; that is, any action in the space around the wires or what happens in the wires themselves should be negligible—in this case the electric field induced by the presence of a time-varying magnetic field around the circuit and in the space occupied by the circuit loop.

A second very important question that arises from an analysis of the simple circuit shown in Figure 3.11 is this: can the result (3.3) be extended also to more complex circuits containing more than a single loop, such as the system in Figure 3.1?

Consider again the system of Figure 3.1, shown again in slightly modified form in Figure 3.13. At any given instant the situation inside the conductors can be imagined as shown in the figure: the field inside generator $\boldsymbol{E}_G$ fulfils the same role as $\boldsymbol{E}_b$ in the previous example; and at any given point of the conductors, field $\boldsymbol{E}_c$ is present. Note that the value of $\boldsymbol{E}_c$ varies with the point of the conductor (it is an electric field with a value at any point of space) and both $\boldsymbol{E}_G$ and $\boldsymbol{E}_c$ vary with time. As in the previous

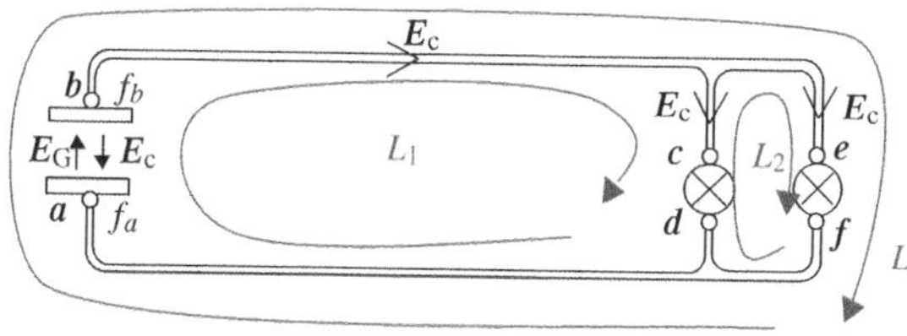

FIGURE 3.13. A different view of the circuital system of Figure 3.1.

example, $\boldsymbol{E}_c$ is due to the surface charges present in the conductors, including the surfaces of the conductive parts f_a and f_b that interface G with the wires. $\boldsymbol{E}_c$ thus has the nature of an electrostatic field, even though it varies with time, and therefore will admit a potential $v(\boldsymbol{p})$, function of the point $\boldsymbol{p}$ of the circuit, variable with time and therefore represented by a lowercase symbol.

Since v is a function of the point, the potential differences in any closed loop sum to zero, which is identical to the process that led to equation (3.3). For instance, for the loops L_1, L_2, L_3 in the system of Figure 3.13, it will be

$$\begin{aligned} &L_1\text{: } u_{ba} + u_{cb} + u_{dc} + u_{ad} = (v_b - v_a) + (v_c - v_b) + (v_d - v_c) + (v_a - v_d) = 0 \\ &L_2\text{: } u_{cd} + u_{ec} + u_{fe} + u_{df} = (v_c - v_d) + (v_e - v_c) + (v_f - v_e) + (v_d - v_f) = 0 \\ &L_3\text{: } u_{ba} + u_{eb} + u_{fe} + u_{af} = (v_b - v_a) + (v_e - v_b) + (v_f - v_e) + (v_a - v_f) = 0 \end{aligned} \tag{3.4}$$

where the equality to zero is due to the fact that all the potential values appear twice, with opposite signs.

The voltages (or potential differences) can be shown in the circuit corresponding to the circuital system, as shown for the generator, the lamps, and the lower conductor in Figure 3.14. Whenever a voltage appears in a circuit, the reference polarities must be indicated: a plus sign or a pair of plus–minus signs. They indicate the references for measuring the corresponding voltages: the number indicate that a voltage is equal to the difference of potentials between the point of the circuit marked "+" and the one marked "–".

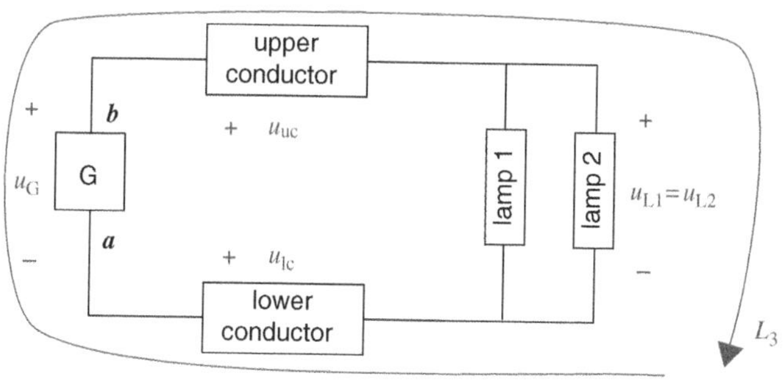

FIGURE 3.14. The circuit of Figure 3.2, showing voltages, along with corresponding polarities.

For instance, in the next figure it is $u_G(t) = v_b(t) - v_a(t)$. Since the negative sign must always be at the side opposite to the positive one, it can be omitted, as in the figure for the conductor voltages u_{uc} and u_{lc}. A time-varying voltage $u(t)$ will normally assume, as time passes, positive and negative values; so the part of circuit marked "+" will actually have a higher potential than the one marked "−" when $u(t)$ has a positive value.

Because of its importance, the voltage sign convention is shown in the following box:

> **Convention:** *Voltage sign*
>
> The number indicating a circuit voltage u is equal to the difference between the potential of the circuit wire marked "+" and that marked "−".

3.4.2 Formulations of Kirchhoff's Voltage Law

It should now be clear that the results obtained in Figures 3.11 and 3.13 can be extended to circuital systems and related circuits of varying complexity and with any number of loops.

These results are summarised by equations (3.3 and (3.4)): in all circuits a function V of terminals exists, called potential, so that the voltages across terminals are expressed as the difference of the corresponding potentials.

The very existence of V implies that the sum of consecutive branch voltages around any loop is equal to zero.

In this section these results are expressed in a more formal and practical way. Prior to this, however, the concepts of *voltage rises* and *voltage drops* must first be introduced. Consider again Figure 3.13. Before writing equations (3.4), possible loops in the circuit were indicated and reference directions were assumed (indicated by the arrows on the lines representing the loop). These reference directions are arbitrary. If the loop is followed, starting from one point and returning to the same point, in the reference direction, branches are traversed. The negatively marked terminal of a traversed branch will be encountered either before the positively marked one or after it. In the former case the branch voltage is considered to be a voltage rise, whereas in the latter case it is considered to be a voltage drop.

For instance, in loop L_3 in Figure 3.14, u_G and u_{lc} are voltage rises, while u_{uc} and u_{L2} are voltage drops. The third of equations (3.4) could be written using the symbols in Figure 3.14, as follows:

$$L_3\text{: } u_G + -u_{uc} + -u_{L2} + u_{lc} = 0$$

or, equivalently,

$$L_3\text{: } u_G + u_{lc} = u_{uc} + u_{L2}$$

Now we can express the rationale behind loop equations like those in equations (3.4) in a general and formal way. This constitutes a fundamental law of circuits called Kirchhoff's Voltage Law (KVL). Similarly to KCL, it can be expressed in three possible forms:

Law: *Kirchhoff's Voltage Law (KVL)*

In any circuit, if any loop is traversed in an arbitrary direction (either clockwise or counter clockwise):

Form 1: The algebraic sum of all voltages across terminals, considered as voltage rises, is zero.

Form 2: The algebraic sum of all voltages across terminals, considered as voltage drops, is zero.

Form 3: The sum of all voltage rises across terminals equals that of all voltage drops across terminals.

KVL applies to any circuit.

For example, the three forms applied to loop L_3 of Figure 3.14 give rise to ($u_{L1} = u_{L2}$ indicated as u_{L12}) the following:

	Form 1	Form 2	Form 3
L_3:	$u_{lc} + u_G - u_{uc} - u_{L12}$	$-u_{lc} - u_G + u_{uc} + u_{L12} = 0$	$u_{lc} + u_G = u_{uc} + u_{L12}$

In Chapter 5 the circuit concept will be expanded, and the possibility of *magnetically coupled circuits*, will be introduced. We will then see that KVL applies individually to each circuit that is magnetically coupled with others.

3.5 SOLUTION OF A CIRCUIT

Solving a circuit normally means finding a value for all node potentials and wire currents. For branch-based circuits, this simply means calculating the values of all branch voltages and currents.

Some of these equations will be KCL and KVL equations; others will give the description of the branches' inner behaviour and will be referred to as "constitutive equations."

This section will provide information on how to write linearly independent Kirchhoff equations and constitutive equations; in addition, we will examine whether or not the resulting equations are sufficient to determine all currents and voltages.

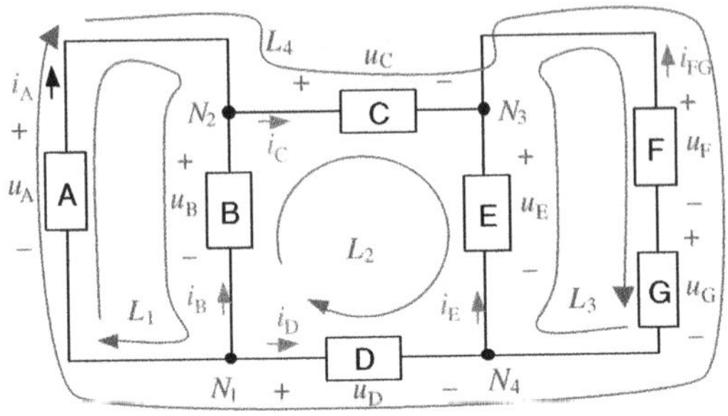

FIGURE 3.15. A circuit highlighting nodes and loops.

3.5.1 Determining Linearly Independent Kirchhoff Equations (Loop-Cuts Method)

Let us consider a branch-based circuit with b branches and n nodes. To fix ideas, consider the circuit shown in Figure 3.15, where names for nodes, reference direction of currents, and some possible loops have already been arbitrarily proposed. Note that the displayed loops do not cover all possibilities (for instance, there could also be a loop sequentially traversing B, C, F, G, and D).

Consider the KCL for all the circuit nodes (written using form 1):

$$\begin{aligned} &N_1\text{: } -i_A - i_B - i_D = 0 \\ &N_2\text{: } i_A + i_B - i_C = 0 \\ &N_3\text{: } i_C + i_E + i_{FG} = 0 \\ &N_4\text{: } i_D - i_E - i_{FG} = 0 \end{aligned} \tag{3.5}$$

It can be immediately verified that these equations are not linearly independent, since

$$N_4 = -(N_1 + N_2 + N_3)$$

It is also easy to verify [using the usual algebraic techniques to solve linear systems, e.g. computing the determinant of subsystems of the system (3.5)] that only three of the equations in (3.5) are linearly independent.

The four KVL equations relating to the loops shown in Figure 3.15 can also be written, again using form 1, as follows:

$$\begin{aligned} &L_1\text{: } u_A - u_B = 0 \\ &L_2\text{: } u_B - u_C - u_E + u_D = 0 \\ &L_3\text{: } u_E - u_F - u_G = 0 \\ &L_4\text{: } u_C - u_F - u_G + u_D + u_A = 0 \end{aligned} \tag{3.6}$$

Again, it is easy to see that these equations are not linearly independent, since

$$L_4 = L_1 + L_2 + L_3$$

and it can be easily verified that only four of the equations (3.6) are linearly independent.

Let us now introduce a technique that allows us to write only linearly independent Kirchhoff equations.

The simple case of a circuit without nodes obviously requires no KCL equations and a single KVL equation. In the other cases the following simple procedure, which in this book will be called the *loop-cuts method,* can be used:

- The KCL is applied to any set of nodes amounting to the total number of nodes n diminished by one.
- The KVL can be determined in the following recursive manner:
  a. Write an equation for an arbitrary circuit loop; a cross can then be applied (or imagined to be applied) to a branch of that circuit. Application of a cross can be imagined to create the effect of "cutting" the branch, hence the name of the method.
  b. If one or more loops are still present in the circuit, return to step a, but loops must not contain branches already crossed.
  c. In the case of branches in parallel, if a single voltage name is used across the branches, a cross must be applied to all branches in parallel but one, before starting to write the KVL equations.

It is apparent that the procedure proposed allows determination of linearly independent equations. If for instance the KVL is considered, crossing a branch ensures that the equation later chosen does not contain the voltage across that branch and is therefore linearly independent of the previous one. Since this reasoning can be repeated recursively, we can conclude that the set obtained is constituted by linearly independent equations. It can also be demonstrated that the number of equations that can be obtained this way is equal to $b-n+1$.

Rule: *Determination of independent Kirchhoff equations (loop-cuts method)*

(1) Independent KCL equations can be obtained by applying the law to all the circuit nodes except one (arbitrarily chosen).

(2) Independent KVL equations can be obtained by crossing ("cutting"), in a loop equation, an arbitrary branch of the loop and choosing the next loop equation in such a way that no crossed branch is traversed.

Example 2. Figure 3.16 shows two possible ways to determine linearly independent KVL sets. After L_1 is determined, crossing C_1 excludes branch A from subsequent loops; and after L_2 is determined, crossing C_2 excludes branches B_1, C, and D from the subsequent final loop L_3, which excludes previously crossed or unconnected branches.

The corresponding sets of linearly independent equations are

$$\text{Set 1: } \begin{cases} L_1\text{: } u_A - u_B = 0 \\ L_2\text{: } u_B - u_C - u_E + u_D = 0 \\ L_3\text{: } u_E - u_F = 0 \end{cases} \qquad \text{Set 2: } \begin{cases} L_1\text{: } u_A - u_B = 0 \\ L_2\text{: } u_A - u_C - u_E + u_D = 0 \\ L_3\text{: } u_E - u_F = 0 \end{cases}$$

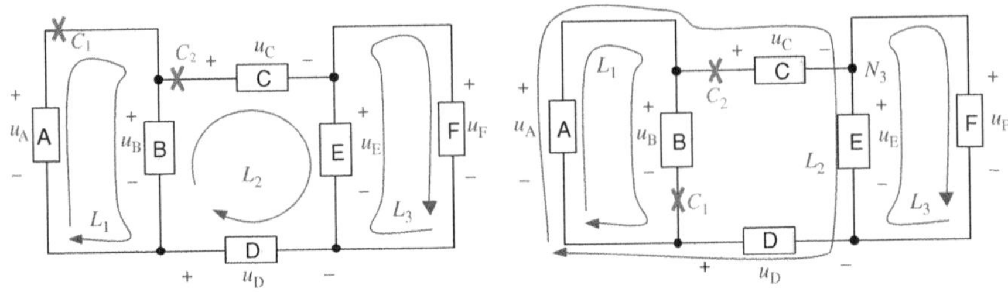

FIGURE 3.16. Graphic procedure to determine linearly independent KVL equations.

The fact that the two sets are equivalent is confirmed by the fact that set 2 can be obtained from set 1 by substituting L_2 with $L_1 + L_2$.

3.5.2 Constitutive Equations

Clearly, Kirchhoff's equations themselves are not sufficient to determine the full behaviour of a circuit (i.e., all the currents and voltages), since information is also necessary on how the circuit elements (i.e., for branch-based circuits, the circuit branches) operate internally.

We may say that Kirchhoff's laws only define the *topology* of the circuit; no information is given about the inner behaviour of the circuit branches. This behaviour is introduced by means of equations, called *constitutive equations*; they will often be referred to by their acronym CE.

Constitutive equations of matter have already been introduced in Chapter 2. Here, instead, we define constitutive equations of lumped circuit components.

The same expression in these two cases indicates different but similar concepts. Indeed, when dealing with the three-dimensional space, at any point of space the existing matter imposes relations between local fields existing in that point: the constitutive equations discussed in Chapter 2. For instance $\boldsymbol{B} = \mu \boldsymbol{H}$ indicates a precise relation between magnetic and flux density fields at the considered point.

When dealing with circuits, the "atomic" element to consider is a single branch, and its constitutive equation characterises how that branch contributes to the definition of circuit variables (i.e., voltage and current). This way a circuit constitutive equation is a relation (algebraic or differential) between branch voltage and current.

Constitutive equations for the most important components will be introduced and discussed in the next two chapters. Here, we will limit ourselves to stating that a constitutive equation is a relation between the branch current and voltage.[6] Examples of constitutive equations are shown in Table 3.4.

[6]Note that for any circuit, it is assumed that the effects on the behaviour of any branch of the potentials at its terminals p_1 and p_2 are only through the potential difference $u = v_{p1} - v_{p2}$. This is coherent with the assumption, valid for any potential, that single potential values cannot have any effect: only potential differences can have measurable effects.

TABLE 3.4. The Constitutive Equation Types Used in this Book[a]

Element	Equation	Description
Resistor	$u_b = R_b i_b$	Branch voltage and current proportional (u_b and i_b measured according to the load sign convention).
Voltage source	$u_b = u_s$	Voltage is equal to u_s (subscript stands for "source") regardless of any other circuit quantity.
Current source	$i_b = i_s$	Current is equal to i_s (subscript stands for "source") regardless of any other circuit quantity.
Inductor	$u_b = L\ \mathrm{d}i_b/\mathrm{d}t$	Branch voltage proportional to time derivative of branch current (the "+" sign must be used if the current is assumed to be entering the branch from the positively marked terminal).
Capacitor	$i_b = C\ \mathrm{d}u_b/\mathrm{d}t$	Branch current proportional to time derivative of branch voltage (the "+" sign must be used if the current is assumed to be entering the branch from the positively marked terminal).
Nonlinear algebraic element	$u_b = f_u(i_b)$ or $i_b = f_i(u_b)$	A component of this type is the diode, which will be discussed in Chapter 8.

[a]In equations containing both current and voltage, the load sign convention is used.

Readers might recognise the equations of typical circuit components already encountered in their previous studies; as regards constitutive equations with a current and voltage, the coefficients are positive (i.e., R, L, C); branch voltage and current must be associated in such a way that the assumed direction of current enters the branch in the positively marked terminal. This combination of reference signs will be later referred to as the load sign convention:

> **Convention:** *Load and generator sign convention*
>
> References for current and voltage in a branch follow the *load sign convention* when the assumed current direction enters the branch from the positively marked terminal.
>
> If the current exits the branch from the positively marked terminal, the references follow the *generator sign convention.*

Not all constitutive equations are possible in all positions in a circuit.

It is impractical to deal with this issue in a general way. As significant examples, however, consider that:

- Since two branches in a series share the same current, they cannot be characterised by current source constitutive equations since even a slight difference between the two currents would contradict the topological constraint imposing perfect equivalence of the two currents.
- Since two branches in parallel share the same voltage, they cannot be characterised by voltage source constitutive equations since even a slight difference between the two voltages would contradict the topological constraint imposing the perfect equivalence of the two voltages.

A quantity which is constant over time is indicated using uppercase symbols. For instance, a constant voltage source can be indicated as equal to U_S. Conversely, lowercase symbols indicate variable quantities, such as $u(t)$, or simply u.

The list of constitutive equations in Table 3.4 does not include branches whose current or voltage is a function of the current or voltage in another branch. Different kinds exist, including, for instance, current-controlled voltage sources or voltage-controlled current sources, which have theoretical and practical significance.

However, their usage is beyond the scope of this book, with one important exception: the *ideal transformer* component. To study this component, we need to expand the definition of circuit given in this chapter, which is the reason why it is not included in Table 3.4. This component, along with these new, expanded circuits (i.e., the *magnetically coupled circuits*) will be discussed in Chapter 5.

3.5.3 Number of Variables and Equations

We have seen that solving a circuit entails finding a value for all node potentials and wire currents; for branch-based circuits, this simply means finding the values of all branch voltages and currents.

A branch-based circuit containing b branches has, in general, b voltages and b currents as independent variables, for a total of $2b$ variables. Solving this kind of circuit, therefore, requires finding a set of $2b$ independent equations that make mathematical connections between all the branch voltages and currents.

It could be demonstrated that finding the solution to a circuit is well-posed since we can write both b Kirchhoff equations and b constitutive equations. However, such a large number of equations is typically written only by algorithms for a computerised solution of the circuits; for a manual solution, this number is typically dramatically reduced by visual analysis of the circuit.

Take, for instance, the circuit shown in Figure 3.17a. It contains 5 branches, and therefore in general we can write a system of 10 equations in 10 unknowns.

If the circuit has to be manually solved, however, it can be immediately understood that branches A, B, and C share the same voltage, for which a unique variable can be used, and that branches D and E share the same current, for which a unique variable

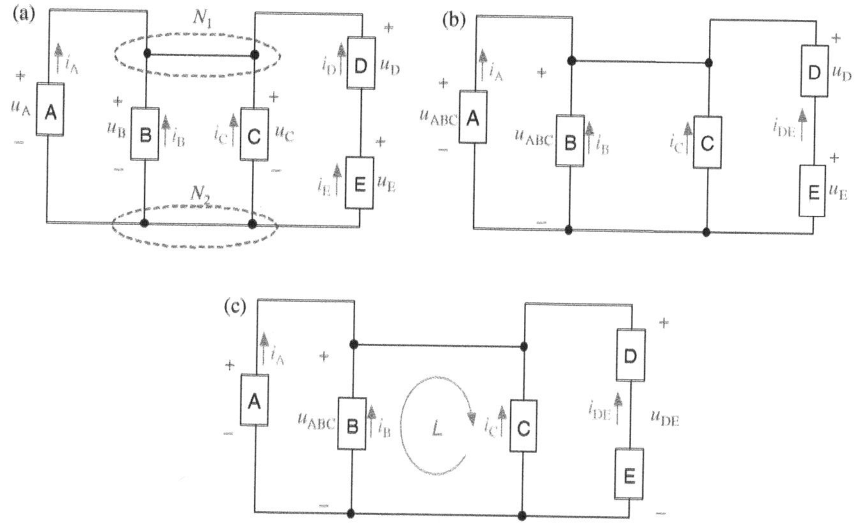

FIGURE 3.17. An example showing circuit simplification and reduction of variables.

can be used, leading to the circuit shown in Figure 3.17b, in which there are now 7 instead of 10 variables (currents and voltages).

Obviously, this reduction in system variables implies a corresponding reduction in Kirchhoff equations, so that the number of equations and variables will be balanced: in this case 7×7.

The circuit can be simplified further, as will be seen in more practical cases in the next two chapters. In a case like the system of Figure 3.17, the system could be reduced to four equations in four variables.

Figure 3.17a also gives us the opportunity to view a particular visual representation of nodes: what is inside of curves N_1 and N_2 is something that connects three or more circuit elements; therefore they are actually nodes. The black circles they contain might be considered nodes as well. However, black circles are connected to each other by wires; it is common and convenient to consider them aggregate into "expanded" nodes such as N_1 and N_2.

Example 3. Consider the circuit shown in Figure 3.18a. For some of the currents, the numerical values are already known, while the others, namely i_1, i_2, and i_3, are unknowns. The circuit contains four nodes, which lead to $4-1=3$ independent KCL equations, for example,

N_1 (form 1): $i_3+2+3=0$ (implying $i_3=-5$ A)
N_3 (form 1): $i_1+1+1=0$ (implying $i_1=-2$ A)
N_2 (form 2): $3+i_2+i_1=0$ (implying $i_2=-i_1-3=-1$ A)

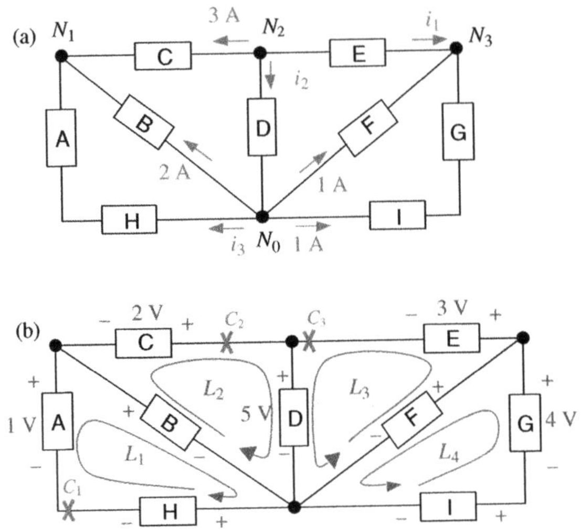

FIGURE 3.18. A basic example showing the use of KCL and KVL.

It can thus be verified that the $n-1$ equations ($n=4$ being the number of circuit nodes) are linearly independent, allowing us to determine the three unknown currents; moreover, the assumed directions for the currents are useful for writing the equations, but in no way indicate the actual currents. In the example of Figure 3.18, the actual flow directions are all opposite the assumed ones.

In Figure 3.18b, the four voltages u_B, u_F, u_H, and u_I are unknowns. Using a somewhat wise choice of loop equations, four linearly independent ones can be determined, for example:

L_2 (form 1): $u_B+2-5=0$ (implies $u_B=3$ V; cut C_2 is made)
L_1 (form 1): $-u_H+1-\ u_B=0$ (implies $u_H=1-u_B=-2$ V; cut C_1 is made)
L_3 (form 2): $-u_F+3+5=0$ (implies $u_F=8$ V; cut C_3 is made)
L_4 (form 3): $u_I+4=u_F$ (implies $u_I=u_F-4=4$ V)

***More in Depth:** Number of Kirchhoff equations*

It could be demonstrated that for any circuit, the number of Kirchhoff equations is equal to the number of branches, computing only once the branches sharing the same current.

In the example of Figure 3.18, A–H, and G–I share the same current (they are called *branches in series*); therefore the total number of independent equations should be equal to the total number of branches, reduced by two, that is, seven.

In fact, we wrote three KCL and four KVL equations.

In this example, some currents and voltages are known, and others, seven in number, remain unknown. In more frequent cases, instead of currents or voltages, some constitutive equations would be known. In these cases, unique constitutive equations must be written for the branches in series, such as A–H and G–I for the example of Figure 3.18.

The circuit can then be solved whenever n_c constitutive equations, n_u branch voltages and n_i branch currents are known, and it is

$$n_c + n_u + n_i = N \qquad \text{(number of branches with unknown current)}$$

(N is 7 in the example of Figure 3.18)

Constitutive equations of branches in series are easily determined by summing the partial constitutive equations expressed in terms of voltages as a function of currents. If, in the example of Figure 3.18 for instance, A, H, G, and I were all resistors, the four equations would be

$$U_A = R_A I_A, \qquad U_B = R_B I_B, \qquad U_G = R_{GI} I_{GI}, \qquad U_I = R_I I_I$$

which allows us to determine the following constitutive equations of the equivalent branches:

$$U_{AB} = (R_A + R_B) I_{AB}, \qquad U_{GH} = (R_G + R_H) I_{GH}$$

3.6 THE SUBSTITUTION PRINCIPLE

Consider a branch-based circuit.

Without loss of generality, we can study the one in Figure 3.17, shown again, slightly simplified, in Figure 3.19.

Any given closed curve such as curve C in the figure, which selects a part of the circuit connected to the remaining part by only two wires, has the structure of a 2-terminal element—that is, a branch.

Is it possible to consider this an actual branch, so that a single box **Eq** can be substituted for the whole circuit section inside C, without changing the solution of the remaining part of the circuit?

Consider the full solution of the left part of the circuit. It allows us to determine, in particular, voltage u and current i.

It is apparent that KCL equations can be written for the part of the circuit outside C, as well as for the part inside it using the value of i instead of the missing part. For instance, the equation for node N will, in both cases, be $i_A + i_B + i = 0$.

It is also apparent that KVL equations can be written for the part of the circuit outside C, as well as for the part inside it using the value of u instead of the missing

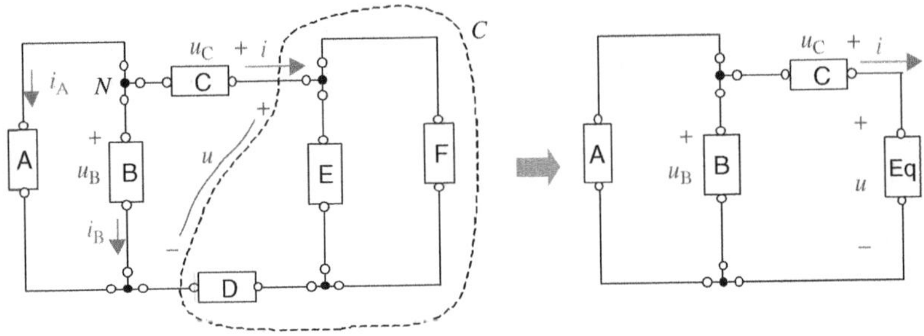

FIGURE 3.19. A sample circuit to show the so-called substitution principle.

part. For instance, the KVL equation for the loop involving **B**, **C**, and u will in both cases be $u_B + u_C - u = 0$.

It should be clear that this conclusion is valid in general for any circuit. The result is therefore as follows:

> **Result:** *Substitution principle of circuits I*
>
> A circuit portion with two terminals can always be substituted by an *equivalent branch*—that is, a branch whose constitutive equation has the same relation between its voltage and current as in the subcircuit it substitutes.

The substitution principle is often used with voltage source branches. Consider, for instance, the circuit shown in Figure 3.20, which is a slightly modified version of

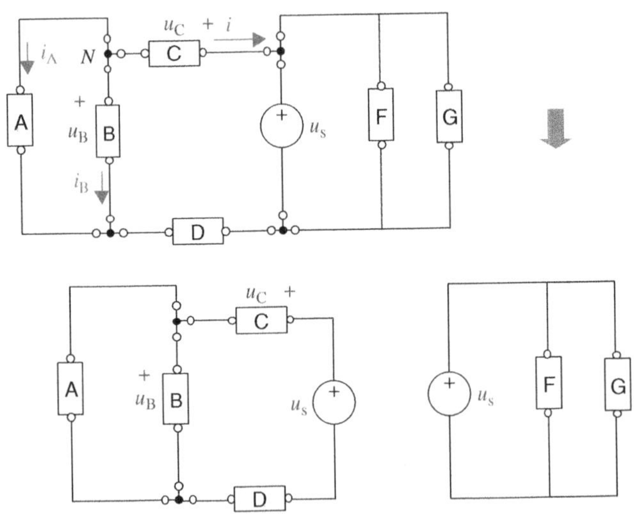

FIGURE 3.20. A simple example showing the substitution principle when a branch is a voltage source element.

the one in Figure 3.19 and in which the constitutive equation for branch E is $u_E = u_s = \text{const}$ (it does not depend on the current flowing through E). Since our knowledge of the voltage across E (i.e., u_s) is sufficient to write all Kirchhoff equations for the left part, this part can be solved without knowing what is in the circuit to the right of u_s. Therefore, the bottom-left circuit can be used. Similarly, the right part of the circuit can be solved without knowing what is in the circuit to the left of u_s. Therefore, the bottom-right circuit can be used.

This way of using the substitution principle is possible even when the component is not a voltage source component but the voltage between two terminals is known. This voltage value will enter the equations as a known value, and the system of equations will be identical to those which are valid when there is a voltage source between the terminals.

Finally, this result is valid also when *the current in a branch* instead of the *voltage across two terminals* is known in advance. These considerations lead to the following second formulation of the substitution principle.

Result: *Substitution principle of circuits II*

If a current in a branch is known in advance, the circuit can be solved by substituting the branch with a source component.

If a voltage between two terminals of a branch is known in advance, the circuit can be solved by substituting the branch with a voltage source component.

This substitution will often enable large circuits to be solved in sequence. Compare Figure 3.20.

3.7 KIRCHHOFF'S LAWS IN COMPARISON WITH ELECTROMAGNETISM LAWS

The reader will already know that there are four basic electromagnetism laws (also called the four Maxwell equations):

- Gauss's law
- Ampère law (with Maxwell's extension)
- Faraday's law
- Gauss's law for magnetism

Those who wish to recall these laws are advised to consult Chapter 2. Maxwell's extension to Ampère's law is in *More in Depth B* of that chapter.

Kirchhoff's laws are not included in the list.

How do Kirchhoff's equations compare to electromagnetism? This is a question that very few people are able to answer immediately. But to the reader of this book the answer should be straightforward, since they will now know that:

> Kirchhoff's laws apply to circuits—that is, to mathematical models of physical systems—while the electromagnetism laws (and therefore Maxwell's equations) apply to actual physical systems.

Therefore, from this perspective, Kirchhoff's laws can be applied to circuits, while their applicability to physical systems is undefined.

In other words, whether or not Kirchhoff's equations are valid for a given circuit or a circuital system is not well-posed: the actual problem is whether and when a given system can be modelled using a given circuit.

It was seen in this chapter that in the so-called DC circuits, where all quantities are constant over time, circuits can be used as models of physical systems whenever the space between wires is perfectly insulating, and this is very often the case.

On the other hand, in the so-called AC circuits, where all quantities vary with time, circuits can be used as models of physical systems, whenever conductive and displacement currents between wires have negligible effects, and voltages induced in the circuit loops, by effect of mutual induction or self-induction, are also negligible.

It will be seen in Chapter 5 that this is not a big limitation and it allows circuits to correctly model a large quantity of systems, but not all. For instance, in long transmission lines, in which conductors remain parallel to each other for kilometres, self-inductance is not negligible. This special case can still be solved if we extend the circuit concept, as will be discussed in Section 5.3.2 and the Appendix.

3.8 POWER IN CIRCUITS

We have mentioned several times that physical systems modelled through circuits have to be independent from other systems, in the sense that no mutual induction with external systems must occur.

In this case, any physical system modelled through a circuit will be energetically autonomous, in the sense that the energy generated by some parts of the given systems will be absorbed by others.

Consider again the basic example proposed throughout this chapter, i.e. the system shown in Figure 3.1. It is apparent that the generator will introduce power into the system from an external source (e.g., a rotating shaft), and that power will be equal to the power which is dissipated in wires and lamps, i.e. converted into heat and luminous energy. This power neutrality is a very important characteristic of systems, and will be retained in any well-made model. It will soon be demonstrated that this power neutrality exists in all circuits; that is, the energy conservation principle applies.

Let us first consider how to mathematically evaluate the power flowing in a section of a circuital system.

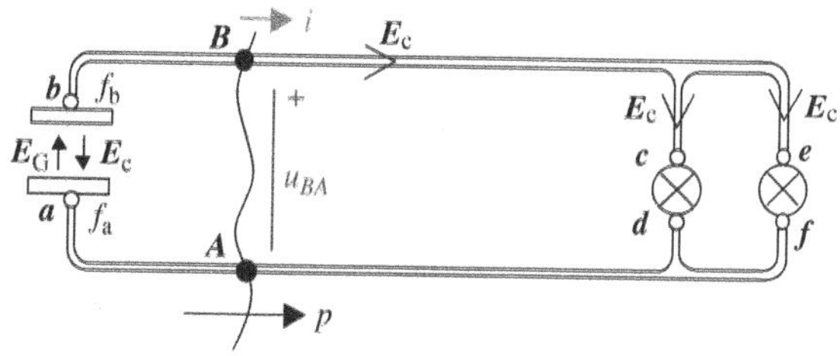

FIGURE 3.21. A physical circuit used to introduce transferred power $p(t)$.

Consider the system of Figure 3.1, again presented in Figure 3.13, and reproduced, with some modifications, in Figure 3.21. The work per unit of charge W that the electric field inside the conductor performs on the conductor charges, is the integral of $\boldsymbol{E}_c \cdot \boldsymbol{dl}$ by an arbitrary path connecting $\boldsymbol{A}$ to $\boldsymbol{B}$. Note that if only $\boldsymbol{E}_c$ is to be accounted for in the following equation, the path should be chosen in the right part of the circuit—for example, ***B–c–d–A*** or ***B–e–f–A***.

$$P = \frac{\mathrm{d}W}{\mathrm{d}t} = \int_A^B \boldsymbol{E}_c \times \mathrm{d}\boldsymbol{l} = v_B - v_A = u_{BA}$$

where v is the potential of electric field $\boldsymbol{E}_c$.

The work per unit of time P—that is, the power that traverses the circuit and goes from the left part of points $\boldsymbol{AB}$ to the right of them—is therefore

$$P = \frac{\mathrm{d}W}{\mathrm{d}t}\frac{\mathrm{d}q}{\mathrm{d}t} = u_{BA} \times i \tag{3.7}$$

Equation (3.7) states that the power flowing through two wires in a circuital system is equal to the product voltage times current. Attention must be paid to the signs, since the voltage $u_{BA} = v_B - v_A$ is the opposite of u_{AB}: the power measured by $p = ui$ is toward the right part of the circuit, when i represents a current that enters the right part of the circuit through the positively marked terminal that gives the reference polarity for u.

When a circuit models a circuital system, it shares the point of interconnection of elements, voltages, and currents with the system to be modelled.

Circuits are power-neutral; that is, the power delivered by branches perfectly equals the power absorbed by the others; the energy conservation principle thus applies:

> **Law:** *Energy conservation in circuits*
>
> In any circuit, at any time interval, the sum of energy absorbed by all branches is zero. This is equivalent to saying that, at any time, the power delivered by the branches acting as generators equals the power absorbed by those acting as loads.

The demonstration of this law is in the next section.

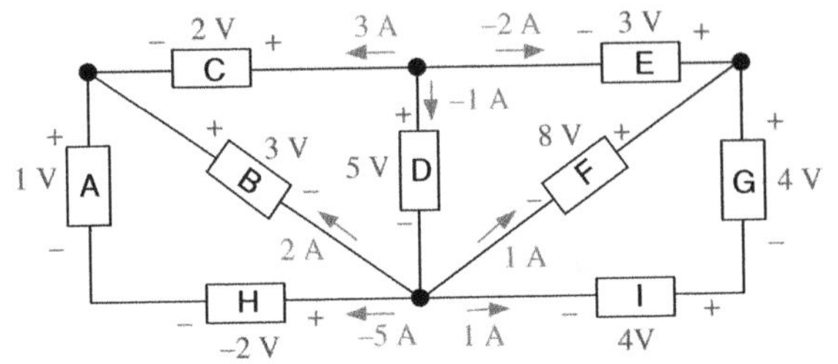

FIGURE 3.22. Currents and voltages computed in Example 3, Section 3.4.3.

TABLE 3.5. Power Balance of the Circuit Shown in Figure 3.22

	u/V	i/A	p_{abs}/W
A	1	5	5
B	3	−2	−6
C	2	3	6
D	5	−1	−5
E	3	2	6
F	8	−1	−8
G	4	−1	−4
H	−2	−5	10
I	4	−1	−4
TOTAL	—	—	**0**

Example 4. As an example consider again the circuit used in Figure 3.18, shown again in Figure 3.22, containing the assumed current signs and voltage polarities used in the same figure, and the corresponding numerical values of voltages and currents as already computed in Section 3.4.3. For each branch, the product ui will be a power produced or absorbed by the branch, depending on the combination of references in the circuit: they are absorbed if the assumed sign for current enters the positively marked terminal.

Voltages, currents (measured as entering into the positively marked terminals) and absorbed powers are shown in Table 3.5, which shows that power conservation applies to the circuit: the algebraic sum of the powers absorbed by all branches is equal to zero.

3.8.1 Tellegen's Theorem and Energy Conservation Law in Circuits

This section is not needed for the comprehension of other parts of this book, except for Section 5.6.3. Therefore its reading can be postponed.

Consider a generic circuit containing n nodes, named by consecutive natural numbers, with reference to just to fix ideas, the five-node circuit shown in Figure 3.23.

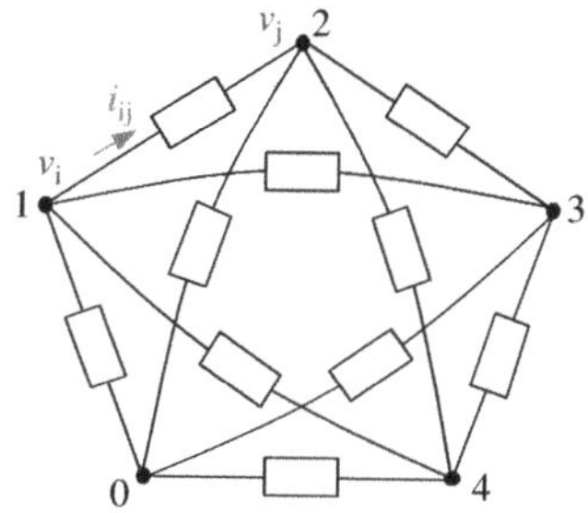

FIGURE 3.23. A generic circuit used for demonstration of Tellegen's theorem.

Without loss of generality let the circuit have only a single branch (possibly equivalent of several branches[7]) between each pair of nodes.

Consider the current between node i and j as having the assumed direction indicated in the figure—that is, from i to j. Tellegen's theorem states the following:

If

- $\{v_0, v_1, \ldots v_n\}$ is a set of node voltages satisfying KVL for the circuit in question
- $\{i_{01}, i_{02}, \ldots i_{0n}, \ldots i_{12}, i_{13}, \ldots i_{1n}, \ldots i_{1n-1,n}, i_{1n}\}$ is a set of currents potentially flowing in the circuit branches (from the first node to the second one) satisfying KCL for the circuit in question

then

$$\sum_{\substack{i=0,n\\ j=0,n\\ i\neq j}} u_{ij} i_{ij} = 0 \tag{3.8}$$

Tellegen's theorem can be very easily demonstrated by writing the branch voltage u_{ij} as the difference of potential v_i and v_j (this can be done because the KVL applies) as follows:

$$\sum_{\substack{i=0,n;\ j=0,n\\ i\neq j}} u_{ij} i_{ij} = \sum_{\substack{i=0,n;\ j=0,n\\ i\neq j}} (v_i - v_j) i_{ij} = \sum_{\substack{i=0,n;\ j=0,n\\ i\neq j}} v_i i_{ij} - \sum_{\substack{i=0,n;\ j=0,n\\ i\neq j}} v_j i_{ij}$$

$$= \sum_{i=0,n} \sum_{\substack{j=0,n\\ j\neq i}} v_i i_{ij} - \sum_{j=0,n} \sum_{\substack{i=0,n\\ i\neq j}} v_j i_{ij} = \sum_{i=0,n} v_i \sum_{\substack{j=0,n\\ j\neq i}} i_{ij} - \sum_{j=0,n} v_j \sum_{\substack{i=0,n\\ i\neq i}} i_{ij} = 0$$

In the latest passage, KCL is used, respectively, in forms 1 and 2: $\forall$ node j, $\sum_{i=0,n} i_{ij} = 0$, $\forall$ node i, $\sum_{j=0,n} i_{ij} = 0$.

[7]Equivalent branches will be discussed in the next chapters. This sentence will thus be clearer at a second reading of this chapter.

There is a special application of Tellegen's theorem, which occurs when u_{ij} and i_{ij} are not only compatible with Kirchhoff's laws, but are also present simultaneously in a circuit. In this case, in accordance with what was stated in Section 3.6, the quantity

$$p_{ij} = u_{ij} i_{ij} = (v_i - v_j) i_{ij} \tag{3.9}$$

is the power absorbed by the branch existing between nodes i and j.

Therefore equation (3.8) can be written as

$$\sum_{\substack{i=0,n \\ j=0,n}} p_{ij} = 0 \tag{3.10}$$

Equation (3.10), which is an immediate consequence of Tellegen's theorem, states exactly what was anticipated—that is, that the sum of all branch powers, assumed to be absorbed, is identically null. This means that the power delivered by branches that deliver power is exactly equal, at any time, to the powers absorbed at that time by branches that absorb power.

3.9 HISTORICAL NOTES

3.9.1 Short Biography of Kirchhoff

Gustav Robert Kirchhoff[8] (1824–1887) was a German physicist and inventor of the two laws that carry his name. He was teacher of physics and later of mathematical physics at the University of Berlin and Breslau.

He worked mainly in the fields of spectroscopy, electric circuit theory, and thermodynamics. In 1860, together with Bunsen, he found the luminous spectra signature of chemical elements. He founded spectroscopy analysis and made it possible to discover new chemical elements.

In circuit analysis he developed the two circuit laws that carry his name—Kirchhoff's Current Law and Kirchhoff's Voltage Law—while updating the results previously obtained by Ohm (whose short biography is in Chapter 4) and eliminating their points of contrast with the already generally accepted theory of electromagnetism.

In thermodynamics he studied radiation emitted by the so-called *black body*.

3.9.2 Short Biography of Tellegen

Bernard D. H. Tellegen (Winschoten, the Netherlands, 1900—Eindhoven, the Netherlands, 1990) was an electrical engineer and inventor of the penthode and

[8]Pronounced (from *Oxford English Dictionary*): /ˈkɪətʃɒf/ or /ˈkɪrxhɔf/

the gyrator. In circuit theory he produced a very important theorem which carries his name.

He obtained his master's degree in electrical engineering in 1923 and then joined the Philips Research Laboratories in Eindhoven. In the period 1946–1966, Tellegen was professor of circuit theory at the University of Delft.

He invented the penthode vacuum tube (adding a fifth electrode to the tetrode) in 1926 and invented the gyrator around 1948. Since its discovery, Tellegen's theorem has attracted increasing interest and is now one of the main pillars of circuit theory.

4

TECHNIQUES FOR SOLVING DC CIRCUITS

For the Instructor

This chapter differs in approach from similar books in that, for a time, it keeps the distinction between Kirchhoff's laws and constitutive equations, as clearly stated in Chapter 3. This is a reminder to students that Kirchhoff's equations are valid not just for DC, but for all circuits.

However, soon we will see how the constitutive equations typical of DC circuits are integrated and combined within Kirchhoff's.

Because of the book's expected use and readership, we have made a selection of the most effective solving techniques available. *Basic usage of combined Kirchhoff-constitutive equations* and *nodal analysis* have been selected here—the first because it naturally completes the knowledge we acquired in chapter three, the second because it allows a dramatic reduction in the number of equations that need to be simultaneously solved.

The mesh current method, which is also very important for experts in circuit analysis, is outlined here.

Fundamentals of Electric Power Engineering: From Electromagnetics to Power Systems, First Edition.
Massimo Ceraolo and Davide Poli.

4.1 INTRODUCTION

In the previous chapter, circuits were introduced as a graphical-mathematical tool, useful for analysing physical systems composed of electric elements connected to each other by means of wires. It was also shown that, under given assumptions, circuits can effectively model physical systems that work with either constant or variable quantities.

In this chapter, we develop our analysis of circuits while working exclusively with constant quantities (voltages, currents, powers, etc.), so as to enable the reader to solve these kinds of circuit.

Circuits that operate with constant quantities are by tradition called direct current circuits or DC circuits.

Examples of DC systems, which can be modelled well by DC circuits, are:

- toys, which are fed by batteries (the most common DC energy source);
- the electrical systems of cars which, when the engine is stopped, are fed by a battery and, when the engine is operating, by a DC generator, consisting of an electromechanical system that draws mechanical power from the engine and converts it into electricity in the DC form.

4.2 MODELLING CIRCUITAL SYSTEMS WITH CONSTANT QUANTITIES AS CIRCUITS

In the previous chapter, in which we explored circuital systems and circuits, we introduced KCL and KVL, referring to examples in which all quantities in the system were constant: currents, voltages, fields, and so on. These systems are traditionally called *direct current systems* or *DC systems*, even though the name is not strictly correct; rather than "direct" the term "constant" would be more appropriate.

Since KCL and KVL were analysed using DC systems, it is quite natural to apply the circuit concept to circuital systems operating with constant quantities.

However, it must always be remembered that circuits are models of physical systems, and modelling always implies making assumptions, which in the systems themselves are satisfied only up to a degree.

This is discussed in the next section.

4.2.1 The Basic Rule

To evaluate the assumptions to make so as to model a circuit system as a circuit, consider again the example discussed in the previous chapter, and in particular the version with constant quantities (voltages, currents, etc., constant over time) in Figure 3.13 and also in Figure 4.1a.

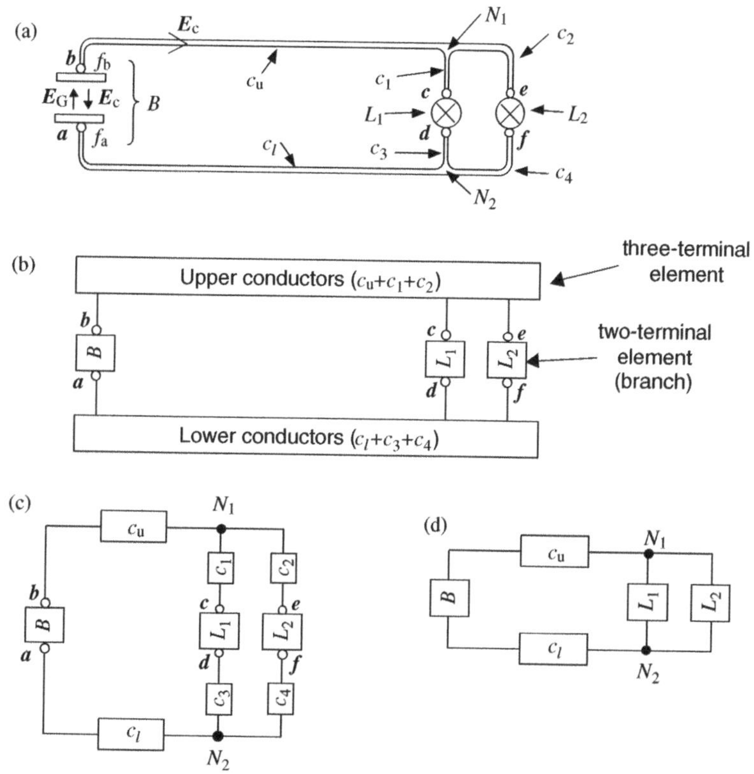

FIGURE 4.1. A sample circuital system (a) and some possible models (circuits: b to d).

Since in the conductor there is electric field $\boldsymbol{E}_c$, caused by the surface charges on the conductors, KVL is satisfied, as already seen in Chapter 3. Moreover, if the charge flows through wires, or wires and elements are negligible, KCL also applies.

Note that this is equivalent to saying that the air surrounding the wires has infinite resistivity, or zero conductivity.

Three different models for the same physical system are shown in Figures 4.1b to 4.1d. The model in Figure 4.1b clearly shows the terminals ***a, b, c, d, e, f,*** as well as the subsystems for which these terminals form borders with other circuit elements. Note that two of the elements have three terminals. The reader might remember that circuits can have multiterminal elements, although circuits with only two terminal elements, called branch-based circuits in this book, are by far the most widely used ones.

The model in Figure 4.1c is equivalent to the one in Figure 4.1b, but the two points N_1 and N_2 of the physical system to be modelled, representing the two nodes N_1 and N_2 in Figure 4.1a, turn it into a branch-based circuit. Finally, the model in Figure 4.1d

TABLE 4.1. Constitutive Equation Types Used in This Book in DC Circuits[a]

Element	Equation	Description
Voltage source	$u_b = U_s$	Voltage is equal to U_s (subscript stands for "source") regardless of any other circuit quantity.
Current source	$i_b = I_s$	Current is equal to I_s (subscript stands for "source") regardless of any other circuit quantity.
Resistor	$u_b = R_b i_b$	Branch voltage and current proportional.[b]

[a]In equations containing both current and voltage, the load sign convention is used.
[b]It will soon be seen that resistors have a positive value of resistance R_b, when u_b and i_b are measured, assuming that the current enters the positively marked terminal.

is a simplified version of the one in Figure 4.1c and is therefore less accurate but easier to deal with.

KCL is assumed to be wholly valid in circuits. Therefore, in the transformation of the system in Figure 4.1a into the circuits of Figures 4.1b to 4.1d, any charge flow in the air surrounding the wires and through the element boundaries is negligible.

For DC circuits, the rule for modelling a given system as a circuit is thus very simple:

Rule: *Modelling of DC systems through circuits*

A circuital system, composed of components that can be modelled as circuit elements connected to each other by conducting wires if all quantities are constant, is a DC system.

It can be modelled by a DC circuit if all charge flow in the air between the wires and around the elements is negligible.

It was earlier stated that in DC circuits all the quantities are constant. Why are they constant? This simply depends on the constitutive equations.

Consider a circuit whose branches have constitutive equations only of the types shown in Table 4.1, a subsystem of those already shown in Table 2.4. In such a circuit, some of the branches impose constant voltages at their terminals, others constant currents (voltage and current sources respectively), while others impose proportionality between current and voltage (resistors).

In this case it can be intuited that all voltages and currents of the circuit will be constant and can be formally demonstrated.

More in Depth: *Equation-only formulation of DC circuit models*

In this chapter the reader will learn that voltages and currents can be determined as the solution of a linear system of equations of the type

$$\mathbf{Ax} = \mathbf{b}$$

where

$\boldsymbol{x}$ is the vector of unknowns (voltages across all branches that are not of the voltage source type and currents through all the branches that are not of the constant current type)

$\boldsymbol{b}$ is the vector of known terms (voltages across all voltage source branches and currents through constant current branches)

$\boldsymbol{A}$ is a matrix of coefficients that will contain constant terms, a function of the resistances of the resistors

Since $\boldsymbol{A}$ and $\boldsymbol{b}$ are both constant, solution $\boldsymbol{x}$ will constitute a vector of constants

Therefore, circuits containing elements having only the constitutive equations shown in Table 4.1 will be DC circuits.

In the following sections, the three basic types of constitutive equations shown in Table 4.1 will be discussed in more detail.

4.2.2 Resistors: Ohm's Law

Consider the load in Figure 4.2. Suppose that if the applied voltage U changes, the current flowing into the load changes in proportion; that is, whenever the voltage doubles, the current doubles; or more generally, when the voltage is multiplied by a given factor, the current varies by the same factor.

This implies direct proportionality between voltage and current:

$$U = kI \tag{4.1}$$

This example is far from rare in nature; indeed this law is valid to a high degree of accuracy for any load composed of conducting material of any shape, such as wire distributing energy along an electric installation (long cylinder of copper) or a spiral (such as in domestic heaters, ovens, and bulb lamps).

The first one that formulated a law that resembles (4.1) was Georg Simon Ohm in 1827 (cf. Section 4.7.1); this equation thus is known today as Ohm's law. The factor

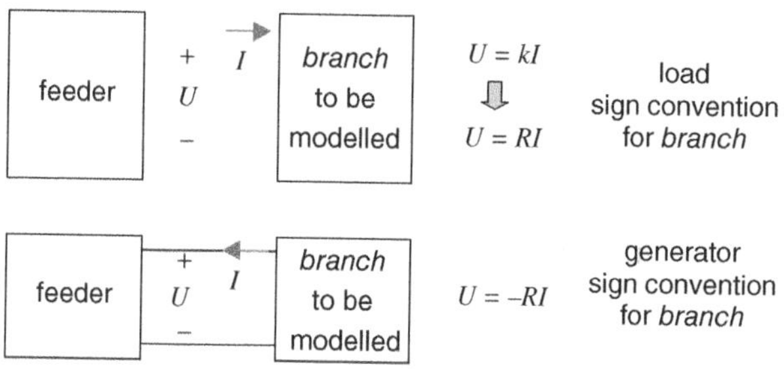

FIGURE 4.2. Proportionality between voltage and current, and Ohm's law for branches.

of proportionality is a quality of the conductor: the higher this factor, the lower the current at a given voltage, or the higher the voltage needed for a given current flow. This factor thus measures the ability of a conductor to *resist* charge transfer. It is therefore called the "resistance" of the load and is consequently normally given the symbol R. Ohm's law is usually written as

$$U = RI \tag{4.2}$$

which not only relates the amplitudes of voltage and current of a given conductor, but also relates the signs. U and I can therefore be positive or negative numbers; however, if the load sign convention is used for the branch to which Ohm's law applies, resistance R is always a positive number (see Figure 4.2).

Resistance R has the dimensions of a voltage times a current; because of its importance, this quantity has its own name in the SI, that is, "ohm" (note the lowercase initial as for all units of SI measures). One ohm is the resistance of a resistor in which, when subjected to a voltage of one volt, a current of one ampere flows.

Definition: *The ohm (the unit of measure of resistance)*

The unit of measure of resistance is the ohm (symbol: Ω).

One ohm is the resistance of a conductor which, under a voltage of 1 V, absorbs a current of 1 A. In formula: $1\,\Omega = (1\text{ V})/(1\text{ A})$

A conductor shows a resistance that is a function of the characteristics of the material used, and of the conductor's geometry. For the most important case of a cylindrical conductor having the length l and the cross-sectional area A, the following relation applies:

$$R = \rho\frac{l}{A} = \frac{l}{\sigma A}$$

where ρ and $\sigma = 1/\rho$ constitute a characteristic of the material with which the conductor is made, and they are called resistivity or conductivity of the material, respectively.

Resistivity (and therefore also conductivity) is rather dependent on temperature. Its dependence is normally considered linear around the value at a reference temperature, according to the law

$$\rho = \rho_0[1 + \alpha(\theta - \theta_0)]$$

where θ_0 is the reference temperature and α is a coefficient characteristic of the material used, defined as

$$\alpha = \frac{\Delta\rho}{\rho_0\Delta\theta} = \frac{(\rho - \rho_0)/\rho_0}{\theta - \theta_0}$$

TABLE 4.2. Some Resistance Parameters (Reference Temperature: 20°C)

Material	Resistivity (nΩm)	Temperature Coefficient (°C^{-1})
Silver	16.4	0.004
Copper	17.5	0.004
Gold	24	0.0035
Aluminium	29	0.004

The numerical values of the best conductors are shown in Table 4.2; among them, widely used in electrical engineering applications, are aluminium and copper.

Ohm's law can be written in a slightly different form from (4.2):

$$I = \frac{1}{R} U = GU \tag{4.3}$$

where G, by definition, is equal to the reciprocal of R, is called conductance of the element with R as resistance, and has a unit of measure of its own, called siemens, whose symbol is S:

Definition: *The siemens (the unit of measure of conductance)*

The unit of measure of resistance is the siemens (symbol: S).

One siemens is the conductance of a conductor which, under a voltage of 1 V, absorbs a current of 1 A. In formula: 1 S = (1 A)/(1 V)

Example 1. A voltage $U = 10\,\text{V}$ is applied to a copper conductor, with length $l = 200\,\text{m}$ and a cross-sectional area $A = 50\,\text{mm}^2$. Calculate the resistance R and current I at 60°C.

From Table 4.2, $\rho_{20°\text{C}} = 17.5\,\text{n}\Omega\text{m}$.

$$R_{20°\text{C}} = \frac{\rho_{20°\text{C}} \cdot l}{A} = \frac{17.5 \times 10^{-9} \cdot 200}{50 \times 10^{-6}} = 0.07\,\Omega$$

$$R_{60°\text{C}} = R_{20°\text{C}}(1 + \alpha \cdot (\vartheta - \vartheta_0)) = 0.07 \cdot (1 + 0.004 \cdot (60 - 20)) = 0.0812\,\Omega$$

$$I = \frac{U}{R} = \frac{10}{0.0812} = 123.1\,\text{A}$$

4.2.3 Ideal and "Real" Voltage and Current Sources

In Chapter 3 we introduced the "battery," referring to the most common and widespread notion of a battery as a source of direct current voltage (and current

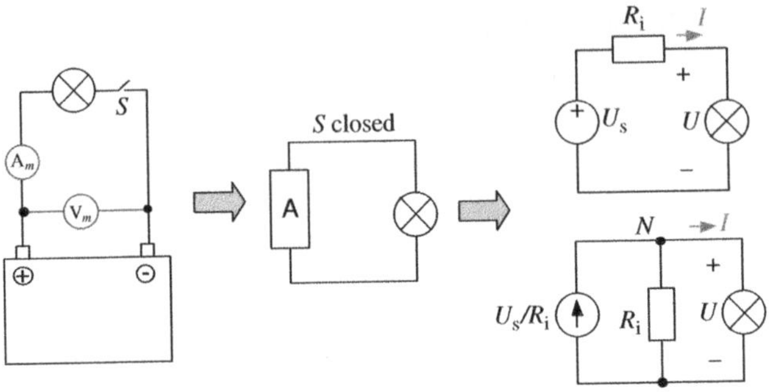

FIGURE 4.3. From a battery to a U_s–R idealised feeder.

and power). A battery shows nonzero voltage (see Figure 4.3), as measured by the voltmeter V_m, even when it is unconnected to any load. When it is connected to the load, it starts delivering current, as measured by the ammeter A_m displayed in the figure; the higher the current delivered, the lower the voltage.

The voltage–current relationship of a battery is very complex and depends on several parameters such as the battery type, the environment temperature, how charged the battery is, and so on. However, it is often sufficient to resort to simplified modelling. The simplest way of modelling this complex behaviour is by resorting to a linear constitutive equation of the type

$$U = U_s - R_i I \tag{4.4}$$

It is interesting to note that this simple constitutive equation allows it to be "translated" into two very basic circuital elements: an ideal, constant voltage source element, in series with an ideal resistor with resistance R_i. Indeed, it is very easy to see that equation (4.4) can immediately be found by applying the KVL to the rightmost upper circuit of Figure 4.3.

Another equivalent circuital interpretation of equation (4.4) is in the lower rightmost circuit of Figure 4.3:

$$U = R_i \cdot I(R_i) = R_i(U_s/R_i - I)$$

where $I(R_i)$ indicates the current flowing through R_i.

The characteristics of a real or ideal generator can also be seen in a U–I plane, as shown in Figure 4.4. Consider the leftmost diagram. It is simply a graphical representation of equation (4.4), taking into account the fact that R_i must be a real positive quantity. However, in the case of real batteries, good practice requires not to load the battery with too high currents; therefore, the operating point $\mathbf{P}$ is near the no-load voltage point $\mathbf{P}_0$ (for which $U = U_s$). The useful part of the left plot of Figure 4.4 will thus correspond to the one shown in the central diagram of this figure. For this part, it is more intuitive to use the generator shown in the top-right part of Figure 4.3, since it highlights the ideal generator U_s, plus the deviation from

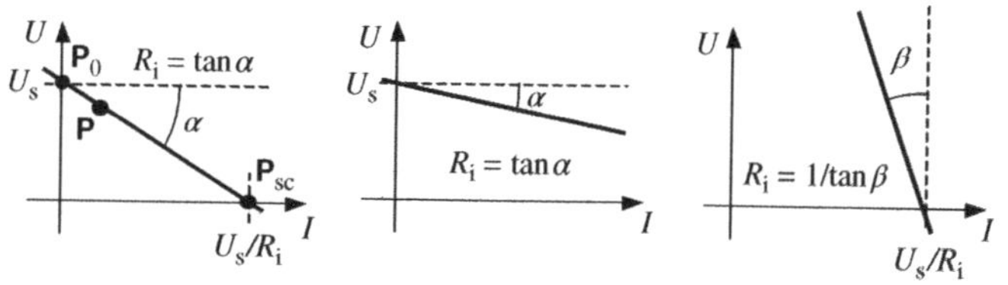

FIGURE 4.4. Ideal and real generators.

ideal behaviour induced by R_i, which causes the actual voltage to be lower than the ideal one.

If the terminals of a real voltage source (i.e., composed of the ideal generator U_s plus the inner resistance R_i) are directly connected to each other, the voltage between them becomes zero and the current flowing is U_s/R_i. This current will usually be very high and must be interrupted quickly to avoid source damage and/or the excessive generation of heat in the conductors, with the risk of fire. These very high currents are normally called *short-circuit currents*; possible interrupting devices are called *breaker* and *fuses*. Some information about them will be supplied in Part IV of this book.

There are some rare cases, however, in which feeders are used in the vicinity of point $\mathbf{P}_{sc}$—that is, when voltage is near zero (subscript stands for "short circuit"). In these cases, the useful part of the leftmost diagram of Figure 4.3 is the one shown in the rightmost diagram of Figure 4.4. For this part, it is more intuitive to use the generator shown in the bottom-right part of Figure 4.3, since it shows the ideal current generator U_s/R_i, plus the deviation from ideal behaviour induced by R_i, which causes the actual current to be lower than the ideal one.

Where the element R_i is omitted from the circuital representations, the actual characteristics of the circuital equivalents shown in Figure 4.3 would be the line of dashes in the middle and right-hand diagrams of Figure 4.4.

4.3 SOLVING TECHNIQUES

In Chapter 3 we stated that solving a circuit means finding a value for all node potentials and wire currents. For branch-based circuits, this simply means finding the values of all branch voltages and currents.

We also stated that we need to write Kirchhoff and constitutive equations, in exactly the number needed to solve the circuit.

However, by restricting the possible constitutive equations for DC circuits to the very limited number shown in Table 4.1, it is possible to greatly simplify this general approach, which allows relatively easy manual computation of circuits.

While computer programmes solve circuits simply by just assembling Kirchhoff and constitutive equations in a mechanical way, electrical engineers (and nonelectrical engineers with some basic knowledge of electrical engineering) use specialised solving techniques for DC circuits. The easiest and more common of these are discussed in the following sections. In particular:

- *Basic usage of combined Kirchhoff-constitutive equations* allows us to limit the system rank to be solved to the number of Kirchhoff equations themselves, since constitutive equations are directly substituted into Kirchhoff's.
- *Nodal analysis* allows a further reduction of equations, resulting in a system whose rank is equal to the number of circuit nodes minus one.
- *Mesh analysis* is another method that allows the number of equations to be reduced to the number of KVL equations.

Because of the introductory nature of this book, mesh analysis is merely mentioned as a possible technique and is not dealt with in detail.

4.3.1 Basic Usage of Combined Kirchhoff-Constitutive Equations

Consider as an example the system shown in Figure 4.5. In this circuit, specialised symbols of the types considered in Table 4.1 (i.e., constant voltage, constant current, and resistors) are used instead of generic branches.

These symbols are commonly used in textbooks. Other symbols are often used, but the ones adopted in this book are those recommended by international standardization bodies (see Chapter 1 for details). Note that the resistor symbol is similar to the generic branch symbol, but it has an aspect ratio of 3:1 instead of 2:1.

The circuit shows all voltage polarities and current flow references (= assumed directions). The reference current directions here always enter positively marked terminals; it goes without saying that any choice of voltage polarities and current references would be acceptable.

A full set of Kirchhoff's equations and constitutive equations (c.e.) would be as follows:

N_1: $I_C - I_1 - I_E = 0$
N_2: $I_E + I_D + I_1 = 0$
N_3: $-I_A - I_C - I_B = 0$
L_1: $U_0 + U_A - U_B = 0$
L_2: $U_B - U_C - U_E + U_D = 0$
c.e.: $U_0 = U_S; U_A = R_A I_A; U_B = R_B I_B; U_C = R_C I_C; U_D = R_D I_D; U_E = R_E I_E; I_1 = I_s$

There are 12 equations in 12 unknowns,[1] which in this form appears lengthy to solve. This system, however, can be simplified immediately by substituting the

[1]In principle the unknowns are all branch voltages and currents—that is, there would be 14 since there are 7 branches; however, the system was written on the basis that U_s is traversed by the same current as R_A and that U_1 shares its current with R_F.

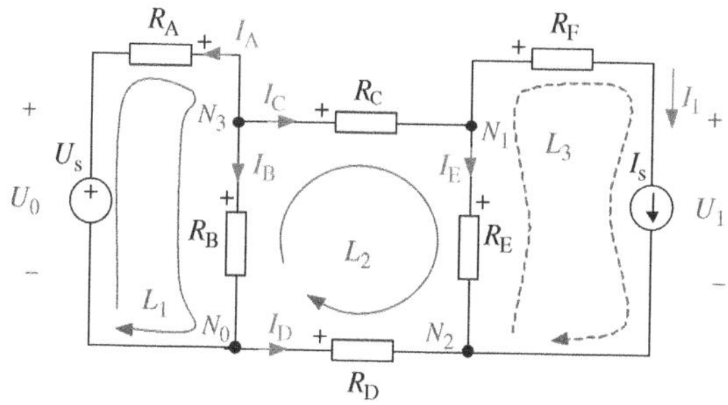

FIGURE 4.5. A sample circuit used for writing solving equations.

constitutive equations into Kirchhoff's, resulting in the following much leaner system:

$$\begin{aligned}
&N_1: \quad I_C - I_s - I_E = 0\\
&N_2: \quad I_E + I_D + I_s = 0\\
&N_3: \quad -I_A - I_C - I_B = 0\\
&L_1: \quad U_s + R_A I_A - R_B I_E = 0\\
&L_2: \quad R_B I_B - R_C I_C - R_E I_E + R_D I_D = 0
\end{aligned} \tag{4.5}$$

It is apparent from this example that, instead of first writing the full set of Kirchhoff and constitutive equations and then doing the substitution, it is much faster to directly write KVL equations in order that voltages are already expressed, whenever resistors are involved, as a function of the corresponding currents:

Rule: *Combined Kirchhoff-Constitutive Equations (KCE)*

When dealing with DC circuits, combined Kirchhoff-constitutive equations are written.

These are Kirchhoff's equations, in which resistor voltages are expressed as products of currents and resistances (with appropriate signs).

This way, the rank of the system to solve is just the number of Kirchhoff's equations.

The reader may have noticed that in the writing of Kirchhoff's equations for the system of Figure 4.5, the loop-cuts rule introduced in Chapter 3 would have caused an additional loop equation to be written

$$L_3: \quad R_E I_E - R_F I_1 - U_1 = 0 \tag{4.6}$$

This equation, however, once added to the system (4.5), will raise its rank by one, due to this additional equation and the new variable U_1, which does not appear anywhere else in the system. It is, therefore, wise to leave this equation out of the system (4.5), which should be then solved. If U_1 is needed, it can be found immediately from eq. (4.6).

This is a general situation whenever current sources are present in the circuit. Therefore, in DC circuits, the loop-cuts rule is applied without considering loops branches containing current sources, in order to reduce the system rank:

Rule: *Kirchhoff-constitutive equations when current sources are present*

When determining combined Kirchhoff-constitutive equations (KCEs) using the loop-cuts method described in Chapter 3, any branch with a current source can be considered already cut.

If the voltage of the current source is needed, the loop in which it is contained can be evaluated after the main system is solved.

The system (4.5) is a linear one with five unknowns I_A, I_B, I_C, I_D, I_E. Once it is solved and all of these have become known, all the branch voltages are easily determined through very simple application of constitutive equations.

This system, as for any circuit in general, can also be represented in the standard form of linear systems:

$$\boldsymbol{A}\boldsymbol{x} = \boldsymbol{b} \tag{4.7}$$

in which

$\boldsymbol{x}$ is the vector of unknowns

$\boldsymbol{b}$ is the vector of known terms

$\boldsymbol{A}$ is the matrix of coefficients

When circuit equations are taken into account, vector $\boldsymbol{b}$ contains the voltages across all voltage source branches and currents through constant current branches; $\boldsymbol{A}$ is a function of the resistances of the resistors; $\boldsymbol{x}$ contains the unknown currents.

For instance, the system (4.5) can be written as follows:

$$\begin{aligned}
&I_C - I_E = I_s \\
&I_D + I_E = -I_s \\
&I_A + I_B + I_C = 0 \\
&R_A I_A - R_B I_B = -U_s \\
&R_B I_B - R_C I_C + R_D I_D - R_E I_E = 0
\end{aligned}$$

that is,

$$Ax = b \quad \text{with} \quad x = \begin{bmatrix} I_A \\ I_B \\ I_C \\ I_D \\ I_E \end{bmatrix}, \quad A = \begin{bmatrix} 0 & 0 & 1 & 0 & -1 \\ 0 & 0 & 0 & 1 & 1 \\ 1 & 1 & 1 & 0 & 0 \\ R_A & -R_B & 0 & 0 & 0 \\ 0 & R_B & -R_C & R_D & -R_E \end{bmatrix}, \quad b = \begin{bmatrix} I_s \\ -I_s \\ 0 \\ -U_s \\ 0 \end{bmatrix}$$

4.3.2 Nodal Analysis

In *basic usage of combined Kirchhoff's and constitutive equations*, constitutive equations are directly inserted into Kirchhoff's equations, resulting in a system whose rank corresponds to the number of the latter.

In reality, this number can be further reduced, often dramatically, using the solution method called the *nodal analysis* technique.[2] This is a somewhat advanced use of combined Kirchhoff and constitutive equations.

The method consists of writing KCL equations in such a way that KVL is automatically satisfied. The order of the obtained system of equations is therefore just $n - 1$, where n is the number of circuit nodes. For instance, the order of the circuit used as an example in Section 4.3.1 can be reduced (after all simplifications described there have been carried out) from five to three.

Indeed, in Chapter 3, we stated that the KVL is equivalent to saying that a potential exists—that is, that it is possible to define a potential function V with a value at each circuit terminal, so that the constitutive equations are a function of potential differences only.

It follows that if the circuit is analysed in terms of terminal potentials instead of branch voltages, KVL is automatically satisfied.

Example 2. Write nodal analysis equations for the circuit reported in Figure 4.5.

Let us assume the potential of nodes N_1, N_2, etc., to be V_1, V_2, *etc.*

Since constitutive equations are a function only of *branch* voltages (i.e., potential differences), the voltage of a terminal, usually one of those in the bottom line of the circuit, can have an arbitrary value and is given the conventional value of zero for simplicity. It can, for instance, be assumed as $V_0 = 0$.

The equations of the circuit are now as follows (we have chosen to write the KCL in form 1):

$$N_1: \quad \frac{V_3 - V_1}{R_C} - \frac{V_1 - V_2}{R_E} - I_s = 0$$

$$N_2: \quad \frac{V_0 - V_2}{R_D} + \frac{V_1 - V_2}{R_E} + I_s = 0$$

$$N_3: \quad \frac{V_3 - U_s}{R_A} + \frac{V_3 - V_0}{R_B} + \frac{V_3 - V_1}{R_C} = 0$$

[2]In other textbooks it is called "node voltage analysis" or "node voltage method."

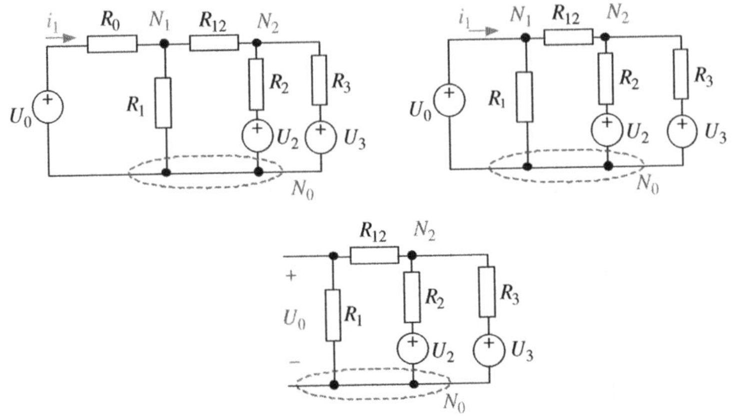

FIGURE 4.6. Reduction of node equations when some node potential is known in advance.

The equations to be written are not always exactly $n-1$. There are cases in which their number is even lower.

Consider, for instance, the circuit shown in the top left part of Figure 4.6. Since it has three nodes (node N_0 is in expanded form), two nodal analysis equations can be written. For instance, taking N_0 as reference, we obtain

$$N_1: \quad \frac{U_0 - V_1}{R_0} - \frac{V_1}{R_0} + \frac{V_2 - V_1}{R_{12}} = 0$$

$$N_2: \quad \frac{V_1 - V_2}{R_{12}} + \frac{U_2 - V_2}{R_2} + \frac{U_3 - V_2}{R_3} = 0 \qquad (^\circ)$$

which allows V_1 and V_2 to be determined; when these potentials are known, every other quantity can be easily computed.

But what happens when R_0 is zero? (top right-hand part of Figure 4.6). We need to write something different from (°).

Circuits like the one shown in the top-right part of Figure 4.6 require special treatment. Indeed, U_1 is exactly equal to V_1 (assuming as usual $V_0 = 0$), and the circuit can be solved referring only to the part in the lower diagram of Figure 4.6, with the following equation:

$$N_2: \quad \frac{U_0 - V_2}{R_{12}} + \frac{U_2 - V_2}{R_2} + \frac{U_3 - V_2}{R_3} = 0$$

which allows immediate computation of V_2; once V_2 is known, all node potentials are known and everything else is then easily computed.

Following these two examples, a general rule for nodal analysis can be written as follows:

> **Rule:** *Nodal analysis equations*
>
> To write a nodal analysis equation, the following procedure applies (let the number of circuit nodes be n):
>
> 1. Choose a reference node arbitrarily whose potential is set to zero.
> 2. Evaluate nodes whose voltage is known in advance. Let their number be m.
> 3. Write a KCL equation for each node, excluding the reference node and the nodes for which the voltage is known in advance.
>
> The total number of equations will thus be $e = n - 1 - m$.

Two further examples clarify the application of nodal analysis in some special situations.

Example 3. Write nodal analysis equations for the following circuit:

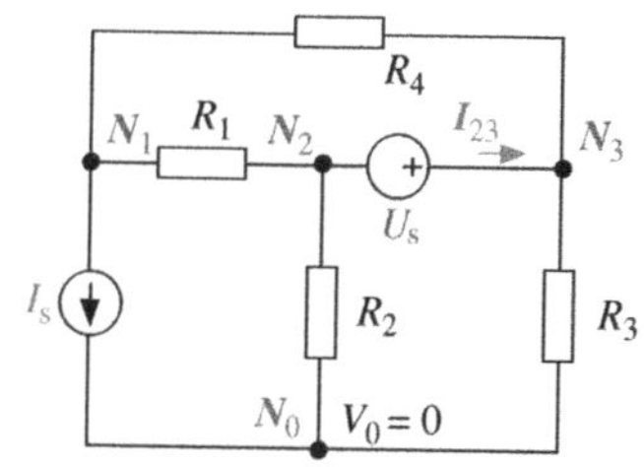

This circuit has two peculiarities:

- It contains a current source.
- It contains a voltage source without a terminal connected to the reference node.

The fact that there is a current source does not create difficulties. The voltage source creates a situation similar to that of the circuit shown in Figure 4.6. Here, however, the voltage source does not determine the potential of a node, but instead determines the mathematical relation between the potentials of two nodes:

$$V_3 = V_2 + U_s \qquad (^\circ)$$

We can still reduce the number of nodal equations to be written. We can arbitrarily decide whether to drop from the KCL equations that of node N_2 or of N_3. If we drop N_3, the following equations can be written:

$$N_1: \quad I_s + \frac{V_1 - V_2}{R_1} + \frac{V_1 - V_3}{R_4} = 0$$

$$N_2: \quad \frac{V_2 - V_1}{R_1} + \frac{V_2}{R_2} + I_{23} = 0$$

in which obviously

$$I_{23} = \frac{(V_2 + U_s)}{R_3} + \frac{(V_2 + U_{23}) - V_1}{R_4}$$

After substituting the expression of I_{23} in the equations N_1, N_2, we produce a 2×2 system. Having solved this system, we know V_1 and V_2; thus, using (°), V_3 is also known.

Example 4. Write nodal analysis equations for the following circuit:

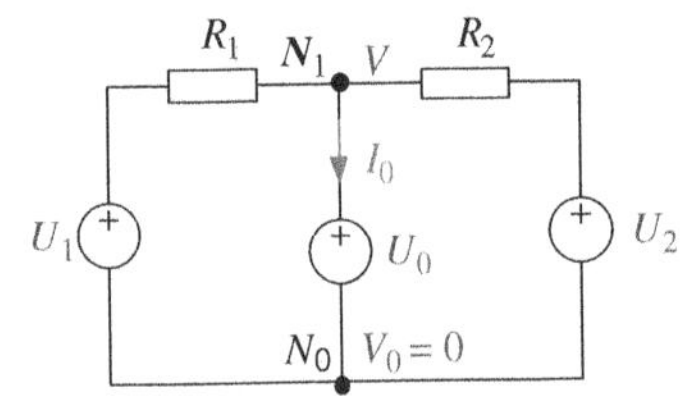

This example is really peculiar since node N_1 potential is known, and therefore the number of nodal analysis equations becomes $n - m - 1 = 2 - 1 - 1 = 0$.

In this case no nodal analysis equation can be written, which is coherent with the fact that the unique circuit node present in the circuit in addition to the reference one has a known voltage.

Since all node potentials are known, from them all branch currents are easily obtained by immediate application of Kirchhoff's voltage law.

Nodal analysis is extremely powerful and *tends to be the preferred method for the manual solution of circuits* because of the very low number of equations it usually generates.

4.3.3 Mesh Analysis

In the previous section we presented nodal analysis, showing how it is able to produce simultaneous equations to be solved in a number as low as $n - 1$, where n is the number of circuit nodes. It is thus the favoured method for the manual solution of circuits.

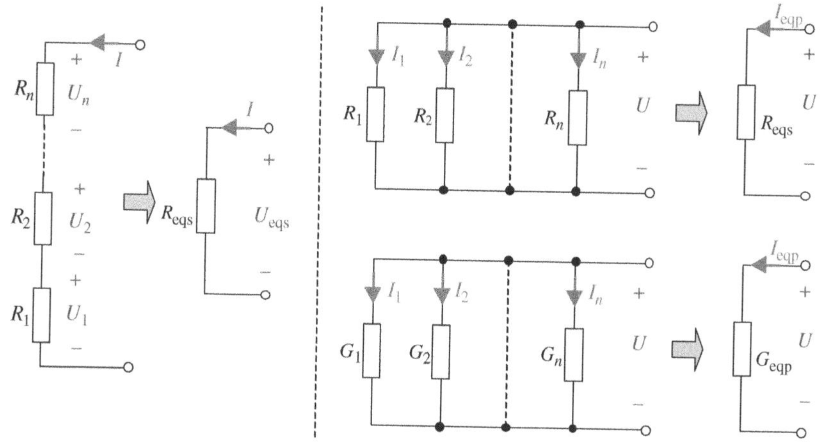

FIGURE 4.7. Series (*left*) and parallel (*right*) composition of circuit branches.

Many other techniques can be used and indeed *are* used by circuit experts; however, to avoid confusion for readers, for whom circuit analysis is simply a tool toward electrical engineering knowledge, these will not be dealt with here.

We wish only to mention *mesh analysis*,[3] which allows us to write, for any given circuit, a number of equations to be solved simultaneously, equal to the number of KVL equations and usually greater than $n-1$.

The interested reader can find details of this technique in references [bc1] and [bc2].

4.3.4 Series and Parallel Resistors; Star/Delta Conversion

Consider the left part of Figure 4.7. Branch currents of resistors $R_1, R_2, \ldots R_n$ have the same value and can therefore be represented by a single variable name: branches sharing the same current are called *branches in series*. The voltage across each resistor is proportional to this current.

By simple mathematical computations, it can be seen that the total voltage is also proportional to this current, and therefore the whole set of resistors in series behave like an equivalent resistor R_{eqs}:

$$U_{eqs} = U_1 + U_2 + \cdots + U_n = R_1 I + R_2 I + \cdots + R_n I = (R_1 + R_2 + \cdots + R_n)I = R_{eqs} I$$

In any circuit containing branches in series, the whole series can be substituted by a single branch with the resistance:

$$R_{eqs} = R_1 + R_2 + \cdots + R_n$$

[3]In other textbooks it is called "mesh current analysis" or "mesh current method".

In the particularly simple case of a series of resistors all having equal resistance R, this is obviously

$$R_{\text{eqs}} = nR$$

Example 5. Calculate the equivalent resistance R_{eqs} of the branch constituted by the series of the following resistances: $R_1 = 20\,\Omega$, $R_2 = 10\,\Omega$, $R_3 = 5\,\Omega$.

$$R_{\text{eqs}} = R_1 + R_2 + R_3 = 20 + 10 + 5 = 35\,\Omega$$

Now consider the right part of Figure 4.7. Branch voltages of resistors R_1, R_2, . . . R_n have the same value and can therefore be represented by a single variable name. Branches sharing the same voltage are called *branches in parallel*. The current entering each resistor is proportional to this voltage.

By simple mathematical computations, it can be seen that the total current is also proportional to this voltage, and therefore the whole set of resistors in parallel behaves like an equivalent resistor R_{eqp}. Computing is easier when, instead of the resistances in parallel, we consider the corresponding conductances (equal to the reciprocal of the resistances):

$$U_{\text{eqp}} = U_1 + U_2 + \cdots + U_n = G_1 I + G_2 I + \cdots + G_n I = (G_1 + G_2 + \cdots + G_n) I = G_{\text{eqp}} I$$

In any circuit containing branches in parallel, the whole parallel can be substituted by a single branch, with a conductance

$$\text{G}_{\text{eqp}} = G_1 + G_2 + \cdots + G_n$$

In everyday work situations, resistances are more common than conductances; it is therefore useful and easy to show this formula in terms of resistors:

$$\frac{1}{R_{\text{eqp}}} = \frac{1}{R_1} + \frac{1}{R_2} + \cdots \frac{1}{R_n} \qquad R_{\text{eqp}} = \left(\frac{1}{R_1} + \frac{1}{R_2} + \cdots \frac{1}{R_n} \right)^{-1}$$

In the simple case of a parallel of resistors all with equal resistances R, this results in

$$R_{\text{eqp}} = R/n$$

Moreover, when only two resistors in parallel are involved, the equivalence formula becomes

$$R_{\text{eqp}} = \frac{R_1 R_2}{R_1 + R_2}$$

Paralleling is usually indicated by the symbol $\|$: $R_{\text{eqp}} = R_1 \| R_2$

Example 6. Calculate the equivalent resistance R_{eqp} of the branch constituted by the parallel of the following resistances: $R_1 = 4\,\Omega$, $R_2 = 2\,\Omega$, $R_3 = 1\,\Omega$.

$$R_{\text{eqp}} = \frac{1}{\frac{1}{R_1} + \frac{1}{R_2} + \frac{1}{R_3}} = \frac{1}{\frac{1}{4} + \frac{1}{2} + \frac{1}{1}} = 0.571\,\Omega$$

where

$$R_{\text{eqp}} = \frac{1}{G_{\text{eqp}}} \quad \text{and} \quad G_{\text{eqp}} = \frac{1}{R_1} + \frac{1}{R_2} + \frac{1}{R_3} = 1.75\,\Omega^{-1}$$

Please note that R_{eqp} is always lower than all the resistances being paralleled.

Following the same approach, other equivalents can be found on particular circuit sections. One important case is the so-called *delta-star* conversion (also named Δ–Y or Δ–Wye conversion).

In circuital terms a *star* is a group of three resistors with one terminal in common and the others free, while a *delta* is a connection of resistors one after the other, in such a way as to form a closed shape resembling a triangle or a Greek uppercase delta (Figure 4.8).

The equivalence formulas must be found in such a way that their behaviour, as seen from the outside terminals (shown in Figure 4.8), is indistinguishable.

The conversion formulas are as follows (demonstration is omitted):

$$R_1 = \frac{R_{12}R_{13}}{R_s}, \qquad R_{12} = \frac{R_1 R_2}{R_p}$$

in which

$$R_s = R_{12} + R_{13} + R_{23}, \qquad R_p = \left(\frac{1}{R_1} + \frac{1}{R_2} + \frac{1}{R_3}\right)^{-1}$$

Since the numbering of terminals is arbitrary, formulas for the other resistors can be obtained by rotating the indexes: for instance, by adding one to each index of the first formula, we obtain the formula on top of next page; in the same scheme there is an

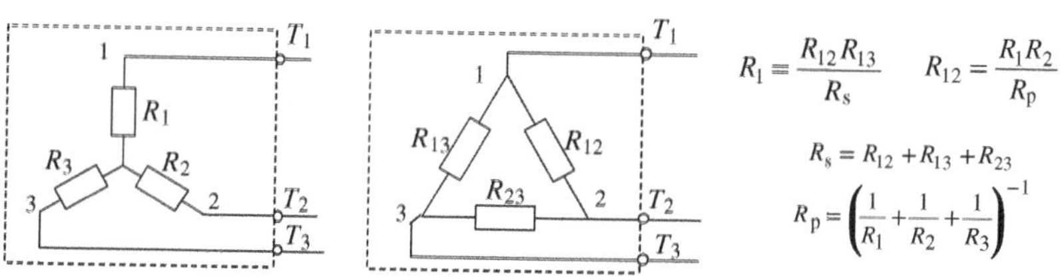

FIGURE 4.8. Delta and star circuital 3-terminal components, and conversion formulas.

intuitive sketch showing how to "add one to an index," where adding one to three restarts the count:

$$R_2 = \frac{R_{23}R_{21}}{R_s}$$

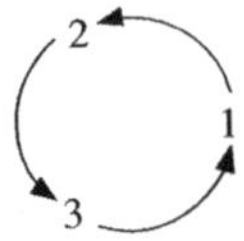

The reader can verify that, in the case of three equal resistances, $R_{\text{delta}} = 3 \cdot R_{\text{star}}$.

Example 7. Referring to Figure 4.7, calculate R_1, R_2, and R_3, if $R_{12} = 10\,\Omega$, $R_{23} = 8\,\Omega$, and $R_{31} = 5\,\Omega$.

$$R_s = R_1 + R_2 + R_3 = 23\,\Omega$$

$$R_1 = \frac{R_{12}R_{31}}{R_S} = \frac{10 \cdot 5}{23} = 2.174\,\Omega$$

$$R_2 = \frac{R_{23}R_{12}}{R_S} = \frac{8 \cdot 10}{23} = 3.478\,\Omega$$

$$R_3 = \frac{R_{31}R_{23}}{R_S} = \frac{5 \cdot 8}{23} = 1.739\,\Omega$$

These results can be verified by recalculating the delta resistances using R_1, R_2 and R_3:

$$R_p = (R_1^{-1} + R_2^{-1} + R_3^{-1})^{-1} = 0.7561\,\Omega$$

$$R_{12} = \frac{R_1R_2}{R_p} = 10\,\Omega, \qquad R_{23} = \frac{R_2R_3}{R_p} = 8\,\Omega, \qquad R_{31} = \frac{R_3R_1}{R_p} = 5\,\Omega$$

Example 8. Calculate the equivalent resistance as "seen" from terminals A and B:

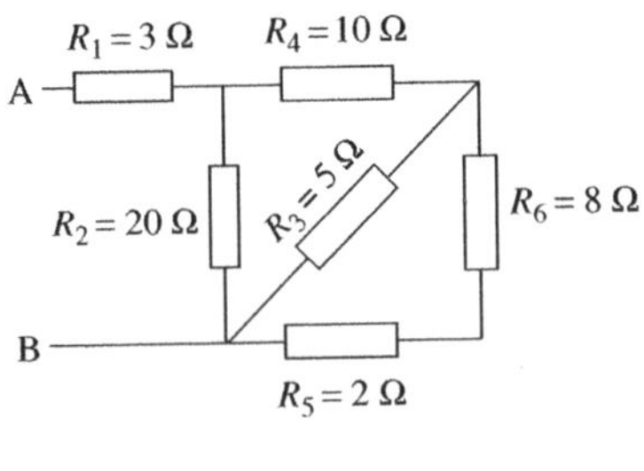

R_5 and R_6 are series connected, thus $R_{56} = R_5 + R_6 = 10\,\Omega$.
R_3 and R_{56} are connected in parallel, thus $R_{356} = R_3 || R_{56} = 3.333\,\Omega$:

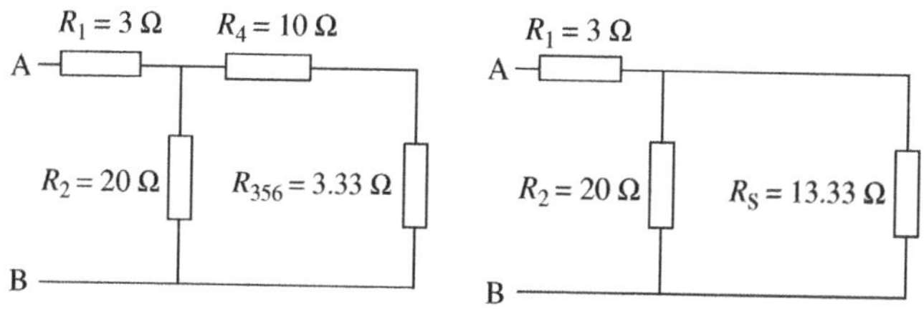

R_{356} and R_4 are series connected: $R_S = R_{356} + R_4 = 13.33\ \Omega$.
Since R_2 and R_s are in parallel, $R_{AB} = R_1 + R_2||R_s = 3 + (20^{-1} + 13.33^{-1})^{-1} = 11\ \Omega$.

4.3.5 Voltage and Current Division

When resistors are in series or in parallel, special relations occur between voltages and currents that are useful to investigate because of their practical implications.

Consider several resistors in series with resistances R_1, R_2, . . . R_n (left part of Figure 4.7).

The ratio of the voltage applied to any of the individual branches, to the total voltage applied to the series is:

$$\frac{U_k}{U_{eqs}} = \frac{R_k I}{R_{eqs} I} = \frac{R_k}{R_{eqs}} \tag{4.8}$$

As trivial as equation (4.8) might appear, it is sufficiently useful to merit a name of its own, i.e. the *voltage division rule* of resistors in series.

Example 9. In the following figure, calculate U_1, U_2, and U_3, if $U = 100$ V, $R_1 = 2\ \Omega$, $R_2 = 4\ \Omega$, and $R_3 = 14\ \Omega$.

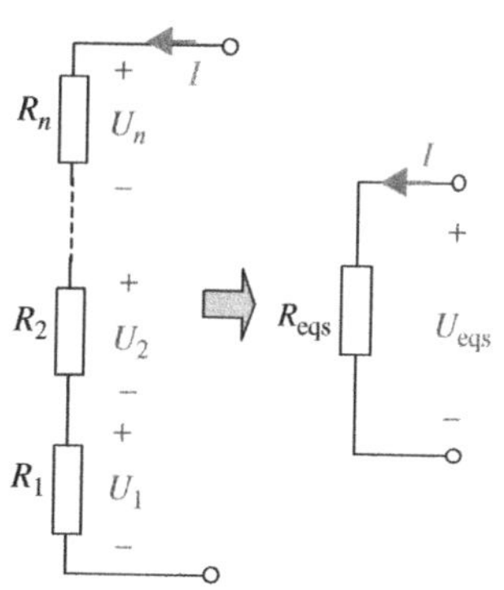

$$U_1 = I \cdot R_1 = U \cdot \frac{R_1}{R_{eqs}} = 100 \cdot \frac{2}{2+4+14} = 10\ \mathrm{V}$$

$$U_2 = I \cdot R_2 = U \cdot \frac{R_2}{R_{eqs}} = 100 \cdot \frac{4}{2+4+14} = 20\ \mathrm{V}$$

$$U_3 = I \cdot R_3 = U \cdot \frac{R_3}{R_{eqs}} = 100 \cdot \frac{14}{2+4+14} = 70\ \mathrm{V}$$

Consider now several branches in parallel with resistances $R_1, R_2, \ldots, R_n$, or, equivalently, conductances $G_1, G_2, \ldots, G_n$ (right part of Figure 4.7).

The ratio of (a) the current flowing into any of the individual branches to (b) the total current flowing into the parallel equivalent is

$$\frac{I_k}{I_{eqp}} = \frac{G_k U}{G_{eqp} U} = \frac{G_k}{G_{eqp}} \tag{4.9}$$

As simple as equation (4.9) might appear, it is sufficiently useful to merit a name of its own—that is, the *current division rule* of resistors in parallel.

Since, as previously mentioned, resistances are more commonly used than conductances, the version of (4.9) for the simpler case of just two branches in parallel can also be written in terms of resistances:

$$\frac{I_1}{I_{eqp}} = \frac{G_k}{G_1 + G_2} = \frac{1/R_1}{\dfrac{1}{R_1} + \dfrac{1}{R_2}} = \frac{R_2}{R_1 + R_2}, \qquad \frac{I_2}{I_{eqp}} = \frac{R_1}{R_1 + R_2}$$

that is, the current flowing into one of two branches in parallel is equal to the total current entering the parallel, multiplied by the resistance of the other branch and divided by the sum of the resistances of the two branches in parallel.

Example 10. Referring to the following figure, calculate I_1, I_2, and I_3, if $I = 10$ A, $R_1 = 2\ \Omega$, $R_2 = 4\ \Omega$, and $R_3 = 3\ \Omega$.

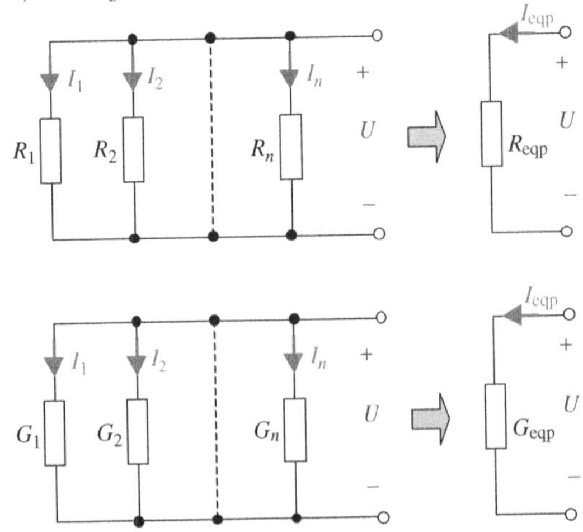

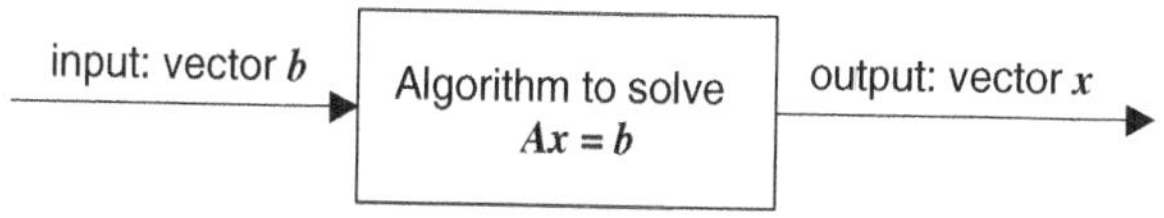

FIGURE 4.9. Circuit solution by means of numerical solution of a linear system.

$$I_1 = U \cdot G_1 = I \cdot \frac{G_1}{G_{\text{eqp}}} = 10 \cdot \frac{2^{-1}}{2^{-1} + 4^{-1} + 3^{-1}} = 4.615 \text{ A}$$

$$I_2 = U \cdot G_2 = I \cdot \frac{G_2}{G_{\text{eqp}}} = 10 \cdot \frac{4^{-1}}{2^{-1} + 4^{-1} + 3^{-1}} = 2.307 \text{ A}$$

$$I_3 = U \cdot G_3 = I \cdot \frac{G_3}{G_{\text{eqp}}} = 10 \cdot \frac{3^{-1}}{2^{-1} + 4^{-1} + 3^{-1}} = 3.077 \text{ A}$$

4.3.6 Linearity and Superposition

It was seen in Section 4.3.1 that sets of Kirchhoff's equations and constitutive equations of DC circuits can be expressed in the standard form of linear systems:

$$\boldsymbol{Ax} = \boldsymbol{b}$$

The vector of known quantities $\boldsymbol{b}$ in circuits contains the numerical values of voltage source voltages and current source currents, while the matrix $\boldsymbol{A}$ contains information on the circuit topology carried by Kirchhoff's equations and on the constitutive equations of the nonsource branches, that is, the resistors.

To solve the system (4.7), any of the existing algorithms to solve linear systems might be used. This can be visualised in Figure 4.9.

If $\boldsymbol{b}$ is composed of a sum of n terms

$$\boldsymbol{b} = \boldsymbol{b}_1 + \boldsymbol{b}_2 + \cdots + \boldsymbol{b}_n$$

it will be

$$\boldsymbol{Ax} = \boldsymbol{b}, \qquad \text{with} \qquad \boldsymbol{x} = \boldsymbol{x}_1 + \boldsymbol{x}_2 + \cdots + \boldsymbol{x}_n \quad \text{where } \boldsymbol{Ax}_1 = \boldsymbol{b}_1, \ldots, \boldsymbol{Ax}_n = \boldsymbol{b}_n$$

Therefore, the solution of the given circuit can be thought of as being the sum of partial solutions $\boldsymbol{x}_1$, $\boldsymbol{x}_2$, $\boldsymbol{x}_n$ obtained by the individual application of a partial set of sources $\boldsymbol{b}_1$, $\boldsymbol{b}_2$, $\boldsymbol{b}_n$:

Result: *Superposition principle*

In a linear system such as a DC circuit, should the input vector be the sum of several partial input vectors, the output vector is equal to the sum of the partial output vectors.

The terminology of this result is explained in Figure 4.9.

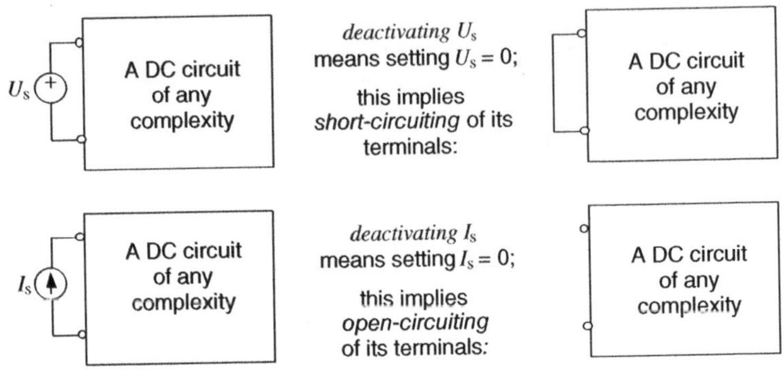

FIGURE 4.10. Deactivating a source.

The graphical nature of circuits facilitates several actions, one of which is the application of the superposition principle. Instead of working with the matrix description (4.7) of the circuit, it appears much more natural to work directly in its graphical representation.

In circuit language, the decomposition of input vector $\boldsymbol{b}$ into several addends is achieved by applying the sources one at a time, or in groups.

When a source is not applied, it is *deactivated*, that is, its value is equalled to zero; this means, for voltage sources, that when it is deactivated, it must be substituted by a wire directly connecting its two terminals (so that the voltage across them is zero); for current sources it means that it must be substituted by the disconnection of any wire between the two terminal ends, so that the current through them is zero (Figure 4.10).

In circuit terminology, when a voltage source is to be deactivated, its terminals must be short-circuited; however, when a current source is deactivated, they must be open-circuited. Two terminals connected to each other are called a short circuit, two terminals without any connections constitute an open circuit. This allows us to summarize superposition in circuits as follows:

Rule: *Superposition principle in circuits*

Linear circuits can be solved by applying voltage and current sources one at a time, and partial summing results. It is possible to use them in groups.

When a voltage or current source is not applied (or is deactivated), the corresponding branch is short- or open-circuited, respectively.

Example 11. The use of the superposition principle can be effectively illustrated if we refer again to the circuit of Figure 4.5, as shown in Figure 4.11. The inputs are U_s and I_s.

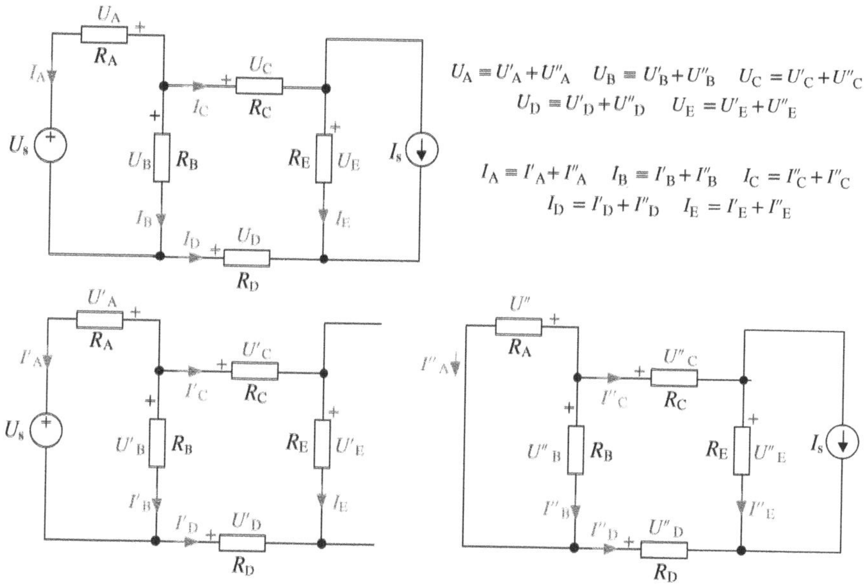

FIGURE 4.11. An example showing circuital application of the superposition principle.

Example 12. Figures 4.12 and 4.13 show possible ways to apply the superposition principle to a more complex circuit. Since this circuit has three sources, either each of them is considered individually, in which case the result includes the three contributions, or they are grouped differently—that is, first the two voltage sources and then the current source.

Any other grouping would be acceptable, so the user can choose whatever is deemed to be more useful or practical.

From the examples shown, it should be clear to the reader that using the superposition principle to solve a circuit is normally not advantageous, since it requires the solution of several circuits, however simplified, rather than one only.

It is, however, very important and useful in understanding how circuits work and can generate simplified computations in several practical situations.

Moreover, it is the basis for demonstrating the very important Thévenin theorem which is very useful and widely used in normal circuit-solving practice.

4.3.7 Thévenin's Theorem

Consider a circuit. If two wires are highlighted (w_1 and w_2), the circuit will be composed of two subcircuits, interfaced by means of them (Figure 4.14a). Any of the two subcircuits can contain an arbitrary number of voltage and current sources and resistors, connected to each other in an arbitrary fashion; it is, however,

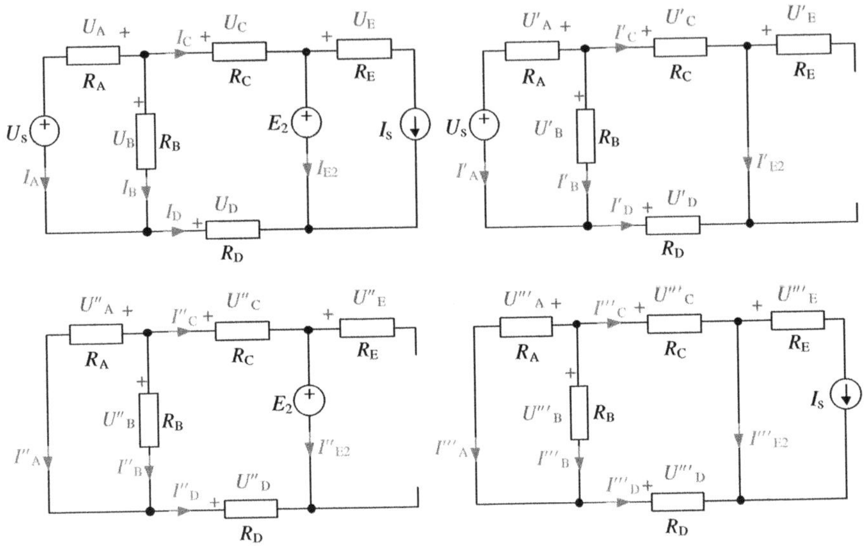

$$U_A = U'_A+U''_A+U'''_A,\quad U_B = U'_B+U''_B+U'''_B,\quad U_C = U'_C+U''_C+U'''_C$$
$$U_D = U'_D+U''_D+U'''_D,\quad U_E = U'_E+U''_E+U'''_E$$

$$I_A = I'_A+I''_A+I'''_A,\quad I_B = I'_B+I''_B+I'''_B,\quad I_C = I'_C+I''_C+I'''_C$$
$$I_D = I'_D+I''_D+I'''_D,\quad I_E = I'_E+I''_E+I'''_E$$

FIGURE 4.12. A first superposition example involving three known quantities (sources).

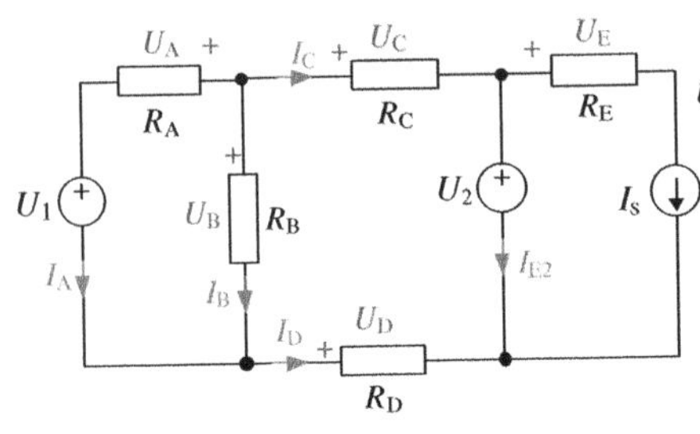

$$U_A = U'_A+U''_A,\quad U_B = U'_B+U''_B,\quad U_C = U'_C+U''_C$$
$$U_D = U'_D+U''_D,\quad U_E = U'_E+U''_E$$

$$I_A = I'_A+I''_A,\quad I_B = I'_B+I''_B,\quad I_C = I'_C+I''_C$$
$$I_D = I'_D+I''_D,\quad I_E = I'_E+I''_E$$

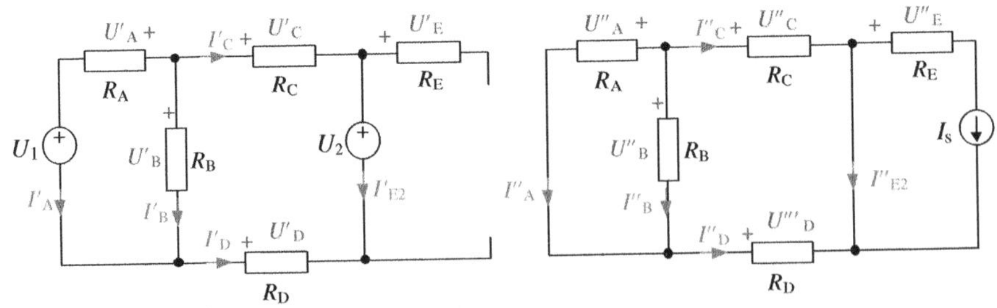

FIGURE 4.13. A second way to apply superposition to the circuit analysed in Figure 4.12.

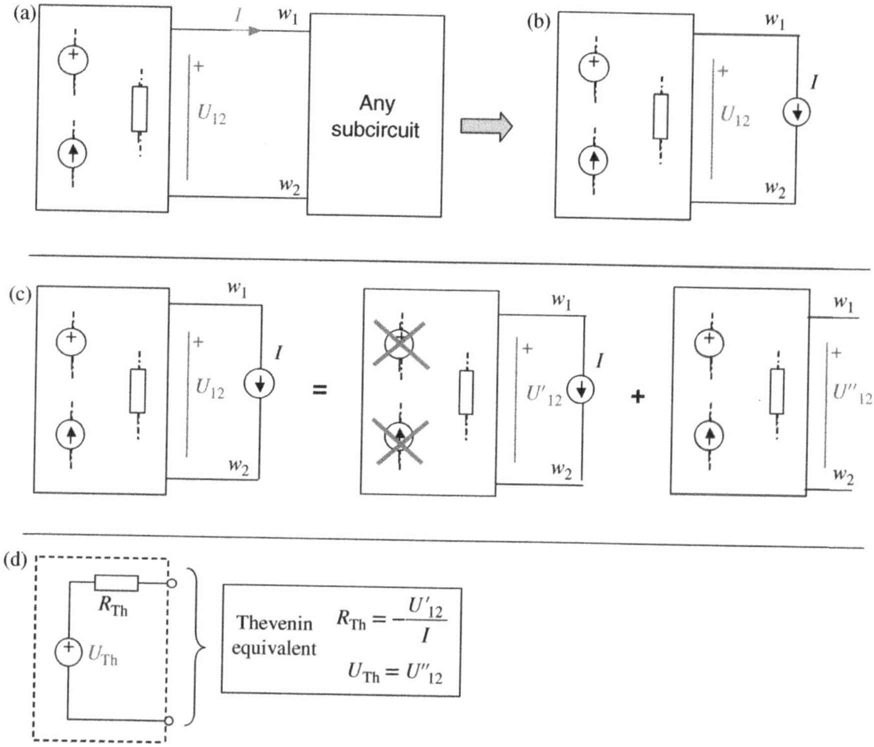

FIGURE 4.14. Thévenin's theorem.

mandatory to suppose that one of the two parts (let this be the left-hand one) is linear.

For a circuit section to be linear, it is sufficient that the constitutive equations of all circuit branches are linear. All the constitutive equations used in this book are linear.

Let the current flowing in the upper wire in the right direction be I, and let the voltage across the two wires be U_{12}. By applying the substitution principle (Section 3.6), the electric equilibrium of the left-hand part of the circuit can be obtained simply by substituting the right-hand, part with the current source I (Figure 4.14b).

Now the superposition principle can be applied; since the hand grouping of sources can be arbitrary, we can decide that in the first circuit all the sources belonging to the left-hand part of the circuit are activated, and I is considered to be inactive (therefore the corresponding branch is eliminated from the circuit); then we can suppose that I is active, but all other sources (i.e., all the sources present in the left-hand part of the circuit) are deactivated.

Therefore (Figure 4.14c),

$$U_{12} = U'_{12} + U''_{12} \tag{4.10}$$

The first term of the sum, U'_{12}, is the voltage appearing across the left-hand circuit terminals when I is applied and internal sources are deactivated; in these conditions, the left circuit is a network of resistances, for which the voltage to current ratio is a constant, a function of the resistor network. This constant, which has the dimensions of a resistance, is the so-called *equivalent resistance* of the left part of the circuit: Therefore we can describe equation (4.9) as follows:

Law: *Thévenin's theorem*

In a circuit composed of two parts, connected to each other by means of two wires, one part, if linear, can be substituted by the series of a voltage source and a resistor.

The voltage source voltage is the voltage appearing across the two terminals of this part when disconnected from the other one; the resistor's resistance is the value "seen" from the terminals, when all the internal sources are deactivated (cf. Figure 4.10).

The Thévenin equivalent of a subcircuit is therefore composed of the series of two simple components (Figure 4.14d). It is apparent that this theorem can greatly simplify circuit analysis: If knowledge of the voltages and currents inside the left part is not necessary, then the whole part, which might be composed of an arbitrary number of branches, loops, and nodes, is substituted by a very simple equivalent, which enables us to analyse all the right part of the circuit.

Example 13. Referring to the following circuit, calculate I_3

(a) using Kirchhoff's laws;
(b) using nodal analysis;
(c) applying Thévenin's theorem at nodes A and B.

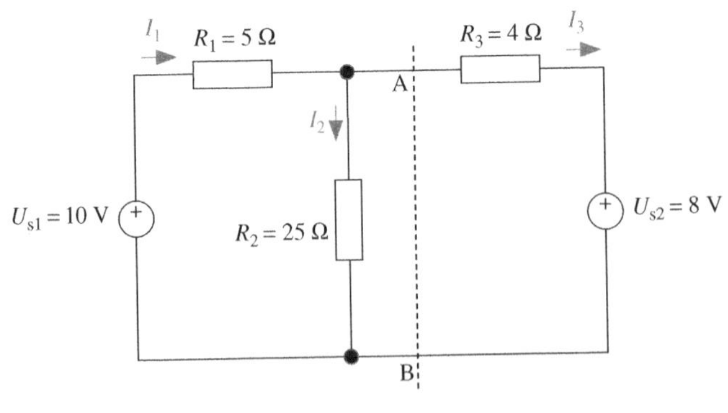

The following set of equations can be written

$$I_1 = I_2 + I_3, \qquad \text{KCL at node A}$$
$$U_{s1} = R_1 I_1 + R_2 I_2, \qquad \text{KVL at left mesh}$$
$$-U_{s2} = R_3 I_3 - R_2 I_2, \qquad \text{KVL at right mesh}$$

$$\begin{aligned} I_1 &= I_2 + I_3 \\ 10 &= 5I_1 + 25I_2 \\ -8 &= 4I_3 - 25I_2 \end{aligned} \quad \Rightarrow \quad \begin{aligned} I_1 &= 0.367\ \text{A} \\ I_2 &= 0.326\ \text{A} \\ I_3 &= 0.041\ \text{A} \end{aligned}$$

The following equation can be written:

$$\frac{U_{s1} - U_{AB}}{R_1} + \frac{0 - U_{AB}}{R_2} + \frac{U_{s2} - U_{AB}}{R_3} = 0, \qquad \text{deriving from KCL at node A}$$

Substituting values, we obtain

$$\frac{10 - U_{AB}}{5} + \frac{0 - U_{AB}}{25} + \frac{8 - U_{AB}}{8} = 0 \quad \Rightarrow \quad U_{AB} = 8.163\ \text{V}$$

Finally,

$$I_1 = \frac{U_{s1} - U_{AB}}{R_1} = \frac{10 - 8.163}{5} = 0.367\ \text{A}$$

$$I_2 = \frac{U_{AB}}{R_2} = \frac{8.163}{25} = 0.326\ \text{A}$$

$$I_3 = \frac{U_{AB} - U_{s2}}{R_3} = \frac{8.163 - 8}{4} = 0.041\ \text{A}$$

Thévenin's voltage U_{Th} at nodes A and B can be easily calculated by disconnecting the right part of the circuit:

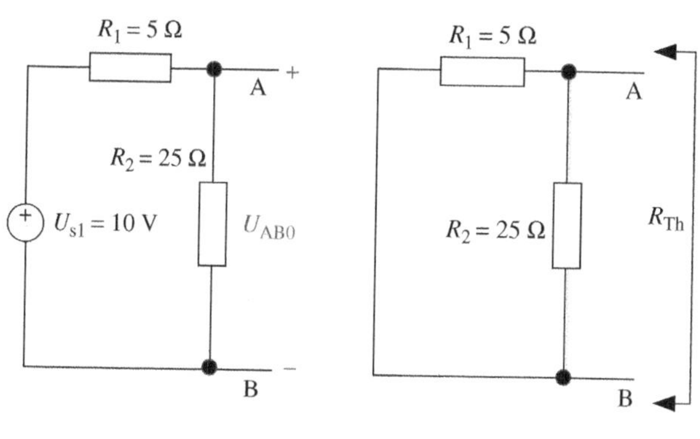

$$U_{Th} = U_{AB0} = (\text{voltage division rule}) = \frac{U_{s1}}{R_1 + R_2} \cdot R_2 = \frac{10}{5 + 25} \cdot 25 = 8.333\ \text{V}$$

Thévenin's equivalent resistance R_{Th} is the resistance "seen" from nodes A and B, when all generators are deactivated (in our case, only E_1 is present):

$$R_{Th} = R_1 \| R_2 = \frac{R_1 R_2}{R_1 + R_2} = \frac{5 \cdot 25}{5 + 25} = 4.166\,\Omega$$

The left side of the circuit can now be substituted by its Thévenin equivalent, in order to calculate I_3:

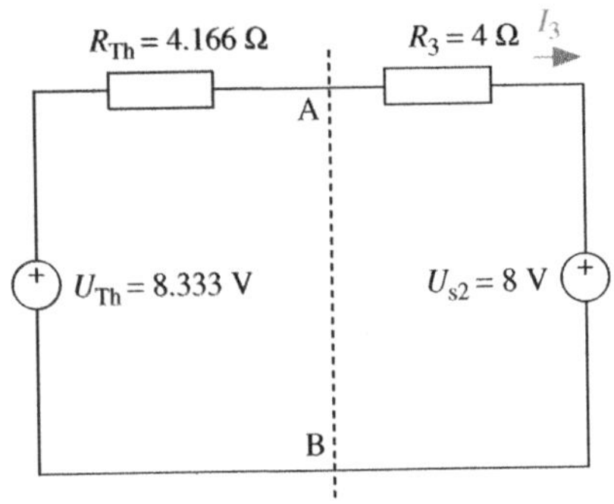

This single-mesh circuit can be easily solved using KVL:

$$U_{Th} - U_{s2} = (R_{Th} + R_3) \cdot I_3$$

$$I_3 = (U_{Th} - U_{s2})/(R_{Th} + R_3) = 0.041\text{ A}$$

Readers should note that $U_{AB0} \neq U_{AB}$:

$U_{AB} = U_{Th} - R_{Th} \cdot I_3 = 8.333 - 4.166 \cdot 0.041 = 8.163$, as calculated in Figure 4.14b.

Comparing the three methods, we can conclude that Thévenin's theorem is very powerful, in particular when a single current value is needed.

4.4 POWER AND ENERGY AND JOULE'S LAW

Power and energy in circuits have already been discussed in Chapter 3. It was shown that the power that flows from one part of a circuit to another is

$$p = ui$$

In DC circuits, everything is constant and uppercase letters are used to advantage to underline this:

$$P = UI$$

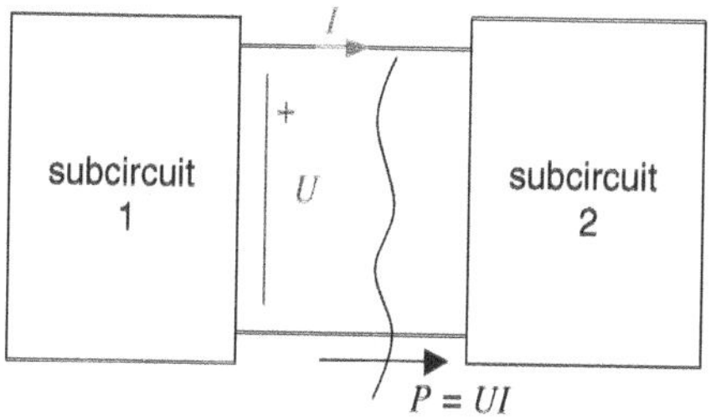

FIGURE 4.15. Power between circuit halves and its reference sign.

Attention must be paid to signs, since P is directed toward the part of the circuit into which the assumed current enters from the positively marked terminal (Figure 4.15). As usual, we must remember that the arrow for I and the "+" for U are only *assumed* flow direction and polarity: actual values of I and U will be positive or negative depending on whether the actual charge flow and voltage polarity[4] are in agreement with the assumed signs or opposite to them.

This also transfers to the value of power. In Figure 4.15, P will be in accordance with the assumed direction, represented by the arrow, if (and only if) its numerical value is positive.

What happens if subcircuit 2 is composed of just a single resistor? Ohm's equation $U = RI$ applies, with U and I again measured according to the load convention—that is, current entering the positive terminal. Therefore,

$$P = UI = (RI) \cdot I = RI^2 \tag{4.11}$$

Thus the resistor always absorbs power, whatever the sign of the applied current and voltage. This power will be converted into heat in the resistor, and the thermal power generated is given by (4.11). The law (4.11) is very important since it establishes a link between the worlds of electricity and thermal phenomena. Since it was first stated by James Prescott Joule, it is called Joule's law.

Law: *Joule's law*

The power absorbed by any resistor is always positive, and equal to $RI^2 = U^2/R$.

This power is converted into electromagnetic radiation, usually at a frequency corresponding to heat.

[4]Furthermore, remember that the current is considered to be composed of a flow of positive charges; electrons (i.e., in reality the negative charges inside conductors) flow opposite the assumed sign for currents, when I is positive.

Example 14. Referring to the circuit of Example 11, calculate the power flowing across section AB, from left to right:

$$P_{AB} = U_{AB} \cdot I_3 = 8.163 \cdot 0.041 = 0.335\ \mathrm{W}$$

P_{AB} can also be calculated by taking into account the power delivered by generators ($P_E = EI$) and the power consumed by resistors ($P_R = RI^2$) and then imposing the power balance:

(a) To the left-hand side of the circuit:

$$\begin{aligned} P_{AB} &= P_{E1} - P_{R1} - P_{R2} = E_1 I_1 - R_1 I_1^2 - R_2 I_2^2 \\ &= 10 \cdot 0.367 - 5 \cdot 0.367^2 - 25 \cdot 0.326^2 = 0.335\ \mathrm{W} \end{aligned}$$

(b) To the right-hand side of the circuit:

$$\begin{aligned} P_{AB} &= -(P_{E2} - P_{R3}) = -(E_3 \cdot (-I_3) - R_3 I_3^2) = -(8 \cdot (-0.041) - 4 \cdot 0.041^2) \\ &= 0.335\ \mathrm{W} \end{aligned}$$

(c) To the left-hand side of the final circuit used in Thévenin's analysis:

$$P_{AB} = U_{Th} I_3 - R_{Th} I_3^2 = 8.333 \cdot 0.041 - 4.166 \cdot 0.041^2 = 0.335\ \mathrm{W}$$

4.5 MORE EXAMPLES

Consider again Example 3 discussed in Section 3.5.3, for reasons of clarity shown again in Figure 4.16.

If the constitutive equations are known, all voltages and currents can be obtained by integrating the latter into Kirchhoff's equations and reducing the set of equations even more by making use of the nodal analysis technique.

Following the procedure in Section 4.3.1, once the branch constitutive equations are known, the circuit is written directly containing specific symbols for the branches. For instance, the circuit in Figure 4.17 is just a specific version of the

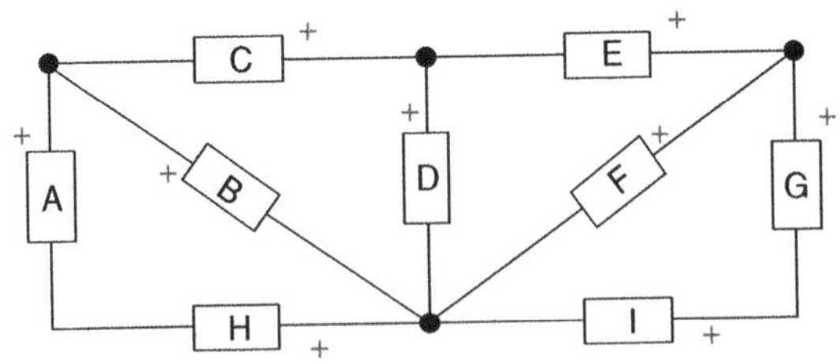

FIGURE 4.16. An example circuit (currents always taken as entering branches from "+" signs).

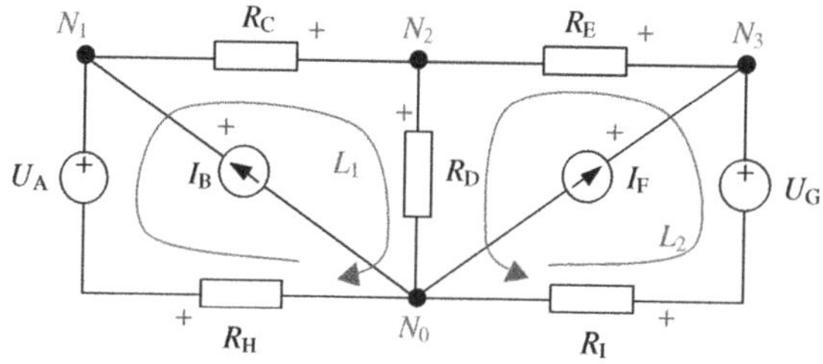

FIGURE 4.17. A specialised version (including constitutive equations) of the circuit in Figure 4.16.

one in Figure 4.16, where A is a voltage source, B is a current source, C is a resistor, and so on. This circuit can be solved using one of the techniques introduced in this chapter:

- direct use of constitutive equations in KCL and KVL (Section 4.3.1);
- nodal analysis (Section 4.3.2).

Other techniques such as mesh analysis can be used, but these are not considered in this book.

The circuit can also be first transformed into a set of simpler circuits using the principle of superposition, and these simpler circuits can be solved using one of the circuit-solving techniques.

It has already been noted in Section 4.3.6 that adoption of the principle of superposition is normally a lengthy process and is not recommended, except in special cases. Therefore, a basic combination of Kirchhoff-constitutive equations, as well as the more advanced nodal analysis, is used here, both to illustrate their practical application and to demonstrate, once again, that node voltage analysis is more effective for manual computations.

Basic Usage of Combined Kirchhoff-Constitutive Equations. This leads us to equations whose structure depends on the choices of nodes for KCL and loops for KVL equations. Choosing, for instance, nodes N_1, N_2, N_3 for KCL and L_1 and L_2 for KVL, the following equations are obtained (remember that currents are taken as entering branches from "+" signs with the obvious exceptions of I_B and I_F):

$$\begin{aligned} N_1&: \quad -I_A + I_B + I_C = 0 \\ N_2&: \quad I_C + I_D - I_E = 0 \\ N_3&: \quad I_E - I_F + I_G = 0 \\ L_1&: \quad R_H I_A + U_A + R_C I_C - R_D I_D = 0 \\ L_2&: \quad R_I I_I + U_G - R_E I_E - R_D I_D = 0 \end{aligned} \tag{4.12}$$

Equations (4.12) are a system of five linearly independent equations in the five unknowns I_A, I_C, I_D, I_E, and I_G. Once solved, voltages across the current sources,

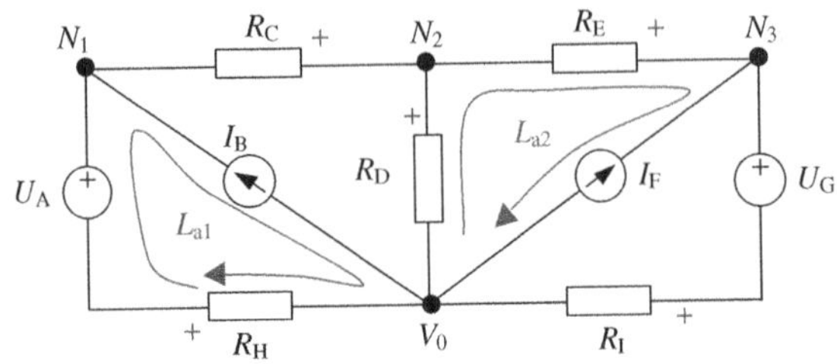

FIGURE 4.18. Additional loops to determine all quantities from those obtained by direct application of Kirchhoff's laws or nodal analysis.

if required, can be immediately determined by loop equations. For instance, referring to auxiliary loops L_{a1} and L_{a2} in Figure 4.18, it will be

$$U_B = R_H I_H + U_A, \qquad U_F = R_D I_D + R_E I_E$$

Nodal Analysis. Nodal analysis will produce a set of $n-1$ equations, n being the number of circuit nodes. Therefore, in this case a set of only three equations in three unknowns is expected. These are the following (V_1 = voltage at node N_1, etc; mixed forms of KCL are used):

$$\begin{aligned} N_1: &\quad -\frac{V_1 - U_A}{R_H} + I_B = \frac{V_1 - V_2}{R_C} \\ N_2: &\quad \frac{V_1 - V_2}{R_C} - \frac{V_2}{R_D} = \frac{V_2 - V_3}{R_E} \\ N_3: &\quad \frac{V_2 - V_3}{R_E} + I_F - \frac{V_3 - U_G}{R_I} = 0 \end{aligned} \tag{4.13}$$

To verify that the two systems (4.12) and (4.13) give the same results, some numerical examples are shown in Table 4.3.

To obtain these results, system (4.12) determines all currents, while if system (4.13) is first used, the node voltages are determined. The remaining quantities can then be obtained using trivial equations.

TABLE 4.3. Numerical Data and Results for Examples

Example	Known Values	I_A/A	I_C/A	I_D/A	I_E/A	I_G/A	V_1/V	V_2/V	V_3/V
1	$R_C = R_E = R_H = R_I = 1\ \Omega$ $I_B = I_F = 1$ A $U_A = U_G = 1$ V.	1/2	−1/2	1	1/2	1/2	3/2	1	3/2
2	$R_C = R_E = R_H = R_I = 1\ \Omega$ $I_B = I_F = 2$ A $U_A = U_G = 1$ V	5/4	−3/4	3/2	3/4	5/4	9/4	3/2	9/4

TABLE 4.4. Power Absorbed by All Circuit Components for the Proposed Example[a]

	A	B	C	D	E	F	G	H	I	TOTAL
U/V	1	3/2	−1/2	1	−1/2	3/2	1	½	1/2	—
I/A	1/2	−1	−1/2	1	−1/2	−1	1/2	½	1/2	—
***P*/W**	**1/2**	**−3/2**	**1/4**	**1**	**1/4**	**−3/2**	**1/2**	**¼**	**1/4**	**0**
U/V	1	9/4	−3/4	3/2	3/4	9/4	1	5/4	5/4	—
I/A	5/4	−2	−3/4	3/2	3/4	−2	5/4	5/4	5/4	—
***P*/W**	**5/4**	**−9/2**	**9/16**	**9/4**	**9/16**	**−9/2**	**5/4**	**25/16**	**25/16**	**0**

[a]Currents *always* assumed to be entering the positively marked terminals.

For instance, the following equations can be written from Ohm's law and/or from auxiliary KVL equations, which allow the node voltages to be determined from the currents once they are computed using the system (4.12):

$$
\begin{aligned}
V_2 &= R_D I_D, && \text{(Ohm's law on } R_D\text{)} \\
V_1 &= R_H I_A + U_A && \text{(KVL in loop } L_{a1} \text{ in Figure 4.18)} \\
V_3 &= V_2 + R_E I_E && \text{(KVL in loop } L_{a2} \text{ in Figure 4.18)}
\end{aligned}
\tag{4.14}
$$

In a similar way, the following trivial equations can be written from Ohm's law and/or KCL equations, which allow the node voltages to be determined from the currents, once they are computed using (4.13):

$$
\begin{aligned}
I_D &= V_2/R_D && \text{(Ohm's law on } R_D\text{)} \\
I_C &= (V_2 - V_1)/R_C && \text{(Ohm's law on } R_C\text{)} \\
I_E &= (V_3 - V_2)/R_E && \text{(Ohm's law on } R_E\text{)} \\
I_A &= I_C + I_B && \text{(node } N_1 \text{ in Figure 4.18)} \\
I_G &= -I_E + I_F && \text{(node } N_3 \text{ in Figure 4.18)}
\end{aligned}
\tag{4.15}
$$

The full numerical solution of the circuit is shown in Table 4.3, with two different sets of values for resistances and for voltage and current sources. From the values of Table 4.3, the power absorbed by all branches can be easily computed and the energy conservation can be easily verified. The numerical data are shown in Table 4.4.

In all cases, resistors absorb power, as they should. In example no. 1 the current generators globally deliver 3 W, while the voltage generators and the resistors absorb 1 W and 2 W, respectively. In example no. 2 the current generators globally deliver 9 W, while the voltage generators and the resistors absorb 2.5 W and 6.5 W, respectively. Clearly, in both cases the power delivered by the current generators perfectly equates that absorbed by the other resistors and the voltage generators.

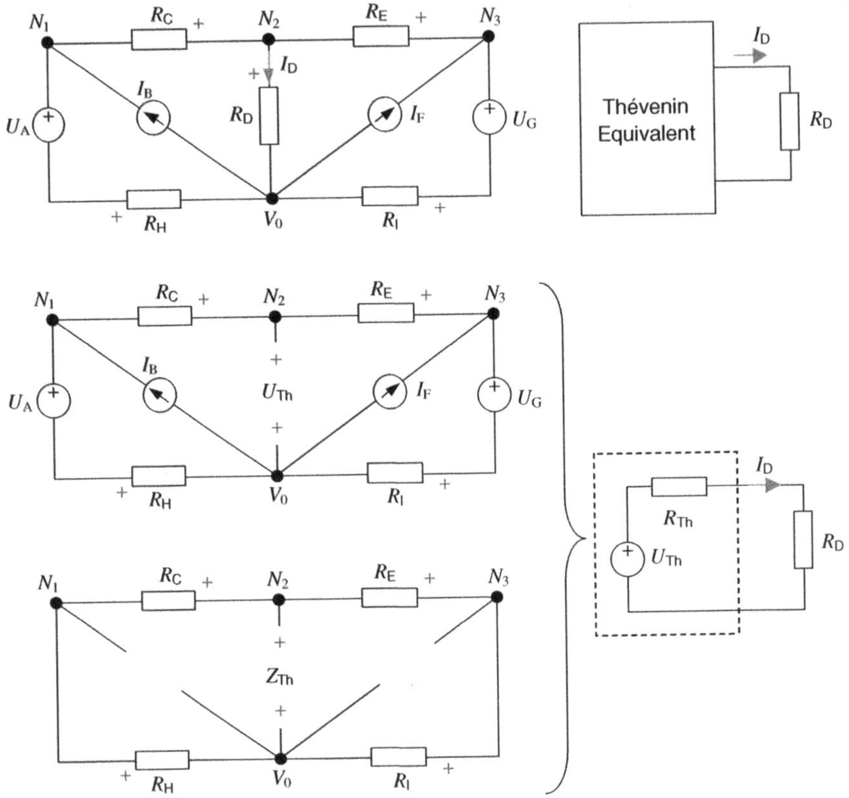

FIGURE 4.19. An example of the application of Thevenin's theorem.

Example 15. Consider Figure 4.19, which shows Thévenin's theorem applied to compute the current I_D. Instead of solving the full circuit, the simple circuit containing U_{Th}, R_{Th}, and R_D can be solved.

The value of U_{Th} is easily computed considering that this quantity is the open-circuit voltage—that is, the voltage that would appear at the R_D terminals if R_D were disconnected from the circuit (see the central circuit of Figure 4.19). R_{Th}, in turn, is the resistance "seen" from the R_D terminals when the circuit to which R_D is connected is made passive, by deactivation of all the inner sources. This is shown in the bottom drawing of Figure 4.19, in which current sources are deactivated and substituted by open circuits, thus forcing the current to zero, while the electromotive force is deactivated and substituted by ideal wires, forcing the corresponding branch voltage to zero.

In order to find U_{Th}, it is much simpler to solve the central circuit of Figure 4.19 than to solve the original (at top of figure). It can be solved using any of the circuit-solving techniques—that is, for the reader of this book, the direct application of Kirchhoff's laws, or nodal analysis.

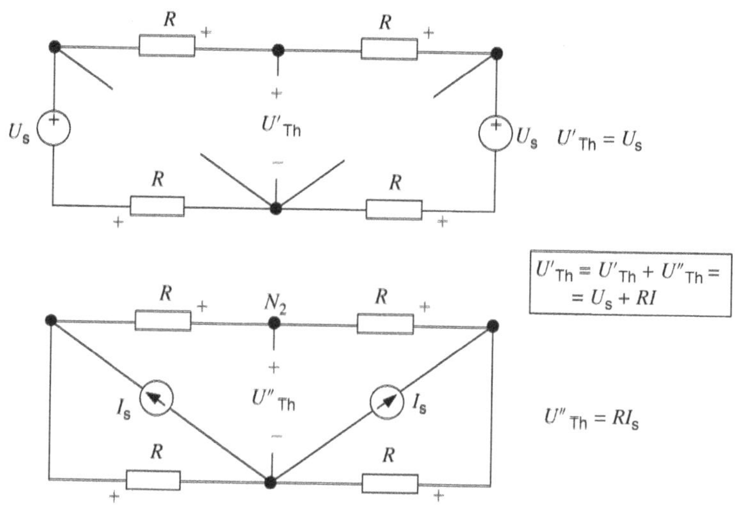

FIGURE 4.20. Calculation of U_{Th} using the superposition principle.

However, the cases considered in the previous examples, shown in Table 4.3, involved special numerical values, for which

$$U_{\mathrm{A}} = U_{\mathrm{G}} = U_{\mathrm{s}}, \quad I_{\mathrm{B}} = I_{\mathrm{F}} = I_{\mathrm{s}}, \quad R_{\mathrm{C}} = R_{\mathrm{E}} = R_{\mathrm{H}} = R_{\mathrm{I}} = R$$

These conditions create a special symmetry that can be exploited, with the help of the superposition technique (Section 4.3.6), to find the final solution with virtually no computations. Indeed, the circuit can be decomposed in two parts to be superposed, as shown in Figure 4.20.

In the upper circuit of Figure 4.20, it is apparent that no current circulates, and therefore $U_{\mathrm{Th}} = U_{\mathrm{s}}$. In the lower part, because of symmetry, the currents in the right part of the circuit must be the mirror version of the ones in the left part. Therefore, the currents flowing in the right and left wires connected to node N_2, measured toward the right part of the circuit must be opposite to each other. But their sums must also be equal, because of the charge conservation in node N_2, and therefore both must be null. Consequently the currents of the current generators flow in the bottom resistances, which implies that U''_{Th} is equal to RI. As regards R_{Th}, it is easy to verify (Figure 4.21) that it is equal to R.

Finally, from Thevenin's equivalent circuit (right-most circuit in Figure 4.19) it is immediately seen that:

$$I_{\mathrm{D}} = \frac{U_{\mathrm{Th}}}{R_{\mathrm{Th}} + R_{\mathrm{D}}} = \frac{U_{\mathrm{s}} + RI_{\mathrm{s}}}{2R}, \qquad I_{\mathrm{D,ex1}} = 1.0\ \mathrm{A}, \qquad I_{\mathrm{D,ex2}} = 1.5\ \mathrm{A}$$

which confirms the results already computed and reported in Table 4.3.

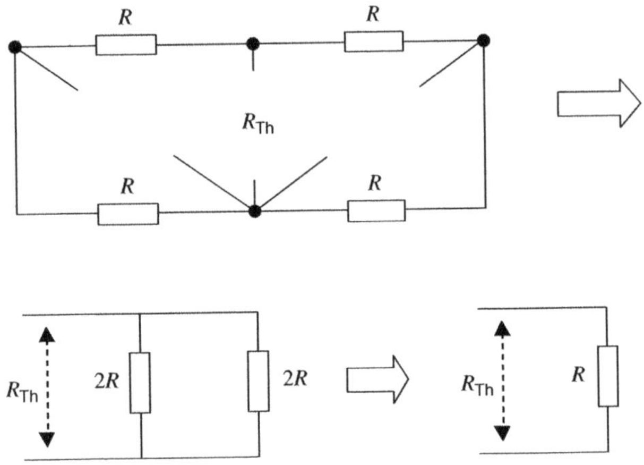

FIGURE 4.21. Calculation of R_{Th} for the example of Figure 4.19 in the event of resistances all being equal to each other.

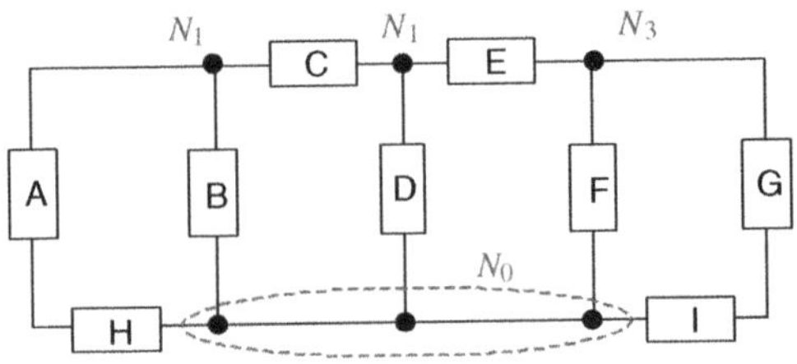

FIGURE 4.22. Representation of the circuit of Figure 4.16 (and Figure 4.17) expanding node N_0.

The example examined in this section contains components (the two current sources) shown inclined. As discussed in Chapter 3, if it is preferable to have all components shown either horizontally or vertically, it is normally sufficient to draw some voltages in an expanded fashion. For instance, the circuit of Figure 4.16 (and therefore also Figure 4.17) can be drawn as shown in Figure 4.22.

4.6 RESISTIVE CIRCUITS OPERATING WITH VARIABLE QUANTITIES

All the developments in this chapter derive from the conclusions of Chapter 3, which are valid for any circuit and for the list of constitutive equations in Table 4.1.

It could be very easy to demonstrate that all the results obtained here—for example, the use of combined Kirchhoff-constitutive equations, nodal analysis, superposition, and Thevenin's theorem—are still valid when the constitutive equations are as shown in Table 4.5, that is, when current and voltage sources are not constant, but all nonsource branches are resistive.

TABLE 4.5. Constitutive Equation Types for Resistive Circuits with Variable Quantities

Element	Equation
(Variable) voltage source	$u_b = u_s(t)$
(Variable) current source	$i_b = i_s(t)$
Resistor	$u_b = R_b i_b$

All the branch voltages and currents will be functions of time, and could be determined using the techniques in this chapter.

4.7 HISTORICAL NOTES

4.7.1 Short Biography of Ohm

The physicist *George Simon Ohm*[5] (Germany) is mainly famous for the law named after him, which relates voltage and potential in a conductor.

He graduated from the University of Erlangen in 1811 and then taught for several years at primary and secondary schools. In 1826 he became physics professor first at the Military Academy of Berlin and then at the universities of Nuremberg and Munich.

His main studies concerned current flow in conductors, which resulted (after further contributions from other scientists) in the law known today as "Ohm's law."

He also conducted studies on light interference and on the ear's perception of complex sounds.

4.7.2 Short Biography of Thévenin

Léon Charles Thévenin (1857–1926) was a French engineer who is mainly famous for his theorem for circuits. Born in Meaux, near Paris, he graduated from the Ecole polytéchnique in 1876. Two years later he joined France's national electrical communication company (Postes et Télégraphes), which he never left.

After studying circuit laws (by Kirchhoff and Ohm), he developed what is now called Thévenin's theorem, which allowed people to reduce complex circuits into simpler ones, containing "Thévenin's equivalent".

Science biographies show that the same theorem had already been developed 30 years earlier, by Hermann von Helmholtz. Credit is given to Thévenin for developing his own version of the same theorem while unaware of Helmholtz's work.

When Thévenin issued his theorem, it was widely disputed. Shortly before his death, however, it became accepted all over the world.

[5]Pronounced (from the *Oxford English Dictionary*): /əʊm/ (Brit.), /oʊm/ (USA).

4.7.3 Short Biography of Joule

James Prescott Joule[6] (1818–1889) was born in Salford, Manchester. He spent his life managing his brewery in Salford and studying physics. Despite his lack of a formal education, his important scientific achievements led to an *honoris causa* degree from the University of Leida.

His most significant achievement was a mathematical relation that links the heat produced by a current-carrying conductor, its resistance, and the current itself. This relationship is today universally referred to as Joule's law. Also very important were his studies leading to his paper on the "mechanical equivalent of heat" which gave vital impetus to the definition of the First Law of Thermodynamics. He also worked with W. Thompson (Lord Kelvin) to develop the absolute scale of temperature.

The S.I. unit of energy (as well as of work and heat), the joule, is named after him.

4.8 PROPOSED EXERCISES

4.1. A voltage $U = 12\,\text{V}$ is applied to an aluminium conductor, with a length of 300 m and a cross-sectional area of $150\,\text{mm}^2$.

Calculate the resistance R, the current I and the absorbed power P, at 20°C and 55°C.

Use Table 4.2.

4.2. Calculate the equivalent resistance "seen" from nodes A and B:

(a)

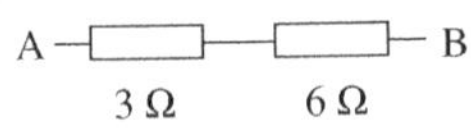

(b)

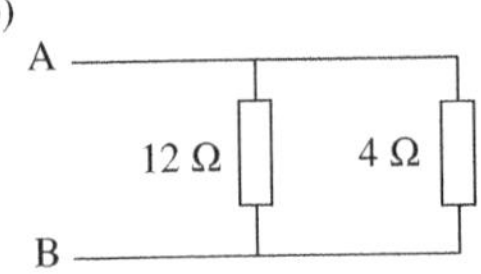

(c)

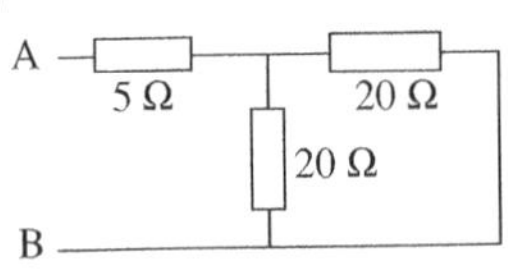

[6]As far as pronunciation is concerned, the *Oxford English Dictionary* states that although some people with this name call themselves /dʒaʊl/ or /dʒəʊl/ it is almost certain that J.P Joule used /dʒu:l/.

(d)

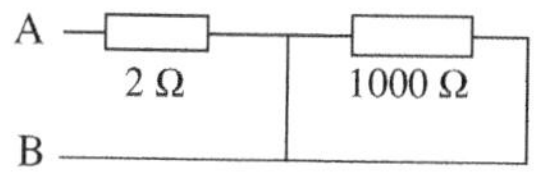

(e)

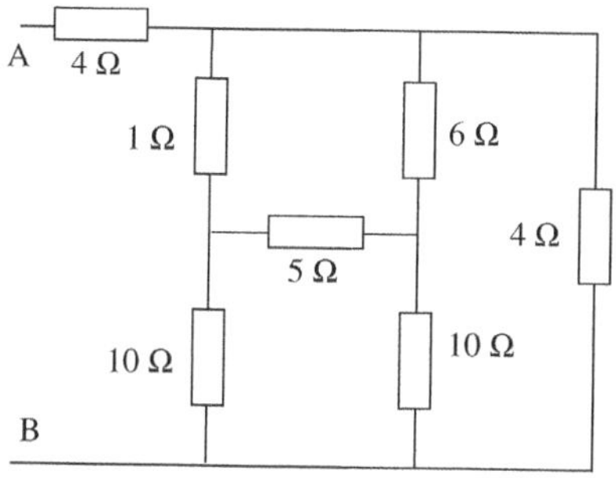

(f)

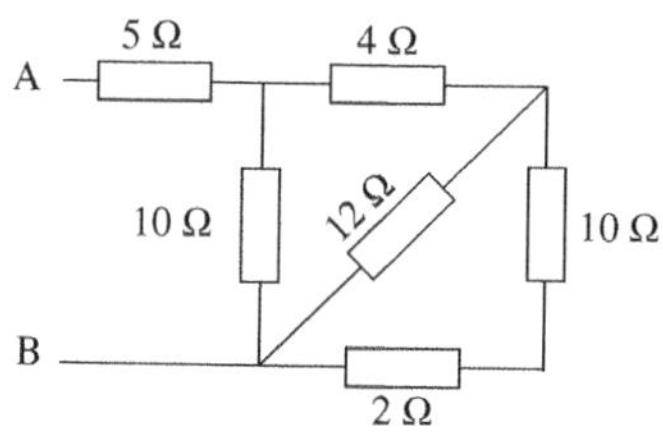

(g)

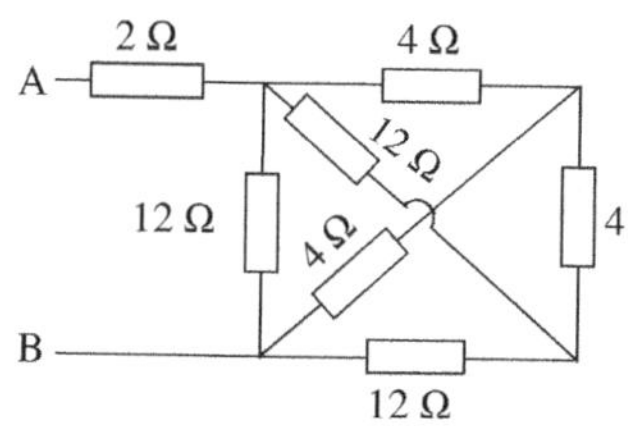

4.3. Using the current division rule, calculate I_1 and I_2, I being 10 A. Verify the solution, calculating U_{AB} as $R_{eqp}I$ and observing that $R_1I_1 = R_2I_2 = U_{AB}$.

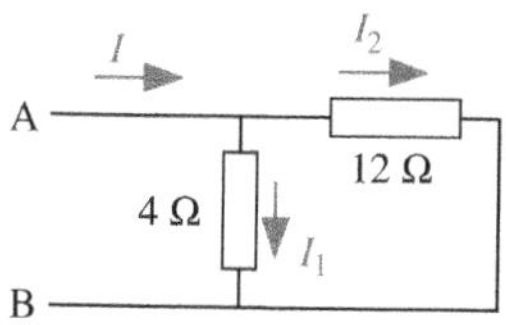

4.4. Determine I and U_{AB}.

If U_{s1} and U_{s2} represent two ideal batteries, which one charges the other?

$$U_{s1} = 120\,\text{V}, \qquad U_{s2} = 90\,\text{V}, \qquad R_1 = R_2 = 10\,\Omega, \qquad R_3 = 40\,\Omega$$

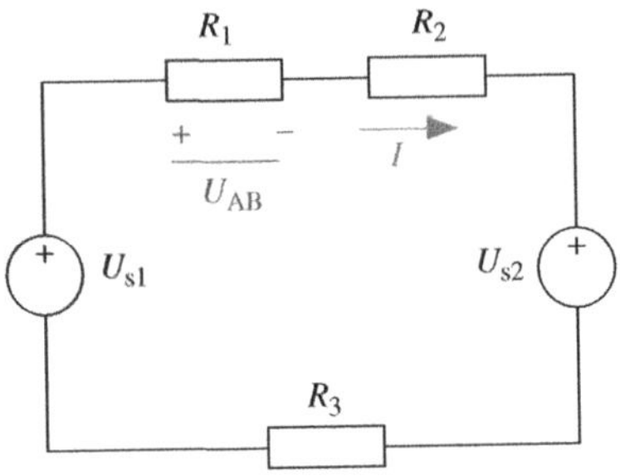

4.5. Calculate the resistance R_G seen by the generator, and I_1. Then, using the voltage division rule, calculate I_2 and I_3.

Check the conservation of power, comparing what is delivered by the generator and what is absorbed by resistors.

$$U_s = 12\,\text{V}, \qquad R_1 = R_2 = 2\,\Omega, \qquad R_3 = 8\,\Omega, \qquad R_4 = 6\,\Omega.$$

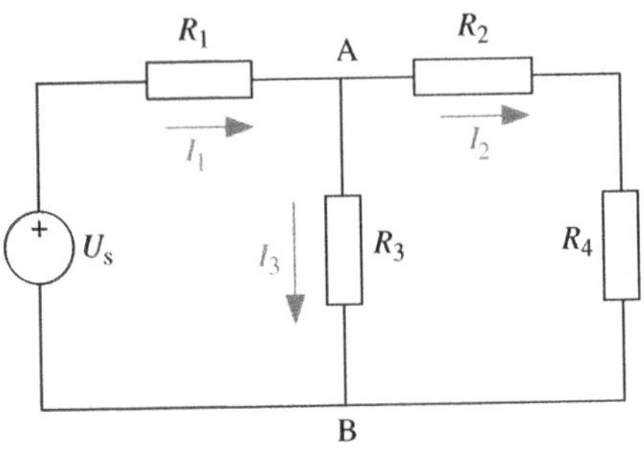

4.6. Solve exercise 4.5 again:

(a) using KCE;

(b) using the nodal analysis in order to calculate U_{AB}.

4.7. By applying Thévenin's theorem between A and B, calculate the equivalent voltage U_{Th} and the equivalent resistance R_{Th}:

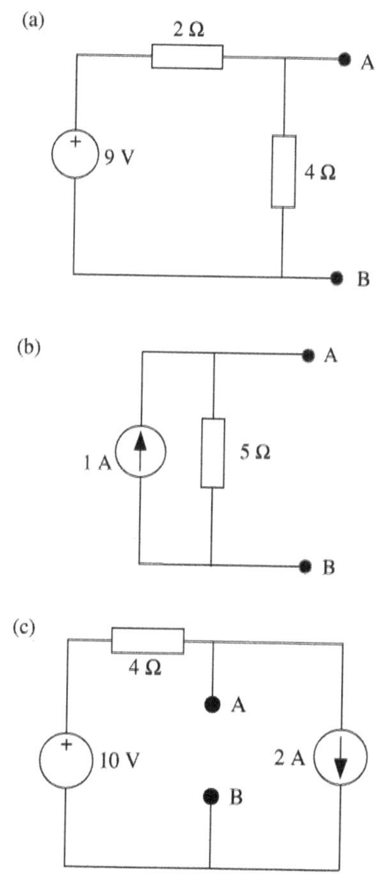

4.8. Referring to exercise 4.5, determine the voltage across R_4, using Thévenin's theorem.

4.9. Solve the following circuit:

(a) using the superposition rule;

(b) using KCE;

(c) using nodal analysis to calculate U_{AB};

(d) using Thévenin's theorem to find an equivalent, left-side or right-side section AB.

$$U_{s1} = 12\,\text{V}, \quad R_1 = 0.5\,\Omega, \quad R_2 = 5\,\Omega, \quad U_{s2} = 9\,\text{V}.$$

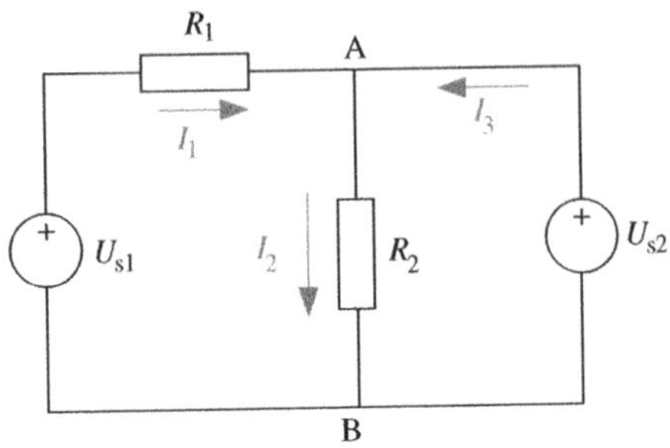

4.10. Solve the following circuit:

(a) using the superposition rule;

(b) using KCE;

(c) using nodal analysis to calculate U_{AB} (write KCL at node A);

(d) using Thévenin's theorem to find an equivalent, left-side section AB.

$$U_s = 100\ \text{V},\quad R_1 = 20\ \Omega,\quad R_2 = 30\ \Omega,\quad I_s = 3\ \text{A}.$$

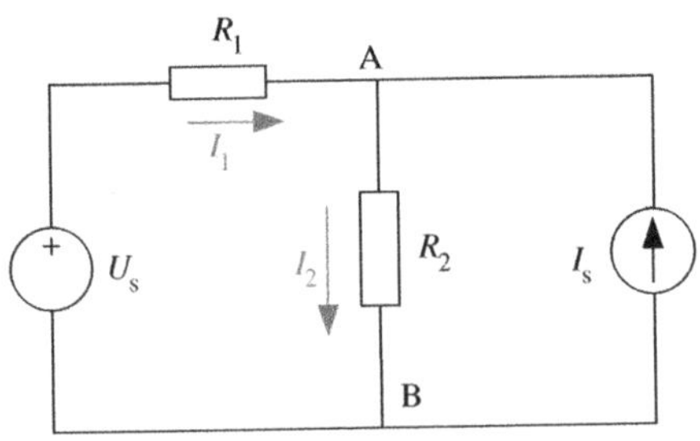

4.11. In the previous exercise, calculate the voltage U_I across the current generator. Then verify the energy balance, comparing the power delivered by the generators and the one absorbed by resistors.

4.12. Find at least three ways to calculate I_2 and I_3. Is R_1 really required to determine currents? And to calculate the power delivered by I_{S1}?

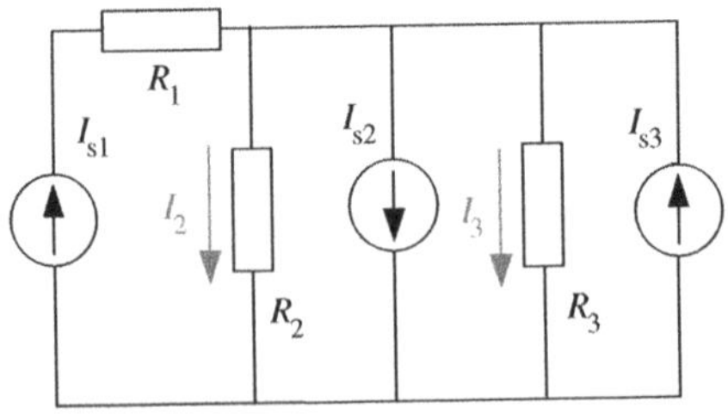

$$I_{s1} = 5\ \text{A},\quad I_{s2} = 1\ \text{A},\quad I_{s3} = 8\ \text{A},\quad R_1 = 5\ \Omega,\quad R_2 = 1\ \Omega,\quad R_3 = 4\ \Omega.$$

4.13. The circuit shown in the figure supplies a lamp connected between nodes A and B and with the following rated values: $U = 12\,\mathrm{V}$, $P = 10\,\mathrm{W}$, $\Phi = 100\,\mathrm{lm}$. Calculate:

(a) the actual luminous flux Φ delivered by the lamp, assuming it is proportional to I^2

(b) the energy $\mathscr{E}$ consumed by the lamp over 100 h.

$$U_s = 15\,\mathrm{V}, \qquad R_1 = 2\,\Omega, \qquad R_2 = 10\,\Omega, \qquad I_s = 0.3\,\mathrm{A}$$

Suggestion: Find Thévenin's equivalent at nodes A and B.

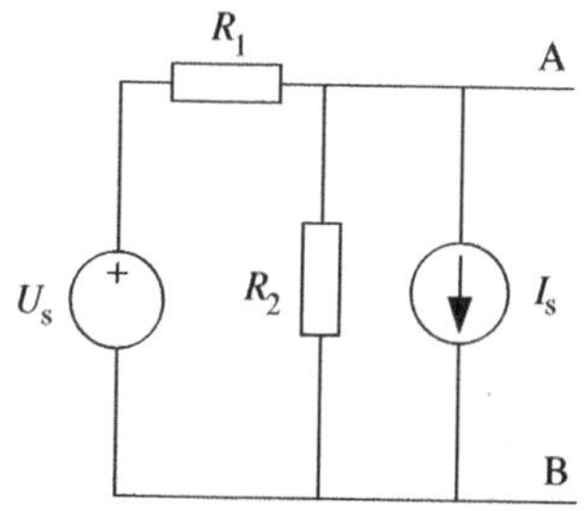

4.14. Calculate current I in the following circuit:

$$U_{s1} = 6\,\mathrm{V}, \qquad U_{s2} = 2\,\mathrm{V}, \qquad R_1 = 1\,\Omega, \qquad R_2 = R_3 = 2\,\Omega, \qquad I_s = 3\,\mathrm{A}$$

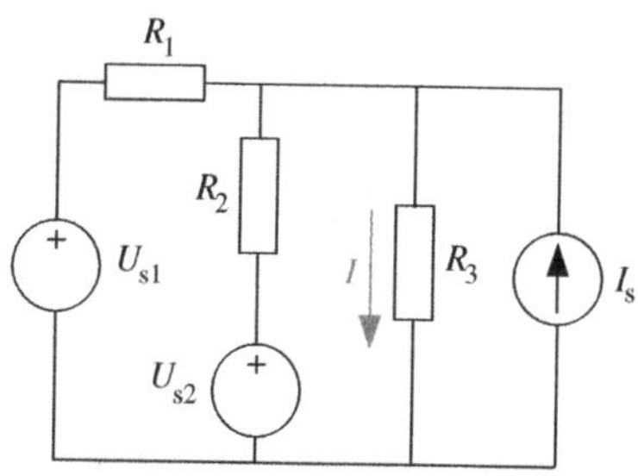

4.15. Calculate the Thévenin equivalent of the following circuit:

$$U_s = 2\,\mathrm{V}, \qquad I_s = 2\,\mathrm{A}, \qquad R_1 = 4\,\Omega, \qquad R_2 = R_3 = 1\,\Omega$$

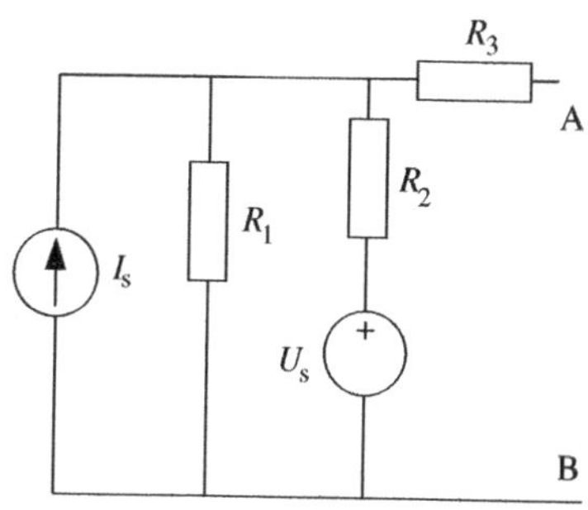

4.16. Calculate the Thévenin equivalent of the following circuit:

$$U_{s1} = 6\,\text{V}, \quad U_{s2} = 2\,\text{V}, \quad R_1 = 4\,\Omega, \quad R_2 = 6\,\Omega, \quad R_3 = 2\,\Omega,$$

$$R_4 = 1\,\Omega, \quad I_s = 1\,\text{A}$$

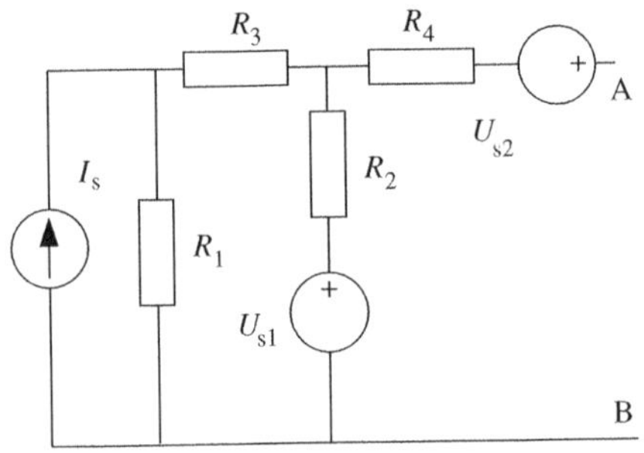

4.17. The previous circuit supplies a 10-Ω resistance. Calculate the power absorbed by the load.

4.18. Express the active power P absorbed by the load, as a function of R, U_s, and R_{line}. Demonstrate that when U_s and R_{line} are fixed, P is maximum when $R = R_{line}$ (*Theorem of maximum power transfer*).

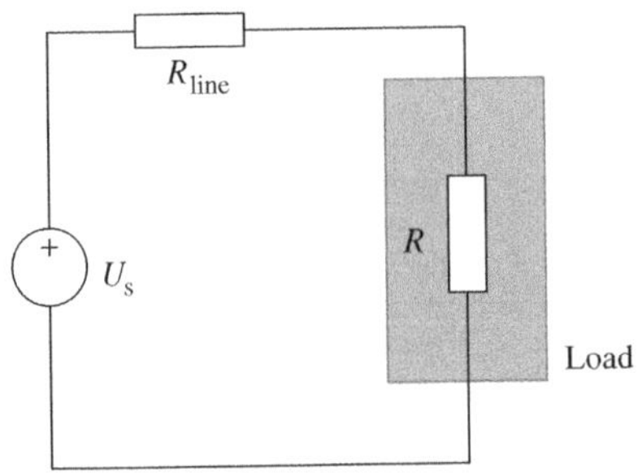

4.19. Find the load resistance R to be supplied by the following circuit, in order to maximize the power transfer to the load. Calculate this power.

$$U_s = 12\,\text{V}, \quad R_1 = 1\,\Omega, \quad R_2 = 15\,\Omega, \quad R_3 = 2\,\Omega, \quad I_s = 2\,\text{A}$$

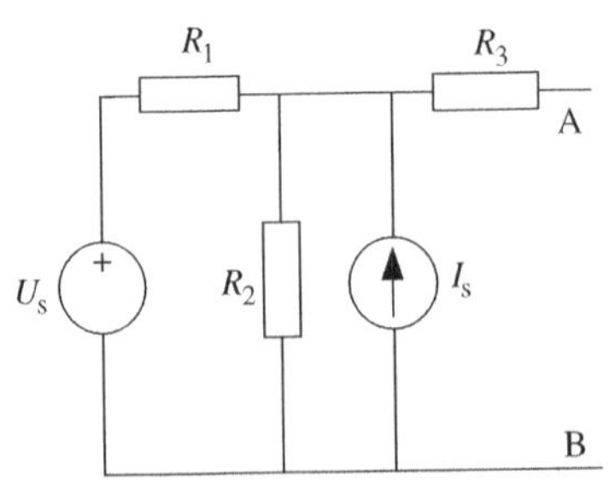

4.20. Calculate I_A and U_{AB} and determine the power flowing through section AB. Verify the result using the principle of power conservation.

$$U_{s1} = 10\,\text{V}, \quad U_{s2} = 4\,\text{V}, \quad R_1 = 1\,\Omega, \quad R_2 = 2\,\Omega,$$
$$R_I = 1\,\Omega, \quad I_s = 1\,\text{A}$$

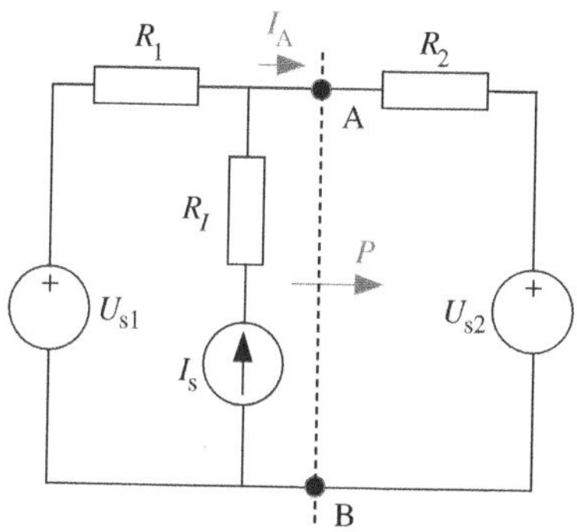

4.21. Find the load resistance R to be supplied by the following circuit, in order to maximize the power transfer to the load. Calculate this power.
(Before solving, read the text of exercise 4.18.)

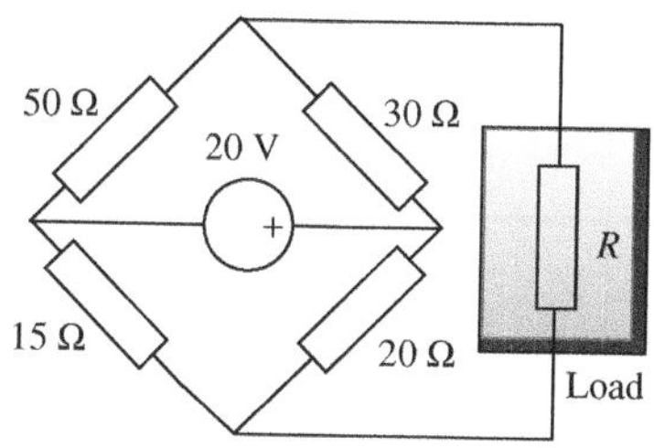

5

TECHNIQUES FOR SOLVING AC CIRCUITS

For the Instructor

This chapter has some special features that should make circuit study clearer.

(1) *Use of R–L models for AC transmission lines.* The use of R–L models for AC transmission lines in circuits is tricky: the user could be induced to consider transmission lines as true "circuits"—that is, to determine voltage between sending-end and receiving-end terminals as a difference of terminal potentials. This is not possible, and only voltages between top and bottom terminals at either ends can be computed. This limitation is normally disregarded in textbooks, thus inducing potential errors. It is, instead, analysed here in a simple but effective way; a deeper analysis can be found in the Appendix.

(2) *Transformer modelling.* In textbooks for electrical engineers, transformers are dealt with both through the T model containing an ideal transformer (and other components such as leakage inductances and coil resistances) and using the mutual inductance concept. To simplify things to our readers, in this book we have adopted only the approach based on the ideal transformer.

(3) *KVL in networks containing transformers.* In this chapter it is explicitly noted that Kirchhoff's voltage equations must be written separately for the parts of the network separated by the ideal transformer. This is so important that the

Fundamentals of Electric Power Engineering: From Electromagnetics to Power Systems, First Edition.
Massimo Ceraolo and Davide Poli.

word "network," often considered to be a synonym of "circuit," is used here to indicate a set of circuits separated from each other by ideal transformers—that is, one or more magnetically coupled circuits.

Some final words regarding the definition of phasors: in some books phasors are defined as having amplitudes equal to the rms values of the corresponding sine waves, instead of their peak values. Other books first use peak-oriented phasors, then, following a presentation of AC power, rms-oriented phasors. The latter is preferable in courses for electrical engineering students, since to some extent it facilitates computations. For the sake of simplicity and clarity, this book uses only peak-oriented phasors—that is, phasors whose amplitude is equal to the peak of the corresponding wave.

5.1 INTRODUCTION

In Chapter 3, "circuits" were introduced as a mathematical–graphical tool, useful for analysing physical systems composed of electric elements connected to each other by means of wires. It was also shown that, under given assumptions, circuits can effectively model physical systems that work with either constant or variable quantities.

In Chapter 4 we focused on DC circuits, in which all quantities (mainly currents, voltages, powers) do not vary over time.

This chapter is mainly devoted to the analysis of alternating current (AC) circuits, in which all the quantities (mainly currents, voltages, powers) vary as sinusoids over time. AC circuits are the most common ones in everyday life, since the energy provided by the mains into all of our homes, industries, and so on, is in this form.

In addition, we will deal with two kinds of circuits in which the quantities do not vary as sinusoids, namely the *R–L* circuit (containing a resistor and an inductor) and the *R–C* circuit (containing a resistor and a capacitor). This is because of their general importance and because they need to be understood in order to fully master the analysis of power electronic components and devices, which is in Chapter 8.

5.2 ENERGY STORAGE ELEMENTS

In Chapter 3 we introduced several possible constitutive equations of circuit branches in their mathematical formulations.

In Chapter 4 we studied circuits with special constitutive equations, which cause all currents and voltages to be constant.

Here we introduce other branches, whose idealized behaviour can be mathematically described by other constitutive equations enclosed in the set in Table 3.4. Since these are differential equations, we will show that the corresponding components have some energy storage capability and are therefore called globally energy storage elements.

5.2.1 Power in Time-Varying Circuits

Consider again the analysis in Chapter 4 (Section 4.4) for developing an expression relating voltage, current, and power in DC circuits. The reader is invited to personally verify that the same considerations developed there can be repeated for any time t. Therefore, the result obtained can be repeated also for circuits working with time-varying quantities:

Result: *Power flowing through two circuit terminals*

In any circuit, operating with either constant or variable quantities, the power exchanged by any two-terminal subcircuit to the other subcircuit is numerically equal to the product of voltage and current.

Power is delivered by the subcircuit in which the current leaves from the positive of the two terminals.

Figure 5.1 shows this result.

5.2.2 The Capacitor

Consider the system depicted in Figure 5.2.

This is composed of two large conductor surfaces, separated by a thin insulating layer (called the dielectric). Conductors are connected to the surface by means of the two terminals: access is given to the voltage across terminals.

Since the surfaces are very large in comparison to the distance between them d (i.e., $h >> d$ and $b >> d$), it can be assumed that the electric field in the space between them is uniform, with the force lines parallel to each other and orthogonal to the surfaces.

This allows us to determine the relation between the charge stored on the two surfaces (assumed to be equal in value but opposite in sign) and the electric field inside

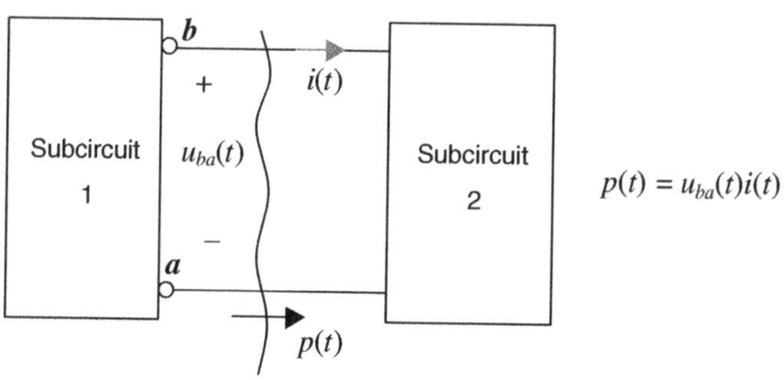

FIGURE 5.1. Power flow between two subcircuits of a generic circuit.

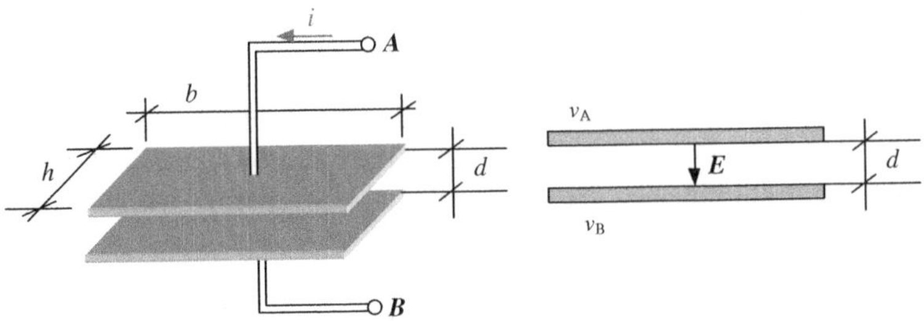

FIGURE 5.2. The basic structure of a capacitor.

using Gauss's law (see Chapter 2):

$$\frac{Q}{\varepsilon} = E \cdot bh \Rightarrow E = \frac{q}{bh\varepsilon} = \frac{q}{A\varepsilon}$$

in which A is the area of any of the two conductor surfaces.

The voltage across terminals is the difference in potential of the electric field and is equivalent to the work done per unit charge to move charges from one terminal toward another:

$$u_{AB} = v_A - v_B = W_{AB} = \int_A^B \mathrm{d}W = E \cdot d$$

We can combine these two equations to draw the relation between voltage U_{AB} and surface charge Q of this system:

$$u_{AB} = E \cdot d = \frac{d}{A\varepsilon} q = \frac{1}{C} q \qquad \text{with } C = \frac{A\varepsilon}{d} \tag{5.1}$$

The quantity C depends on the geometry of the system and not on the applied charge. The system represented in Figure 5.2 is called a capacitor.

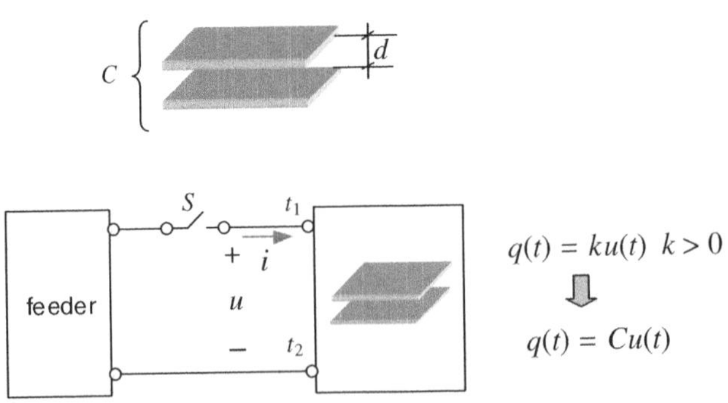

FIGURE 5.3. Proportionality between charge and voltage: the law of capacitors.

Any systems constituted by two large conductors separated by a thin layer of insulating material have the same basic characteristics as the system shown in Figure 5.2 and are thus capacitors, whose capacitance has a fixed value which can be computed with formulas that are very similar to (5.1).

Equation (5.1) is the *constitutive equation* of a capacitor. For a discussion of what a constitutive equation is, see Section 3.5.2.

The constitutive equation is useful to analyse the behaviour of capacitors inside circuits.

Consider the scheme in Figure 5.3. Imagine the two terminals t_1 and t_2 as being connected internally to the two plates, in such a way that current can flow freely between the outside wires and the plates. The testing circuit, this time, also contains the *ideal switch S*.

An ideal switch is a circuit component that can be either in the so-called *closed* state, in which case it behaves like an ideal wire, or in the so-called *open* state, in which it behaves as if nothing was present between its terminals—that is, as an *open circuit*, as it is usually termed.

Note that the terminology of electric switches is opposite to that of hydraulics circuits: when a hydraulic valve or a water tap is open, it allows fluid to flow; when it is closed, no flow can occur.

Suppose the voltage u across the two plates is zero at $t=0$, and that at $t=0$ the switch S, having been previously open, is closed. If the integral of the current i flowing in the circuit, is measured over time, along with the voltage across the plates u, equation (5.1) says that

$$q(t) = \int_0^t i(t)\mathrm{d}t = Cu(t) \tag{5.2}$$

Attention must be paid to the various signs: As clearly indicated in Figure 5.3, the proportionality coefficient is $C>0$ if the current is assumed to be entering the component through the positively marked terminal.

It goes without saying that if the choice of reference signs for voltage and current is different—that is, if the current is assumed to be entering the pin with an assumed negative potential—the equation should be written with the minus sign, as already stated in Table 3.4.

Equation (5.2) indicates that the physical system shown in the figure between terminals t_1 and t_2 can be mathematically described in a given circuit, with a close approximation, using the "capacitive element" constitutive equation. This component is very important in electrical engineering and is widely used. The actual physical construction of a capacitor can be different from the basic structure shown in Figure 5.3. For instance, a capacitor can be physically built in a cylindrical shape, so as to occupy less space.

When a capacitor is being charged or discharged, it stores or delivers energy to the circuit. When it is charged, the energy can be imagined to accumulate in the electric field in the space around the capacitor conducting plates.

Evaluating the energy transfer during capacitor charge is quite simple. Consider a charge starting at $t = t_0$ (with a null capacitor voltage, and therefore null stored charge q) and ending at $t = t_f$. It is

$$\mathscr{E} = \int_{t_0}^{t_f} u(t)i(t)\mathrm{d}t = \int_{q_0}^{q_f} q(t)\mathrm{d}q = \frac{1}{2C}\left(q_f^2 - q_0^2\right) = \frac{1}{2C}q_f^2$$

The energy accumulated in the capacitor, whose capacitance is C, charged at a voltage u is therefore $\mathscr{E} = \frac{1}{2}Cu^2$; this charge is sent back to the circuit when the capacitor is discharged.

Note the evident similarity with the energy accumulated in a translating mass m moving at a velocity v: $\mathscr{E} = \frac{1}{2}mv^2$

The charging transient of a capacitor will be dealt with in Section 5.4.

Example 1. The energy $\mathscr{E}$ accumulated in a capacitor charged at $u = 100$ V, and whose capacitance is $C = 1\ \mu$F, can be calculated as follows:

$$\mathscr{E} = \frac{1}{2}Cu^2 = \frac{1}{2} \cdot 10^{-6} \cdot 100^2 = 5 \text{ mJ}$$

5.2.3 Inductors and Magnetic Circuits

Consider the system depicted in Figure 5.4. It is composed of a coil with N turns of conductive material around a toroid made of magnetic material (e.g., iron). Since the toroid constitutes a loop in which the magnetic field circulates, it resembles a mesh in an electric circuit and is therefore also called *magnetic circuit.* If current i flows through the coil, a magnetic field is generated around it. It can be demonstrated that, as a consequence of iron's much higher permeability compared to air, most magnetic flux flows through iron; the flux traversing the coil but flowing at least partially in air is called leakage flux. For simplicity's sake, let us disregard the leakage flux for a while.

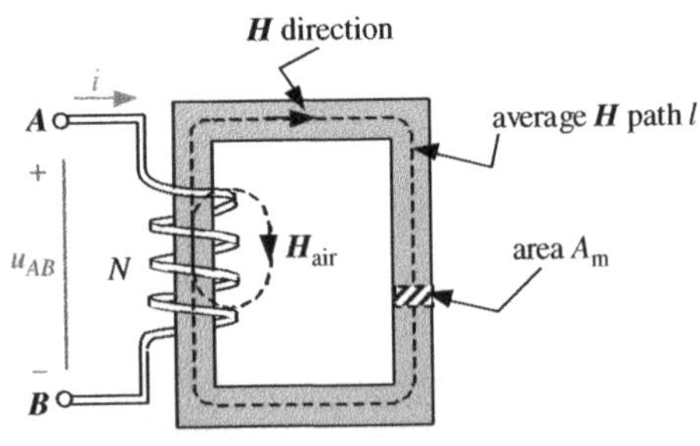

FIGURE 5.4. A coil around an iron toroid, constituting a magnetic circuit.

In this case, all the force lines of magnetic field inside the iron coil share the same length l, and therefore Ampère's law (Section 2.4) can be applied to the system in Figure 5.4 as follows:

$$H = \frac{N\,i}{l}$$

This magnetic field inside the iron torus produces a flux density $\boldsymbol{B}$ expressed as

$$\boldsymbol{B} = \mu \boldsymbol{H}$$

Throughout this book, iron nonlinearity is disregarded; as a consequence the permeability μ is assumed to be constant in the previous equation.

This flux density produces the flux ($\boldsymbol{B}$ being assumed to be constant along the area A_{m}) in the iron toroid

$$\phi = BA_m = \mu BA_m$$

and the flux linkage

$$\psi = N\phi = \mu NBA_m$$

All the constants introduced can be combined in a global constant L, called inductance, which is the coefficient between current and flux linkage:

$$\psi = \mu NBA_m = \mu NHSA_m = \mu \frac{N^2 A_m}{l} i = Li \tag{5.3}$$

The inductance L is composed of a characteristic of the coil (the number N) and other variables which are characteristic of the magnetic circuit. This is clearer if it is written

$$L = \frac{N^2}{\mathcal{R}}, \qquad \text{where} \quad \mathcal{R} = \frac{l}{\mu A_m} \tag{5.4}$$

The quantity $\mathcal{R}$ is called *reluctance* of the magnetic circuit.

The system shown in Figure 5.4 can be interfaced with other systems by means of its terminals ***A–B***. Therefore, it is very useful to describe the behaviour of the whole system by means of an equation that relates the potential difference across its terminals $u_{\mathrm{AB}} = v_{\mathrm{A}} - v_{\mathrm{B}}$ with the current i entering terminal $\boldsymbol{A}$. This is easily done by combining Faraday's law (Chapter 2) with result (5.3):

$$E_f = -\frac{\mathrm{d}\psi}{\mathrm{d}t} = -L\frac{\mathrm{d}i}{\mathrm{d}t} \tag{5.5}$$

Equation (5.5) is the constitutive equation of an inductor.

The sign "–" in this equation depends on how the reference signs for E_{f} and i are set in a circuit. Compare the discussion in Section 2.5, and the circuital representations of inductors and reference signs immediately before *More in Depth* box D of Chapter 2.

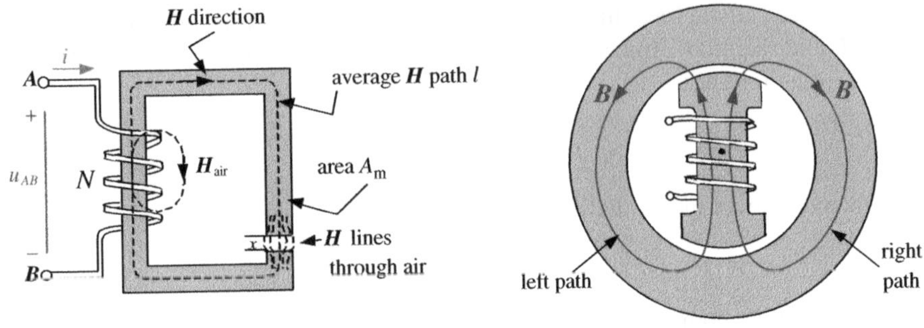

FIGURE 5.5. Magnetic circuits containing air gaps.

Equation (5.5) remains valid also in the event that leakage flux is not neglected: The formulation of L is different from that obtained earlier when all the flux flowing through the iron toroid was taken into account.

A particularly significant case is that of a circuit which mainly traverses iron (i.e., material with very high permeability), but has a small part constituted by air. Two examples are shown in Figure 5.5.

Consider first the magnetic circuit shown in the left-hand part of Figure 5.5. The force lines of magnetic field $\boldsymbol{H}$ are closed loops that mostly traverse iron, but also traverse the air gap whose width is indicated as x in the figure.

It could easily be demonstrated that equations (5.3), (5.4) and (5.5) are still valid, with an appropriate evaluation of the circuit reluctance as (subscripts "ag" stands for "air-gap")

$$\mathcal{R} = \mathcal{R}_{fe} + \mathcal{R}_{ag}, \quad \mathcal{R}_{fe} = \frac{l_{fe}}{\mu_{fe}A_{fe}} \cong \frac{l}{\mu_{fe}A_m}, \quad \mathcal{R}_{ag} = \frac{x}{\mu_{air}A_{ag}} \cong \frac{x}{\mu_{air}A_m}$$

Since iron permeability is thousands of times that of air, in many cases it can be more simply assumed as

$$\mathcal{R} \cong \mathcal{R}_{ag}$$

The same considerations apply to the more complex magnetic circuit shown in the right-hand part of Figure 5.5. Here iron reluctance must be computed by taking into account the two paths in parallel and summing two air gaps to the iron path, as follows:

$$\mathcal{R} = \mathcal{R}_{fe,tot} + \mathcal{R}_{ag,tot} = \mathcal{R}_{fe,path}/2 + 2 \cdot \mathcal{R}_{ag}$$

and, approximately

$$\mathcal{R} \cong \mathcal{R}_{fe,tot}$$

The constitutive equation is useful to analyse the behaviour of inductors inside circuits.

Consider the scheme in Figure 5.6. Imagine the two terminals t_1 and t_2 as being connected internally to the two coil terminals. The testing circuit this time also contains the ideal switch S, whose concept was introduced in Section 5.2.2.

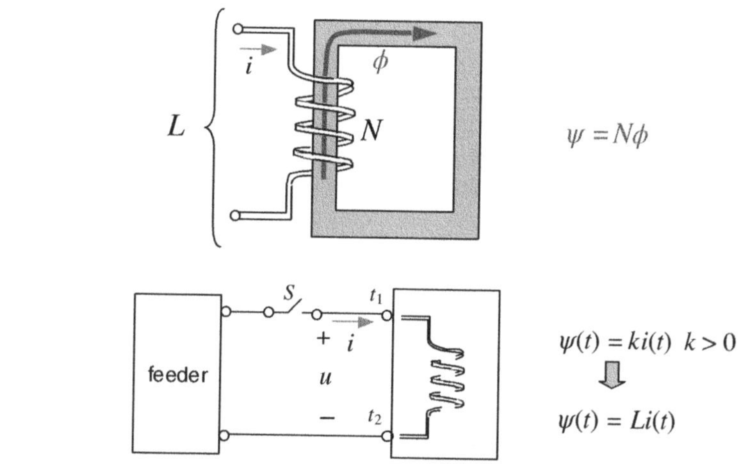

FIGURE 5.6. Proportionality between current and flux linkage: the law of the inductor.

Suppose that the current i flowing into terminal t_1 is zero at $t=0$, and that at $t=0$ the switch S, which was previously open, is closed. If the coil flux linkage and the current i flowing in the circuit are measured over time, it can be concluded from the laws of electromagnetism (Chapter 2) and experimentally verified that flux linkage is proportional to the current:

$$\psi(t) = Li(t)$$

A practical way to experimentally evaluate the flux linkage is by using Faraday's law (see Chapter 2 again), coupled with Ohm's law:

$$u(t) = Ri(t) + \frac{d\psi(t)}{dt}$$

Therefore, the equation defining the proportionality coefficient L between flux linkage and current is

$$u(t) = Ri(t) + L\frac{di(t)}{dt} \tag{5.6}$$

As with the capacitor, attention must be paid to signs. Signs are to be determined using Lenz's rule: a rising current generates a positive polarity in the terminal where the current is entering the coil. The result is the positive sign shown in equation (5.6).

Moreover, to allow flux linkage to be related to the current that generates it by a positive proportionality coefficient, its reference direction must be combined with the current reference direction according to the right-hand rule, as shown in Figure 5.3.

The rule is applied as follows. Using the right hand, curl the fingers and align them with the coil windings in such a way that fingertips are oriented as arrowheads of the reference direction of the coil current; the thumb indicates the reference direction of flux (the arrowhead is its tip).

The sign rule for inductors can be concisely expressed in the following way: if current and voltage reference signs are paired in such a way that the current enters the positively marked terminal, then Faraday's electromotive force has the same sign as the corresponding ohmic voltage drop, as in (5.6).

When an inductor is being charged or discharged, it stores or delivers energy to the circuit. When it is charged, the energy can be imagined as accumulating in the magnetic field in the space around the inductor coil.

As for the capacitor, evaluating the energy transfer during capacitor charge is quite simple: consider a charge starting at $t = t_0$ and ending at $t = t_f$. This is shown here, for simplicity sake, for an ideal inductor—that is, an inductor for which the effects or resistance R are negligible. It is

$$\mathscr{E} = \int_{t_0}^{t_f} u(t)i(t)\,dt = \int_{t_0}^{t_f} L\frac{di}{dt}i(t)\,dt = \int_{i_0}^{i_f} Li(t)\,di = \frac{1}{2}L\left(i_f^2 - i_0^2\right)$$

By convention, it is assumed that the energy stored in the inductor is null when the current is also null (i.e., zero). Therefore, the energy accumulated in the inductor, whose inductance is L and in which the current i flows is $\mathscr{E} = \frac{1}{2}Li^2$; this charge is sent back to the circuit when the inductor is discharged. Again, note the evident similarity with the energy accumulated in a translating mass m moving at a velocity v: $\mathscr{E} = \frac{1}{2}mv^2$.

The charging transient of an inductor will be dealt with in Section 5.4.

Example 2. The energy $\mathscr{E}$ accumulated in an inductor whose inductance is $L = 0.1$ H, and in which the current $i = 2$ A flows, can be calculated as follows:

$$\mathscr{E} = \tfrac{1}{2}Li^2 = \tfrac{1}{2}\cdot 0.1 \cdot 2^2 = 0.2\ \text{J}$$

5.3 MODELLING TIME-VARYING CIRCUITAL SYSTEMS AS CIRCUITS

In Chapter 3, Kirchhoff's laws were introduced for circuits. In Chapter 4 it was shown that *circuital systems* (physical systems with circuital shape, i.e., composed of subsystems connected to each other by wires), operating with all constant quantities—commonly known as "DC circuits"—can be modelled through *circuits* (i.e., lumped-component models of circuital systems) containing a branch between each pair of points in space through which non-negligible current can flow.

Modelling a physical system as a circuit gives us the power of using Kirchhoff's laws.

The hypotheses which in DC allowed a physical circuital system to be modelled by a circuit, relied on the fact that all quantities were constant (this in turn implied that the branches were either resistors or constant voltage or constant current sources).

These hypotheses are not valid when time-varying circuital systems are considered. Analysis of these systems will be dealt with in the following sections.

5.3.1 The Basic Rule

Consider the physical system shown in Figure 5.7a. To understand this, we must postulate that it is possible to build a "generator"—that is, a device able to generate a

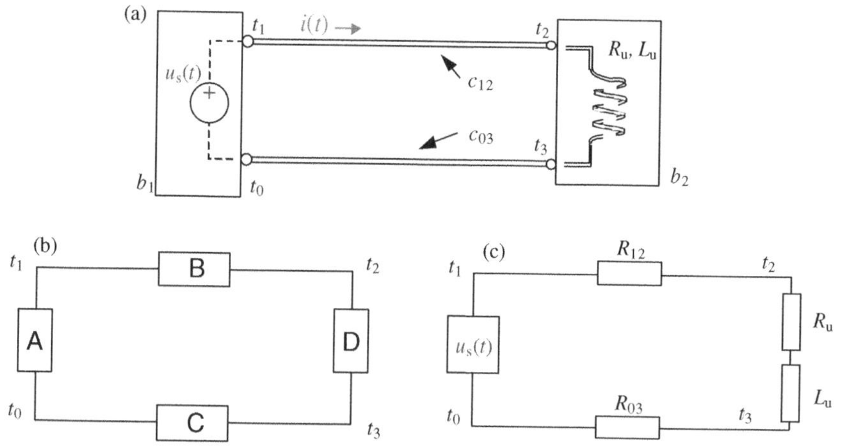

FIGURE 5.7. A physical system containing an inductor as a load (a) and a lumped-components model candidate (b and c).

time-varying voltage at its left-hand side (i.e., between terminals t_0 and t_1). From engineering practice, this is known to be true and should be part of the student's previous knowledge.

This device is postulated as having a border b_1, which isolates it from the external world, in such a way that no electric phenomena can cross its borders: no electric charge, no displacement current, and no effects of electric and magnetic fields inside this source influence the external world, and vice versa, except for what occurs through the component terminals.

This characteristic of interacting with the external world through the terminals is also postulated for the load.

The remainder of the circuit is constituted by a two-conductor line (imagine them to be constituted by two copper cylinders), connecting the generator with the load constituted by an inductor, whose behaviour was summarised in Section 5.2.3.

It appears natural to build, as a candidate circuit for modelling the physical system of Figure 5.7a, a circuit with four terminals at the connection points of the two line conductors with source and load, as shown in Figure 5.7b.

To do this, the actual system behaviour should be mainly confined to the four boxes, meaning that:

- Conduction and displacement currents outside the circuit wires are negligible.
- Any other influence of what happens in the space around the conductors on what occurs inside them must be negligible. Therefore, the effects of induced electromotive force on the loop constituted by the two wires, the source and the load must be negligible.

Under these assumptions, the conductor wires connecting source and load can be modelled by their ohmic resistance, and therefore the equation governing the system in Figure 5.7a, is

$$u_s(t) = R_{12}i(t) + R_u i(t) + R_{03}\, i(t) + L_u \frac{di(t)}{dt}$$

This equation can be written in terms of branch voltages as

$$u_{10}(t) = u_{12}(t) + u_{23}(t) + u_{30}(t)$$

with

$$u_{12}(t) = R_{12}i(t)\, u_{12}(t) = R_{12}i(t)$$

$$u_{01}(t) = u_s(t), \qquad u_{23}(t) = R_u i(t) + L_u \frac{di(t)}{dt}$$

Therefore

$$u_{10}(t) + u_{21}(t) + u_{32}(t) + u_{03}(t) = 0$$

It is possible to define, at any given time t, a potential function of the circuit terminal T $v(t, T)$, thus allowing branch voltages to be computed as the difference between the values this function assumes, at any given time, at the corresponding terminals; in fact, if such a function exists, it can be written as

$$(v_1 - v_0) + (v_2 - v_1) + (v_3 - v_2) + (v_3 - v_0) = 0$$

The result of the above sum is zero because each term occurs once with a positive sign and once with a negative one.

This mathematical analysis allows the system of Figure 5.7a to be modelled, under the given hypotheses, by the circuit in Figure 5.7b, with the following constitutive equations:

$$\mathsf{A}: u_A(t) = e(t), \qquad \mathsf{B}: u_B(t) = R_B i_B(t), \qquad \mathsf{C}: u_C(t) = R_C i_C(t),$$

$$\mathsf{D}: u_D(t) = R_u i_D + L_u (di_D(t)/dt)$$

The constitutive equation of block D is not in one of the forms considered in Table 3.4; to obtain this, the component's behaviour needs to be split into two parts, one purely ohmic (modelled by R_u) and another purely inductive (modelled by L_u) as shown in Figure 5.7c.

It should now be clear that this analysis could be repeated for any circuital system with components that can be modelled as circuit elements, connected to each other by

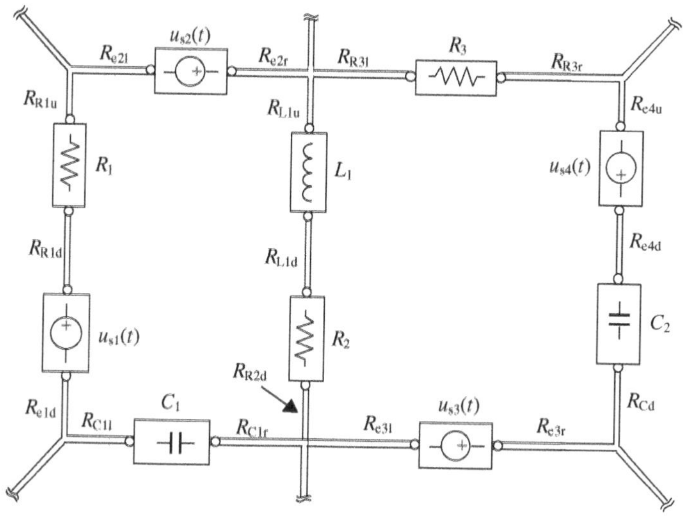

FIGURE 5.8. A complex example of a time-varying circuital system.

their terminals, and a unique potential function v can be defined to evaluate circuit terminal potentials.

For instance, Figure 5.8 shows the example of a circuital system containing multiple loops. It should be clear that also in this example the boundaries around the circuit devices define the latter as circuit elements. If, in addition, conduction and displacement currents around the conductors are disregarded, along with the effects of the time-varying magnetic field around the wires, then the system can be modelled using the circuit in Figure 5.9 and, possibly, if the wire resistances are disregarded, the one in Figure 5.10.

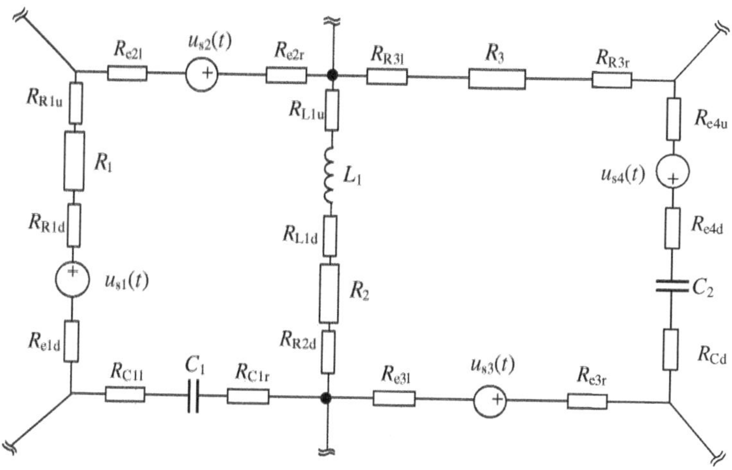

FIGURE 5.9. A circuit modelling, under given assumptions, the system of Figure 5.8.

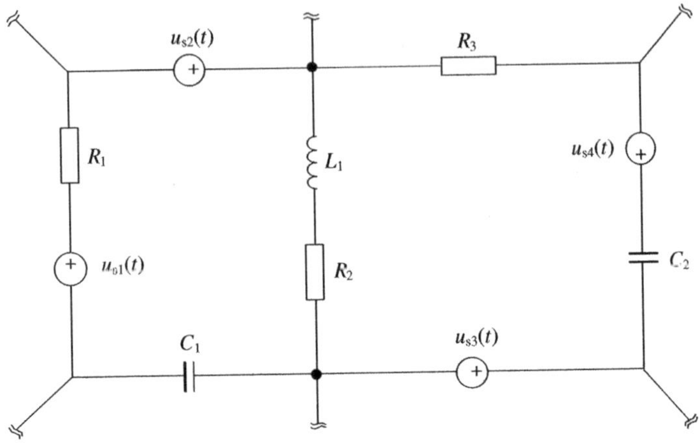

FIGURE 5.10. A simplified version of the circuit of Figure 5.9, which can be drawn when the resistance of the conductors is disregarded.

Example 3. Write the KVL equations for the two loops of the previous circuit. Let's assume the following conventional signs for currents and voltages:

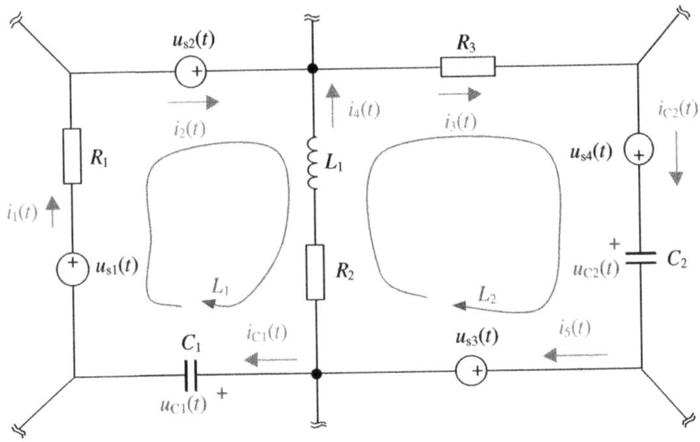

Loop L_1:

$$u_{s1}(t) + u_{s2}(t) = R_1 i_1(t) - L_1 \frac{di_4(t)}{dt} - R_2 i_4(t) + u_{C1,0} + \frac{1}{C_1}\int_0^t i_{C1}(t)\,dt$$

Loop L_2:

$$u_{s4}(t) - u_{s3}(t) = R_2 i_4(t) + L_1 \frac{di_4(t)}{dt} + R_3 i_3(t) + u_{C2,0} + \frac{1}{C_2}\int_0^t i_{C2}(t)\,dt$$

The systems proposed in this section should make it clear under which hypotheses circuital systems can be modelled as circuits. They can be expressed as follows.

Rule *Modelling of time-varying systems through circuits*

In a circuital system made of components that can be modelled as circuit elements, connected to each other by conducting wires, the following statement is true:

If conduction and displacement currents in the air surrounding circuit wires are negligible, as are voltages induced according to Faraday's law in the areas between wires,

then the conductor wires can be represented by their equivalent resistance, and KVL applies to the system, which can therefore be modelled as a circuit. Branch voltages can be written as the difference of corresponding terminal potentials.

Time-varying systems to be modelled as circuits will contain elements whose constitutive equations are those reported in Table 3.4. It is fairly obvious that in order for at least some of the quantities to have actual variation over time, at least one of the sources (either voltage or current source) will have to be a function of time.

5.3.2 Modelling Circuital Systems When Induced EMFs Between Wires Cannot Be Neglected

There are some cases in which magnetic induction by Faraday's law is not negligible and thus we need to extend the analysis of the previous section to accommodate this need.

Consider again Figure 5.7a. Let us now analyse it without ignoring the effect of induction in the area between the upper and lower conductors. This figure is shown in Figure 5.11 with minor modifications.

If the two wires cover a long distance, it may be of interest to compute only the voltage between terminals t_1 and t_0, and between t_2 and t_3. For instance, the line could be a transmission line and only the voltage at the sending and receiving ends are of interest.

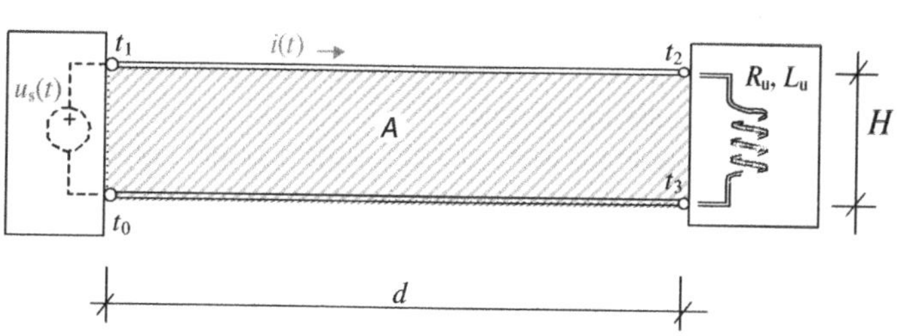

FIGURE 5.11. System containing two components connected by a long transmission line.

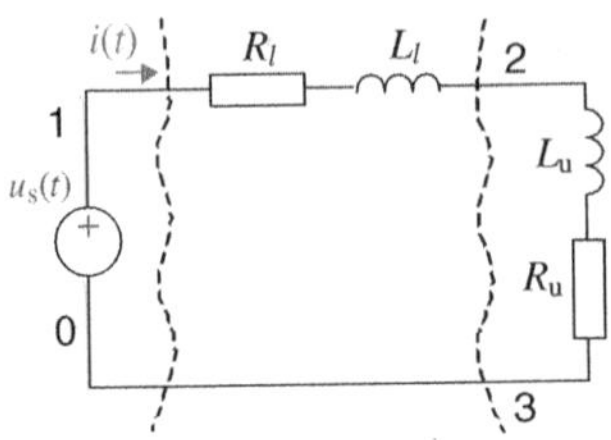

FIGURE 5.12. A "metacircuit" representing the circuital system shown in Figure 5.11.

In this case, even though a potential cannot be defined for each of the shown terminals, we can derive the mathematical relation between them.

Indeed, the two constitutive equations of sending-end and receiving-end blocks:

$$u_{10}(t) = u_s(t), \qquad u_{23}(t) = R_u i(t) + L_u \frac{di}{dt} \tag{5.7}$$

can be associated with the equation of the line containing Ohm's and Faraday's laws:

$$u_{10}(t) - u_{23}(t) = R_l i(t) + \frac{d}{dt}\psi_A(t) = R_l i(t) + L_l \frac{d}{dt} i(t) \tag{5.8}$$

in which R_l and L_l are line resistance and inductance respectively, and ψ_A is the magnetic flux linked with area A.

Equation (5.8) is similar to a constitutive equation since it relates electrical quantities as seen from the line terminals; however, it is not exactly the same, since it does not allow us to determine "cross-voltages" such as $u_{13}(t)$ or $u_{20}(t)$ since they are undefined.

The set of (5.7 and (5.8)) is able to describe the behaviour of the *metacircuit* shown in Figure 5.12.

The word "metacircuit" is introduced here. The dashed lines indicate that this is not a true circuit, since it cannot be used to compute potential differences between a terminal at the sending end of the transmission line and one at the receiving end, but only potential differences between terminals at the same end of the line.

In common engineering practice, the two dashed lines are not drawn, and circuits containing transmission lines (i.e., systems in which the effects of induction are not negligible) are written as normal circuits; any electrical engineer, however, knows that in these cases it is not correct to evaluate potential differences between terminals situated at two opposite ends of the line.

For those interested, the topic is further developed in the Appendix.

5.3.3 Mutual Inductors and the Ideal Transformer

From their background of electromagnetism, readers should already have some knowledge of mutually coupled windings. These are rapidly revised here.

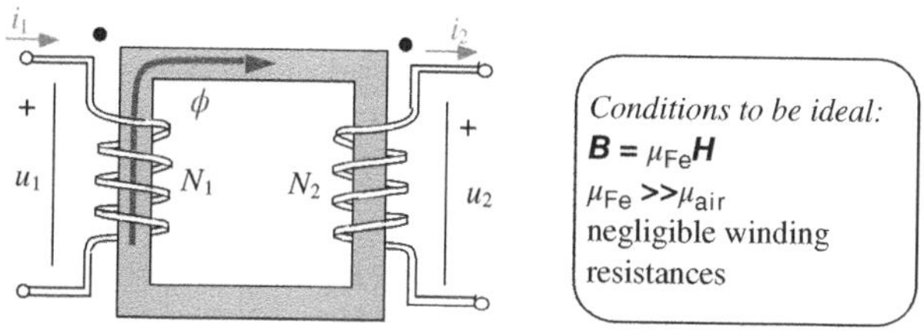

FIGURE 5.13. The principle structure of a transformer, and the conditions to be met for ideal behaviour.

Consider a component constituted as shown in Figure 5.13. It is composed of two conductors, each of them wound several times around a toroid of ferromagnetic material. The very high magnetic permeability of ferromagnetic materials, compared to air, causes the magnetic field produced by any current flowing in the conductor to be confined, for the most part, within the toroid itself.

A simplified description of the behaviour of the system in Figure 5.13 can be found under the hypothesis of proportionality of electric field and magnetic flux density within the iron core $\boldsymbol{B} = \mu_{\mathrm{Fe}} \boldsymbol{H}$ when iron magnetic permeability is much greater than the surrounding air permeability: $\mu_{\mathrm{Fe}} >> \mu_{\mathrm{air}}$.

In Chapter 7, where the system in Figure 5.13 will be analysed in much more detail, it will be shown that under these hypotheses the following equations apply (ϕ is the magnetic flux flowing in the torus):

$$\begin{aligned} u_1(t) &= R_1 i_1(t) + N_1 \frac{\mathrm{d}\phi(t)}{\mathrm{d}t} \\ u_2(t) &= -R_2 i_2(t) + N_2 \frac{\mathrm{d}\phi(t)}{\mathrm{d}t} \\ N_1 i_1(t) &- N_2 i_2(t) = 0 \end{aligned} \tag{5.9}$$

Different ways of winding the two conductors around the torus are possible. In the left coil of the figure, the wire winds from top to bottom in a counterclockwise direction (as seen from the upper part of the drawing); the opposite occurs in the right coil. Different winding orientations have effects on the signs of the different terms of equations (5.9).

Dots "•" reported on the figure give useful information about the coil winding orientation. They indicate that the second terms of equations (5.9)—that is, those expressing Faraday's law—are in phase, when voltages u_1 and u_2 are expressed as potentials at the dot-marked minus potentials at the unmarked terminals.

In the third equation, we make an algebraic sum of two magnetomotive forces. The minus sign is a consequence of how the coils are wound and the directions assumed for i_1 and i_2. In fact, applying the right-hand rule it is easily seen that i_1 creates a magnetomotive force that tends to "push" the magnetic field (thus the magnetic flux) in the direction of the symbol ϕ shown; i_2, on the contrary, creates a magnetomotive force that acts against the one created by i_1.

The dots "•" reported in the drawing also indicate this. They can be interpreted as follows: If both currents enter or leave the dot-marked terminals, the magneto motive forces created by the two coils contribute together to the flux and they must be summed using plus signs; if on the contrary *one* of them leaves a dot, one of the two magnetomotive forces receives minus signs, as in the third of (5.9).

The sign of the term of Faraday's law is such that the induced electromotive forces tend to create currents in the circuits outside the transformer, and these tend to oppose the flux change, according to Lenz's law.

Consider for instance the left coil. When ϕ is increasing, it must produce a voltage that, when applied to an external resistor, would circulate a current opposite in direction to ϕ. Such a current, according to the right-hand rule, would have the opposite direction to i_1. Therefore the Faraday's law induced voltage must have the plus sign in the upper terminal, as u_1 in figure. The same reason can be repeated for the right coil and voltage u_2.

If the further assumption that the first terms of the right-hand sums (i.e., Ohm's law terms) are negligible in comparison to the second ones is correct, the following can be written:

$$u_1(t) \cong u_{10}(t) = N_1 \frac{\mathrm{d}\phi(t)}{\mathrm{d}t}$$

$$u_2(t) \cong u_{10}(t) = N_2 \frac{\mathrm{d}\phi(t)}{\mathrm{d}t}$$

which, combined with the third equation of (5.9), give the following set of equations (equations of the ideal transformer):

$$\begin{aligned} u_1(t) &\cong \frac{N_1}{N_2} u_2(t) \\ i_2(t) &\cong \frac{N_1}{N_2} i_1(t) \end{aligned} \tag{5.10}$$

An *ideal transformer*—that is, a device characterised by the two equations in (5.10)—transfers to one side, for instance the right side (called the secondary side), exactly the same power that it receives from the other side (called the primary side). So *power conservation* applies to ideal transformers (see Figure 5.14b)

Note that if $\phi(t)$ varies slowly with time, or even if it does not vary at all, (5.10) cannot be inferred from (5.9), because the first terms of the sums on the right of (5.9) are not negligible, which was the assumption that enabled us to obtain (5.10).

Therefore, the ideal transformer concept is used in physical transformers only in AC, since the commonly used frequencies of 50 Hz or 60 Hz allow us to infer (5.10) from (5.9) with reasonable approximation for basic analysis. In no way can physical transformers based on Faraday's laws be used in DC.

The device shown in Figure 5.13, as seen from its exterior, has four terminals, two on its left side, two on the right. When inserted into a circuit, it should be represented

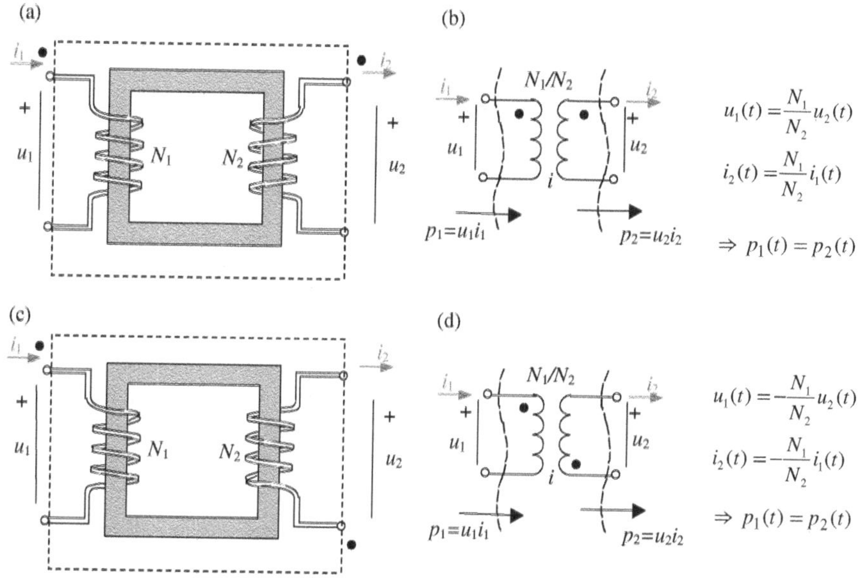

FIGURE 5.14. The ideal transformer: physical scheme (a and c), circuital symbols (b and d) and corresponding equations.

as "a box" with two terminals on the left and two on the right. Indeed, the importance of the component is such as to merit a specific symbol rather than a generic box, as shown in Figure 5.14b.

The symbol must be such that the electric behaviour must be determined even though the inner structure of the windings and magnetic circuit are not known. This is obtained by means of the two dots "•". The interpretation of these dots is immediate after looking at Figure 5.13 or at the left part of Figure 5.14b: the ratio of voltages measured with positive signs both at the dot-marked terminals is a positive number (equal to the turns ratio); similarly, the ratio of currents supposedly entering the dot-marked terminals is a positive value and is equal to the reciprocal turns ratio.

Whatever the signs, however, the power entering one side of an ideal transformer is equal to the power leaving the other side. It is very important to understand that an ideal transformer is characterised by equations which relate voltage and current at the two sides (windings), but does not imply any relation between the voltages at the primary and secondary side.

Therefore, they operate as a link between circuits, since, according to the developments in Chapter 3, in a single circuit a unique potential function must be known for the whole circuit, and the potential difference between all circuit terminals must be determined as a difference of the values the potential function assumes in correspondence to terminals.

The consequences of the inclusion of ideal transformers in circuits will be discussed in the next section. The ideal transformer does not accumulate any energy, since the power entering the primary coil is identically equal to the power leaving the secondary one: $p_1(t) = p_2(t)$.

However, we will see in Chapter 7 that a nonideal transformer in addition to transferring energy from one of its sides to the other, accumulates a little energy in the magnetic field around the coils.

5.3.4 Systems Containing Ideal Transformers: Magnetically Coupled Circuits

Consider a system composed of two circuital systems, connected to each other by an ideal transformer; a very simple paradigm of this system can be seen in Figure 5.15.

If we substitute the physical systems at the two ends of the ideal transformer with their circuit counterparts, we build the object represented in Figure 5.15b. What kind of object is it? It is not a physical system, nor is it a circuit of the types encountered up to now.

Indeed, no relation exists between the potentials of terminals of the left-hand part of the circuit and those of the right-hand part. For these "expanded circuits" the following definitions are issued:

Definition: *Magnetically coupled circuits*

Two circuits, each connected to the coil of an ideal transformer, are called magnetically coupled circuits.

Definition: *Network*

A network is a set of magnetically coupled circuits.

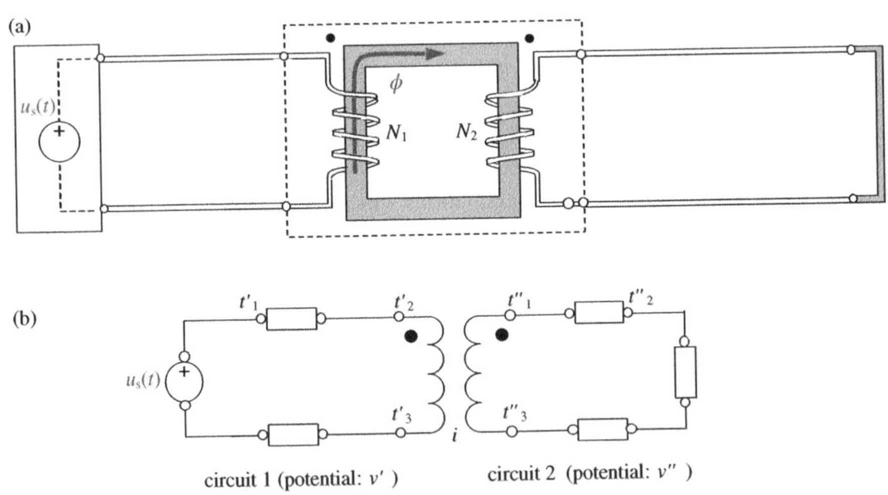

FIGURE 5.15. A system containing a transformer and its magnetically coupled circuits counterpart.

More in Depth: *Terminology*

The terminology used in books and papers for magnetically coupled circuits tends to vary. They are sometimes called *mutually coupled circuits.*

More confusing still, however, is the fact that often, in books and papers, the set of the two circuits connected to each other by the ideal transformer is called a single "magnetically coupled circuit." This is to be avoided since it suggests that KVL equations can be written for loops containing branches of both circuits, which is not possible.

Finally, the words "network" and "circuit" are often used as synonyms. In this book a *network* is a set of one or more circuits, separated from each other by ideal transformers. A circuit, therefore, cannot contain ideal transformers.

Each of the two magnetically coupled circuits has a potential function v of its own, and thus there is no way of determining cross voltages, that is, "differences of potentials between two terminals." Similarly, no current can flow between the two sides of an ideal transformer.

Therefore the following applies:

Result: *Kirchhoff's laws in magnetically coupled circuits*

In a network of two magnetically coupled circuits, each of the circuits has a set of KCL and KVL equations of its own.

Now we verify that magnetically coupled circuits allow all branch voltages and currents to be determined, just like ordinary circuits.

Let's first consider the simple network shown in Figure 5.15b and again, with some more information, in Figure 5.16.

This can be analysed by writing the loop equations of the two individual circuits and the constitutive equations of the ideal transformer as follows ($\alpha = N_1/N_2$):

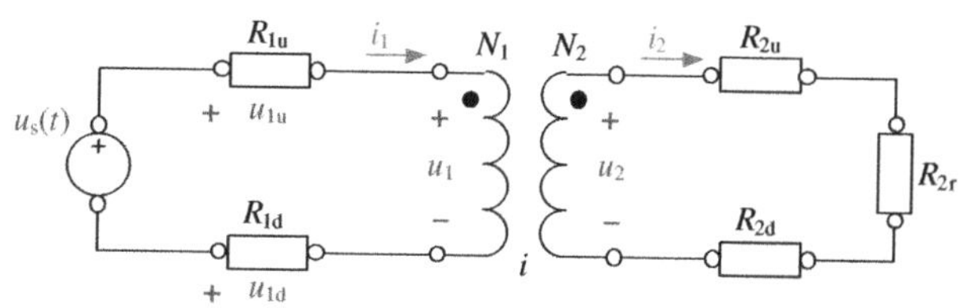

FIGURE 5.16. The coupled circuits of Figure 5.15b, with resistance names for analysis.

$$\begin{aligned} &u_s - R_{1u}i_1 - R_{1d}i_1 = u_1 \\ &u_2 = R_{2u}i_2 + R_{2r}i_2 + R_{2d}i_2 \\ &u = \alpha u_2 \\ &i_2 = \alpha i_1 \end{aligned} \tag{5.11}$$

which allows easy solution of the circuits:

$$u_s - R_{1u}i_1 - R_{1d}i_1 = \alpha(R_{2u} + R_{2r} + R_{2d})\alpha i_1 \Rightarrow i_1 = \frac{u_s}{R_{1u} + R_{1d} + \alpha^2(R_{2u} + R_{2r} + R_{2d})}$$

Branch voltages come immediately from Ohm's law: $u_{1u} = R_{1u}\ i_1$, $u_{1d} = -R_{1d}\ i_1$, etc.

Generalising this procedure, it can be shown that in networks containing magnetically coupled circuits the number of linearly independent Kirchhoff equations is equal in number to the unknowns, and therefore the network be solved.

5.4 SIMPLE *R–L* AND *R–C* TRANSIENTS

The introductory nature of this book does not allow sufficient scope for the inclusion of complex transients. However, because simple transients are given importance elsewhere, namely in Chapter 8, they need to be dealt with here.

Consider the transient in the circuit shown in Figure 5.17, activated by closing the ideal switch at time $t = 0$. The equations describing the behaviour of the current are also included in the figure. The differential equation to be solved is a first-order, constant-coefficient linear equation of the type

$$y' + ay = b \tag{5.12}$$

where y is the dependent variable, x the independent variable, and the apostrophe indicates the x derivative of y. The solution of this simple equation is

$$y(x) = ke^{-ax} + \frac{b}{a}$$

as can be verified with direct substitution.

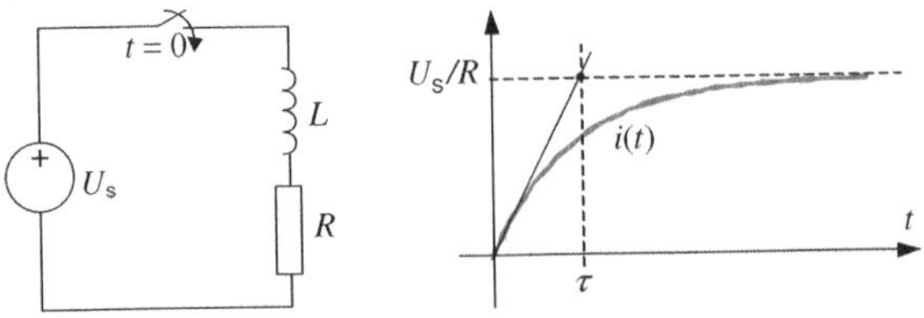

$$\begin{cases} U_s = Ri + L\dfrac{di}{dt} \\ i(0) = 0 \end{cases}$$

FIGURE 5.17. A simple *R–L* transient: circuit and equations.

If the symbols of the circuit of Figure 5.17 are substituted into (5.12), the equation to be solved is written as

$$\frac{\mathrm{d}i}{\mathrm{d}t} + \frac{R}{L} i(t) = \frac{U_\mathrm{S}}{L}$$

and its solution is

$$i(t) = k e^{-t/\tau} + \frac{U_\mathrm{S}}{R} \qquad \text{with } \tau = \frac{L}{R}$$

the condition $i(0) = 0$ allows the determination of constant k, giving

$$i(t) = \frac{U_\mathrm{S}}{R}(1 - e^{-t/\tau})$$

The behaviour of the current over time is therefore as reported in Figure 5.17.

The intersection of the horizontal current asymptote with the geometrical tangent at the $i(t)$ curve for $t = 0$ is equal to the time constant $\tau = L/R$. In fact,

$$\frac{\mathrm{d}i}{\mathrm{d}t} = \frac{U_\mathrm{S}}{\tau R} e^{-t/\tau}, \qquad \frac{\mathrm{d}i}{\mathrm{d}t}(0) = \frac{U_\mathrm{S}}{\tau R} = \frac{U_\mathrm{S}}{L}$$

The intersection t^* with the asymptote is given by

$$\frac{\mathrm{d}i}{\mathrm{d}t}(0) \cdot t^* = i(\infty) = \frac{U_\mathrm{S}}{R}, \qquad \frac{U_\mathrm{S}}{L} t^* = \frac{U_\mathrm{S}}{R} \Rightarrow t^* = \frac{L}{R} = \tau$$

In Chapters 11 and 12 a solution to a problem similar to the one shown in Figure 5.17—that is, the problem shown in Figure 5.18—will need to be found.

The solution, in this case, can be found by determining the right k constant:

$$i(t) = k e^{-t/\tau} + \frac{U_\mathrm{S}}{R} \Rightarrow I_0 = i(0) = k + \frac{U_\mathrm{S}}{R} \Rightarrow k = I_0 - \frac{U_\mathrm{S}}{R}$$

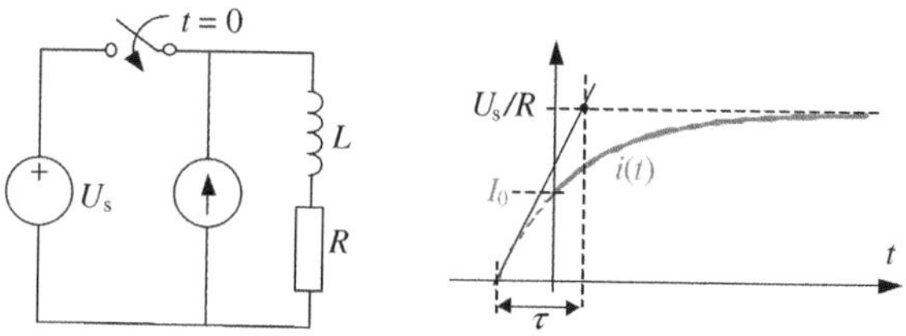

$$\begin{cases} E = Ri + L\dfrac{\mathrm{d}i}{\mathrm{d}t} \\ i(0) = I_0 \end{cases}$$

FIGURE 5.18. A simple R–L transient with a starting current different from 0.

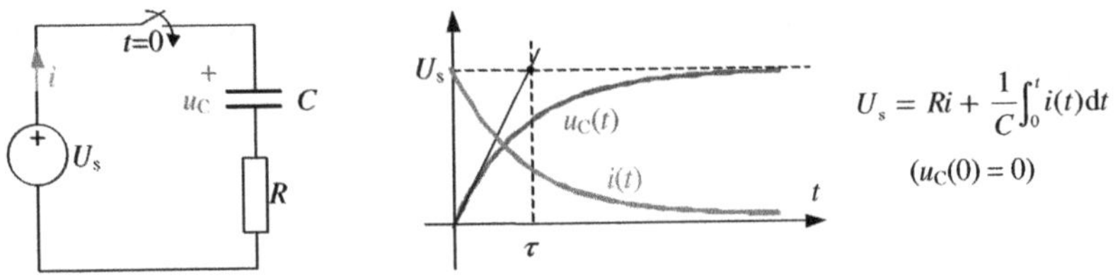

FIGURE 5.19. A simple *R–C* transient: circuit and equations.

that gives

$$i(t) = \left(I_0 - \frac{U_S}{R}\right)e^{-t/\tau} + \frac{U_S}{R}$$

Another simple transient which is important to know even at an introductory level is a simple *R–C* transient, as shown in Figure 5.19. The equation of the circuit, also shown in Figure 5.19, has an integral form in accordance with the characteristic equation of a capacitor, that is,

$$u_C(t) = \frac{q}{C} = \frac{1}{C}\left(q(0) + \int_0^t i(t)\mathrm{d}t\right) = u_C(0) + \frac{1}{C}\int_0^t i(t)\ \mathrm{d}t$$

The full circuit equation is therefore

$$U_s - u_C - Ri = 0; \qquad U_s = Ri + u_{C0} + \frac{1}{C}\int_0^t i(t)\ \mathrm{d}t \tag{5.13}$$

and already contains the initial condition constituted by the voltage on the capacitor at time $t=0$, that is, $u_C(0) = u_{C0}$.

This is due to the writing of the whole equation, instead of the usual differential form.

Equation (5.13) is equivalent to the following Cauchy problem, obtained by derivation and exposition of the initial condition:

$$0 = R\frac{\mathrm{d}i}{\mathrm{d}t} + \frac{i(t)}{C}$$

$$i(0) = \frac{U_S - u_{C0}}{R}$$

Again, the solution of the differential equation is easily found by using the right-hand symbols when applying the general solution of (5.12):

$$\frac{\mathrm{d}i}{\mathrm{d}t} + \frac{i(t)}{RC} = 0 \Rightarrow i(t) = ke^{-t/\tau}$$

$$i(0) = (U_s - u_{C0})/R \Rightarrow k = (U_s - u_{C0})/R \Rightarrow i(t) = [(U_s - u_{C0})/R]e^{-t/\tau}$$

The transient has some similarity to the R–L transient, but the current in this case starts from a finite value and tends to zero.

Once the current is known, the circuit voltages can easily be computed:

$$u_C(t) = U_s - Ri(t) = U_s - (U_s - u_{C0})e^{-t/\tau}$$

5.5 AC CIRCUIT ANALYSIS

5.5.1 Sinusoidal Functions

By and large, the majority of applications of electricity rely on the so-called "alternating current." It is a rather confusing name, whose actual meaning is "applications in which currents and voltages are sine waves."

Indeed, the sinusoidal variation of voltages and currents over time makes the actual behaviour of electrical systems very smooth; it therefore constitutes an ideal behaviour very often used in applications.

A generic sinusoidal function of time, having a given *frequency* f and a given *angular frequency* $\omega = 2\pi f$, has the following formulation:

$$x(t) = A\sin(\omega t + \alpha) = A\sin(2\pi f t + \alpha) \tag{5.14}$$

where

- A is the amplitude of the function.
- α is its phase angle.

Therefore, all possible sinusoidal functions of given frequency or angular frequency differ from each other by their amplitude or phase angle.

They are *periodic*; that is, whatever the time, we have

$$x(t) = x(t + T) \tag{5.15}$$

where $T = 1/f$ is their *period*.

The validity of (5.15) can be immediately verified by substituting it into (5.14).

The meaning of period, amplitude, and phase of a sinusoidal time function are graphically represented in Figure 5.20:

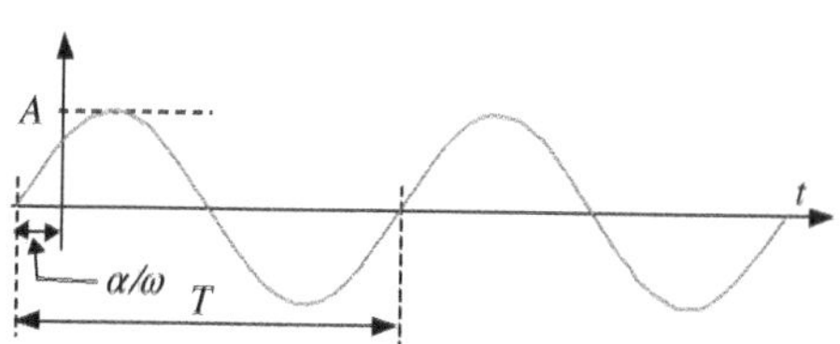

FIGURE 5.20. A sinusoidal time function and its parameters.

5.5.2 Steady-State Behaviour of Linear Circuits Using Phasors

In AC circuits all the sources—that is, branches having constitutive equations of the voltage source or current source type—are sinusoids with the same frequency, which is, therefore, the frequency of the circuit.

Consider a sinusoid, with angular frequency ω, as in equation (5.14). Its instantaneous value is equal to the projection of a vector, rotating counterclockwise, having its tail constantly positioned in the origin of a pair of Cartesian axes x–y, and its amplitude equal to the sinusoid amplitude A. This rotating vector is represented in the x–y plane at time $t=0$, and is therefore called *phasor* (its position at $t=0$ clearly indicates its phase, Figure 5.21).

It is apparent that the phasor indicates not only the value of the sine wave at the time in which it is represented (i.e., $t=0$), but also the previous and subsequent times.

Figure 5.22 shows two different phasors, as well as the corresponding time functions. Again, the rotating vectors effectively show all the functions with a single time shot.

The phasor corresponding to a sinusoidal function $a(t) = A\sin(\omega t + \alpha)$, according to international standards, is to be indicated as $\underline{A}$, that is, an underlined uppercase letter. This convention is followed throughout this book. The phasor can also be represented in amplitude and phase using one of the following graphical ways: $\underline{A} = Ae^{j\alpha} = A\angle\alpha$.

When circuits are operating in AC, current and voltages will be sine waves. Their KVL equations will often contain derivatives or integrals. For instance, the current derivative is needed to solve the circuit in Figure 5.17, as expressed in the circuit

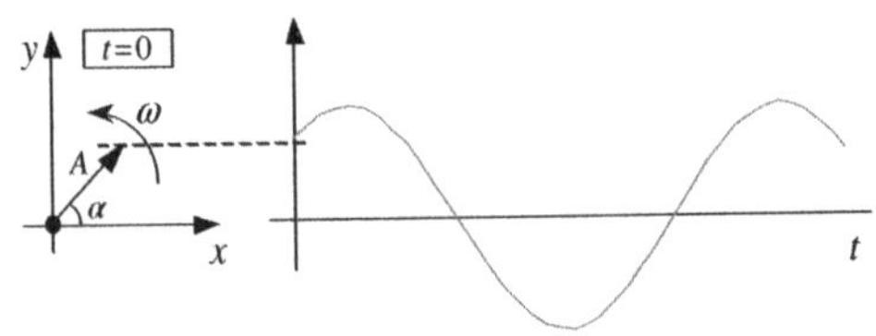

$$\mathbf{A} \equiv \begin{pmatrix} A\cos\alpha \\ A\sin\alpha \end{pmatrix} \leftrightarrow a(t) = A\sin(\omega t + \alpha)$$

FIGURE 5.21. Correspondence between a phasor and a sinusoidal function.

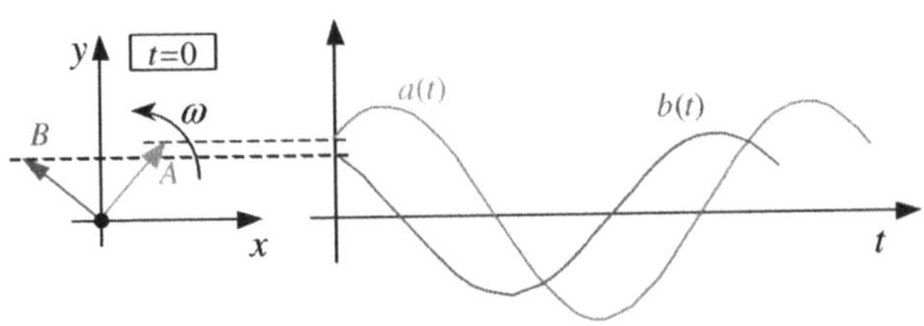

FIGURE 5.22. Two phasors, with their phases and amplitudes, and the corresponding sinusoidal functions.

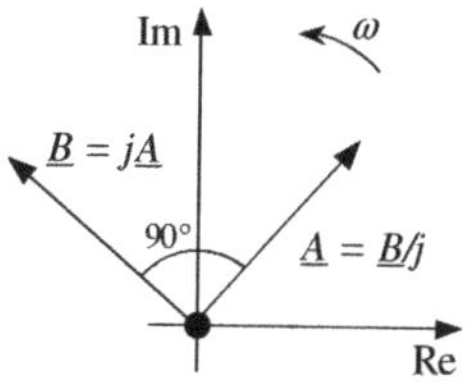

FIGURE 5.23. The number j, used as a factor, advances the phase of a phasor by 90°.

equation shown in the same figure; similarly, the integral of the capacitor current is used to solve the circuit reported in Figure 5.19.

The time derivative and integral of a sine wave have a special form as follows:

$$x(t) = A\sin(\omega t + \alpha) \Rightarrow \frac{dx}{dt} = \omega A\cos(\omega t + \alpha) = \omega A\sin(\omega t + \alpha + \pi/2)$$
$$\int x(\tau)\,d\tau = -\frac{A}{\omega}\cos(\omega t + \alpha) + C = \frac{A}{\omega}\sin(\omega t + \alpha - \pi/2) + C \tag{5.16}$$

Therefore, the time derivative of a sine wave corresponds to multiplication by ω and phase advance by $\pi/2$ radians, while the integral (here the presence of the constant C is negligible) corresponds to division by ω and phase reduction by $\pi/2$ radians.

Is it possible to enhance our representation of sine waves by phasors so that derivation and integration (i.e., phase shift of $\pi/2$), are easily taken into account? The answer is yes: in mathematics an operator already exists which adds $\pi/2$ to the phase of a vector. This is the imaginary unit j, and axes x and y would be the real and imaginary axes of Gauss's plane (Figure 5.23): multiplication by j corresponds to phase advance by $\pi/2$ radians, or 90° degrees, division by j to phase reduction. If $\underline{X}$ is a phasor representing the sine wave

$$x(t) = A\sin(\omega t + \alpha)$$

the derivation rule of sinusoids recalled in (5.16) becomes

$$\frac{d\underline{X}}{dt} = j\omega\underline{X}, \qquad \int \underline{X}\,dt = \frac{\underline{X}}{j\omega}$$

Because of this easy treatment of derivation and integration of sinusoidal time functions, the preferable plane in which to represent the phasors, rather than a generic one, is Gauss's plane,[1] and therefore the phasors are numerically represented by complex numbers.

[1]In a generic Cartesian plane, no internal vector product is defined: dot product gives a scalar (not a vector) as output, while cross product gives a vector which does not belong to the original plane. On the other hand, a product between complex numbers is defined that gives as output a complex number. Therefore Gauss's plane can be imagined as a specialised version of a Cartesian plane, in which an internal product between vectors is defined.

Example 4. Calculate the phasors of the following sinusoids:

$$u(t) = 10\sin(200t + 40°)\,\text{V}$$

$$i(t) = 2\cos(100t - 20°)\,\text{A}$$

$\underline{U} = 10\angle 40° = 10\cos 40° + j\,10\sin 40° = 7.660 + j\,6.428$ V

$i(t) = 2\sin(100t - 20° + 90°) \rightarrow \underline{I} = 2\angle 70° = 2\cos 70° + j\,2\sin 70° = 0.6840 + j\,1.879$ A

Consider now a first example of application to circuits, the one-loop circuit shown in Figure 5.24.

The symbol of the voltage source $u_s(t)$ has been integrated, as often happens, with a sinusoid symbol ~, in order to make explicit the sinusoidal shape of the source (cf. Section 1.3.3). A single KVL loop equation can be written, with block constitutive equations already taken into account:

$$u_s(t) = Ri + u_C(0) + \frac{1}{C}\int_0^t i(t)\mathrm{d}t \tag{5.17}$$

This equation is equivalent to

$$\frac{\mathrm{d}u_s}{\mathrm{d}t} = \frac{1}{C}i(t) + R\frac{\mathrm{d}i}{\mathrm{d}t} + L\frac{\mathrm{d}^2 i}{\mathrm{d}t^2} \tag{5.18}$$

where the condition that the capacitor voltage at $t = 0$ must be equal to $u_C(0)$, as built into equation (5.17), is explicitly added to (5.18).

From mathematics, we know that (5.18), as a linear differential equation, has a solution with both a transient term, which vanishes with time, and a permanent term, the only term that stays in the circuit when the transient term has become zero.

Very often in electrical engineering we are interested in the *steady-state* behaviour of circuits—that is, how the circuit behaves when initial transients have vanished.

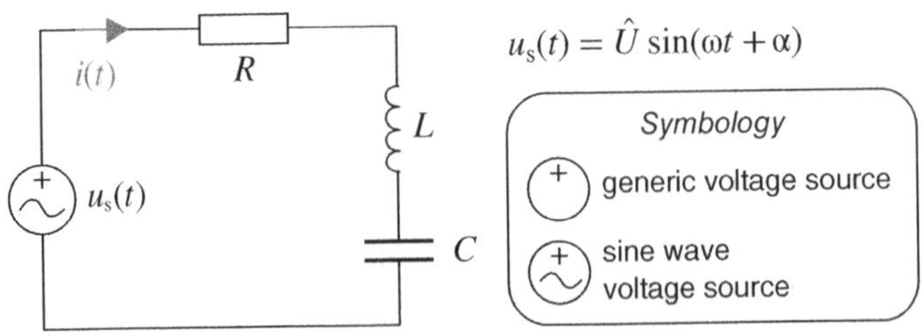

FIGURE 5.24. A simple RLC circuit.

If only the steady-state part of the solution is of interest, the term $u_C(0)$ can be omitted from equation (5.17). Moreover, if $u_s(t)$ is a sinusoid and has ω radian frequency, the solution of the equation—that is, the current $i(t)$—is also a sinusoid and has the same frequency ω.

Equation (5.17), therefore, can be written in phasor form, in which phasors represent sinusoids with a given frequency ω. The circuit equation (5.17), without the term $u_C(0)$, thus becomes

$$\underline{E} = R\underline{I} + L\frac{\mathrm{d}I}{\mathrm{d}t} + \frac{1}{C}\int \underline{I}\,\mathrm{d}t$$

Applying the derivative-integral rule of phasors it becomes

$$\underline{U}_s = R\underline{I} + j\omega L\,\underline{I} + \frac{1}{j\omega C}\underline{I} = \underline{Z}\,\underline{I} \tag{5.19}$$

where the complex number

$$\underline{Z} = R + j\omega L + \frac{1}{j\omega C} = R + j\left(\omega L - \frac{1}{\omega C}\right)$$

is a "complex operator," called *impedance*, that does not represent a phasor, just a ratio of phasors.

The great advantage of using (5.19) for solving the circuit instead of (5.17) or (5.18) should now be apparent: instead of solving a differential equation, only an algebraic equation needs to be solved, albeit in the domain of complex numbers. It must be remembered, however, that this works only in finding the steady-state solution of the starting equation.

Example 5. Calculate the impedance corresponding to the following values of R, L, C, and ω:

$$R = 10\,\Omega, \qquad L = 0.5\,\mathrm{H}, \quad \omega = 10\,\mathrm{rad/s}$$

$$R = 2\,\Omega, \qquad C = 5\,\mathrm{mF}, \quad \omega = 100\,\mathrm{rad/s}$$

$$R = 30\,\Omega, \qquad L = 0.5\,H, \quad C = 1\,\mu\mathrm{F}, \quad \omega = 1000\,\mathrm{rad/s}$$

$$\underline{Z} = R + j\omega L = 10 + j(0.5\cdot 10) = 10 + 5j$$

$$\underline{Z} = R - \frac{j}{\omega C} = 2 - \frac{j}{100\cdot 5\cdot 10^{-3}} = 2 - 2j$$

$$\underline{Z} = R + j\left(\omega L - \frac{1}{\omega C}\right) = 30 + j\left(0.5\cdot 1000 - \frac{1}{1000\cdot 10^{-6}}\right) = 30 - 500j$$

Example 6. Calculate $i(t)$ in the following circuit, knowing that

$$u_s(t) = 6\sin(2t + 45^\circ)$$
$$R = 3\,\Omega, \quad L = 5\,\text{H}, \quad C = 1/8\,\text{F}$$

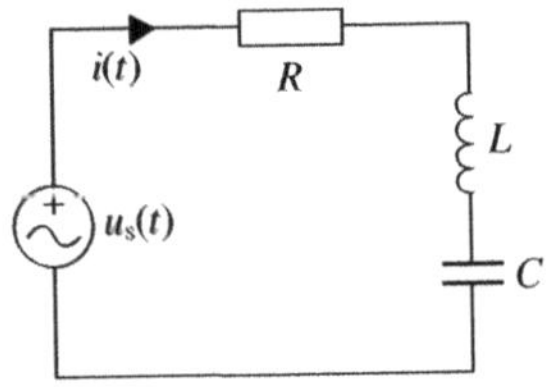

$$\underline{U}_s = 6\angle 45^\circ = 4.24 + j4.24\ \text{V}$$

$$\underline{Z} = R + j\left(\omega L - \frac{1}{\omega C}\right) = 3 + j\left(5\cdot 2 - \frac{1}{2/8}\right) = 3 + 6j$$

$$\underline{I} = \underline{U}_s/\underline{Z} = \frac{4.243 + 4.243j}{3 + 6j} = 0.8485 - 0.2828j = 0.8944\angle -18.43^\circ$$

$$i(t) = 0.8944\sin(2t - 18.43^\circ)$$

Is this process applicable to more complex circuits? The answer is yes: any linear circuit—that is, a circuit with linear constitutive equations of its components (such as those specified in Table 3.4), in which all the sources are sinusoidal functions with the same frequency—can be treated this way.

This is a direct consequence of the fact that, as demonstrated in books on differential equations, systems of linear differential equations, whose forcing functions are sinusoids sharing the same frequency, have a steady-state solution in which all the variables are again sinusoids with the same frequency as the forcing functions.

Since circuits with branches that have linear constitutive equations, whose sources are all sinusoids with the same frequency, are described by systems of linear differential equations, it can be concluded that the phasor method can also be applied to all these circuits.

As a more complex application of the phasor analysis, consider the two-loop circuit in Figure 5.25. Note that if a voltage source has a sinusoidal shape, this can be reflected in the symbol, thus giving more graphical information, as in the example in Figure 5.25b.

Kirchhoff's equations, written with phasors, are

$$N\colon \underline{I}_3 = \underline{I}_1 + \underline{I}_2$$

$$L_1\colon \underline{U}_A = \frac{1}{j\omega C}\underline{I}_1 + R_3\underline{I}_3 \qquad (^\circ)$$

$$L_2\colon \underline{U}_B = j\omega L\underline{I}_2 + R_2\underline{I}_2 + R_3\underline{I}_3$$

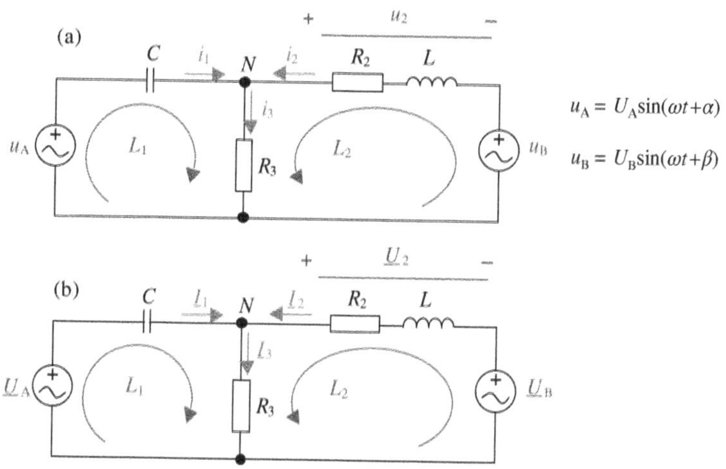

FIGURE 5.25. Example of a two-loop AC circuit (a) with variables named as time functions; (b) with variables named as phasors.

where

$$\underline{U}_A = U_A(\cos\alpha + j\sin\alpha) \quad \underline{U}_B = U_B(\cos\beta + j\sin\beta)$$

Once system (°) is solved, $\underline{I}_1$, $\underline{I}_2$, and $\underline{I}_3$ are found. Thus if

$$\underline{I}_1 = I_1(\cos\alpha_1 + j\sin\alpha_1)$$

then

$$i_1(t) = I_1(\sin\omega t + \alpha_1)$$

and similarly for $i_2(t)$ and $i_3(t)$. Should other circuit voltages be required, they can easily be found from the now known currents. For instance,

$$\underline{U}_2 = -\underline{I}_2(R_2 + j\omega L) = U_2(\cos\alpha_{u2} + j\sin\alpha_{u2}) \Rightarrow u_2(t) = U_2\sin(\omega t + \alpha_{u2})$$

Phasor analysis of an AC circuit involves a three-step procedure:

1. The circuit is converted into its complex-plane form: the voltage and current sources, all sinusoidal functions with a given frequency ω, are converted into phasors using the correspondence rule in Figure 5.21, and the circuit elements are written in their complex operator form: R–L–C elements are written as a $\underline{Z} = R + j\,[\omega L - 1/(\omega C)]$.
2. The complex form of circuit equations have the same structure as DC circuits in which, however, currents voltages and factors between voltages and currents are complex numbers instead of real numbers. This can be solved as an ordinary algebraic system of complex equations.

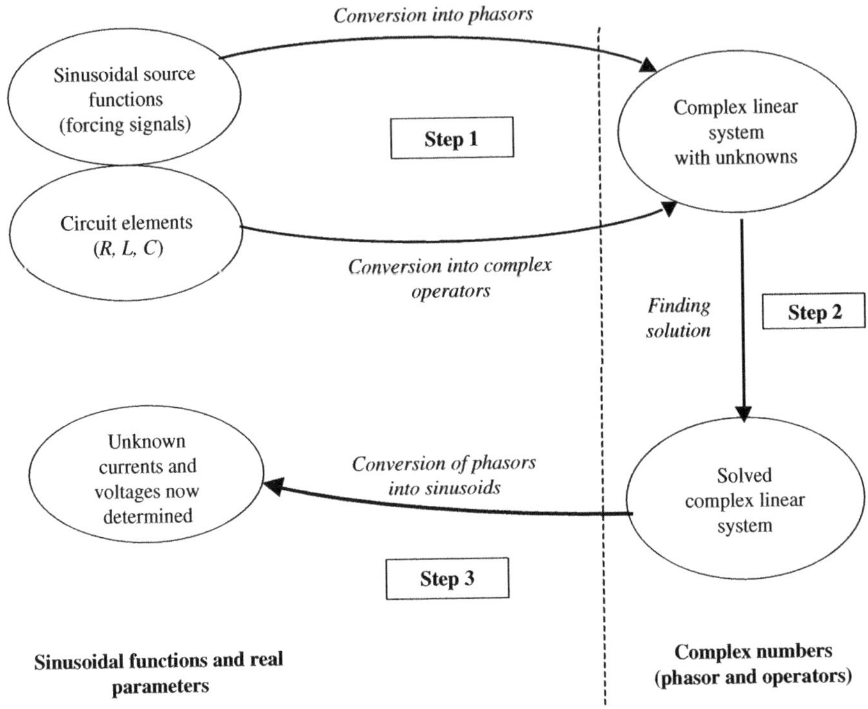

FIGURE 5.26. Schematic description of the phasor analysis technique for solving linear circuits.

3. Once all the phasors (or all the phasors corresponding to all the required electric quantities) are computed, they are converted back into sinusoidal functions of time.

The process is summarised in the scheme of Figure 5.26. It shows that instead of finding the direct solution while staying in the domain of sinusoidal functions and real parameters, it is simpler to first transfer ("transform") the problem into the domain of complex numbers and then solve the problem, afterwards converting the results back into the domain of sinusoidal functions and real parameters ("untransform"). This is an example of a general technique used in all fields of engineering called the "use of transformations": the problem is first translated into another domain, where it is much simpler to solve, and then the result is untransformed.

More in Depth *Phasors and sinusoidal functions correspondence options*

The rule suggested by Figure 5.21 for phasors and sinusoidal functions is

$$\underline{A} = A(\cos\alpha + j\sin\alpha) \leftrightarrow a(t) = A\sin(\omega t + \alpha) \qquad (^\circ)$$

It should be very easy to show that, as an alternative, it could be

$$\underline{A} = A(\cos\alpha + j\sin\alpha) \leftrightarrow a(t) = A\cos(\omega t + \alpha) \qquad (^{\circ\circ})$$

The two choices are almost exactly equivalent. When solving a circuit with a single source containing all functions expressed as sinusoids—for example, of the type $x(t) = X\sin(\omega t + \alpha)$—the correspondence (°) is the more immediate, since it does not require conversion of a sine into a cosine; vice versa, the correspondence (°°) is more immediate if all the source functions are expressed as cosinusoids—for example, are of the type $x(t) = X\cos(\omega t + \alpha)$.

Some books adopt the correspondence (°), others (°°), while others suggest adopting one or another on a case-by-case basis. In the examples developed in this book, the correspondence (°) is used.

5.5.3 AC Circuit Passive Parameters

The concept of impedance was introduced in the previous section; this is the ratio of voltage to current phasors measured at the ends of any circuit branch as follows:

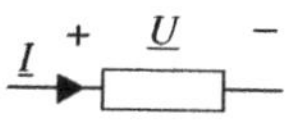

$$\underline{U} = \underline{Z}\underline{I} \qquad (5.20)$$

where the complex number:

$$\underline{Z} = R + jX = Ze^{j\varphi}$$

impedance across voltage $\underline{U}$, is a *complex operator*, that is, a ratio of phasors.

Since the impedance is not a phasor it cannot be converted into a sine wave. Its modulus Z is the ratio of moduli of $\underline{U}$ and $\underline{I}$, while its phase angle φ is the phase difference between them: the phase angle of $\underline{U}$ is the phase angle of $\underline{I}$ plus φ.

Because of its importance, the imaginary part of $\underline{Z}$ has a symbol and name of its own: it is normally represented by the symbol X (or x) and called reactance. Reactance is a factor describing the behaviour of inductors or capacitors and has a value which is respectively positive or negative:

- Inductor with inductance $L \Rightarrow X_L = \omega L$
- Capacitor with capacitance $C \Rightarrow X_C = -1/(\omega C)$

The reactance of a generic passive branch may be positive or negative, depending on whether the prevailing effect is that of inductive or capacitive parts.

The ratio of current to voltage is often of interest and has a new symbol and name:

$$\underline{I} = \underline{Y}\underline{U}, \qquad \underline{Y} = G + jB = Ye^{-j\varphi} = 1/\underline{Z}$$

where, for the branch in question,

- $\underline{Y}$, the ratio of current to voltage phasors, is a complex number called *admittance*;
- G, the real part of $\underline{Y}$, is called *conductance* (in the case of real impedance, i.e. with zero imaginary part, it coincides with the reciprocal of the resistance);
- B, the imaginary part of $\underline{Y}$, is called *susceptance*.

Since the phase angle of the admittance is the opposite of the phase angle φ of the impedance of the same branch, when the reactance of a branch is positive, the corresponding susceptance is negative, and vice versa.

Often the impedance parameters are represented as a triangle, shown here, called "triangle of impedance". It is useful to find ratios between quantities, such as $X/Z = \sin\varphi$ or $X/R = \tan\varphi$.

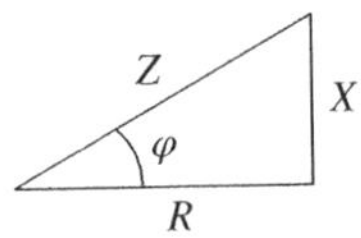

5.5.4 The Phasor Circuit

Consider again Figure 5.25. It contains Figure 5.25a, showing a circuit in which voltages and currents are sine waves, and Figure 5.25b, in which they are phasors. The latter is called *phasor circuit*. It is structurally identical to the original one; just the quantities are complex numbers (i.e., phasors) instead of time functions.

The introduction of complex impedance $\underline{Z}$ enables the writing phasor circuits in a more compact way: series of resistors, inductors, and/or capacitors are thought of as single entities, which correspond to circuit impedances.

For instance, the circuit in Figure 5.25a can be drawn as indicated in Figure 5.27.

Circuits represented using the impedance symbol for components containing resistors, inductors and capacitors are called *phasor circuits*. They are structurally similar to DC circuits: instead of constant real quantities, however, constant complex quantities are used.

Phasor diagrams of different circuits are shown in Figure 5.28. To draw the diagrams, it is important to pay due attention to the reference signs in the circuit. In the case, for instance, of the central diagrams of Figure 5.28, referring to R–L loads,

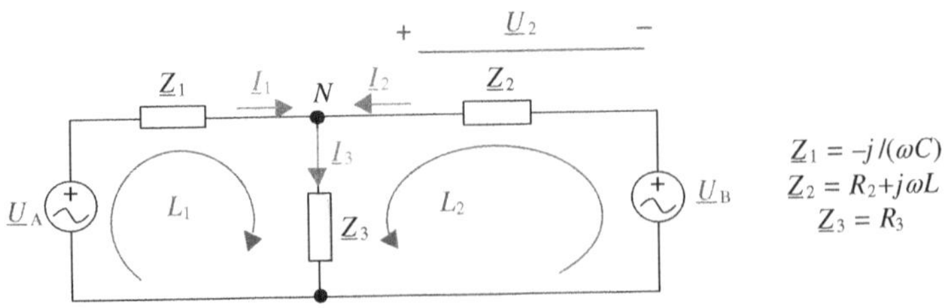

FIGURE 5.27. The circuit of Figure 5.25, using symbols for impedance elements: phasor circuit.

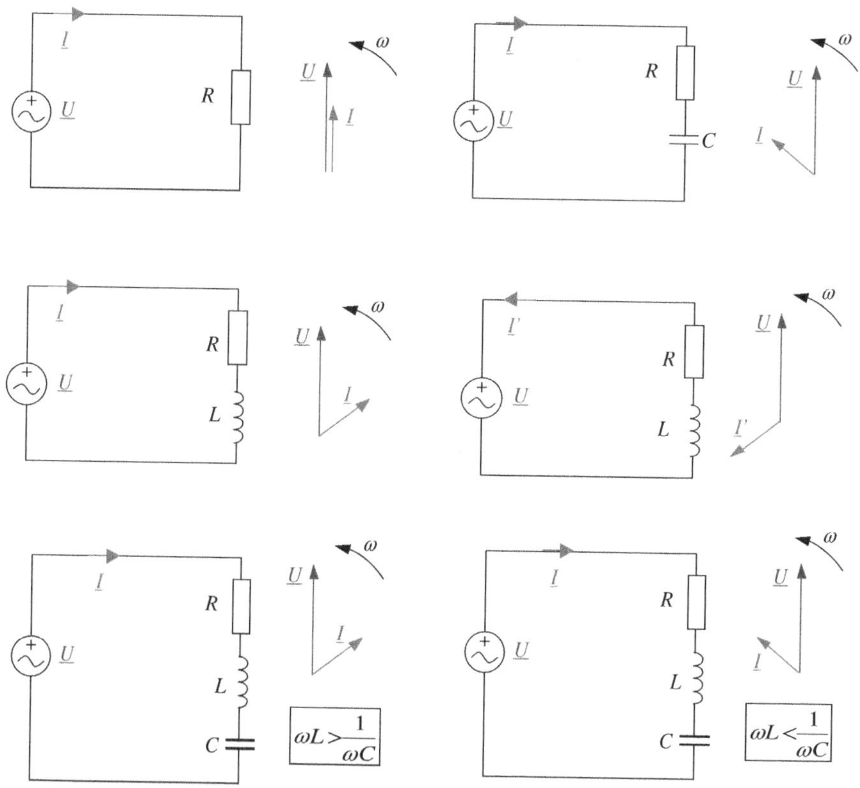

FIGURE 5.28. Phasor diagrams for different kinds of load and reference signs.

the current was assumed to have two opposite reference directions. Two opposite values of current correspond (i.e., $\underline{I}' = -\underline{I}$), and the corresponding phasor diagrams must be changed accordingly, as shown in the figure.

If the load is composed of a pure reactance (i.e., its resistance is zero), the phase displacement between current and voltage phasors is 90 degrees. The current lags the voltage if the reactance is positive (it is constituted by an inductor), while it leads the voltage if the reactance is constituted by a capacitor, and therefore is negative.

In the case of a generic *R–X* load, the angle between voltage and current phasors (voltage angle minus current angle), when the current is assumed to be entering the load from the positively marked terminal, is between $-90°$ and $+90°$. Negative values refer to negative reactances.

Finally, *R–L–C* load is taken into account. As seen from the voltage source, it behaves in a way that could be *R–L*-like or *R–C*-like in $\omega L > 1/(\omega C)$ or $\omega L < 1/(\omega C)$, respectively—that is, whether the effects of inductance prevail over the effects of the capacitor or vice versa.

Phasor circuits also effectively describe circuits containing mutually coupled inductive elements. In fact, the mathematical behaviour of any time-varying circuit in which

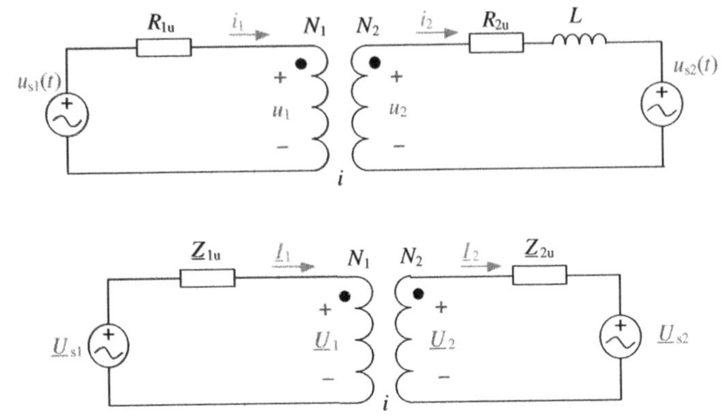

FIGURE 5.29. A circuit containing mutually coupled inductors and the corresponding phasor circuit.

the branches have constitutive equations of the types shown in Table 3.4 is described by a system of linear differential equations. Should all the sources be sinusoidal with the same frequency, the solution of this system is composed entirely of sinusoidal quantities (voltages and currents) and therefore the phasor circuit can be used.

An example of a circuit containing an ideal transformer (two mutually coupled inductors) is shown in Figure 5.29, along with its phasor circuit.

Phasor circuits can be seen as circuits having branches with constitutive equations expressed as complex numbers. The only equations of interest are those seen in Table 5.1. Note that the presence of the component ideal transformer, which does not exist in DC, requires a special row which is not a constitutive equation of a branch; this can be imagined to be a system of constitutive equations of a four-terminal circuit element.

TABLE 5.1. The Constitutive Equation Types of Phasor Circuits[a]

Element	Equation	Description
Linear passive element	$\underline{U}_b = \underline{Z}_b\,\underline{I}_b$	Branch voltage and current phasors proportional (the "+" sign must be used when load sign convention is used, i.e. when the current is assumed to enter the branch from the positively marked terminal).
Voltage source	$\underline{U}_b = \underline{U}_s$	Voltage phasor is equal to the complex constant $\underline{U}_s$ (subscript stands for "source") regardless of any other circuit quantity.
Current source	$\underline{I}_b = \underline{I}_s$	Current phasor is equal to the complex constant $\underline{I}_s$ (subscript stands for "source") regardless of any other circuit quantity.
Ideal transformer	$\underline{U}_2/\underline{U}_1 = \alpha$ $-\underline{I}_2/\underline{I}_1 = 1/\alpha$	α is the turns ratio N_2/N_1. The equations are written for positive polarities corresponding to dots, and load sign convention for both coils.

[a]In equations containing both current and voltage, load sign convention is used.

Because of its great similarity to DC circuits (see Table 3.4: the only structural difference, in addition to the use of complex numbers instead of real numbers, is the presence of ideal transformers), nearly all the formal results obtained in Chapter 4 for DC circuits can be replicated in AC circuits, studied using phasors and complex impedances.

Here they are shown for clarity's sake.

Nodal Analysis. Nodal analysis can be applied to AC circuits exactly in the same way as in DC circuits, except for the use of complex variables. For instance, the circuit in Figure 5.27 can be solved using a unique equation, written using nodal analysis:

$$\frac{\underline{U}_A - \underline{U}_N}{\underline{Z}_1} + \frac{\underline{U}_B - \underline{U}_N}{\underline{Z}_2} = \frac{\underline{U}_N}{\underline{Z}_3}$$

whose unique unknown is voltage $\underline{U}_N$.

Series, Parallel, and Star/Delta Conversion of Impedances. The formulas seen in Chapter 3 for resistor conversions are immediately extended to AC circuits, simply by substituting resistances with impedances:

- Equivalent impedance of n impedances in series: $\underline{Z}_{eqs} = \underline{Z}_1 + \underline{Z}_2 + \cdots + \underline{Z}_n$
- Equivalent impedance of two impedances in parallel: $\underline{Z}_{eqp} = \dfrac{\underline{Z}_1 \underline{Z}_2}{\underline{Z}_1 + \underline{Z}_2}$
- Equivalent impedance of n impedances in parallel: $\underline{Z}_{eqp} = \left(\dfrac{1}{\underline{Z}_1} + \dfrac{1}{\underline{Z}_2} + \cdots + \dfrac{1}{\underline{Z}_n}\right)^{-1}$
- Star-delta conversion formulas: $\underline{Z}_1 = \dfrac{\underline{Z}_{12}\underline{Z}_{13}}{\underline{Z}_s}$, $\underline{Z}_{12} = \dfrac{\underline{Z}_1 \underline{Z}_2}{\underline{Z}_p}$

in which $\underline{Z}_s = \underline{Z}_{12} + \underline{Z}_{13} + \underline{Z}_{23}$ and $\underline{Z}_p = \left(\dfrac{1}{\underline{Z}_1} + \dfrac{1}{\underline{Z}_2} + \dfrac{1}{\underline{Z}_3}\right)^{-1}$

Finally, it is to be noted that since any phasor circuit is described by linear-algebraic (though complex) equations, linearity applies, then superposition and Thévenin's theorem also apply.

Voltage and Current Division. Again, the formulas found in Chapter 4 are immediately extended to AC circuits substituting resistances with impedances:

- Voltage division. Consider the following impedances in series: $\underline{Z}_1, \underline{Z}_2, \ldots, \underline{Z}_n$. The ratio of the voltage applied to any of the individual branches to the total voltage applied to the series is

$$\frac{\underline{U}_k}{\underline{U}_{eqs}} = \frac{\underline{Z}_k \underline{I}}{\underline{Z}_{eqs} \underline{I}} = \frac{\underline{Z}_k}{\underline{Z}_{eqs}} \tag{5.21}$$

- Current division. Consider branches in parallel with impedances $\underline{Z}_1, \underline{Z}_2, \ldots,$ $\underline{Z}_n$, or, equivalently, as admittances $\underline{Y}_1, \underline{Y}_2, \ldots, \underline{Y}_n$. The ratio of the current flowing into any of the individual branches to the total current flowing into the parallel equivalent is

$$\frac{\underline{I}_k}{\underline{I}_{eqp}} = \frac{\underline{Y}_k \underline{I}}{\underline{Y}_{eqs} \underline{I}} = \frac{\underline{Y}_k}{\underline{Y}_{eqp}} \tag{5.22}$$

- Simplified current division. When only two impedances in parallel are involved, the current division formula can be simplified as follows:

$$\frac{\underline{I}_1}{\underline{I}_{eqp}} = \frac{\underline{Y}_k}{\underline{Y}_1 + \underline{Y}_2} = \frac{1/\underline{Z}_1}{\dfrac{1}{\underline{Z}_1} + \dfrac{1}{\underline{Z}_2}} = \frac{\underline{Z}_2}{\underline{Z}_1 + \underline{Z}_2}, \qquad \frac{\underline{I}_2}{\underline{I}_{eqp}} = \frac{\underline{Z}_1}{\underline{Z}_1 + \underline{Z}_2}$$

Example 7. Calculate $i_1(t)$, $i_2(t)$, and $i_3(t)$ in the following circuit, where

$$u_s(t) = 100 \sin(200t + 20°)\text{ V}$$

$$R_1 = 8\,\Omega, \quad R_2 = 4\,\Omega, \quad R_3 = 25\,\Omega, \quad \text{and} \quad L_2 = 0.2\text{ H}$$

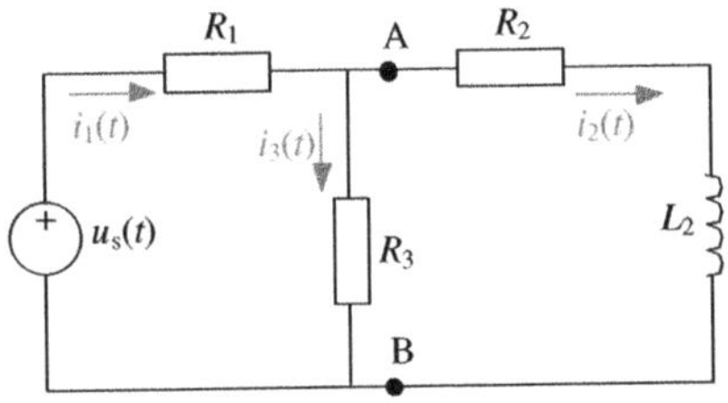

Let us first determine the phasor circuit:

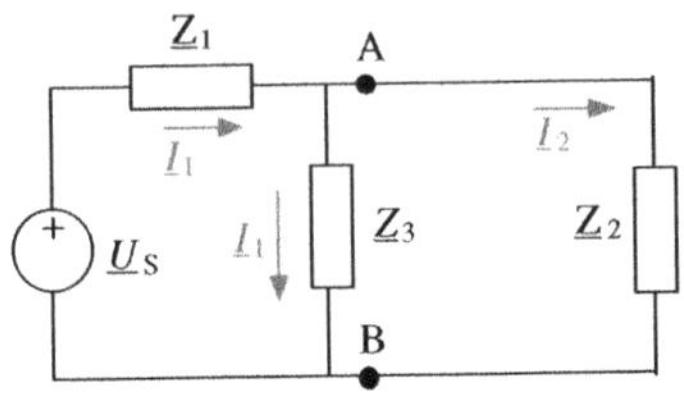

where

$$\underline{U}_s = 100\angle 20° = 93.97 + j34.2\text{ V}$$

$$\underline{Z}_1 = 8\,\Omega, \qquad \underline{Z}_2 = 4 + 40j\,\Omega, \qquad \underline{Z}_3 = 25\,\Omega$$

Then

$$\underline{Z}_{\parallel} = \underline{Z}_2 \| \underline{Z}_3 = \frac{\underline{Z}_2 \cdot \underline{Z}_3}{\underline{Z}_2 + \underline{Z}_3} = 17.57 + j10.24\,\Omega$$

$$\underline{Z}_{tot} = \underline{Z}_1 + \underline{Z}_{\parallel} = 25.57 + j10.24\,\Omega$$

$$\underline{I}_1 = \underline{U}_s / \underline{Z}_{tot} = 3.628 - j0.1155\text{ A} = 3.630\angle{-1.824°}\text{A}$$

Considering the current division rule:

$$\underline{I}_2 = \underline{I}_1 \cdot \frac{\underline{Z}_3}{\underline{Z}_2 + \underline{Z}_3} = 1.030 - j1.521\ \text{A} = 1.836\angle{-55.88}°\text{A}$$

$$\underline{I}_3 = \underline{I}_1 \cdot \frac{\underline{Z}_2}{\underline{Z}_2 + \underline{Z}_3} = 2.598 + j1.405\ \text{A} = 2.953\angle 28.41°\text{A}$$

then

$$i_1(t) = 3.630 \sin(200t - 1.824°)\text{A}$$

$$i_2(t) = 1.836 \sin(200t - 55.88°)\text{A}$$

$$i_3(t) = 2.953 \sin(200t + 28.41°)\text{A}$$

The same results can be found by applying Thévenin's theorem between terminals A and B:

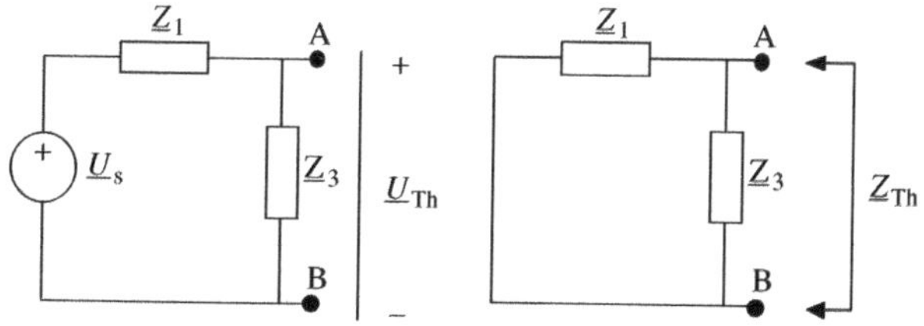

$$\underline{U}_{\text{Th}} = \underline{U}_s \cdot \frac{\underline{Z}_3}{\underline{Z}_1 + \underline{Z}_3} = 71.19 + j25.91\ \text{V}, \quad \underline{Z}_{\text{Th}} = \underline{Z}_1 \| \underline{Z}_3 = 6.061\ \Omega$$

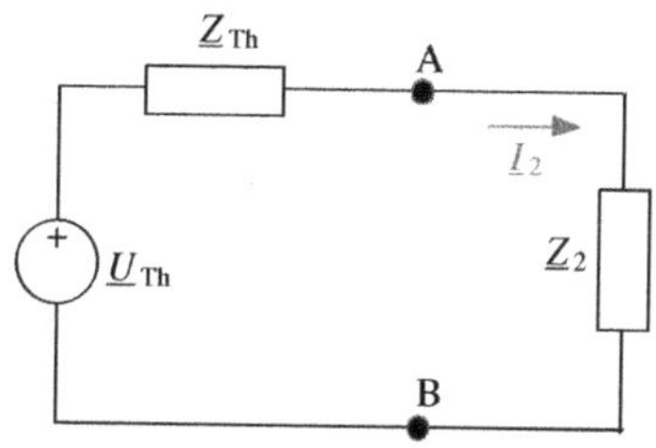

$$\underline{I}_2 = \frac{\underline{U}_{\text{Th}}}{\underline{Z}_{\text{Th}} + \underline{Z}_2} = 1.030 - j1.521\ \text{A}$$

$$\underline{U}_{\text{AB}} = \underline{Z}_2 \cdot \underline{I}_2 = 64.92 + j35.12\ \text{V}$$

$$\underline{I}_3 = \underline{U}_{\text{AB}} / \underline{Z}_3 = 2.598 + j1.405\ \text{A}$$

$$\underline{I}_1 = \underline{I}_2 + \underline{I}_3 = 3.628 - j0.1155\ \text{A}$$

5.5.5 Circuits Containing Sources with Different Frequencies

In Section 5.5.2 we stated that the phasor circuit concept can be used only when all the sources are sinusoidal functions sharing the same frequency.

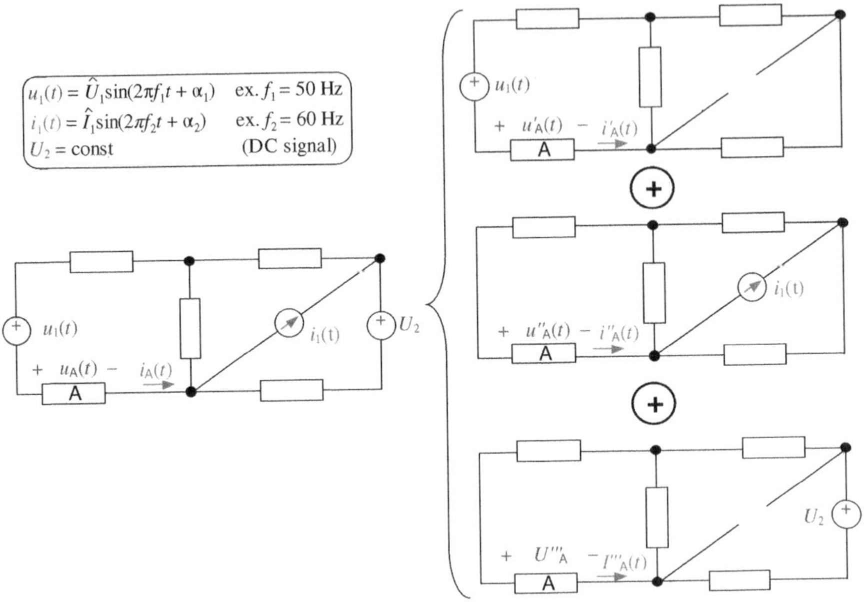

FIGURE 5.30. Use of superposition with linear circuits with sources at different frequencies and/or in DC.

It often happens in practice, however, that circuits contain sources with different frequencies, or else some sources are constant while other are sinusoids. This case will be treated in this book only for linear systems, for which superposition applies.

Superposition was dealt with in Chapter 3, in which it was shown that when the branches are either known voltages (constant or variable) or known currents (constant or variable), or are characterised by a linear relation between voltage and current, the circuit can be solved by applying the different known voltages and currents in sequence and then summing the results.

The process is exemplified using the example in Figure 5.30.

The solution of the circuit shown on the left-hand side of the figure can be found by solving the three circuits on the right side and then summing all the electrical quantities. Just as an example, it can be said that $u_A(t) = u_A'(t) + u_A''(t) + U_A'''$, $i_A(t) = i_A'(t) + i_A''(t) + I_A'''$.

Naturally, $u_A'(t)$ and $i_A'(t)$ can be found using the phasor circuit at frequency f_1, $u_A''(t)$ and $i_A''(t)$ can be found using the phasor circuit at frequency $f_2 \neq f_1$, and U_A''' and I_A''' can be found using the solving techniques of DC circuits. It is important to remember that phasors computed in the circuit operating at frequency f_1 *cannot be summed* to phasors computed in the circuit operating at f_2: the sum can only be done when the phasor circuit solutions have already been converted into time functions.

Example 8. Calculate $i_1(t)$ and $i_2(t)$ in the following circuit, where

$$u_s(t) = 10\sin(100t)\text{V}, \quad i(t) = 2\text{ A}, \quad R = 1\,\Omega, \quad L = 0.05\text{ H}, \quad C = 1\text{ mF}$$

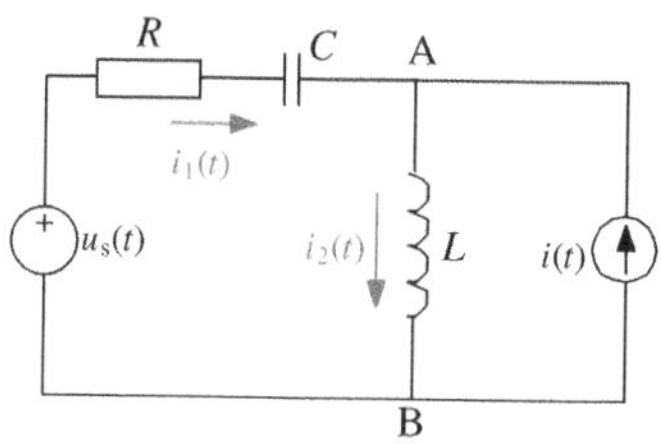

Applying the superposition principle, the circuit becomes

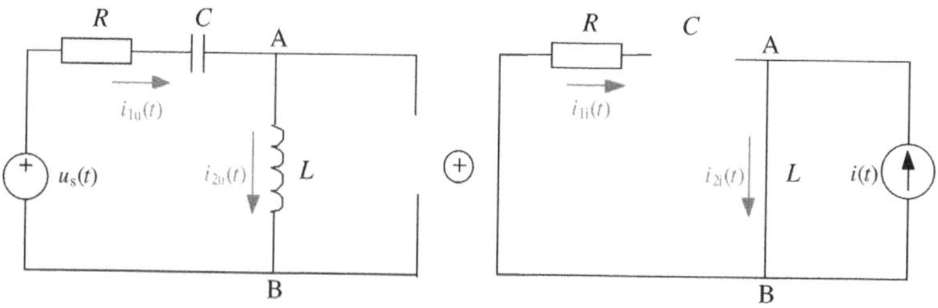

Please note that in the right-hand circuit, where $f = 0$, the inductor corresponds to a short circuit and the capacitor to an open branch.

In the left-hand circuit, we have

$$\underline{U}_s = 10\text{ V}$$

$$\underline{Z} = R + j\left(\omega L - \frac{1}{\omega C}\right) = 1 - j5\,\Omega$$

$$\underline{I}_{1u} = \underline{I}_{2u} = \underline{U}_s/\underline{Z} = 0.3846 + j1.923\text{ A} = 1.961\angle 78.69°$$

$$i_{1u}(t) = i_{2u}(t) = 1.961\sin(100t + 78.69°)\text{ A}$$

In the right-hand circuit, we have $i_{1i}(t) = 0$ and $i_{2i} = i(t) = 2$ A. Hence,

$$i_1(t) = 1.961\sin(100t + 78.69°)\text{A}$$

$$i_2(t) = 1.961\sin(100t + 78.69°) + 2\text{ A}$$

5.6 POWER IN AC CIRCUITS

5.6.1 Instantaneous, Active, Reactive, and Complex Powers

Consider the simple circuit shown here, for which we assume

$$i(t) = \hat{I}\sin\omega t$$

The power $p(t)$ transferred to the resistor is

$$p = ui = R\hat{I}\,\sin^2\omega t = R\hat{I}^2\,\frac{1-\cos 2\omega t}{2}$$

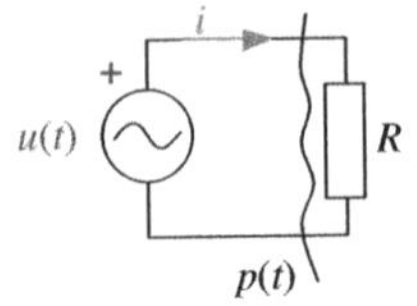

This function of time is called the "instantaneous power" absorbed by the resistor.

This instantaneous power is composed of an average value and a fluctuating term around it, as shown in Figure 5.31.

For the majority of applications, it is sufficient to consider the average value of $p(t)$. Take, for instance, a home heater. The fact that heat is produced at a rapidly variable speed (with periodical cycling every 10 ms for a standard 50-Hz frequency, or 8.33 ms for a frequency of 60 Hz) is of no practical significance: the user feels only the average of this power. The same applies for the light of an incandescence bulb: luminance fluctuations are very small due to the thermal inertia of the bulb filament and are hardly discerned by the human eye.

Therefore, the time average of the instantaneous power transferred to the resistor has a name and a symbol of its own: it is called *power* (without any attributes) and its symbol is an uppercase P:

$$P = avg(p(t)) = \frac{R\hat{I}^2}{2} = RI^2 \tag{5.23}$$

The quantity I, uppercase and without any additional sign, is called *root mean square* (rms) of $i(t)$.

The rms value of a generic periodic quantity $x(t)$—that is, a function $x(t)$ having the characteristic that $x(t) = x(t - T)$, where t is *any* generic time and T is a fixed quantity—is defined as

$$X = \sqrt{\frac{1}{T}\int_{t}^{t+T} x^2(t)\,\mathrm{d}t}$$

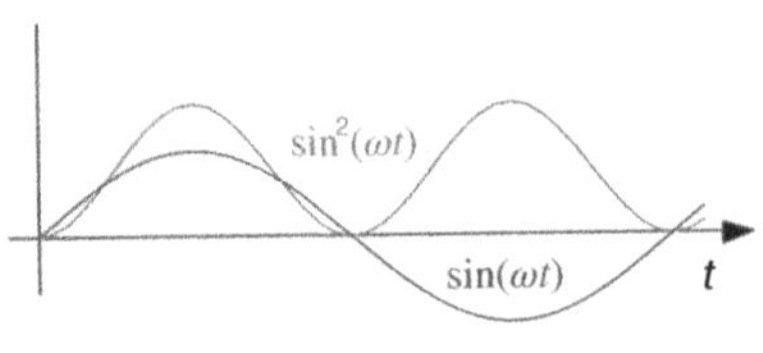

FIGURE 5.31. The shape of a sine function compared with its square.

In particular, for a sinusoid quantity $x(t) = \hat{X}\sin(\omega t + \alpha)$ it is

$$X = \sqrt{\frac{1}{T}\hat{X}^2\int_t^{t+T}\sin^2(\omega t + \alpha)\,dt} = \hat{X}\sqrt{\int_t^{t+T}\frac{1-\cos(2\omega t + 2\alpha)}{2}\,dt} = \frac{\hat{X}}{\sqrt{2}}$$

because the integral extended to the period T of the quantity $\cos(2\omega t + \alpha)$ which is periodic with a period $T/2$ is zero.

The expression $P = RI^2$ of the (average) power absorbed by resistor R is exactly the same as that which applies to DC circuits. This justifies the definition adopted for the rms value of the current: it is the value of a constant current that has the same thermal effect of the actual $i(t)$ which is sinusoid in shape.

The equation (5.23) can now be written as follows:

$$p = P(1 - \cos 2\omega t)$$

which shows in a synoptic way its average and fluctuating components.

Now we evaluate what happens to the instantaneous power entering an energy storage element.

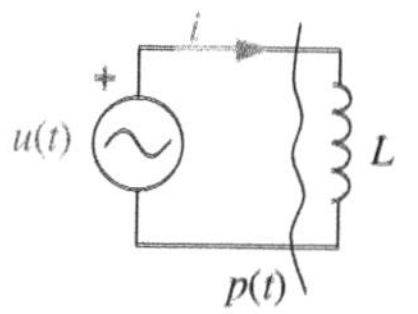

Consider the simple circuit shown here, for which it is assumed:

$$i(t) = \hat{I}\sin\omega t$$

The power $p(t)$ transferred to the inductor is

$$p = ui = L\frac{di}{dt}i = L\omega\sin(\omega t + \pi/2)\sin\omega t = L\omega\hat{I}^2\cos\omega t\,\sin\omega t$$
$$= X\hat{I}^2\,\frac{\sin 2\omega t}{2} = XI^2\sin 2\omega t$$

Therefore, the instantaneous power transferred to the inductor has a null average, and this is coherent with the fact, demonstrated in Section 5.2.3, that it is a storage component that stores energy whenever the absolute value of current rises, and delivers it back whenever it reduces.

The instantaneous power absorbed by an inductor has a maximum whose importance is such as to merit a name of its own: the so-called *reactive power Q*:

$$Q = XI^2, \qquad p(t) = Q(\sin 2\omega t)$$

Now it should be clear that capacitors behave similarly to inductors. As before, let the current be $i(t) = \hat{I} \sin \omega t$.

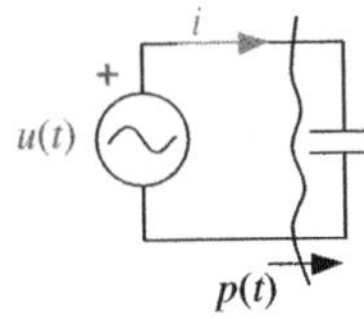

Consider the simple circuit shown here. The power $p(t)$ transferred to the capacitor is:

$$p = ui = \frac{1}{C}\int u(t)\mathrm{d}t \cdot i = \frac{1}{\omega C}\sin(\omega t - \pi/2)\hat{I}\sin\omega t = \frac{-1}{\omega C}\hat{I}^2\cos\omega t \sin\omega t$$

$$= X\hat{I}^2\frac{\sin 2\omega t}{2} = XI^2 \sin 2\omega t$$

Therefore, the instantaneous power transferred to the capacitor has a null average, and this is coherent with the fact, demonstrated in Section 5.2.2, that it is a storage component that (a) stores energy whenever the absolute value of voltage rises and (b) delivers it back whenever it reduces.

The instantaneous power absorbed by a capacitor has a maximum which, as in the case of the inductor, is called reactive power Q:

$$Q = XI^2, \quad p(t) = Q(\sin 2\omega t)$$

Remember that for a capacitor the reactance X is a negative number. Indeed the fact that the reactive power absorbed by a capacitor is negative in value can be understood by thinking of a capacitor as a component that delivers positive reactive power:

$$Q_{absorb} = XI^2, \qquad Q_{deliv} = -Q_{absorb} = \left|XI^2\right|$$

Consider now the general case of a connection between two networks **A** and **B** shown in Figure 5.32.

If the whole network is linear and contains, inside **A** and **B**, only sinusoidal sources all sharing the same frequency f, then all the voltages and currents in the network are

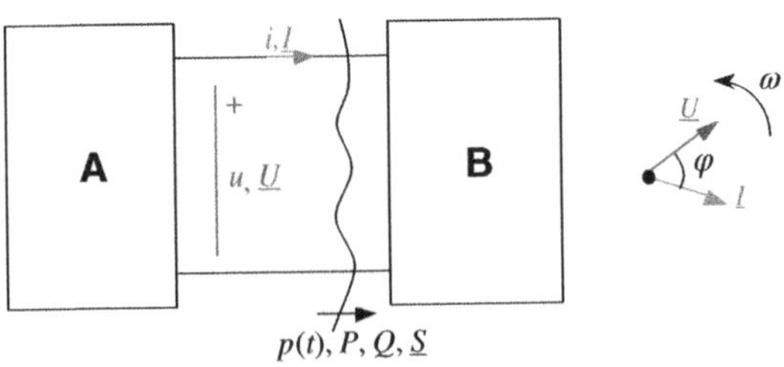

FIGURE 5.32. Instantaneous, active, reactive, and apparent powers exchanged by two networks.

sinusoidal quantities with the same frequency f and the same angular frequency ω and can be represented by the corresponding phasors.

The power transferred from **A** to **B** is still $p(t) = u(t)i(t)$. It can be developed as follows:

$$p = \hat{U}\hat{I}\sin(\omega t + \varphi)\sin\omega t = \hat{U}\hat{I}(\sin\omega t\cos\varphi + \cos\omega t\sin\varphi)\sin\omega t$$
$$= \hat{U}\hat{I}(\sin^2\omega t\cos\varphi + \sin\omega t\cos\omega t\sin\varphi) = UI\cos\varphi(1-\cos 2\omega t) + UI\sin\varphi\sin 2\omega t$$

Should network **B** be passive, (i.e., its Thévenin equivalent has a null electromotive force), it can be considered to be equivalent to an impedance $\underline{Z}_{eq} = R + jX$. In this case, angle φ is the characteristic angle of the impedance $\underline{Z}_{eq}$ and it is

$$p = RI^2(1 - \cos 2\omega t) + XI^2\sin 2\omega t = P(1 - \cos 2\omega t) + Q\sin 2\omega t$$

in which P is the power absorbed by the real part R of $\underline{Z}_{eq}$—that is, since reactors absorb no active power by the whole $\underline{Z}_{eq}$. In turn, Q is the reactive power absorbed by the reactor X—that is, since resistors absorb no reactive power by the whole $\underline{Z}_{eq}$.

It can be concluded that, at least when the receiving network is passive, this is

$$P = UI\cos\varphi, \qquad Q = UI\sin\varphi \tag{5.24}$$

It could be easily seen that also in the more general case of a generic (active or passive) receiving network, identities (5.24) apply.

Result: *Active and reactive power*

When two subnetworks **A** and **B** are connected to each other by a pair of wires, the active and reactive powers going from **A** to **B** are $P = UI\cos\varphi$ and $Q = UI\sin\varphi$, where U and I are the rms values of voltage and current exchanged (I entering **B** through the positively marked terminal), and φ is the phase difference between voltage and current phasors, measured positive when current lags voltage.

The term $\cos\varphi$ reduces the power transferred with the maximum available when instantaneous voltage and current are proportional to each other, or, equivalently, voltage and current phasors have the same phase.

The quantity $\cos\varphi$ is of sufficient importance to merit a name of its own: it is called *power factor*. The importance for power factors to be near unity will be discussed in Section 5.6.4.

Note that angle φ is, by definition, the difference between phase angle of voltage phasor and phase angle of current phasor. A rapid look at Figure 5.32 immediately tells us that this angle can be either positive (e.g., in the case of an R–L load) or negative (e.g., for an *R–C* load). Two opposite values of φ imply the same power factor, but a different positioning of phasors. Because of its importance, specific terms are used to avoid ambiguity. These terms refer to the position of current phasor with respect to voltage phasor: the power factor is said to be *lagging* when current lags voltage (and φ is positive), but *leading* when current leads voltage. For instance, in the situation represented in Figure 5.32 there is a lagging power factor. In circuits feeding *R–C* loads the power factor is of the leading type. This is remembered in the following box:

Definition: *leading and lagging power factor*

If the difference between voltage phasor angle and current phasor angle is called φ, the *power factor* is, by definition, $f = \cos\varphi$.

A power factor is said to be *leading* when the current phasor leads the voltage phasor (the angle φ is negative) and *lagging* in the opposite situation (the angle φ is positive).

Note that we refer to the power factor of a load. When we consider the $\underline{U}$–$\underline{I}$ couple to determine the displacement angle φ, we consider that it is defined with reference to the load with the load sign convention (see Section 3.5.2)—that is, with $\underline{I}$ assumed to be entering the positively marked load terminal. This is important in order to evaluate whether the power factor is leading or lagging.

When a phasor circuit is solved, phasors of voltages and currents are computed. It is very important to also have active and reactive powers expressed in terms of these phasors, using minimal computations. This is easily done by using the concept of *complex power*. The complex power $\underline{S}$, flowing through a pair of wires, is defined as half the product of the phasor of voltage across the wires and the conjugate of the phasor of the flowing current:

$$\underline{U} = \hat{U}e^{j\alpha} \qquad \underline{I} = \hat{I}e^{j\alpha-\varphi} \Rightarrow$$

$$\underline{S} = \frac{\underline{U}\,\underline{I}^*}{2} = \frac{\hat{U}\hat{I}^*}{2}e^{j(\alpha-\alpha+\varphi)} = UIe^{j\varphi} = UI\cos\varphi + jUI\sin\varphi = P + jQ$$

In some books, phasors are defined as having amplitudes equal to the rms values of the corresponding sine waves, instead of their peak values. In other books, peak-oriented phasors are used first and then, after the introduction of AC power, rms-oriented ones.

The latter is the preferred choice in courses for electrical engineering students, since it makes computations somewhat easier. In this book, for the sake of the greatest simplicity and clarity, just peak-oriented phasors—that is, phasors whose amplitude is equal to the peak of the corresponding wave—are used.

Therefore, in its real and imaginary parts respectively, complex power contains active and reactive powers, and therefore it summarises them in a unique complex number.

The modulus of complex power,

$$|\underline{S}| = UI = \sqrt{P^2 + Q^2}$$

is called apparent power.

All the "powers" considered in this section—active, reactive, complex, and apparent—are homogeneous with physical power and should in principle be measured in the unit of measure of power (i.e., watt) in the S.I.

However, it appears slightly confusing to attribute the same unit of measure to Ps, which are really powers in terms of derivative of energy flows, and Qs, which have a totally different definition; these have a conservation law of their own (see Section 5.6.3), but are never obtainable by conversion of real power. Apparent power is even trickier, since no conservation law applies to it. The latter statement will be clear once Section 5.6.3 has been read.

By international convention, therefore, the following units of measure have to be used (care should be taken with uppercase and lowercase letters and with roman and italic styles):

Quantity		Unit of Measurement	
Name	Symbol	Name	Symbol
Power, active power	P	Watt	W
Reactive power	Q	Var	var
Apparent power	S	Volt-ampere	VA

To conclude this section, it should be noted that, similarly to the triangle of impedances, a triangle of powers can also be used as shown here. Note that the two triangles are *similar*, since their angles are equal.

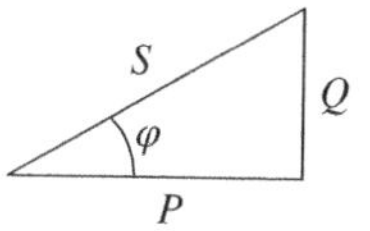

5.6.2 Circuits Containing Sources Having Different Frequencies

It was shown earlier that the solution to linear circuits containing different frequencies can be found using superposition.

Once the current $i(t)$ and voltage $u(t)$ entering a branch are found—that is, measured using the load convention (current entering the positively marked terminal)—the instantaneous power entering that branch will obviously be

$$p(t) = u(t)i(t)$$

whose shape will depend on the shapes of $u(t)$ and $i(t)$, but will be *periodic*, in the sense that it will replicate itself after a period T.[2]

It is therefore useful to compute the average value of $p(t)$ over the period T. This average will be referred to, in a manner similar to the case of AC circuits, with the symbol P:

$$P = avg[p(t)] = \frac{1}{T}\int_{t}^{t+T} p(t)\mathrm{d}t \tag{5.25}$$

The instant t in (5.25) is arbitrary since, given the periodic nature of $p(t)$, the integral in this equation does not depend on the instant t.

By expressing both $u(t)$ and $i(t)$ as the sum of contributions at different frequencies (see Section 5.5.5), it is possible to demonstrate that P simply equals the sum of active powers calculated at each frequency separately:

$$P = avg[p(t)] = avg[u(t)\cdot i(t)] = avg\left[\sum_{\text{freq j}} u_j(t)\cdot \sum_{\text{freq k}} i_k(t)\right]$$

$$= avg\left[\sum_{j=k} u_j(t)i_k(t)\right] + avg\left[\sum_{j\neq k} u_j(t)i_k(t)\right] = avg\left[\sum_{j=k} u_j(t)i_k(t)\right] + 0 = \sum_{j} P_j$$

In fact, each term $u_j(t)i_k(t)$, $j \neq k$, has a null average, since it represents the product between two sinusoidal functions with different frequencies.

In other terms, *the mutual interaction between voltages and currents at different frequencies affects p(t), but adds nothing to P.*

Reactive power has much less importance in periodic, nonsinusoidal circuits and will not be dealt with in this book.

5.6.3 Conservation of Complex, Active, and Reactive Powers

In the previous section the reactive power in an AC circuit was defined in a mathematical way, as the peak value of the alternating instantaneous power flowing through a reactive element, specifying that reactive power absorbed by a reactor is the same as its reactance.

This, however, in no way justifies the use of the word *power* in its name. The real importance and usefulness of reactive power comes from *Boucherot's theorem*, which can be very rapidly derived as an application of Tellegen's theorem.[3]

In Section 3.8.1, Tellegen's theorem was stated as follows. Consider a generic circuit. *If*

- $\{v_0, v_1, \ldots, v_n\}$ is a set of potential values for the circuit node, satisfying KVL for the circuit in question;
- $\{i_{01}, i_{02}, \ldots i_{0n}, \ldots i_{12}, i_{13}, \ldots i_{1n}, \ldots i_{1n-1,n}, i_{1n}\}$ is a set of currents flowing in the circuit branches (from the first to the second node) satisfying KCL for the circuit in question,

[2]More information about periodic waveforms and how they are to be treated in circuits will be given in Chapter 8.
[3]The theorem is named after Paul, Boucherot, who first discovered and demonstrated it. Later, Tellegen's theorem made it faster to demonstrate.

then

$$\sum_{i=0,n;\ j=0,n;\ i\neq j} u_{ij}\, i_{ij} = 0 \tag{5.26}$$

In this chapter the phasors circuit was introduced, showing that Kirchhoff's equations can be written directly in terms of phasors.

Since (5.26) is a direct consequence of Kirchhoff's laws, and Kirchhoff's laws apply to phasors in the phasor circuit, it can also be written in phasor form:

$$\sum \underline{U}_{ij}\underline{I}_{ij} = 0$$

Consider that if a given set of phasor currents,

$$\left\{\underline{I}_{01}, \underline{I}_{02}, \ldots \underline{I}_{0n}, \ldots \underline{I}_{12}, \underline{I}_{13}, \ldots \underline{I}_{1n}, \ldots \underline{I}_{1n-1,n}, \underline{I}_{1n}\right\}$$

satisfies the KCL equations for the given circuit, then also a set containing their conjugate satisfies the same equations. Tellegen's theorem can then be applied, using the actual branch voltages and the conjugate of the actual branch currents:

$$\sum \underline{S}_{ij} = \frac{1}{2}\sum \underline{U}_{ij}\underline{I}^{*}_{ij} = 0$$

or, equivalently, $\sum P_{ij} + jQ_{ij} = 0$,

which implies

$$\sum P_{ij} = 0, \qquad \sum Q_{ij} = 0 \tag{5.27}$$

Result (5.27) shows the conservation of active power in a circuit, which is also a fairly obvious consequence of the conservation of instantaneous power, plus the conservation of reactive power. The conservation of reactive power in circuits is a very important law.

It allows conservation techniques to be applied in circuit analysis: for example, if the Q absorbed by some of the branches is known or computed, this equals the Q globally delivered by the set of the remaining branches. This is summarised as follows:

Law: *Power conservation in AC circuits*

This law can be expressed in one of the following equivalent formulations:

1. In any AC circuit, the sum of complex power values absorbed by all its branches is identically null.
2. In any AC circuit, the sum of active power values absorbed by all its branches and the sum of reactive power values absorbed by all its branches are identically null.

Note that the fact that the sum of all complex $\underline{S}_k$ is zero does not imply that the sum of their moduli is zero, and in fact this is not generally true.

The conservation of reactive power [second of equations (5.27)] in a circuit is often called Boucherot's theorem.

Example 9. Calculate the active and reactive power absorbed by any passive component of Example 7. Calculate the active and reactive power provided by the generator and verify Boucherot's theorem.

$$P_{R1} = R_1 \cdot I_1^2 = R_1 \cdot \frac{\hat{I}_1^2}{2} = 8 \cdot 3.630^2/2 = 52.70\ \text{W}$$

$$P_{R2} = R_2 \cdot I_2^2 = R_2 \cdot \frac{\hat{I}_2^2}{2} = 4 \cdot 1.837^2/2 = 6.747\ \text{W}$$

$$P_{R3} = R_3 \cdot I_3^2 = R_3 \cdot \frac{\hat{I}_3^2}{2} = 25 \cdot 2.953^2/2 = 109.0\ \text{W}$$

$$Q_{L2} = X_2 \cdot I_2^2 = X_2 \cdot \frac{\hat{I}_2^2}{2} = 40 \cdot 1.837^2/2 = 67.47\ \text{var}$$

$$\underline{S}_{gen} = \frac{\underline{E}\,\underline{I}^*}{2} = \frac{(93.97 + j34.2)(3.63 + j0.11)}{2} = 168.5\text{W} + j67.47\text{var} = P_{gen} + jQ_{gen}$$

It can be verified that $P_{R1} + P_{R2} + P_{R3} = P_{gen}$ and that $Q_{L2} = Q_{gen}$.

5.6.4 Power Factor Correction

Consider the system represented in Figure 5.33. It contains an ideal AC generator—that is, a device which produces at its terminals a perfectly sinusoidal voltage and feeds through conductor wires a linear, passive system which operates as a load. Under the hypotheses, normally satisfied, that displacement currents and self-induced voltage are negligible between wires, this system can be represented by the equivalent circuit shown in the central part of the figure.

Should only the steady-state behaviour be of interest, the load can be represented by its Thévenin equivalent which in our case, since the load is passive, is reduced to the simple impedance $\underline{Z}_{eq}$. The line wires are modelled through resistors R_{lu} and R_{ld} (bottom part of Figure 5.33).

In general it is not known whether the reactive part of $\underline{Z}_{eq}$ is positive or negative, although in the large majority of cases it is positive. In Figure 5.34, it is modelled as positive, and therefore $\underline{Z}_{eq}$ can be represented as an equivalent series R_{eq}–L_{eq}. The sum of R_{lu} and R_{ld} is reported as a single resistor, named R_0.

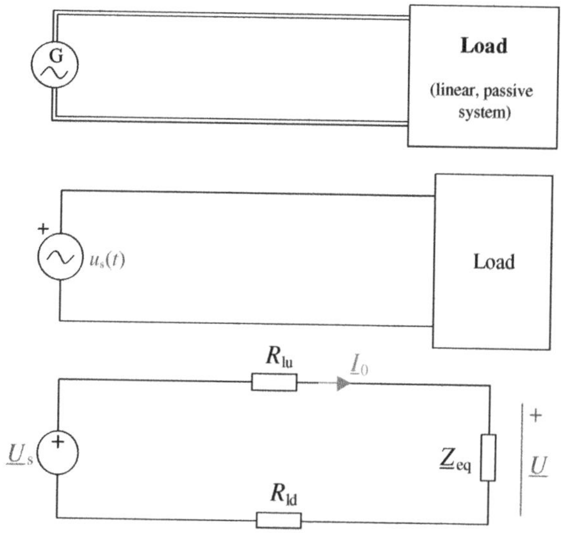

FIGURE 5.33. A sample physical circuital system and a possible circuit modelling it.

Firstly, consider the circuit in Figure 5.34a, in which the switch T is open (i.e., no current can flow through it). In this case, the phasor diagram is of the type shown alongside the same figure. It is clearly seen that the current $\underline{I}_0$ flowing through the line, and causing a net power loss equal to $R_0\, I_0^2$, is composed of a part which is in phase with the user voltage $\underline{U}$, and a part that is in quadrature[4] with it.

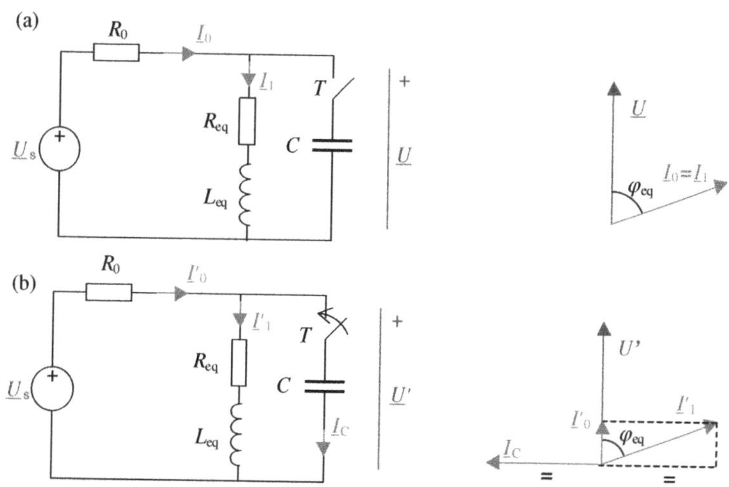

FIGURE 5.34. A circuit before and after power factor correction and the corresponding phasor diagrams.

[4]In electrical engineering, to be "in quadrature" means having a phase difference of 90°.

While the component in phase is associated with a flow of active power, the one in quadrature can be generated locally by a capacitor, since capacitors absorb currents in quadrature leading the voltage phasor, which can offset the quadrature component of the $\underline{Z}_{eq}$ current.

The concept is illustrated in Figure 5.34b, which considers the steady-state circuit operation when switch T is closed. Now the line transfers only the part of the load current in phase with the voltage. Naturally, this implies an advantage in terms of power lost in the line, since the modulus of $\underline{I}_0$ is reduced. The reduction can be substantial, as in the case shown in the figure, when the initial phase difference between load voltage and current (i.e., the characteristic angle of the impedance $\underline{Z}_{eq}$) is large (around 90°).

The quantity $\cos\varphi$, as discussed in Section 5.6.1, is the power factor of the load. The closer the power factor of a load gets to unity, the more effectively the power can be transferred to it. Which is why, when the power factor is too small (e.g., below 0.8), it might be advisable to correct it by bringing it nearer to unity.

The example of Figure 5.34 shows that this can be done by absorbing, in the vicinity of the load, a current that is the same size (or nearly) as the quadrature component of the load current, and opposite in phase. Since, as already noted, typical loads have positive φ angles, this operation normally requires placing a capacitor with suitable capacity in parallel to the load.

The operation of installing a reactive component near the load, which is able to eliminate, or at least strongly reduce, the phase difference between the voltage applied to the load and the current that it absorbs, is called *power factor correction*. Should the displacement be reduced to zero, the correction is total, otherwise it is partial.

Let us now compute the value of the capacitance of the capacitor which has to perform a total power factor correction. The formula can be derived by reasoning in terms of powers: the reactive power delivered by the correcting capacitor must match the reactive power absorbed by the R–L load,

$$Q_{absorbed}(Z_{eq}) = UI_1 \sin\varphi_{eq} = \frac{U^2}{Z_{eq}}\sin\varphi_{eq} = \frac{XU^2}{Z_{eq}^2} = \frac{\omega L_{eq}U^2}{R_{eq}^2 + X_{eq}^2}$$

since from the triangle of impedances it is immediate that for any impedance $X = Z\sin\varphi$.

As regards the capacitor, it is

$$Q_{delivered}(C) = -UI_C \sin\varphi_C = \frac{U^2}{|X_C|} = U^2\omega C$$

because the angle φ for the capacitor is −90°.

Equating the two powers the result is

$$C = \frac{L_{eq}}{R_{eq}^2 + X_{eq}^2}$$

When a total power factor correction is carried out, the line feeding the parallel of $\underline{Z}_{eq}$ and C "sees" a load that absorbs a current that is in phase with the voltage—that is, an impedance having null (= zero) reactance or, equivalently, a pure resistor.

The reader can verify this by computing the equivalent impedance of $\underline{Z}_{eq}$ and $\underline{Z}_C = -j/(\omega C)$.

Example 10. Measure the capacitor required to correct the load power factor to 1, if $R = 4\ \Omega$, $L = 0.01$ H, $\omega = 314$ rad/s.

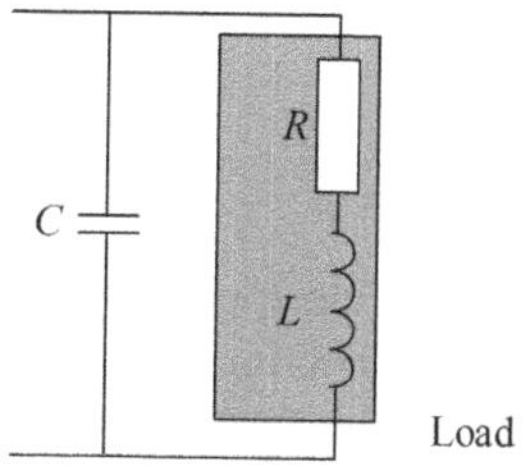

$$C = \frac{L}{R^2 + X^2} = \frac{0.01}{4^2 + (314 \cdot 0.01)^2} = 386.7\ \mu\text{F}$$

In fact,

$$\underline{Z}_{\parallel} = \frac{j}{\omega C} \| (R + j\omega L) = 6.47\ \Omega \qquad \text{(pure resistance, } \cos\varphi = 1\text{)}$$

Please note that the equivalent resistance is different from R.

Example 11. In the previous example, calculate C to correct the power factor to 0.9:

$$\underline{Z} = R + jX = 4 + j3.14$$
$$\underline{Y} = 1/\underline{Z} = 0.1546 - j0.1214 = G + jB$$
$$\underline{Y}_{\parallel} = \underline{Y} + \underline{Y}_c, \text{ where } \underline{Y}_c = jB_c = j\omega C$$
$$\underline{Y}_{\parallel} = G + j(B + B_c) = G + jB'$$

The desired B' equals:

$$B' = -G \cdot \tan\varphi', \text{ being } \cos\varphi' = 0.9$$
$$B' = -0.1546 \cdot \tan(\cos^{-1} 0.9) = -0.07487\Omega^{-1}$$
$$B_c = B' - B = -0.07487 + 0.1214 = 0.04652\Omega^{-1} = \omega C \rightarrow C$$
$$= 0.04652/314 = 148.1\ \mu\text{F}$$

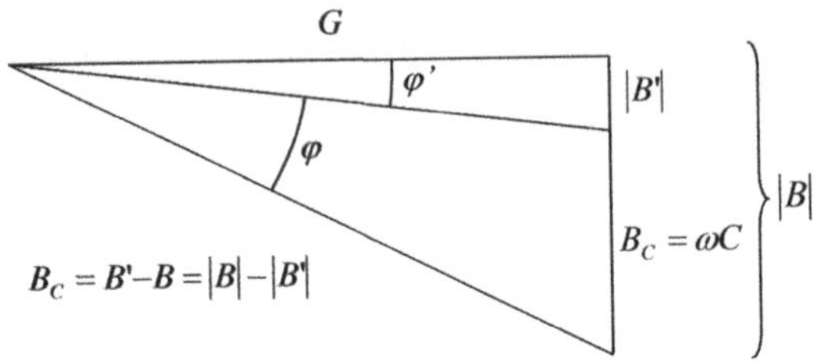

5.7 HISTORICAL NOTES

5.7.1 Short Biography of Boucherot

Paul Boucherot (1869–1943) was an electrical engineer, known for several important studies in the field of electrical engineering.

In circuit theory, he introduced the distinction of active, reactive, and apparent power and demonstrated the theorem that today carries his name (discussed in Section 5.6.3).

In the field of electrical machines, he carried out important studies on asynchronous motors, thus contributing greatly to their increased use.

In power system engineering, he built an innovative power station able to exploit the ocean's thermal energy.

5.8 PROPOSED EXERCISES

5.1. Determine A, ω and θ in the following functions:
- **(a)** $10\sin(5t + 30°) = A\cos(\omega t + \theta)$
- **(b)** $2\cos(3t - 20°) = A\sin(\omega t + \theta)$
- **(c)** $\sqrt{2}\cdot 230\sin(2\pi 50t + 10°) = A\cos(\omega t + \theta)$
- **(d)** $\sqrt{2}\cdot 127\cos(2\pi 400t + 70°) = A\sin(\omega t + \theta)$

5.2. Determine A and B, or M and θ, in the following functions:
- **(a)** $2\sin(10t + 30°) = A\cos 10t + B\sin 10t$
- **(b)** $5\cos(t - 15°) = A\cos t + B\sin t$
- **(c)** $3\cos 2t + 4\sin 2t = M\sin(2t + \theta)$
- **(d)** $5\cos t + \sin t = M\cos(t + \theta)$

5.3. Calculate the phasor corresponding to the following voltages or currents:
- **(a)** $u(t) = 100\sin(\omega t + 20°)$
- **(b)** $u(t) = 5\cos(\omega t - 60°)$
- **(c)** $i(t) = 3\sin(\omega t)$
- **(d)** $i(t) = 4\cos(\omega t - 15°)$

5.4. Calculate the impedance associated with the following R, L, C, and ω:

(a) $R = 5\,\Omega,\ L = 2\,\text{H},\ \omega = 10\,\text{rad/s}$

(b) $R = 10\,\Omega,\ C = 10\,\mu\text{F},\ \omega = 1000\,\text{rad/s}$

(c) $R = 10\,\Omega,\ L = 100\,\text{mH},\ C = 100\,\mu\text{F},\ \omega = 1000\,\text{rad/s}$

(d) $R = 1\,\Omega,\ L = 5\,\text{mH},\ C = 100\,\text{mF},\ \omega = 200\,\text{rad/s}$

(e) $R = 30\,\Omega,\ L = 0.5\,\text{H},\ C = 1\,\mu\text{F},\ \omega = 1000\,\text{rad/s}$

5.5. Calculate the equivalent impedance between A and B:

(a)

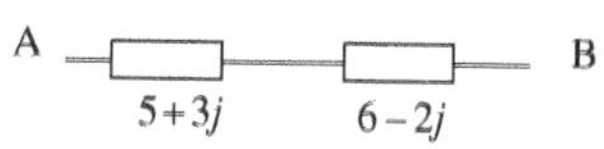

(b)

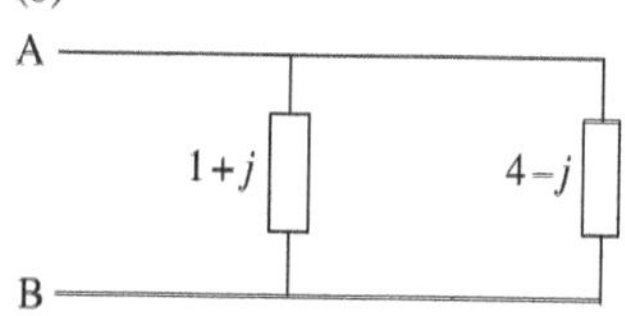

(c)

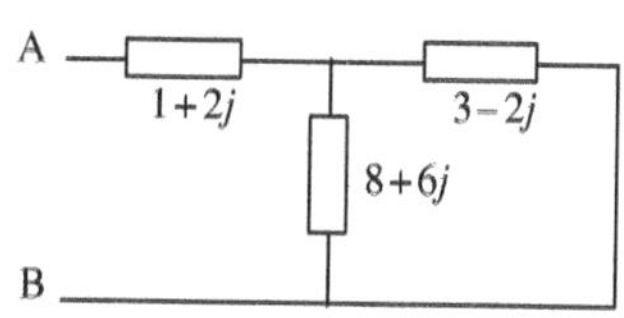

(d)

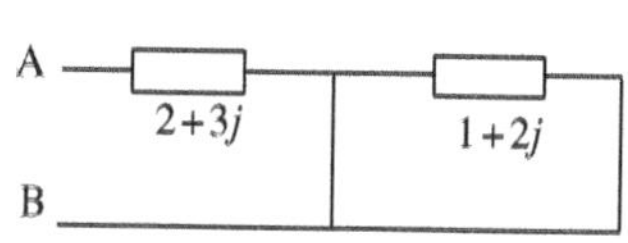

(e)

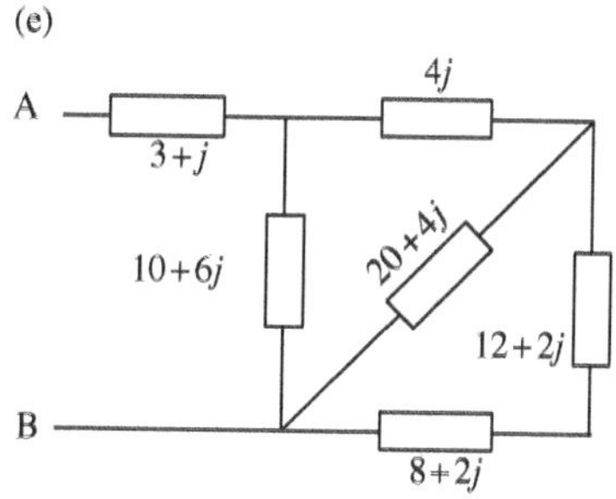

5.6. Using the rule of current divider, calculate $\underline{I}_1$ and $\underline{I}_2$, the total current $\underline{I}_{tot}$ being $10+5j$. Verify the solution calculating $\underline{U}_{AB}=\underline{Z}_{eq}\,\underline{I}_{tot}$ and observing that $\underline{U}_{AB}$ is applied to each impedance.

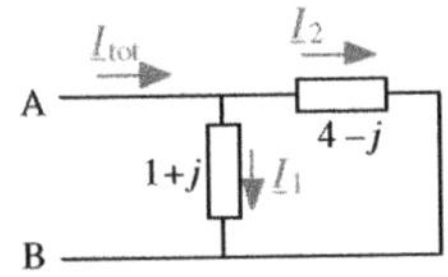

5.7. Using the rule of voltage divider, calculate $\underline{U}_1$ and $\underline{U}_2$, the total voltage $\underline{U}_{tot}$ being $6+3j$. Verify the solution by calculating $\underline{I}=\underline{U}_{tot}/\underline{Z}_{eq}$ and observing that such a current flows through each impedance.

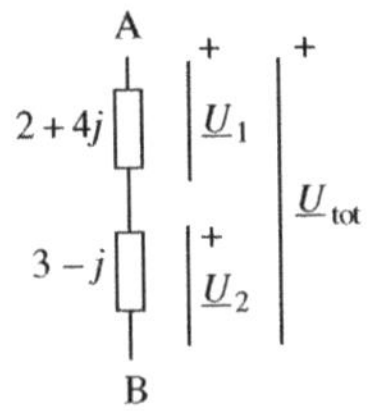

5.8. Applying Thévenin's theorem between A and B, calculate the equivalent voltage $\underline{U}_{Th}$ and the equivalent impedance $\underline{Z}_{Th}$:

(a)

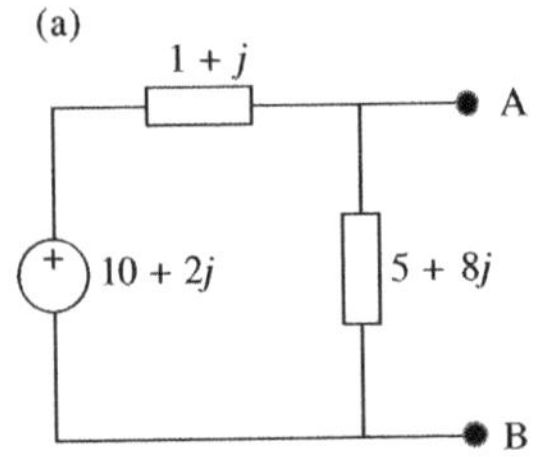

(b)

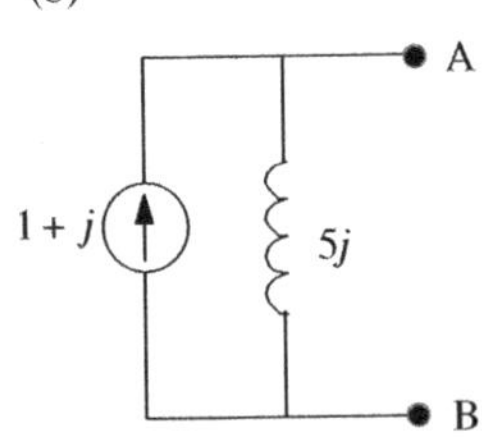

(c)

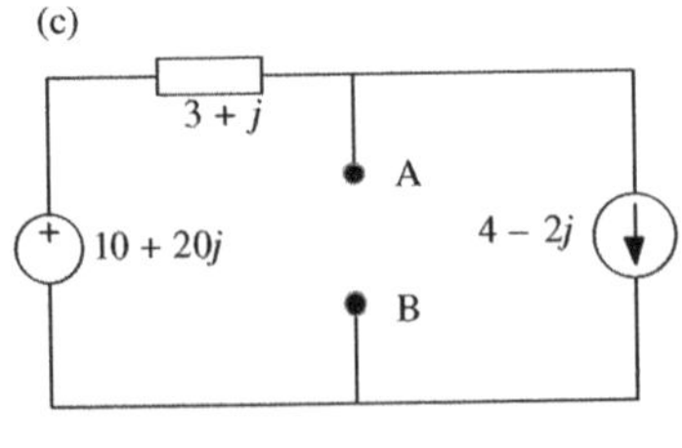

5.9. Calculate $i_1(t)$, $i_2(t)$, and $i_3(t)$ in the following circuit, where

$$u_s(t) = 100\sin(2\pi 50t)\text{V}$$

$$R_1 = 10\,\Omega,\ R_2 = 5\,\Omega,\ R_3 = 20\,\Omega \text{ and } L_2 = 0.1\text{ H}.$$

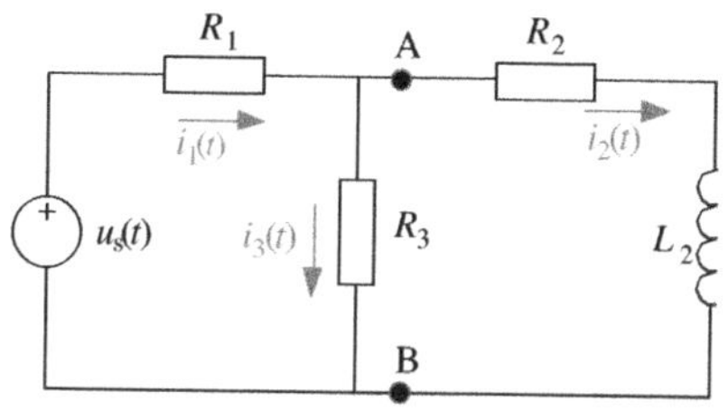

5.10. Calculate once again $i_2(t)$ of the previous exercise, this time applying Thévenin's theorem between A and B.

5.11. Calculate $i_1(t)$, $i_2(t)$ and $u_{AB}(t)$ in the following circuit, where

$$u_s(t) = 10\sin(100t)\text{V}$$

$$i_s(t) = 5\sin(314t + 30°)\text{A}$$

$$R = 1\,\Omega, \qquad L = 0.02\text{ H}$$

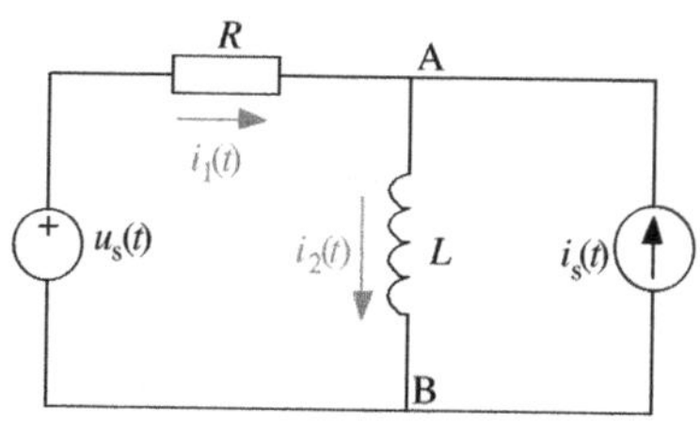

5.12. Calculate $i(t)$ in the following circuit:

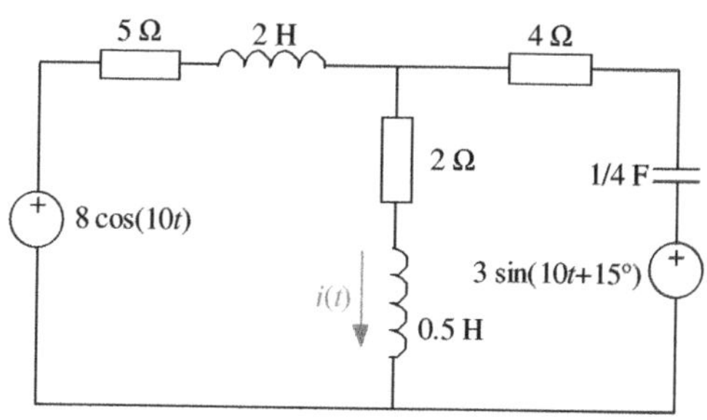

The reader is invited to compare the use of the following methods:

- the use of KCL and KVL equations.
- the superposition principle
- Thévenin's theorem
- nodal analysis

5.13. Calculate the current across the capacitor in the following circuit:

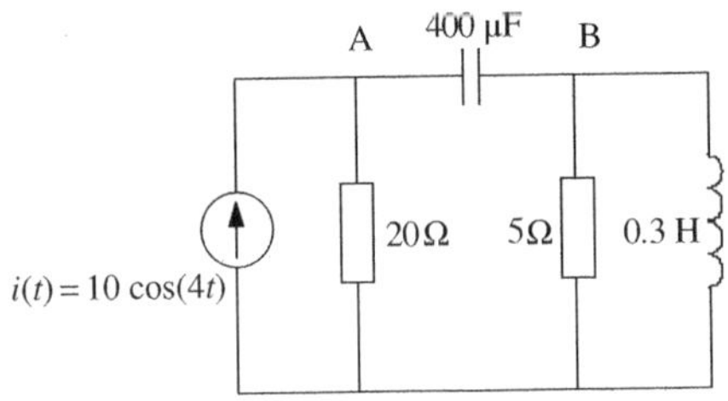

5.14. Calculate the real and reactive power delivered by the generator in exercise 5.9. Calculate the active and reactive power absorbed by resistors R_1, R_2, R_3 and by the inductor L_2. Verify Boucherot's theorem.

5.15. Calculate the active power delivered by the generators in exercise 5.11 and the active power absorbed by resistor R. Verify the conservation of active and reactive powers.

5.16. Consider the circuit in exercise 5.13. Knowing that the rms value of the current through the capacitor is 0.226 A, calculate the reactive power provided by the generator.

5.17. Calculate the average and maximum value of $p(t)$, where

$$u(t) = \sqrt{2}\,10\sin(200t + 65^\circ)$$

$$i(t) = \sqrt{2}\,3\sin(200t + 5^\circ)$$

5.18. Express the active power P absorbed by the load, as a function of R, L, U_s, and R_{line}. Demonstrate that, U_s, L, and R_{line} being fixed, P is maximum when $R = R_{line}$ (*Theorem of maximum power transfer*).

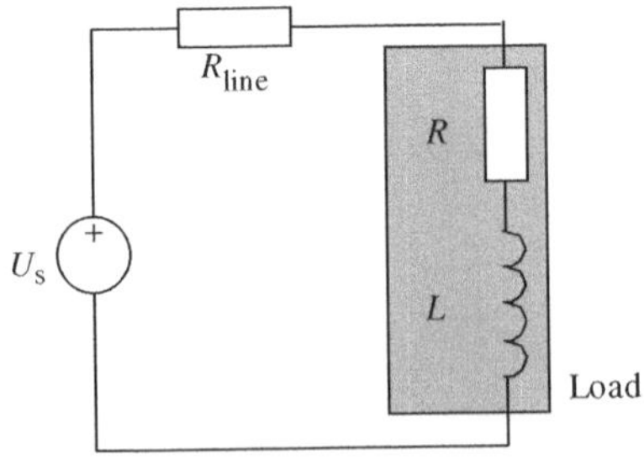

5.19. Calculate I_A and U_{AB} and determine the active and reactive power flowing through the section AB. Verify the result applying Tellegen's theorem at the right-side or left-side AB.

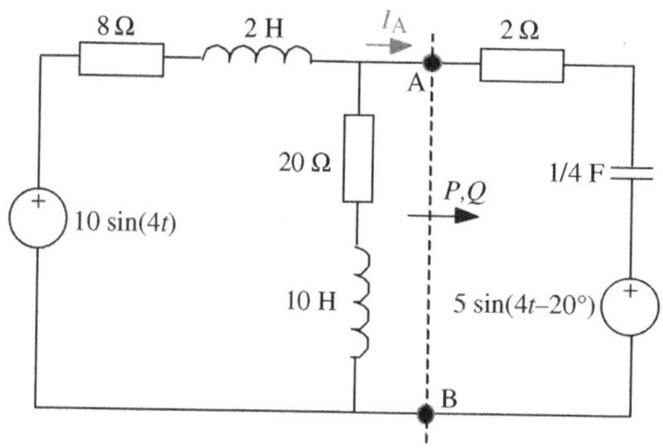

5.20. Calculate $u(t)$ in the following circuit, $i(t)$ being known:

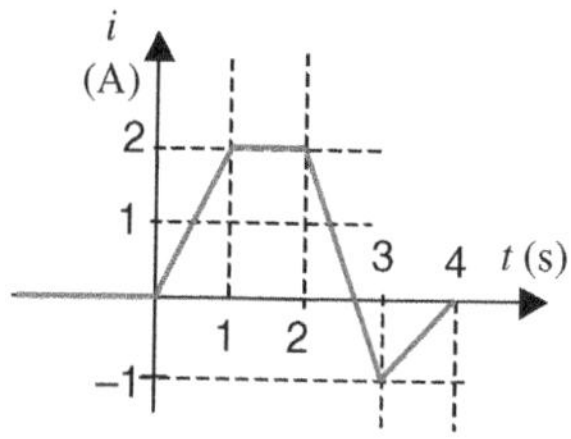

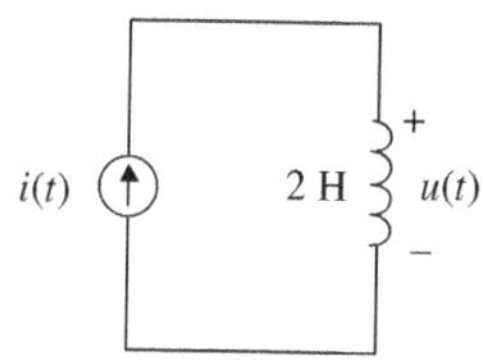

5.21. Calculate the capacity C_1 and $C_{0.8}$ of the capacitor required to correct the load power factor respectively to 1 and 0.8, if $R = 1\Omega$, $L = 0.01$ H, $\omega = 200$ rad/s.

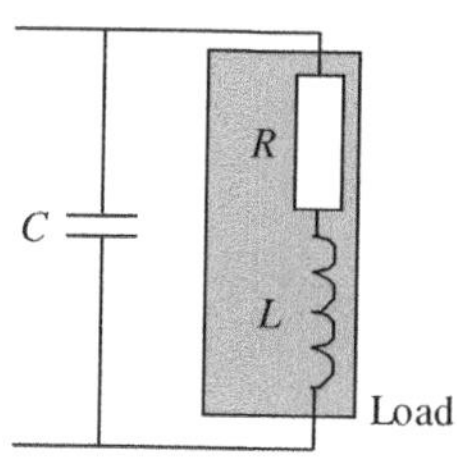

6

THREE-PHASE CIRCUITS

For the Instructor

In this chapter, consistently with the approach adopted in Chapter 5, the three-phase system is introduced by initially modelling the lines connecting generators and loads as purely resistive branches; that is, we ignore cross conduction between wires (both conductive and displacement currents) as well as the effect of EMF induced in meshes containing the lines due to the alternating current circulating in the line conductors.

The inductive effects are introduced later, even though it is remarked that in low voltage lines, line reactances are normally negligible in comparison to resistances.

Details on transmission line issues when they are modelled as circuits are included in the Appendix.

6.1 INTRODUCTION

Every reader of this book has surely seen overhead high-voltage transmission lines, such as the ones depicted in Figure 6.1. It can be seen from the pictures that energy is transferred using three-wire lines: the left-hand pole has three wires, the right-hand

Fundamentals of Electric Power Engineering: From Electromagnetics to Power Systems, First Edition.
Massimo Ceraolo and Davide Poli.

FIGURE 6.1. Two three-phase overhead lines (*left*: medium voltage; *right*: high voltage).

tower holds four, but the highest wire just has the purpose of protecting the line from lightning and does not contribute to carrying electric energy. This is highly visible evidence of the fact that electric energy is used and transferred, when power surpasses a few kilowatts, by means of so-called *three-phase systems* that are based on three wires for transferring electric power.

This chapter shows the genesis of three-phase lines and systems and provides information on how to make computations with these systems in the particularly important (and easy) case of *balanced* three-phase systems.

6.2 FROM SINGLE-PHASE TO THREE-PHASE SYSTEMS

Imagine we want to feed three load impedances $\underline{Z}_u$ by means of a voltage source, situated relatively far from the load, at distance d (Figure 6.2a). The electrical behaviour of this system, if induced EMF between wires is negligible, can be analysed as represented in Figure 6.2b.

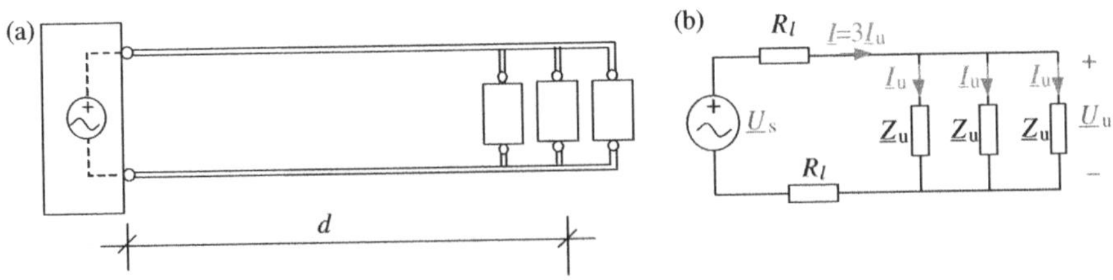

FIGURE 6.2. Single-phase transmission to three loads.

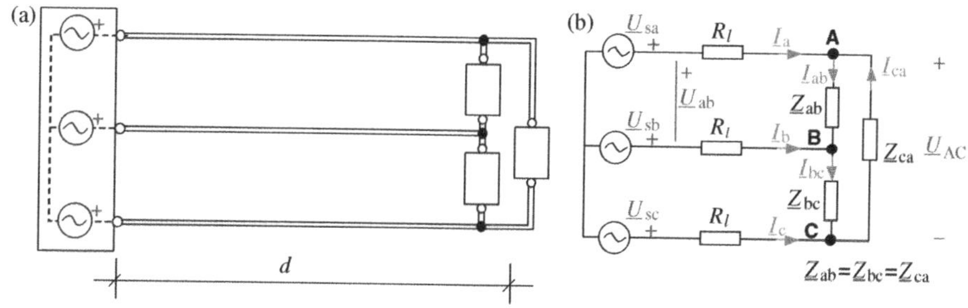

FIGURE 6.3. Three-phase transmission to three loads.

Consider that the total load requires a power $\underline{S} = P + jQ$ to be delivered at a voltage $U_u = U$ ([1]). The current flowing through the line is $I = S/U = 3I_u$ and the power lost through the line is

$$P_{jl} = 2R_l I^2 = 2R_l(3I_u)^2 = 2 \cdot 9R_l I_u^2$$

Now we will show a technique that allows the same power to be transferred under the same voltage to the three loads while reducing the line losses, if the quantity of copper for the lines is the same, or reducing the quantity of copper involved, at equal line losses.

Consider the arrangement shown in Figure 6.3. The three voltage sources have equal amplitudes, and they are displaced in phase with each other by $2\pi/3$ rad. Such sets of voltages are called *balanced three-phase sets of voltages*. It will be shown in the next chapter that these sets are very easily created by means of electrical machines.

The amplitude of voltages applied to single load impedances will have to be the same as $\underline{U}_u$ in the scheme of Figure 6.2.

This implies that the voltage sources $\underline{U}_{sa}$, $\underline{U}_{sb}$, $\underline{U}_{sc}$ are equal in amplitude to $|\underline{U}_u|/\sqrt{3}$, as will be demonstrated later.

The load impedances, as all sets of three equal impedances, is called a *balanced three-phase* load.

Figure 6.3b shows a circuit modelling the system of Figure 6.3a, valid if induced EMFs between wires are negligible. Using this arrangement, it can be easily demonstrated that the three loads are fed with the same active and reactive power as the arrangement of Figure 6.2, but with reduced losses in the lines.

Consider Figure 6.4. The left part of the figure shows the balanced set of voltage composed of $\underline{U}_{sa}$, $\underline{U}_{sb}$, $\underline{U}_{sc}$, as well as the voltages between wires at the source terminals, that are determined using the usual rule for vector differences (Figure 6.4a); the voltages between the corresponding phases at the load terminals (indicated in figures with uppercase subscripts) will be similar to them, and they are shown with a

[1]Remember that when a voltage value is quoted in an alternating current, if there is no other specification, reference is made to the rms value of the sinusoidal waveform. Therefore U, I, and I_u are rms values.

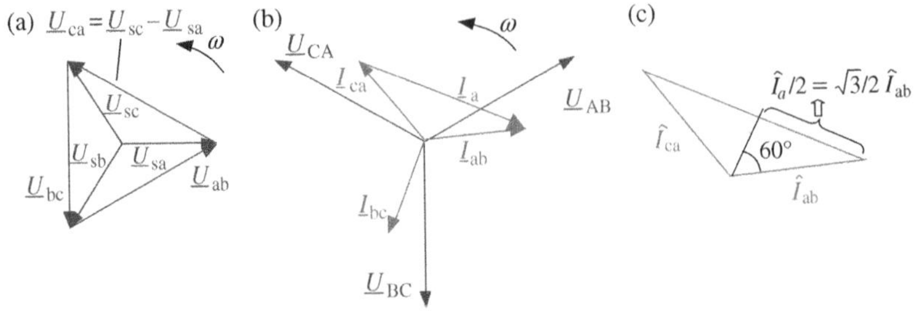

FIGURE 6.4. Phasors relative to the three-phase transmission shown in Figure 6.3.

common tail in Figure 6.4b. Considering that, by hypothesis, the three load impedances are equal to each other, the currents absorbed by the three loads will have the same amplitudes and phase displacements with respect to the corresponding voltages. Therefore, they will be of the type shown in Figure 6.4b. Considering KCL at nodes **A**, **B**, **C**, the corresponding line currents are found by the difference of the currents flowing in the load impedances, and they will be of the type shown in Figure 6.4b for the first wire ($\underline{I}_a$). The amplitudes of the line currents will therefore be equal to those of load currents, multiplied by $\sqrt{3}$, as shown by the construction in Figure 6.4c. The conclusion is that the total line losses will be

$$P'_{jl} = 3R_l(\sqrt{3}I_u)^2 = 9R_lI_u^2$$

compared with the solution of Figure 6.2:

$$P_{jl} = 2R_l(3I_u)^2 = 2 \cdot 9R_lI_u^2$$

The ratio of the two powers is therefore

$$P'_{jl}/P_{jl} = 0.5$$

which demonstrates that the transmission of electric power shown in Figure 6.3, called three-phase transmission, is more effective than single-phase transmission).[2] Naturally, if the power loss is required to be unchanged, the total copper used for the three-phase transmission can be reduced. In this case, instead of having a reduction in power loss, we would have a reduction in line cost.

Another very important advantage of three-phase systems is the fact that, differently from single-phase transmission, the instantaneous power delivered by

[2]The analysis carried out considered only the amount of copper at equal line-to-line voltages. A more detailed analysis should also have considered the cost of insulating three wires from each other instead of two. However, it can be seen that even with this more detailed analysis, the advantages of three-phase transmission remain sound.

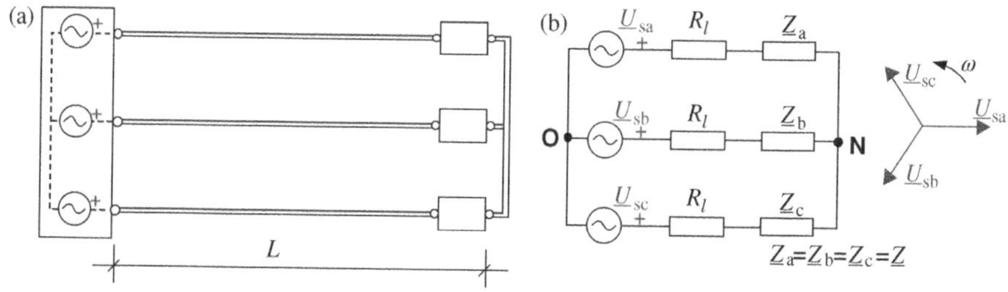

FIGURE 6.5. Three-phase transmission to a star-connected balanced load.

three-phase generators and transferred to three-phase loads is constant. This will be demonstrated in the next section.

The three load impedances discussed earlier were connected between the line wires, forming a triangle (“delta connection”). This was because equality of the voltage at the load terminals was required to compare with the single-phase transmission.

Consider now the diagram in Figure 6.5. In this case the load is constituted by three “star-connected” impedances.[3]

As usual, the left-hand part of the figure represents a physical system whose mathematical model is the circuit in the right-hand part of the figure.

This circuit, as any three-phase circuit, can be studied using the usual rules for solving AC circuits. In particular, nodal analysis can be used, assuming that the reference node is **O**. The following equation can thus be written:

$$\frac{\underline{U}_{sa}-\underline{V}_{N}}{R_l+\underline{Z}_a}+\frac{\underline{U}_{sb}-\underline{V}_{N}}{R_l+\underline{Z}_b}+\frac{\underline{U}_{sc}-\underline{V}_{N}}{R_l+\underline{Z}_c}=0$$

or

$$\frac{\underline{U}_{sa}+\underline{U}_{sb}+\underline{U}_{sc}}{R_l+\underline{Z}}=3\frac{\underline{V}_{N}}{R_l+\underline{Z}} \qquad (°)$$

It must now be observed that any balanced three-phase set of quantities (voltages, currents) sums to zero, as can be immediately inferred from the diagram shown here, evidently:

$$-\underline{U}_{sa}=\underline{U}_{sb}+\underline{U}_{sc}\Rightarrow \underline{U}_{sa}+\underline{U}_{sb}+\underline{U}_{sc}=0$$

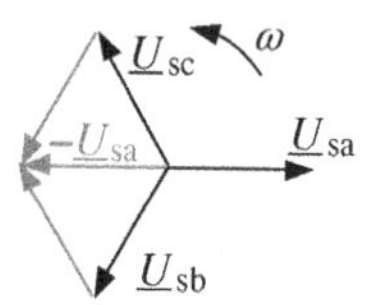

[3]Star connection is also referred to as “wye” connection.

As a consequence, equation (°) yields

$$0 = 3\frac{\underline{V}_N}{R_l + \underline{Z}} \Rightarrow \underline{V}_N = 0$$

This shows that, in the circuit of Figure 6.5, nodes **O** and **N** have the same potential.

Points **O** and **N** are called *star centres* or *neutral points* of the three-phase system. It is easy to convince ourselves that this is a rather general conclusion:

Result: *Potential of neutral points in three-phase balanced systems*

In a three-phase system fed by a balanced set of voltages, if all the impedances present in the three phases are equal to each other, all the neutral points of the star-connected impedances and voltage sources share the same potential.

Therefore, entire three-phase networks can be built, where several three-phase loads might be present—for example, constituted by star- or delta-connected impedances. In the case of balanced networks, in which the impedances present in the three phases are equal to each other and the sources are all constituted by balanced three-phase sets, all the star centres have the same potential.

In the example of Figure 6.6, for instance, this common potential is $\underline{V}_{N1} = \underline{V}_{N2} = \underline{V}_{N3} = \underline{V}_O$. When impedances are delta-connected, they can be replaced by three star-connected equivalents. The star-delta conversion formulas are the same as those used in Chapter 5 for generic AC circuits. In the case considered here, where impedances are equal to each other, such formulas become simply

$$\underline{Z}_\Delta = 3\underline{Z}_\lambda$$

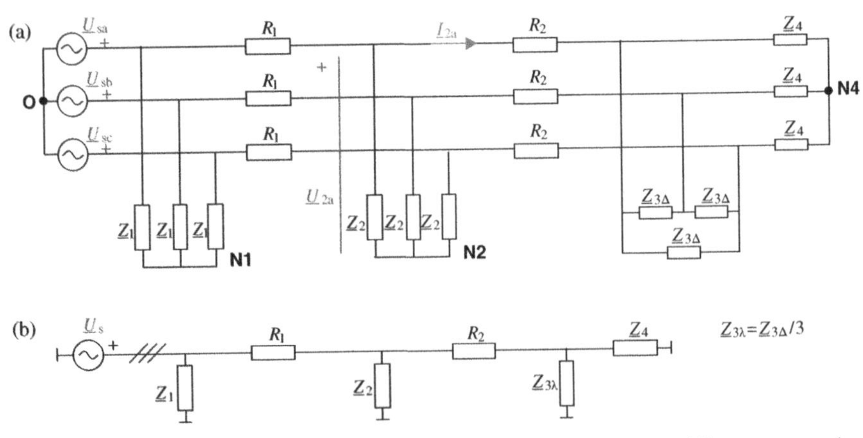

FIGURE 6.6. A three-phase system with several components and its unifilar representation.

where $\underline{Z}_\Delta$ and $\underline{Z}_\lambda$ are respectively the delta-connected and the star-connected impedances.

It is apparent that when large balanced three-phase networks are to be represented, it is superfluous to always replicate circuit parameters for the three phases. Therefore, very often in these cases, we resort to the *unifilar representation*, such as that shown in Figure 6.6b, where a single line is shown to indicate three elements by the use of three small slanting lines: ///. Three-phase systems are characterised by two kinds of voltages: *line-to-line* and *line-to-neutral voltages*.

Line-to-line voltages can be measured between two of the three wires, while line-to-neutral are the voltages between lines and the neutral point. Both are normally expressed as rms values.

There is always a very simple and important relation between line and phase voltages of a three-phase system:

$$U_{\text{line}} = \sqrt{3} \cdot U_{\text{phase}} \tag{6.1}$$

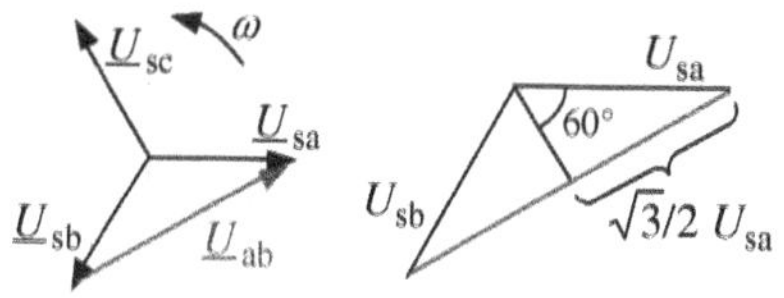

This can be easily seen looking at the phasors of these voltages as in the diagram above. For instance, it is

$$\underline{U}_{ab} = \underline{U}_{sa} - \underline{U}_{sb}; \quad |\underline{U}_{\text{line}}| = |\underline{U}_{ab}| = \sqrt{3}|\underline{U}_{sa}| = |\underline{U}_{\text{phase}}|$$

The same relationship applies to rms values that are amplitudes divided by $\sqrt{2}$. The duality of line and phase voltages is so important that it is very common for three-phase systems to be named after two numbers, representing the two voltages.

For instance, in European Union countries the voltage used for distributing energy to users needing slightly more power than for domestic use homes is the following:

$$400/230\ \text{V}$$

The first number represents the line-to-line nominal voltage of this system, while the second one represents the phase voltage. Note that the ratio between the two is only approximately $\sqrt{3}$; this was chosen to avoid decimal numbers in the representation of any of the two nominal values.

When a three-phase system is named using just a single value, it always refers to the rms value of the line-to-line voltages; for instances some standard voltage levels are 10 kV, 15 kV, 20 kV, 130 kV, and 220 kV. All these values are line rms voltages.

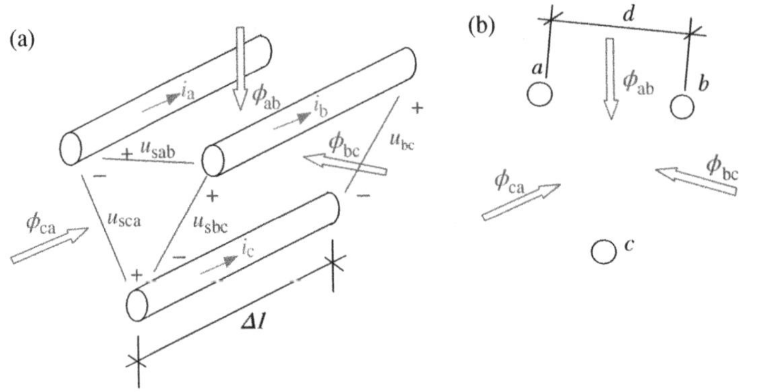

FIGURE 6.7. A three-dimensional view of the conductors constituting a transmission line (*left*) and a cross-sectional representation, to better show the mesh-coupling fluxes (*right*). A short segment Δl is shown; actual line length $l \gg d$.

6.2.1 Modelling Three-Phase Lines When Induced EMFs Between Wires Are Not Negligible

Consider again the system in Figure 6.5a. Let us now discuss what happens when the EMFs between wires are not negligible.

To understand the situation, we consider a three-phase wire of the line such as that shown in Figure 6.7a. There are fluxes linked with the three different meshes constituted by the different pairs of conductors. The following equations can be written, in general, for time-varying functions (left) and for phasors (right):

$$u_{sab} - u_{ab} = R_a i_a - R_b i_b + \frac{d\phi_{ab}}{dt}, \qquad \underline{U}_{sab} - \underline{U}_{ab} = R_a \underline{I}_a - R_b \underline{I}_b + \frac{d\underline{\Phi}_{ab}}{dt}$$

$$u_{sbc} - u_{bc} = R_b i_b - R_c i_c + \frac{d\phi_{bc}}{dt}, \qquad \underline{U}_{sbc} - \underline{U}_{bc} = R_b \underline{I}_b - R_c \underline{I}_c + \frac{d\underline{\Phi}_{bc}}{dt}$$

$$u_{sca} - u_{ca} = R_c i_c - R_a i_a + \frac{d\phi_{ca}}{dt}, \qquad \underline{U}_{sca} - \underline{U}_{ca} = R_c \underline{I}_c - R_a \underline{I}_a + \frac{d\underline{\Phi}_{ca}}{dt}$$

In this book we only consider balanced three-phase systems. They are fed by balanced sets of voltages, and loaded by balanced loads. Similarly, we assume that the three conductors have the same resistances and that the rectangular areas between conductors *a–b*, *b–c*, *c–a*, are equal to each other.[4] It will constitute a *three-phase balanced line*.

Consider now a three-phase system of voltages (phase voltages $\underline{U}_{sa}$, $\underline{U}_{sb}$, $\underline{U}_{sc}$) feeding a balanced load (equal impedances $\underline{Z}_{ld}$) through a balanced line (resistance

[4]For instance the considered area for conductors *a–* is $d \times l$ *(Δl is just a small part of l)*. For the line in Figure 6.7 to be balanced, in the cross-sectional representation the conductors must be at the vertexes of an equilateral triangle.

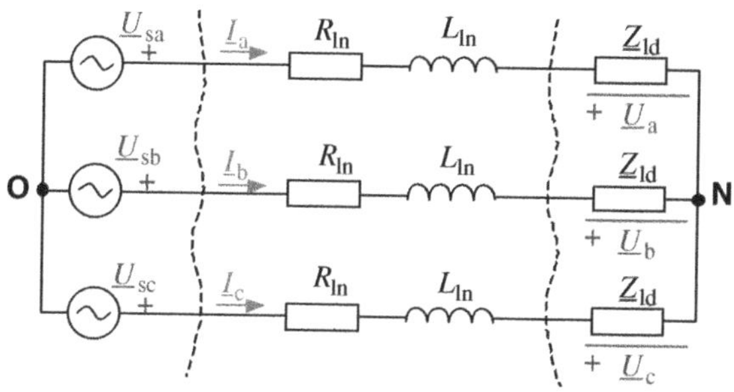

FIGURE 6.8. Metacircuit representing the system of Figure 6.5a when the EMFs between wires are not negligible ("ln" indicates "line," "ld" indicates "load").

R_{ln}). It can be demonstrated that the following equations apply:

$$\underline{U}_{sa} - \underline{U}_a = \underline{U}_{sa} - \underline{Z}_{ld}\underline{I}_a = R_{ln}\,\underline{I}_a + j\omega L_{ln}\,\underline{I}_a$$
$$\underline{U}_{sb} - \underline{U}_b = \underline{U}_{sb} - \underline{Z}_{ld}\underline{I}_b = R_{ln}\,\underline{I}_b + j\omega L_{ln}\,\underline{I}_b$$
$$\underline{U}_{sc} - \underline{U}_c = \underline{U}_{sc} - \underline{Z}_{ld}\underline{I}_c = R_{ln}\,\underline{I}_c + j\omega L_{ln}\,\underline{I}_c$$

The parameter L_{ln} is a function of line geometry and is called line inductance.

The system can therefore be studied by the *metacircuit*[5] shown in Figure 6.8.

The line inductance can be expressed as

$$L_{ln} = kl \ln \frac{d}{k_r r} \tag{6.2}$$

where l is the line length, d is the distance between two conductors of the line, r is their radius, and k and k_r are constant values. More details on the value of L_{ln} are not supplied here, because they are well beyond the scope of this book and because even electrical engineers tend to read about them in manuals rather than compute them directly. Equation (6.2) is given just to show that L_{ln} is proportional to the line length, increases when d increases, and decreases when r increases; the dependence on d and r is a logarithmic one.

As a final comment on this section, it is worthwhile noticing that in low voltage lines (up to 1 kV), inductances L_{ln} are very small and designers very often ignore them. In other terms, at the usual frequency, in low voltage lines it can often be assumed that $\omega L_{ln} \ll R_{ln}$. Only in the case of very large LV cables, having a cross section of copper larger than 95 square millimetres, L_{ln} has some importance.

On the other hand, in medium voltage lines (say between 1 and 35 kV)[6] this line inductance becomes important, and the corresponding reactance $X_{ln} = \omega L_{ln}$ usually reaches the same order of magnitude of line resistances R_{ln}.

When much higher voltage lines are considered line reactances tend to become dominant with respect to line resistances.

[5] For an explanation of the concept of metacircuit the reader is suggested to consult Section 5.3.2.

[6] The threshold of 35 kV has been set here because it is considered in the International Standard IEC 60038. Individual countries might have slightly different thresholds; for instance, in Italy this value is set to 30 kV.

6.3 THE SINGLE-PHASE EQUIVALENT OF THE THREE-PHASE CIRCUIT

As we saw in the previous chapter, in a balanced three-phase system whatever occurs in phase *a* also occurs in phase *b*, only with a phase lag of 120°; it then occurs in phase *c*, with another phase lag of 120°.

Thus, in order to understand the full behaviour of the three-phase system, it is sufficient to evaluate what happens in one of the three phases: all voltages and current phasors in the other phases will have the same amplitudes and will just be rotated counterclockwise by 120 or 240 degrees.

By convention, the phase considered is the first of the three: phase *a*, if phases are named *a*, *b*, *c*; phase *1* if they are named *1*, *2*, *3*; phase *r* if the naming convention dictates *r*, *s*, *t*. The circuit constituted by only the first phase of a three-phase system is called the single-phase equivalent of the three-phase circuit.

Just to make this approach more explicit, consider again the circuit shown in Figure 6.6. Its behaviour can be easily evaluated by referring to the single-phase equivalent shown in Figure 6.9. The process for determining the single-phase equivalent of a three-phase circuit should now be clear: since the potential of all neutral points are the same, they are connected to each other, creating the lower common node of the circuit. The upper nodes of the single equivalent represent the corresponding nodes in phase *a* of the three-phase circuit. To make things clearer, in both figures voltage $\underline{U}_{2a}$ and current $\underline{I}_{2a}$ are shown.

The same process can be used when EMFs between wires of transmission lines are not negligible. For instance, the metacircuit shown in Figure 6.8 can be studied by its single-phase equivalent, in Figure 6.10. The currents and voltages flowing in this single-phase equivalent are the ones flowing in conductor *a* of the actual three-phase system.

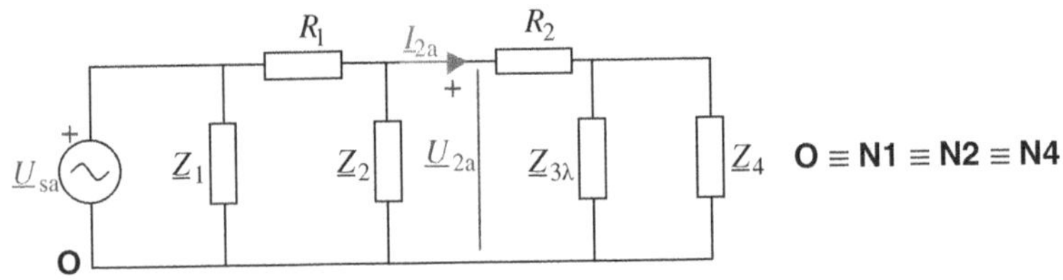

FIGURE 6.9. Single-phase equivalent of the circuit of Figure 6.6.

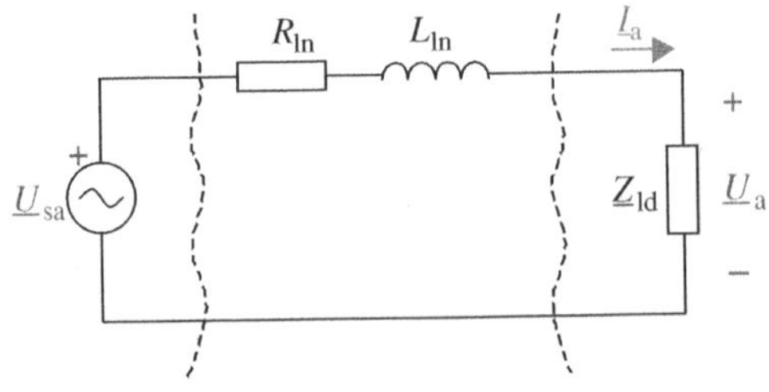

FIGURE 6.10. Single-phase equivalent of the circuit of Figure 6.8.

Example 1. Determine $\underline{I}_{1\Delta}$, $\underline{I}_1$, $\underline{I}_2$, $\underline{I}_3$, and $\underline{U}_{AB}$ in the following circuit, where

$u_{s1\Delta}(t) = \sqrt{2} \cdot 400 \cdot \sin(314t + 62°)$ V (balanced system of voltages)
$u_{s2}(t) = \sqrt{2} \cdot 230 \cdot \sin(314t + 30°)$ V (balanced system of voltages)
$\underline{Z}_1 = 1 + 2j\,\Omega$
$\underline{Z}_2 = 1.5 + 2.5j\,\Omega$
$\underline{Z}_3 = 120 + 40j\,\Omega$

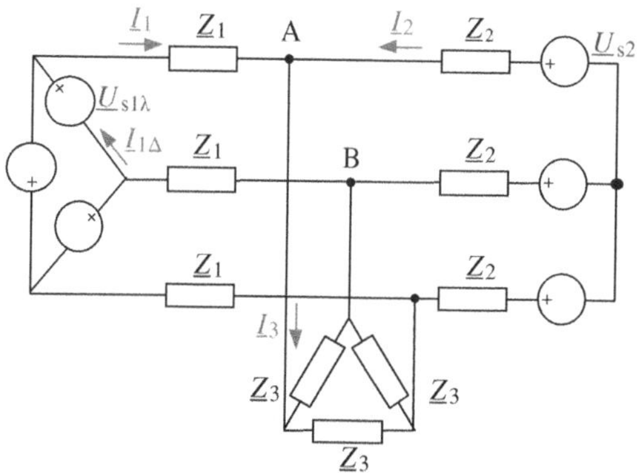

Let us first determine the phasors of sources:

$$\underline{U}_{s1} = \sqrt{2} \cdot 400 \text{ V}\angle 62°$$
$$\underline{U}_{s2} = \sqrt{2} \cdot 230 \text{ V}\angle 30°$$

Now transform all delta-connections into equivalent star-connections:

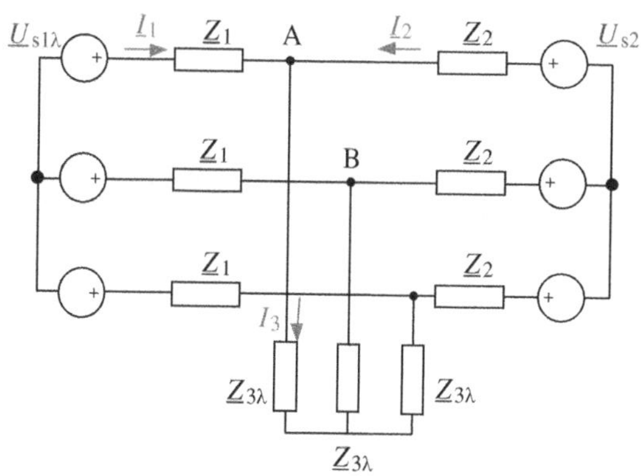

where

$$\underline{U}_{s1\lambda} = \underline{U}_{s1\Delta}/\sqrt{3} \cdot e^{-j30°} = 326.6 \text{ V}\angle 32° = 277.0 + j173.1 \text{ V}$$
$$\underline{Z}_{3\lambda} = \underline{Z}_3/3 = 40 + j13.33\,\Omega$$

Since all the neutral points have the same potential (balanced generators and loads), the following single-phase equivalent circuit can be used:

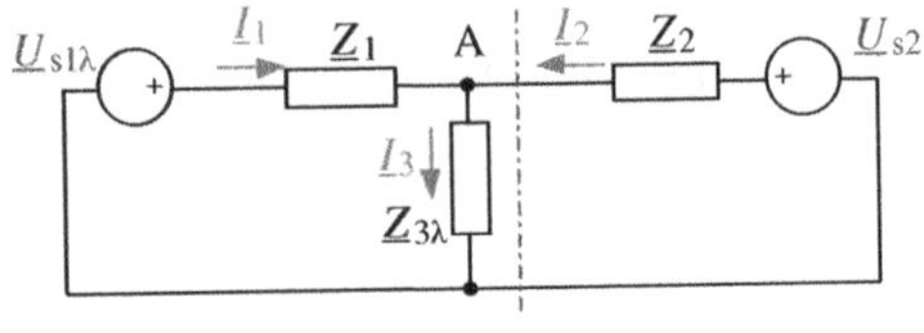

To solve this circuit, Thévenin's theorem can be applied to the side of the circuit on the left of the dot–dashed line:

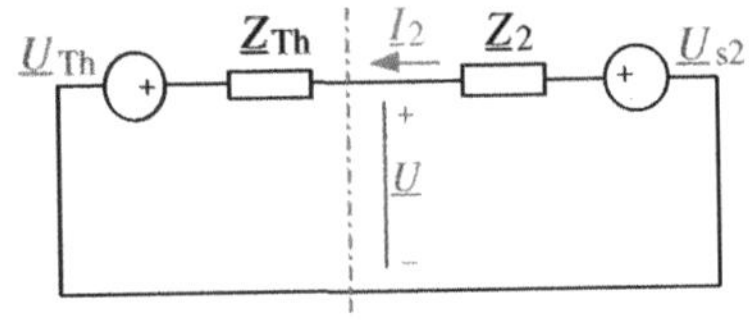

where

$$\underline{U}_{\text{Th}} = \frac{\underline{U}_{\text{s}1\lambda}\underline{Z}_3}{\underline{Z}_1 + \underline{Z}_3} = 272.6 + j157.0\ \text{V}$$

$$\underline{Z}_{\text{Th}} = \underline{Z}_1 \| \underline{Z}_3 = 1.032 + j1.890\ \Omega$$

Hence,

$$\underline{I}_2 = \frac{\underline{U}_{\text{s}2} - \underline{U}_{\text{Th}}}{\underline{Z}_{\text{Th}} + \underline{Z}_2} = 1.863 - j0.989\ \text{A}$$

$$\underline{U} = \underline{U}_{\text{s}2} - \underline{I}_2 \cdot \underline{Z}_2 = 276.43 + j159.5 = 319.1\ \text{V}\angle 29.97° = \sqrt{2} \cdot 225.6\ \text{V}\angle 29.97°$$

$$\underline{I}_3 = \underline{U}/\underline{Z}_{3\lambda} = 7.415 + j1.515\ \text{A}$$

$$\underline{I}_1 = (\underline{U}_{\text{s}1\lambda} - \underline{U})/\underline{Z}_1 = 5.553 + j2.503 = 6.091\ \text{A}\angle 24.27°$$

To calculate the phase-to-phase voltage $\underline{U}_{\text{AB}}$ and the phase current $\underline{I}_{1\Delta}$, $\underline{U}$ and $\underline{I}_1$ must be modified in amplitude and angle, in accordance with Section 6.2.

$$\underline{U}_{\text{AB}} = \underline{U} \cdot \sqrt{3} \cdot e^{j30°} = 276.5 + j478.6 = 552.7\ \text{V}\angle 59.98° = \sqrt{2} \cdot 390.8\ \text{V}\angle 59.98°$$

$$\underline{I}_{1\Delta} = (\underline{I}_1/\sqrt{3}) \cdot e^{j30°} = 2.054 + j2.855 = 3.517\ \text{A}\angle 54.27° = \sqrt{2} \cdot 2.487\ \text{A}\angle 54.27°$$

6.4 POWER IN THREE-PHASE SYSTEMS

First of all consider a generic circuital system constituted by two parts, connected to each other by means of several wires, and the corresponding circuit (Figure 6.11a).

The charges flowing through the wires are moved by the electric field present inside the conductor, whose potential values, written with reference to node **O**, are

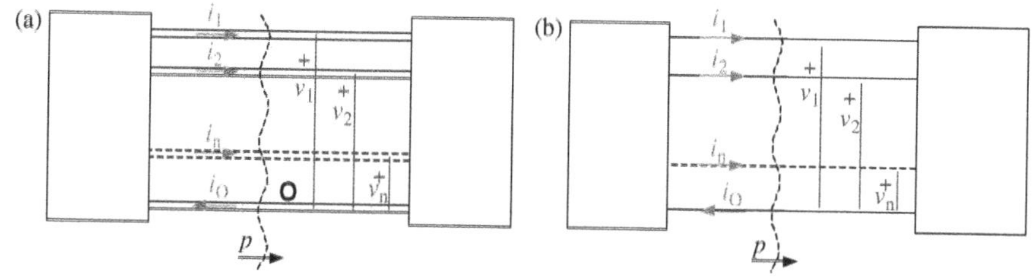

FIGURE 6.11. Power flowing between two circuit sections between multiconductor interface.

$v_1(t)$, $v_2(t)$, $v_3(t)$ (the two circuit sections are imagined to be so near to each other that the potentials of all points of the lower conductor are assumed to be equal to each other).

Therefore, the work that the electric field does inside the conductors onto the charges flowing in the conductors is

$$dW_{AB} = v_1 dq_1 + v_2 dq_2 + \cdots + v_n dq_n \Rightarrow p_{AB} = \frac{dW_{AB}}{dt} = \sum_k v_k i_k$$

With the signs assumed in Figure 6.11, the electric fields inside the conductors transfer energy into the charges assumed to be moving from the left to the right part of the system. Therefore, this energy is assumed to be generated at the left part and received (absorbed) at the right part. This justifies the reference signs assumed for $p(t)$ in the same figure.

Consider now the three-phase circuit shown in Figure 6.5 and Figure 6.12, where the three loads are generic linear boxes, and the electric quantities are indicated as time functions. Since the voltage sources are sinusoidal, also the currents are sinusoidal; since the three loads are equal (they share the same constitutive equation $u = f(i)$) and the three voltage sources constitute a three-phase balanced system, the currents also constitute a three-phase balanced system.

Similar to the process in Chapters 4 and 5, we can evaluate the power $p(t)$ transferred across the curved line of Figure 6.12.

The charges flowing through the three phases are moved by the electric field present inside the conductor, whose potential values, written with reference to node **O**, are $v_1(t)$, $v_2(t)$, $v_3(t)$.

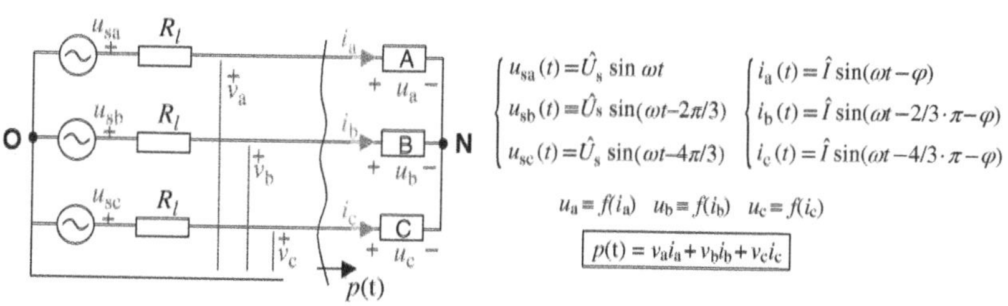

FIGURE 6.12. Circuit showing computation of instantaneous three-phase power.

Therefore, the power transferred to the load is the sum of the powers transferred though each wire and it is

$$p(t) = p_a(t) + p_b(t) + p_c(t) = v_a i_a + v_b i_b + v_c i_c = u_a i_a + u_b i_b + u_c i_c \tag{6.3}$$

the latter equality being justified by the fact that potentials of node **N** and **O** are the same.

Consider the expression for the instantaneous power entering a single linear circuit branch:

$$p = UI \cos \varphi(1 - \cos 2\omega t) + UI \sin \varphi \sin 2\omega t = UI \cos \varphi + UI\, f_\varphi(2\omega t)$$

where U and I are rms values of branch voltage and current respectively, and

$$f_\varphi(x) = \cos \varphi \cos x + \sin \varphi \sin x.$$

$f_\varphi(x)$ has the following property:

$$f_\varphi(x) + f_\varphi(x - 4/3 \cdot \pi) + f_\varphi(x + 4/3 \cdot \pi) = 0 \qquad \text{whatever } x \text{, whatever } \varphi$$

In fact:

$$f_\varphi(x \pm 4/3 \cdot \pi) = f_\varphi(x \mp 2/3 \cdot \pi)$$

because

$$\sin(x \pm 4/3 \cdot \pi) = \sin(x \mp 2/3 \cdot \pi) \quad \text{and} \quad \cos(x \pm 4/3 \cdot \pi) = \cos(x \mp 2/3 \cdot \pi)$$

and

$$\begin{aligned} f_\varphi(x) + f_\varphi(x - 4/3 \cdot \pi) + f_\varphi(x + 4/3 \cdot \pi) &= f_\varphi(x) + f_\varphi(x + 2/3 \cdot \pi) + f_\varphi(x - 2/3 \cdot \pi) \\ &= \cos \varphi \left[\cos x + \cos(x + 2/3 \cdot \pi) + \cos(x - 2/3 \cdot \pi)\right] \\ &\quad + \sin \varphi \left[\sin x + \sin(x + 2/3 \cdot \pi) + \sin(x - 2/3 \cdot \pi)\right] = 0 \end{aligned}$$

(the latter equality because the sum of three sinusoidal functions displaced from each other $2\pi/3$ is always zero).

Therefore it is

$$p(t) = 3UI \cos \varphi \tag{6.4}$$

In fact:

$$\begin{aligned} p(t) &= 3UI \cos \varphi + UI\{f_\varphi(2\omega t) + f_\varphi[2(\omega t - 2/3 \cdot \pi)] + f_\varphi[2(\omega t + 2/3 \cdot \pi)]\} \\ &= 3UI \cos \varphi + UI\{f_\varphi(2\omega t) + f_\varphi[2\omega t - 4/3 \cdot \pi] + f_\varphi[2\omega t + 4/3 \cdot \pi]\} = 3UI \cos \varphi \end{aligned}$$

Equation (6.4) is a very important result:

> **Result:** *Instantaneous power in three-phase balanced systems*
>
> The instantaneous power flowing in three-phase balanced systems is constant over time and is expressed by the simple formula (6.4).

This is different from what happens in single-phase circuits and is therefore another big advantage of three-phase over single-phase circuits.

It will be seen in the next chapters that this implies a very uniform transfer of power to mechanical shafts when three-phase machines are involved, since when the machine shafts rotate at constant speed, they receive constant torque, therefore with no vibration induced.

As far as active, reactive and apparent powers are concerned, their values can be obtained trivially considering that a three-phase phasor circuit is still an AC circuit:

$$P_{3\text{ph}} = 3UI\cos\varphi,\quad Q_{3\text{ph}} = 3UI\sin\varphi,\quad \underline{S}_{3\text{ph}} = (\underline{U}_{a}\underline{I}_{a}^{*} + \underline{U}_{b}\underline{I}_{b}^{*} + \underline{U}_{c}\underline{I}_{c}^{*})/2,\quad S_{3\text{ph}} = 3UI \tag{6.5}$$

It is again remembered that in (6.5), U is the rms value of the line-to-neutral voltages. The power transferred though a three-phase circuit is often expressed in terms of the line-to-line voltage U_l. From the fundamental relation $U_l = \sqrt{3}U$, we immediately obtain

$$P_{3\text{ph}} = \sqrt{3}U_l I\cos\varphi,\quad Q_{3\text{ph}} = \sqrt{3}U_l I\sin\varphi,\quad S_{3\text{ph}} = \sqrt{3}U_l I$$

Example 2. Consider the circuit of Example 1 and determine the active and reactive power:

- delivered by the two three-phase generators;
- absorbed by the two lines;
- absorbed by the load.

Assuming the frequency to be 50 Hz, determine the capacity of the capacitors (firstly star-connected and then delta-connected) required to correct the load power factor to 1.

Remembering that the three-phase powers can be obtained by multiplying by 3 the powers calculated in the single-phase circuit, we obtain

$$\begin{aligned}
\underline{S}_{\text{gen1}} &= 3 \cdot \underline{U}_{\text{s1}\lambda} \cdot \underline{I}_1^*/2 = 2957\ \text{W} + j401.5\ \text{var}\\
\underline{S}_{\text{gen2}} &= 3 \cdot \underline{U}_{\text{s2}} \cdot \underline{I}_2^*/2 = 545.8\ \text{W} + j872.2\ \text{var}\\
\underline{S}_{\text{gen3}} &= 3 \cdot I_1^2 \cdot \underline{Z}_1 = 55.65\ \text{W} + j111.3\ \text{var}\\
\underline{S}_{\text{gen4}} &= 3 \cdot I_2^2 \cdot \underline{Z}_2 = 10.00\ \text{W} + j16.68\ \text{var}\\
\underline{S}_{\text{load}} &= 3 \cdot I_3^2 \cdot \underline{Z}_{3\lambda} = 3437\ \text{W} + j1145\ \text{var}
\end{aligned}$$

where $I_1 = |\underline{I}_1|/\sqrt{2}$ (phasors were calculated considering the peak value of quantities), and so on, for the other phases. Please note that this accords with Tellegen/Boucherot's theorem:

$$\underline{S}_{\text{gen1}} + \underline{S}_{\text{gen2}} = \underline{S}_{\text{line1}} + \underline{S}_{\text{line2}} + \underline{S}_{\text{load}}$$

The power factor correction can be carried out on the single-phase circuit, directly obtaining the capacity of each star-connected capacitor:

$$C_\lambda = \frac{L_{3\lambda}}{R_{3\lambda}^2 + X_{3\lambda}^2} = \frac{13.33/(2\pi 50)}{40^2 + 13.33^2} = 23.88\ \mu\text{F}$$

Also,

$$C_\lambda \frac{Q_{\text{single-phase } Z_{3\lambda}}}{\omega \cdot U^2} = \frac{1145/3}{2\pi 50 \cdot 225.6^2} = 23.88\ \mu\text{F}$$

The capacitance of the three delta-connected capacitors can be easily calculated considering that in a balanced load we have

$$\underline{Z}_\Delta = 3\underline{Z}_\lambda, \quad \text{thus } C_\Delta = C_\lambda/3 = 7.962\ \mu\text{F}$$

6.5 SINGLE-PHASE FEEDING FROM THREE-PHASE SYSTEMS

In Part IV of this book, there will be a description of how practical power systems are designed and built. The analysis of these systems makes use of circuit knowledge acquired in this part.

It is worthwhile, at this point, to examine the general usage of three-phase and single-phase AC systems.

It was already mentioned at the beginning of this chapter that large power systems are three-phase, and the rationale behind this was also discussed. However, inside houses and homes it is much more common to find electricity distributed through just two wires—that is, single-phase AC. This is a natural consequence of the fact that the lower powers of low-voltage installations do not justify the adoption of the more complex three-phase structure for the installation themselves.

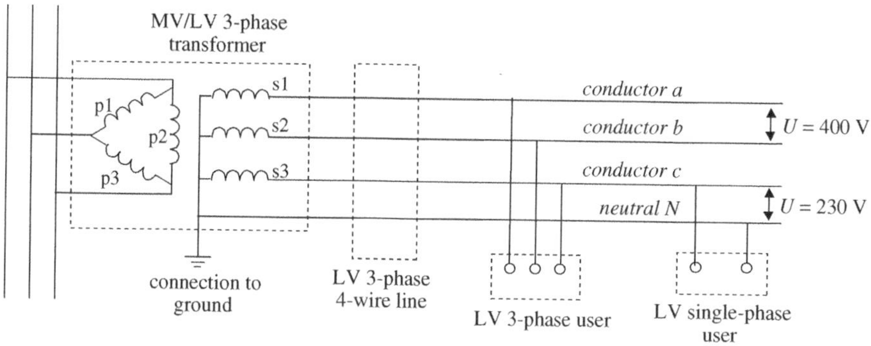

FIGURE 6.13. Connection between a three-phase MV line and LV users using the three-phase four-wire LV distribution system.

How do we generate single-phase systems from three-phase systems? It must be remembered here that the domestic installations are supplied by low-voltage systems. The low-voltage supply comes from MV/LV transformers.

> A very basic discussion of what transformers are has been made in Section 1.1; a more detailed idea of transformers was introduced in Chapter 5, where ideal transformers were dealt with. More comprehensive analysis of this subject is in Chapter 7.
>
> Finally, more information about the structure of power systems and why MV/LV transformers are needed is discussed in Chapter 15.

It is often in this transformer that the generation of single-phase systems from a three-phase system occurs.

This is explained in Figures 6.13 and 6.14 in the two most widely used cases. Here, just for graphical simplicity, wires are indicated by unique lines, but these figures represent physical systems, not circuits. So, the line named "conductor a" is a real conductor, with its own conductor material and insulation, resistance, and so on.

The three-phase transformer is constituted by mutually coupled windings: for instance, the secondary winding s1 is mutually coupled with p1, s2 with p2, and s3 with p3.

As a consequence, the voltage at the s1 terminals is proportional to the voltage at p1 terminals, and the same applies to s2 and s3, the proportionality coefficients being the same [in ideal transformers they are exactly equal to the turn ratios (see Chapter 5), and in real transformers they are near to them (Chapter 7)].

Equation (6.1) indicates a fundamental mathematical relation between line and phase voltages. In Europe, the standard LV levels for these two voltages are 400 and 230 V.[7] Finally, note that the neutral point of the LV transformer is connected to ground, so that its potential is maintained in all situations very near to the ground potential. Since the

[7]The ratio in not exactly $\sqrt{3}$ but is much closer to this value; the approximation introduced by using these round numbers instead of more precise numbers with some decimal digits is 0.4%, much lower the usual tolerance on LV voltage levels, which is around 10%.

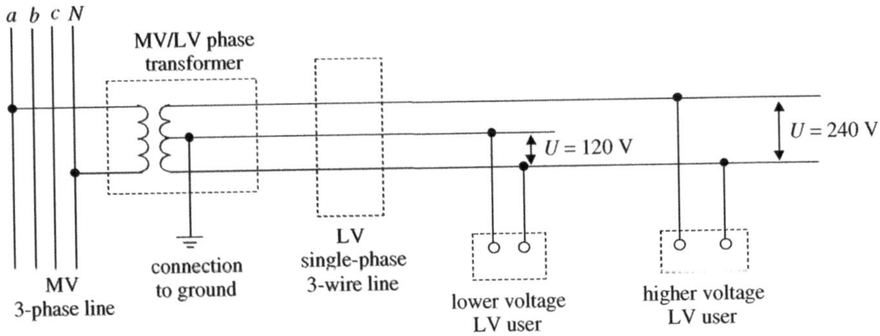

FIGURE 6.14. Connection between a three-phase MV line and LV users using the single-phase three-wire LV distribution system.

ground is where human beings place their feet, having this connection to ground is valuable for safety. This topic will be dealt with in detail in Chapter 15.[8]

Figure 6.13 shows the four-wire distribution system, which allows true three-phase users to be fed from the three conductors *a–b–c*, at a line-to-line voltage level of 400 V, while single-phase users are fed through neutral and one phase conductor. Naturally, since an important target is to achieve balanced operation of the three-phase systems, the electricity distributor allocates individual LV users along the three conductors, in such a way that their combined effect is a three-phase system which is as balanced as possible. For instance, since the single-phase user of Figure 6.14 draws its current from conductor *c*, additional users who must be served will be connected to conductors *b* and *c*, also taking into account the amount of current each of them draws.

The four-wire distribution system is usually used in Europe.

Figure 6.14 shows a single-phase three-wire distribution system that allows single-phase users to be fed from the three conductors of the LV side of a distribution transformer.

As for the scheme of Figure 6.14, to pursue the target of drawing balanced currents from the three-phase network, the connection of different transformers is allocated by rotating between conductors. For instance, since the single-phase MV/LV transformer of Figure 6.14 draws its current from conductor *a*, additional users will be served through transformers connected to conductors *b* and *c*, also taking into account the amount of current each of them draws. If the three currents are equal to each other (and their phase displacement with respect to the corresponding voltage is equal to each other), they will constitute a balanced set of currents and therefore the current globally flowing through the MV neutral conductor will be null. In real life situations it will be very small in comparison with the currents flowing through the phase conductors.

Also in this case, the connection to ground guarantees that in all the systems fed by the transformer involved, all potential differences with reference to the ground will be limited, which is useful for safety.

[8]There are some cases, however, in which this neutral is not connected to ground. This occurs in the so-called IT systems that will be discussed in Chapter 14.

The single-phase three-wire distribution system is usually used in North and Central America.

6.6 HISTORICAL NOTES

6.6.1 Short Biography of Tesla

Nicolas Tesla (Croatia, 1865–New York, USA, 1943) was a scientist and engineer who gave an outstanding contribution to the development and diffusion of electrification in the United States in the early years of the twentieth century.

Born in the Austro-Hungarian Empire, he moved to the United States in 1884. There, he participated in the so-called "war of the currents" between those promoting DC current for commercial electricity distribution, led by Edison, and those promoting AC, led by Westinghouse.

One of the big advantages of AC is the existence of polyphase systems, especially three-phase systems, developed by Tesla, whose patents were bought by Westinghouse.

Tesla first worked under Edison but then, because he felt underrated, moved to Westinghouse, where he was able to deploy all his ideas and knowledge on AC and polyphase AC systems.

Tesla is also credited with discovering the *rotating magnetic field* (to be introduced and discussed in the next chapter), independently of Galileo Ferraris, who made the same discovery in the same period.

The S.I. unit of magnetic flux density is the *tesla* in his honour.

6.7 PROPOSED EXERCISES

Note: To solve the following problems, remember that phasor amplitudes represent the amplitudes of the corresponding sine waves (peak values).

6.1. Determine $\underline{I}_a$, $\underline{I}_b$, and $\underline{I}_c$ in the following circuit, where

$$\underline{U}_{sa} = \sqrt{2} \cdot 100\ \text{V}\angle 20^\circ \qquad \text{(balanced set)}$$
$$\underline{Z}_l = 2 + 3j\ \Omega$$
$$\underline{Z} = 30 + 10j\ \Omega$$

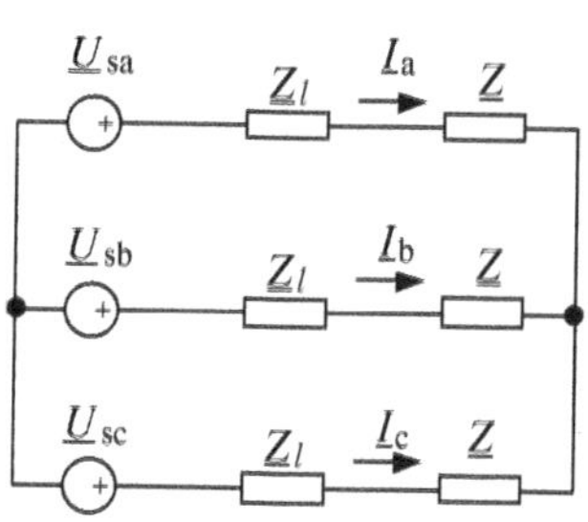

6.2. Determine $\underline{I}_a$, $\underline{I}_b$, $\underline{I}_c$, and $\underline{I}_{ab}$ in the following circuit, where

$$\underline{U}_{sab} = \sqrt{2} \cdot 400\ \mathrm{V} \angle 30° \qquad \text{(balanced set)}$$

$$\underline{Z}_l = 5 + 10j\ \Omega$$

$$\underline{Z} = 18 + 6j\ \Omega$$

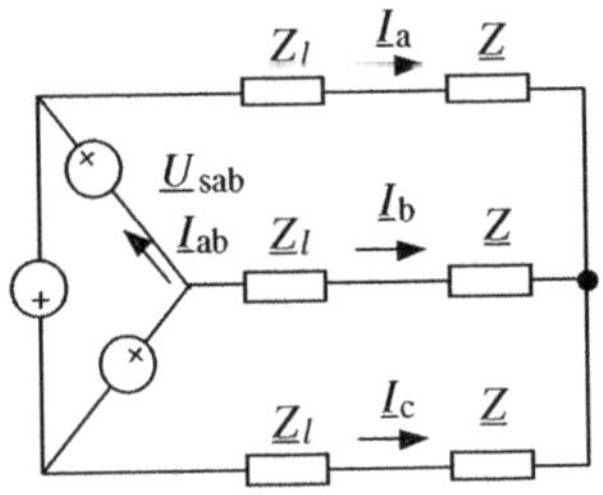

6.3. Determine $\underline{I}_a$, $\underline{I}_b$, $\underline{I}_c$, and $\underline{I}_{ab}$ in the following circuit, where

$$\underline{U}_{sab} = \sqrt{2} \cdot 400\ \mathrm{V} \angle 0° \qquad \text{(balanced set)}$$

$$\underline{Z}_l = 2 + 4j\ \Omega$$

$$\underline{Z} = 12 + 3j\ \Omega$$

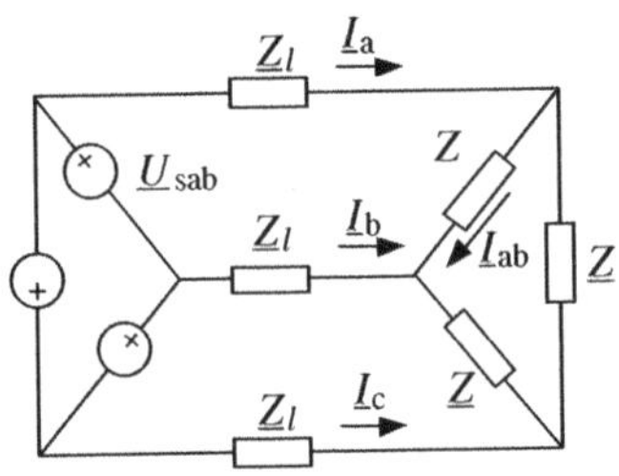

Hint: first convert into star the voltage sources and the loads, and compute $\underline{I}_a$, $\underline{I}_b$, $\underline{I}_c$. To compute $\underline{I}_{ab}$, and so on, consider that, as a consequence of the circuit's symmetry, $\underline{I}_{ab}$, $\underline{I}_{bc}$, $\underline{I}_{ca}$ are a balanced set of currents and that $\underline{I}_a = \underline{I}_{ab} - \underline{I}_{ca}$: this allows determination of relations between amplitude and phase of $\underline{I}_{ab}$ and $\underline{I}_{ca}$, relative to $\underline{I}_a$.

6.4. Determine $\underline{I}_a$, $\underline{I}_1$, $\underline{I}_2$, and $\underline{U}_{AB}$ in the following circuit, where

$$\underline{U}_{sa} = \sqrt{2} \cdot 230\ \mathrm{V} \angle 15° \qquad \text{(balanced set)}$$

$$\underline{Z} = 1 + 3j\ \Omega$$

$$\underline{Z}_1 = 18 + 12j\ \Omega$$

$$\underline{Z}_2 = 5 + j\ \Omega$$

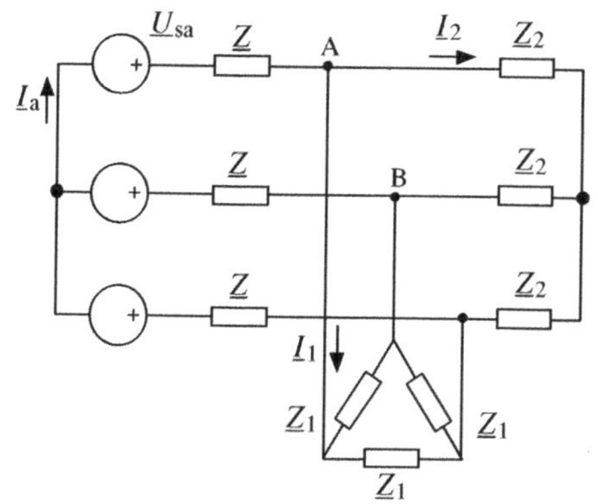

6.5. Determine $\underline{I}_1$, $\underline{I}_2$, $\underline{I}_3$, and $\underline{U}_{AB}$ in the following circuit, where

$$\underline{U}_{s1} = \sqrt{2} \cdot 230\ \text{V}\angle 0° \qquad \text{(balanced set)}$$

$$\underline{U}_{s2} = \sqrt{2} \cdot 230\ \text{V}\angle 30° \qquad \text{(balanced set)}$$

$$\underline{Z}_1 = 2 + 1j\,\Omega$$

$$\underline{Z}_2 = 2 + 3j\,\Omega$$

$$\underline{Z}_3 = 15 + 9j\,\Omega$$

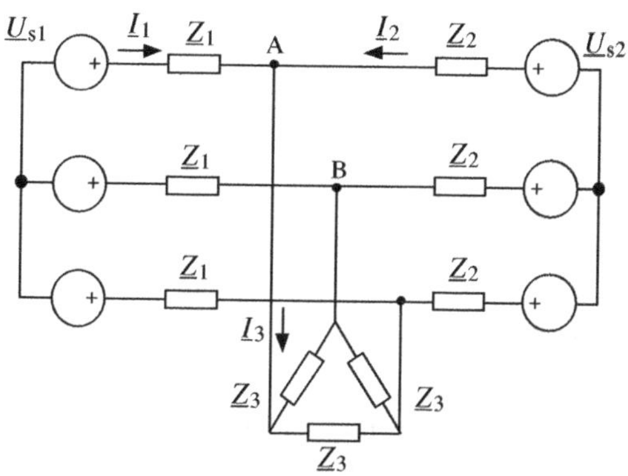

6.6. Using the method of nodal analysis, determine $\underline{I}_a$, $\underline{I}_b$, and $\underline{I}_c$ in the following unbalanced system, where

$$\underline{U}_{sa} = \sqrt{2} \cdot 220\ \text{V}\angle 0°$$

$$\underline{U}_{sb} = \sqrt{2} \cdot 210\ \text{V}\angle - 120°$$

$$\underline{U}_{sc} = \sqrt{2} \cdot 215\ \text{V}\angle 115°$$

$$\underline{Z} = 1 + 3j\,\Omega, \quad \underline{Z}_a = 20\,\Omega, \quad \underline{Z}_b = 5 + 25j\,\Omega, \quad \underline{Z}_c = 30 - 8j\,\Omega$$

Then calculate the current $\underline{I}_{NO}$ flowing in a resistor $R = 10\,\Omega$, positioned between O and N (use Thévenin's theorem).

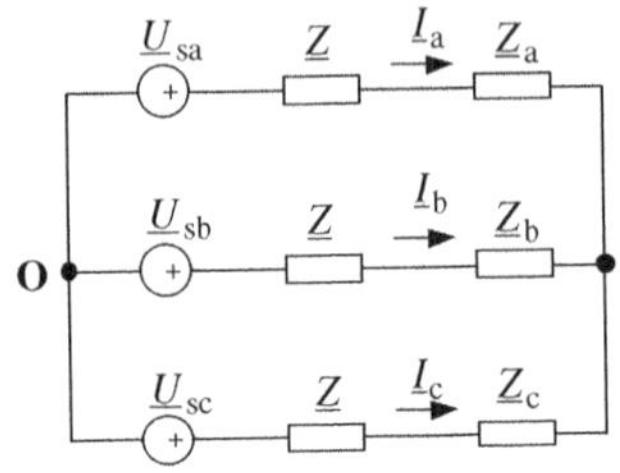

6.7. Using the method of nodal analysis, determine $\underline{I}_a$, $\underline{I}_b$, $\underline{I}_c$, and $\underline{I}'$ in the following unbalanced system, where

$$\underline{U}_{sa} = \sqrt{2} \cdot 127 \text{ V}\angle 30°$$

$$\underline{U}_{sb} = \sqrt{2} \cdot 120 \text{ V}\angle -90°$$

$$\underline{U}_{sc} = \sqrt{2} \cdot 130 \text{ V}\angle 145°$$

$$\underline{Z} = 1 + j\,\Omega, \quad \underline{Z}_a = 10 + 5j\,\Omega, \quad \underline{Z}_b = 8j\,\Omega, \quad \underline{Z}_c = 12j\,\Omega,$$

$$\underline{Z}' = 15j\,\Omega$$

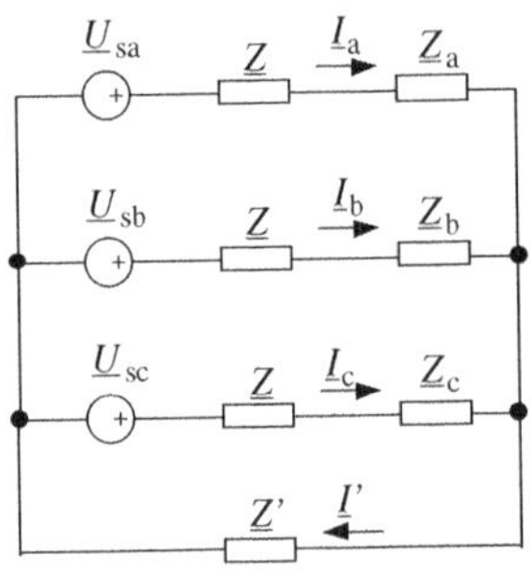

6.8. Consider exercise 6.1. Calculate the active and reactive power absorbed by the three-phase load and by the line; determine the active and reactive power delivered by the generating system and verify Boucherot's theorem.

6.9. Consider exercise 6.2. Calculate the active and reactive power absorbed by the three-phase load and by the line; determine the active and reactive power delivered by the generating system and verify Boucherot's theorem.

6.10. Consider exercise 6.3. Calculate the active and reactive power absorbed by the three-phase load and by the line; determine the active and reactive power delivered by the generating system and verify Boucherot's theorem.

6.11. Calculate the active and reactive power delivered by the two three-phase generators of exercise 6.5 and the active power losses of the two lines.

6.12. Verify Boucherot's theorem in exercise 6.7.

6.13. Calculate the power factor of the load of exercise 6.4 ($\underline{Z}_1$ and $\underline{Z}_2$). Determine the capacity of the three star-connected capacitors required to correct the power factor to 1. Repeat if capacitors are delta-connected.

PART III

ELECTRIC MACHINES AND STATIC CONVERTERS

Fundamentals of Electric Power Engineering: From Electromagnetics to Power Systems, First Edition.
Massimo Ceraolo and Davide Poli.

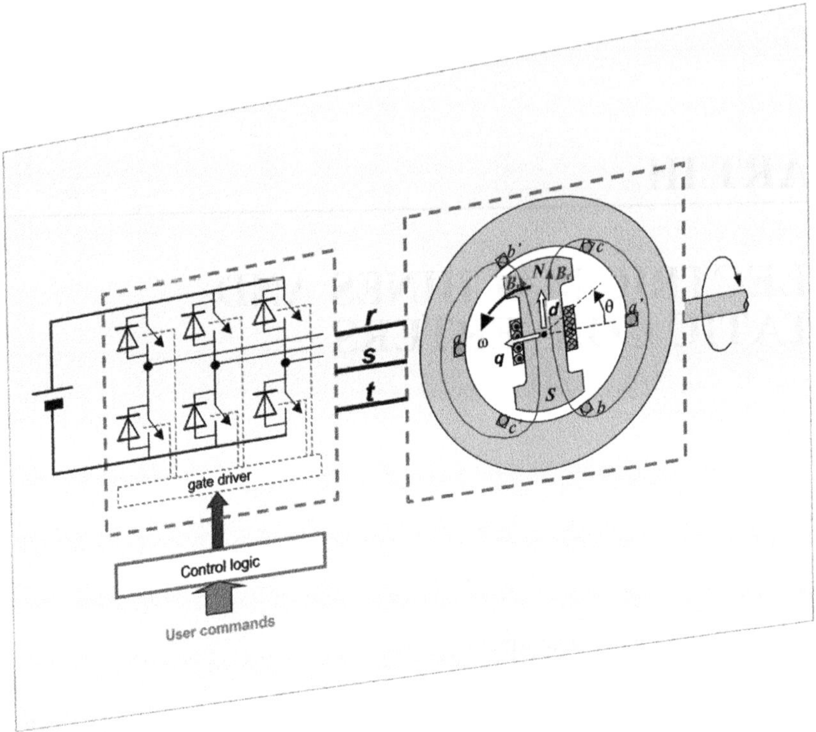
r
s
t
gate driver
Control logic
User commands
N
d
q
ω
θ
S
a
a'
b
b'
c
c'

7

MAGNETIC CIRCUITS AND TRANSFORMERS

For the Instructor

This short chapter contains basic concepts that can be used for all rotating machines and for power transformers. This, along with Chapter 9, avoids the repetition of common topics in chapters relating specifically to electric machines (Chapters 10, 11 and 12).

7.1 INTRODUCTION

In Chapter 5 we introduced the ideal transformer. Here we present a more complete analysis and more detailed circuit models of real-life transformers.

7.2 MAGNETIC CIRCUITS AND SINGLE-PHASE TRANSFORMERS

Consider the basic correspondence between magnetic flux density $\boldsymbol{B}$ and magnetic field $\boldsymbol{H}$, as known from the study of electromagnetism:

$$\boldsymbol{B} = \mu \boldsymbol{H} \tag{7.1}$$

Fundamentals of Electric Power Engineering: From Electromagnetics to Power Systems, First Edition.
Massimo Ceraolo and Davide Poli.

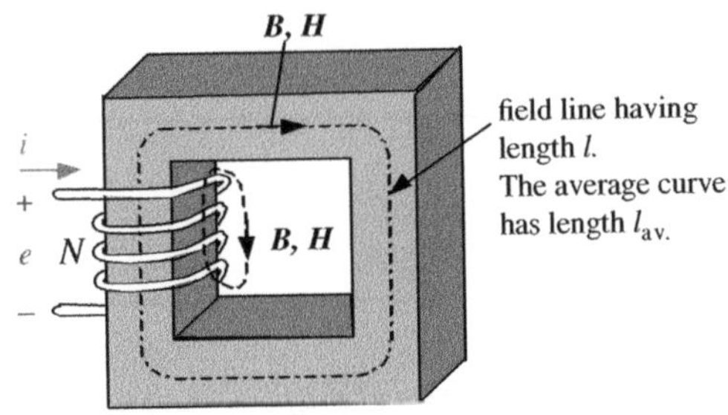

FIGURE 7.1. The structure of a simple magnetic circuit with a coil.

where μ is the permeability of the material at the point in which $\boldsymbol{H}$ and $\boldsymbol{B}$ are evaluated.

In a subsequent section we will briefly examine magnetic hysteresis; here, however, we can disregard it, thus assuming pure proportionality between the two fields indicated by equation (7.1).

A *magnetic circuit* is composed of loops of ferromagnetic material containing magnetic flux, which is conveyed by means of coils made of conducting material, wound around some parts of the loops. The simplest and most important magnetic circuit has a single loop, usually with a rectangular cross section, like that illustrated in Figure 7.1. More complex shapes are used in some three-phase transformers which, however, are not dealt with in this book.

Consider what happens in the system of Figure 7.1 when a current i flows through the coil, which consists of N turns wound along the magnetic core. As a result of this current, a magnetic field is produced in all points of space. However, it can be demonstrated that, as a consequence of the much higher permeability of iron compared to air, most of the magnetic flux flows through iron; the flux that traverses the coil but which passes, at least partially, through air is called *leakage flux* (this is also discussed in Section 5.2.3).

For simplicity's sake, let us ignore the leakage flux. Under this hypothesis, the magnetic field flows only inside iron. The magnetic flux flowing through a cross section of the iron having a cross-sectional area A is

$$\phi = \iint_A \boldsymbol{B} \bullet \mathrm{d}\boldsymbol{A} = B_{av} A$$

in which B_{av} is the average value of $\boldsymbol{B}$ over the whole cross-sectional area A, and d$\boldsymbol{A}$ is the versor of the infinitesimal area element dA.

Since the magnetic field is supposed to flow only inside iron, ϕ is constant throughout the magnetic circuit. Moreover, because the area A is constant along the circuit (without considering the effects of singularities near the circuit corners), B_{av} also has the same value all around the circuit.

Moreover, it is

$$\phi = B_{av}A = \mu H_{av}A$$

The magnetic field H_{av} is related to the current flowing in the coil around the magnetic circuit by Ampère's law (the path l of the integral may be any path inside iron, not just l_{av}):

$$\phi = \oint_{l} \boldsymbol{H} \bullet \mathrm{d}\boldsymbol{l} = H_{av}l_{av} = Ni$$

Finally it is

$$\phi = \iint_{A} \boldsymbol{B} \bullet \mathrm{d}\boldsymbol{A} = \mu NiA/l_{av} = Ni/\mathcal{R} \tag{7.2}$$

where the quantity

$$\mathcal{R} = l_{av}/\mu A$$

is a characteristic of the magnetic circuit and is called circuit *reluctance*. The corresponding linked flux is

$$\psi = N\phi = N^2 i/\mathcal{R} = Li$$

in which

$$L = N^2/\mathcal{R} \tag{7.3}$$

is the circuit self-inductance, that is, the proportionality coefficient between current and linked flux, as introduced in Chapter 5 in our study of inductors.

> **Result:** *Simplified behaviour of a one-coil magnetic circuit*
>
> In a magnetic circuit having a constant cross section, current and linked flux are proportional, the proportionality coefficient being the inductance L, whose expression is (7.3).

The computation performed here is practically identical to those made in the Appendix, when computing the inductor constitutive equation.

Consider now a magnetic circuit with two different coils, as shown in Figure 7.2. Now both currents, i_1 flowing in coil 1 and i_2 flowing in coil 2, contribute to

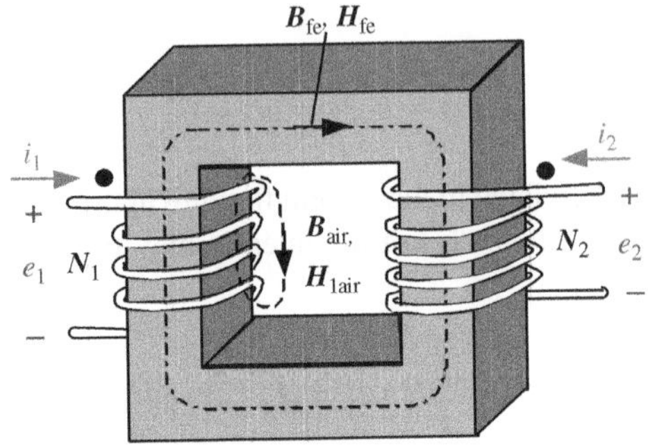

FIGURE 7.2. Basic construction of a two-winding transformer.

generating the magnetic field $\boldsymbol{H}$. Again, because $\mu_{fe} \gg \mu_{air,}$ the force lines of magnetic field $\boldsymbol{H}$ (and therefore $\boldsymbol{B}$) are much denser inside the iron core than in air.

As a result of these currents, a magnetic field is produced in all points of space. However, as before, because iron has a much higher permeability than air, most of the magnetic flux is through iron; again the fluxes traversing the coils but flowing at least partially in the air (field lines containing $\boldsymbol{B}_{air}$, $\boldsymbol{H}_{air}$ in Figure 7.2) are called leakage fluxes.

Let us for a moment disregard the leakage fluxes. In this case the total flux ϕ traversing coils 1 and 2 is the same and can be computed using Ampère's law as follows:

$$\oint \boldsymbol{H} \bullet \mathrm{d}\boldsymbol{l} = H_{av} l_{av} = N_1 i_1 + N_2 i_2$$

Therefore,

$$\phi = \mu A H_{av} = \mu A (N_1 i_1 + N_2 i_2)/l_{av} = (N_1 i_1 + N_2 i_2)/\mathcal{R}$$

or

$$N_1 i_1 + N_2 i_2 = \mathcal{R}\phi \qquad (7.4)$$

The quantity $\mathcal{R}$ is the reluctance of the magnetic circuit as before.

If, instead, we take the leakage fluxes into account, Faraday's law as applied to the two coils is

$$\begin{aligned} e_{1tot}(t) &= \frac{\mathrm{d}\psi_1}{\mathrm{d}t} = \frac{\mathrm{d}(\psi_{l1} + \psi)}{\mathrm{d}t} = L_{l1}\frac{\mathrm{d}i_1}{\mathrm{d}t} + N_1\frac{\mathrm{d}\phi}{\mathrm{d}t} \\ e_{2tot}(t) &= \frac{\mathrm{d}\psi_2}{\mathrm{d}t} = \frac{\mathrm{d}(\psi_{l2} + \psi)}{\mathrm{d}t} = L_{l2}\frac{\mathrm{d}i_2}{\mathrm{d}t} + N_2\frac{\mathrm{d}\phi}{\mathrm{d}t} \end{aligned} \qquad (7.5)$$

where ψ_1 is the flux linked to coil 1, equal to N_1 times the iron flux ϕ plus ϕ_{l1}—that is, primary leakage flux. In a similar manner, ϕ_2 is also composed of the iron flux ϕ and its leakage flux ϕ_{l2}, and ψ_2 equals N_2 times ϕ_2.

The leakage fluxes are also proportional to the corresponding coil currents, because all points of space surrounding the iron are supposed to be linear. This is written in equation (7.5). Using the same terminology as equation (7.4), it can, in fact, be written as follows:

$$\phi_{l1} = N_1 i_1/\mathcal{R}_{l1}$$

$$\phi_{l2} = N_2 i_2/\mathcal{R}_{l2}$$

and

$$\psi_{l1} = L_{l1} i_1$$

$$\psi_{l2} = L_{l2} i_2$$

where $L_{l1} = N_1^{\,2}/\mathcal{R}_{l1}$ and $L_{l2} = N_2^{\,2}/\mathcal{R}_{l2}$ can be interpreted as equivalent inductances of the leakage path through air.

Equation (7.4) describes the basic behaviour of transformers, which, however, needs some correction when leakage inductances are taken into account. Equations (7.5) complete the description of the transformer.

These two equations allow a more detailed description of transformers than that offered by the ideal transformer discussed in Chapter 5. If we ignore the term on the right-hand side in equation (7.4) and leakage fluxes in (7.5), we immediately obtain the following equations of ideal transformers:

$$e_{1\mathrm{tot}}(t) \cong e_1(t) = \frac{N_1}{N_2} e_2(t) \cong e_{2\mathrm{tot}}(t), \qquad i_2(t) \cong \frac{N_1}{N_2} i_1(t)$$

Remember that e_1 and e_2 are in phase, since they indicate voltages between the dot-marked terminals and the others (unmarked).

A complete analysis of transformer behaviour is beyond the scope of this section. Here we present only the final result, under the hypothesis that magnetic linearity, stated by equation (7.1) with constant μ.

When the transformer operates in sinusoidal steady state, the corresponding phasor circuit can be used to analyse it. A fairly complete version of this circuit, which takes into account the leakage circuits, the coil resistances, and the term $\mathcal{R}\phi$ in equation (7.4), is shown in Figure 7.3; this circuit is the classic equivalent of a single-phase transformer, used (with minor variations) in all textbooks and also when a fairly high level of detail in the transformer modelling is required. Only one component is still missing: a resistor simulating iron losses, which will be introduced in the final section of this chapter.

The circuit shown in Figure 7.3 contains the following elements:

- an ideal transformer between $\underline{E}_1$ and $\underline{E}_2$;
- resistors R_1 and R_2 corresponding to the ohmic resistance of primary and secondary winding;

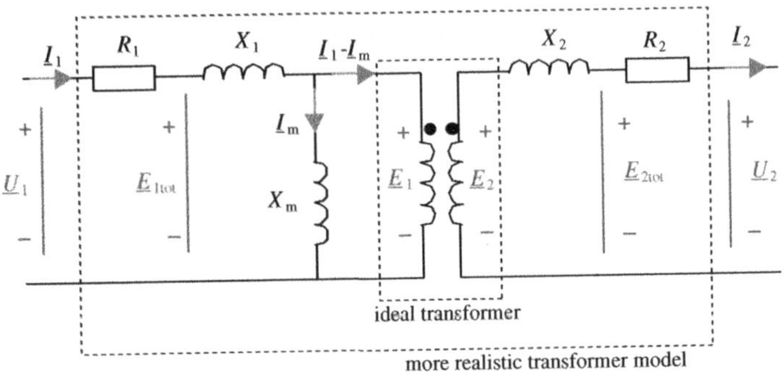

FIGURE 7.3. Ideal and more realistic transformer models.

- reactances $X_1 = \omega L_{l1}$ and $X_2 = \omega L_{l2}$ which model the effects of the leakage inductances;
- reactance $X_m = \omega L_m$ which absorbs a fictitious current $\underline{I}_m$ defined from (7.4) as follows:

$$N_1 \underline{I}_1 + N_2 \underline{I}_2 = \mathscr{R}\underline{\Phi} = N_1 \underline{I}_m$$

This circuit can be used like any phasor circuit proposed in the previous chapters.

Whenever possible, the transformer model can be made simpler than the one in Figure 7.3. For instance, often the inductance X_m is removed from the circuit, since the current it absorbs, except when the secondary is disconnected from any load, is very small with respect to the other currents $\underline{I}_1$ and $\underline{I}_2$.

When this is done, the equations of the circuit of Figure 7.3 become

$$\underline{U}_1 = R_1 \underline{I}_1 + jX_1 \underline{I}_1 + \underline{E}_1, \qquad \underline{E}_1 = \frac{d\underline{\Psi}_1}{dt} = N_1 \frac{d\underline{\Phi}}{dt}$$

$$\underline{U}_2 = \underline{E}_2 - R_2 \underline{I}_2 - jX_2 \underline{I}_2, \qquad \underline{E}_2 = \frac{d\underline{\Psi}_2}{dt} = N_2 \frac{d\underline{\Phi}}{dt}$$

As usual, here uppercase underlined quantities indicate phasors corresponding to sinusoidal quantities. Compare with equations (7.5).

If we further ignore all the leakage fluxes, the equations of the circuit of Figure 7.3 become

$$\begin{aligned} \underline{U}_1 &= R_1 \underline{I}_1 + \underline{E}_1, \qquad \underline{E}_1 = \frac{d\underline{\Psi}_1}{dt} = N_1 \frac{d\underline{\Phi}}{dt} \\ \underline{U}_2 &= \underline{E}_2 - R_2 \underline{I}_2, \qquad \underline{E}_2 = \frac{d\underline{\Psi}_2}{dt} = N_2 \frac{d\underline{\Phi}}{dt} \end{aligned} \qquad (°)$$

Equations (°), except for the phasor notation, are the equations of mutually coupled inductors introduced in Chapter 5, that is, equations (5.9).

The presence of the ideal transformer in the circuit introduces an additional difficulty into the analysis. This is, therefore, often "eliminated." The process can be

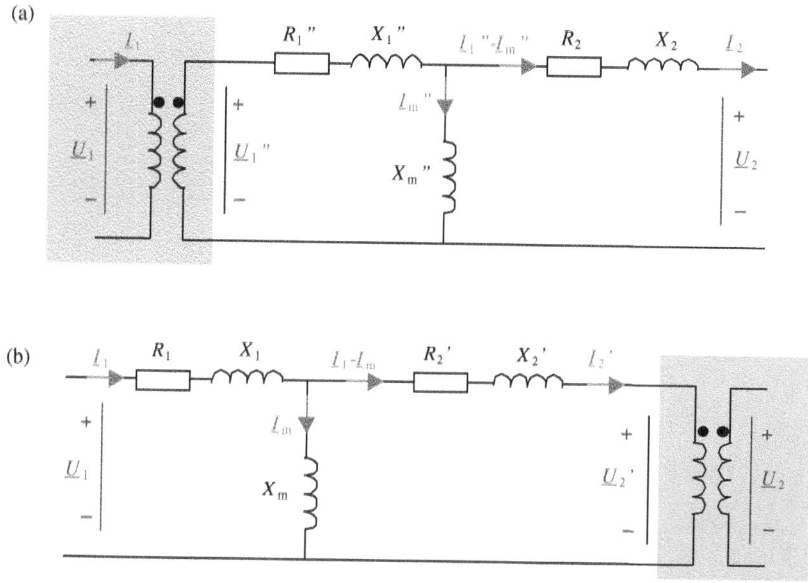

FIGURE 7.4. Models of transformer of Figure 7.3 with impedances referred to secondary (a) and to primary (b).

described with reference to the drawings of Figure 7.4. In the case of the model of Figure 7.4a, all the primary voltages, currents, and impedances are *referred* or *moved* to secondary: for instance, R''_1 is the resistance R_1 referred to secondary. In the case of Figure 7.4b) all the secondary impedances and voltages are referred to primary: R'_2 is R_2 referred to primary; and so on.

Indeed the ideal transformer is still present in the shaded areas of the circuits, but they are often not considered (i.e., they are omitted) since the circuits are studied "as seen from" the secondary or primary winding, respectively. What mathematical operation is performed on voltages, currents, and impedance to transfer them from one side of the ideal transformer to the other?

For simplicity, the general rule is drawn here in the case in which the reactance X_m is disregarded (Figure 7.5a). The part of the circuit containing the primary and the ideal transformer can be substituted by its Thévenin equivalent. Thévenin voltage is obtained under the condition $\underline{I}_2 = 0$ (Figure 7.5b).

Posing as usual $\alpha = N_2/N_1$, it is

$$\underline{U}_{Th} = \underline{E}_2 = \alpha \underline{E}_1 = \alpha \underline{U}_1$$

The equality of $\underline{E}_1$ and $\underline{U}_1$ is a consequence of $\underline{I}_2 = 0$ and therefore $\underline{I}_1 = 0$, since $\underline{I}_1 = N_2/N_1 \cdot \underline{I}_2$.

Thévenin's inner impedance is obtained by deactivating the primary voltage $\underline{U}_1$ and feeding the circuit from the secondary (Figure 7.5c).

$$\underline{Z}_{Th} = \underline{U}_{s2}/\underline{I}_2 = \underline{E}_2/\underline{I}_2 = \alpha \underline{E}_1/(\underline{I}_1/\alpha) = \alpha^2 \underline{E}_1/\underline{I}_1 = \alpha^2 (R_1 + jX_1)$$

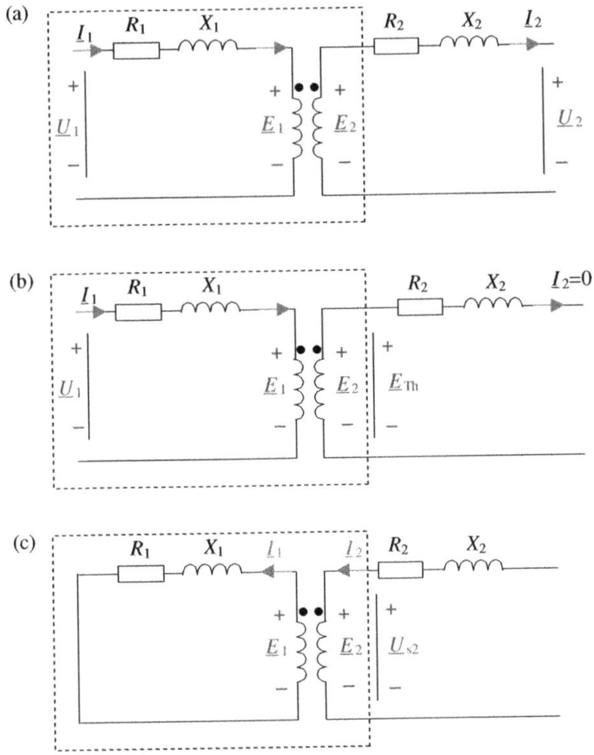

FIGURE 7.5. Thevenin's equivalents to determine the primary circuit as seen from the secondary.

Therefore, the circuit shown in Figure 7.5a can be transformed using Thévenin's equivalent of the part in the box, thus obtaining the circuit in Figure 7.6. The total impedance $(\alpha^2 R_1 + R_2) + j(\alpha^2 X_1 + X_2)$ is also called "secondary short-circuit impedance" ($\underline{Z}_{sc}''$). In fact, dividing by $\underline{Z}_{sc}''$ the secondary no-load voltage $\alpha \underline{U}_1$, we obtain the secondary short-circuit current (when the transformer is fed at its full voltage).

Now the result can be generalised as follows:

> **Rule:** *Referring (or moving) quantities from one side of an ideal transformer to the other*
>
> Quantities can be transferred from one side of the ideal transformer to the other, multiplying them by a suitable factor. If $\alpha = N_d/N_s$ indicates the ratio destination turns/source turns):
>
> - Voltages referred to destination are the original voltages multiplied by α.
> - Currents referred to destination are the original currents divided by α.
> - Impedances referred (or moved) to destination are the original impedances multiplied by α^2.

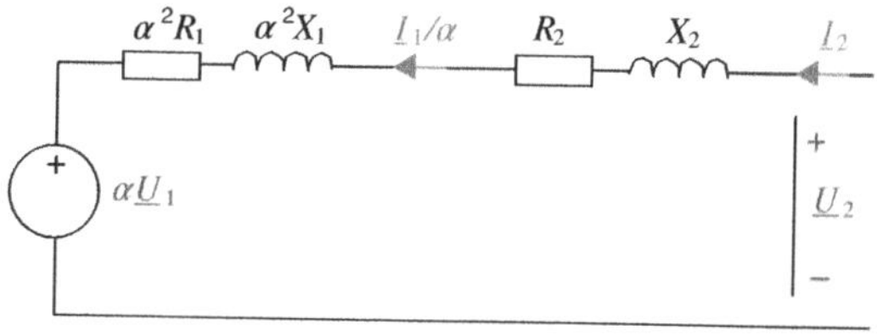

FIGURE 7.6. Transformer model with primary referred to secondary, and ideal transformer omitted.

To be more explicit, therefore, some of the quantities shown in Figure 7.4 are as follows ($\alpha = N_2/N_1$):

$$\begin{aligned} \underline{U}_1'' &= \alpha \underline{U}_1, & \underline{U}_2' &= \underline{U}_2/\alpha \\ \underline{I}_1'' &= \underline{I}_1/\alpha, & \underline{I}_2' &= \alpha \underline{I}_2 \\ R_1'' &= \alpha^2 R_1, & R_2' &= R_2/\alpha^2 \end{aligned} \tag{7.6}$$

It was already shown in Chapter 5 that the ideal transformer, in terms of energy, is transparent, that is, $p_1 = u_1 i_1 = p_2 = u_2 i_2$ (using reference flows and polarities shown in Figure 7.2), which, in AC (sinusoidal) circuits, implies conservation of active and reactive power, that is, conservation of the complex power:

$$\underline{S}_1 = \underline{S}_2 \Rightarrow \underline{U}_1 \underline{I}_1^* = \underline{U}_2 \underline{I}_2^* \tag{7.7}$$

It can be easily verified that the moving rule summarised in the above box and exemplified in (7.6) maintains this power invariance:

$$\underline{U'' I''^*} = \alpha \underline{UI^*}/\alpha = \underline{UI^*}$$

For single-phase transformers, IEC[1] considers the two symbols shown in Figure 7.7, respectively in the compact (unifilar) and in the detailed (full) form. According to IEC graphical standards, these forms are also called form 1 and form 2, respectively.

FIGURE 7.7. Standard symbols of single-phase transformers.

[1]IEC id an International standardization body. Compare Section 1.3.1.

Example 1. The primary winding of an ideal single-phase transformer is fed by a 120-V 50-Hz voltage source. The turn ratio is $N_1/N_2 = 1{:}2$. The secondary winding feeds an impedance of 32 Ω, 0.9 power factor (lagging). Calculate:

(a) the impedance seen by the primary voltage source;
(b) the primary and secondary current;
(c) the active power delivered to the load.

The load impedance Z_2 can be referred to the primary, multiplying it by $(N_1/N_2)^2 = 1/4$:

$$\begin{aligned} Z_2' &= Z_2/4 = 8\,\Omega\ (0.9\ \text{power factor}) \\ I_1 &= U_1/Z_2' = 120/8 = 15\ \text{A} \\ I_2 &= I_1/(N_2/N_1) = 7.5\ \text{A} \end{aligned}$$

I_2 can also be calculated as $U_2/Z_2 = (U_1 \cdot N_2/N_1)/Z_2 = 120 \cdot 2/32 = 7.5$ A.

The power delivered to the load can be calculated as $P_2 = U_2 I_2 \cos\varphi_2 = 240 \cdot 7.5 \cdot 0.9 = 1620$ W ($P_1 = U_1 I_1 \cos\varphi_1 = 120 \cdot 7.5 \cdot 0.9 = 1620\ \text{W} \equiv P_2$, the transformer being ideal).

Example 2. An ideal single-phase transformer is fed by a 100-V sinusoidal voltage source, thus providing 25 V at the secondary winding. If N_1 is 40 turns, $f = 50$ Hz, and the net cross-sectional area of the core is $A = 100\ \text{cm}^2$, calculate N_2, the flux ϕ, and $\hat{B}$.

If the load absorbs 12 A and $\cos\varphi_2 = 0.9$, calculate the primary current and the active and reactive power provided to the transformer by the electric grid.

The turn ratio N_1/N_2 equals $U_1/U_2 = 100/25 = 4$, thus $N_2 = N_1/4 = 40/4 = 10$ turns. Since $u_1(t) = e_1(t) = \mathrm{d}/\mathrm{d}t(N_1\ \phi)$, $\underline{U}_1 = j\omega N_1 \underline{\Phi}$ and $U_1 = \omega N_1 \phi$:

$$\begin{aligned} \phi &= \frac{U_1}{\omega N_1} = \frac{100}{2\pi 50 \cdot 40} = 7.958\ \text{mWb} && \text{(rms value, like } U_1) \\ B &= \phi/A = 7.958 \cdot 10^{-3}/10^{-2} = 0.7958\ \text{T} && \text{(rms value)} \\ \hat{B} &= B \cdot \sqrt{2} = 1.125\ \text{T} && \text{(peak value)} \\ I_1/I_2 &= N_2/N_1, \quad \text{thus } I_1 = I_2/4 = 3\ \text{A} \end{aligned}$$

The active and reactive power absorbed by the load are, respectively,

$$\begin{aligned} P_2 &= U_2 I_2 \cos\varphi_2 = 25 \cdot 12 \cdot 0.9 = 270\ \text{W} \\ Q_2 &= U_2 I_2 \sin\varphi_2 = 25 \cdot 12 \cdot \sqrt{1 - \cos^2\varphi_2} = 130.8\ \text{var} \end{aligned}$$

Since an ideal transformer has no losses, according to Boucherot's theorem the primary and secondary complex powers always correspond: $S_1 = S_2$, meaning $P_1 = P_2$ and $Q_1 = Q_2$.

I_1 can also be calculated as S_1/U_1, where $S_1 = S_2 = \sqrt{P_2^{\,2} + Q_2^{\,2}} = 300$ VA.

Example 3. A single-phase transformer has the following parameters (see Figure 7.3): $R_1 = 2.8\,\Omega$, $R_2 = 30\,\text{m}\Omega$, $X_1 = 5\,\Omega$, $X_2 = 60\,\text{m}\Omega$, $N_1/N_2 = 10$. The transformer is fed by a primary voltage of 230 V. Calculate the equivalent circuit referred to the secondary.

The equivalent voltage source seen by the secondary is, disregarding I_m with respect to I_1, $230/10 = 23$ V.

The primary impedance referred to the secondary can be easily calculated as $R_1'' + jX_1'' = (R_1 + jX_1)/(N_1/N_2)^2 = 28 + j50\,\text{m}\Omega$.

The secondary short-circuit impedance is then $R_1'' + jX_1'' + R_2 + jX_2 = 58 + j110\,\text{m}\Omega = 124\,\text{m}\Omega \angle 62.2°$.

The secondary short-circuit current, when the primary is rated fed, is $23/0.124 = 185$ A.

7.3 THREE-PHASE TRANSFORMERS

Three-phase transformers can have different construction modes. Here we briefly present only the transformers created as a three-phase bank of single-phase transformers.

The three-phase bank is constituted by a simple connection of single phase transformers.

Possible connections are star[2]/star, star/delta, and delta/delta. For example, star/star and star/delta are shown in Figure 7.8.

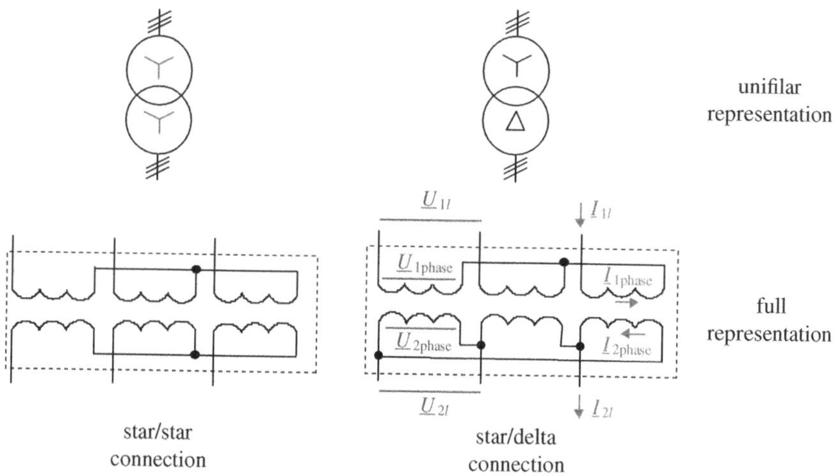

FIGURE 7.8. Three-phase transformers obtained interconnecting single-phase transformers.

[2]As already discussed in Chapter 6, star connection is also referred to as "wye" connection.

It easy to verify in the previous figure that in a three-phase star connection the three single-phase windings share one terminal, and the other terminals connect with the transformer exterior. In the delta connection, on the other hand, the terminals are connected in sequence, and the connection points constitute the points of junction with the transformer exterior.

The principal difference between the star-connected side and the delta-connected side of a three-phase transformer is that in the star-connected side it is possible to access the neutral point. Therefore both phase and line voltages can be accessed.

However, this is not very important in perfectly balanced systems, such as those dealt with in this book, since the phase voltages are easily reproduced in the phase of any star-connected impedances connected to the three-phase system in question.

Numerical relationships between line and phase voltage and currents are derived directly as we saw in Chapter 6:

$$\text{in a star connection}: U_l = \sqrt{3}\, U_{\text{phase}}, \quad I_l = I_{\text{phase}} \tag{7.8}$$

$$\text{in a delta connection}: U_l = U_{\text{phase}}, \quad I_l = \sqrt{3}\, I_{\text{phase}} \tag{7.9}$$

where U_l is the line (phase-to-phase) voltage and I_l is the line current, that is, the current flowing in each conductor of the three-phase power line (see Figure 7.8). Phase voltage and currents are instead applied to each single-phase transformer which constitutes the three-phase bank.

Once the phase voltage and currents are determined, models made for single-phase transformers can be used.

Regardless of the winding connections, in an ideal three-phase transformer the complex power $\underline{S}_1$ entering the machine equals the complex power $\underline{S}_2$ provided to the load by the secondary winding:

$$\underline{S}_1 = P_1 + jQ_1 = P_2 + jQ_2 = \underline{S}_2 \tag{7.10}$$

Using (7.8), (7.9), and (7.10), the reader can easily demonstrate that, regardless of the winding connections, the three-phase bank powers are three times those of each single-phase transformer.

Example 4. The primary and secondary winding of a three-phase transformer are respectively delta- and wye-connected. The rated power is 100 kVA and the rated primary voltage is 10 kV (phase-to-phase). If each phase has $N_1/N_2 = 50$, calculate:

(a) the secondary rated voltage (both phase and phase-to-phase);
(b) the primary rated current (both line and phase);
(c) the secondary rated current (both line and phase);
(d) $\hat{B}$ if $f = 50$ Hz and the magnetic circuit section is 250 cm^2, if $N_1 = 2000$ turns.

(a) $U_{1\text{phase}} = U_{1l} = 10\,\text{kV}$ (delta connection)

$$U_{2\text{phase}} = U_{1\text{phase}} \cdot N_2/N_1 = 10000/50 = 200\,\text{V} \text{ (as in a single-phase transformer)}$$
$$U_{l2} = \sqrt{3}\,U_{2\text{phase}} = 346.4\,\text{V} \text{ (wye connection)}$$

(b) $I_{1l} = S_{\text{rated}}/(\sqrt{3}\,U_{1l}) = 10^5/(\sqrt{3}\cdot 10^4) = 5.77\,\text{A}$ (any connection)

$$I_{1\text{phase}} = I_{1l}/\sqrt{3} = 3.333\,\text{A} \text{ (delta connection)}$$

(c) $I_{2\text{phase}} = I_{1\text{phase}} \cdot N_1/N_2 = 3.333 \cdot 50 = 166.6\,\text{A}$ (as in a single-phase transformer)

$$I_{2l} = I_{2\text{phase}} = 166.6\,\text{A} \qquad \text{(wye connection)}$$

(d) $$\phi = \frac{U_{1\text{phase}}}{\omega N_1} = \frac{10^4}{2\pi 50 \cdot 2000} = 15.92\,\text{mWb}$$

$$B = \phi/A = 15.92 \cdot 10^{-3}/(250 \cdot 10^{-4}) = 0.637\,\text{T} \qquad \text{(rms value)}$$
$$\hat{B} = B \cdot \sqrt{2} = 0.9\,\text{T} \qquad \text{(peak value)}$$

7.4 MAGNETIC HYSTERESIS AND CORE LOSSES

In the previous sections we examined a simple linear relation between magnetic flux density and field:

$$B = \mu H$$

The ferromagnetic material of which the transformer core is made (like other machines, as will be seen in the following chapters) does not show a linear correspondence between these two quantities. The two vectors share the same direction, but the amplitude ratio is not constant.

The real behaviour shows a hysteresis loop like the one shown in Figure 7.9. While the general shape depicted in the figure can be attributed to any ferromagnetic material, the numerical values are typical of ordinary polycrystalline metal, such as fairly pure iron.

The user can verify that the ratio B/H on the parts of the curve far from the axes is much larger than μ_0, the value valid for nonferromagnetic materials.

If the magnetic field varies as a sine, with an amplitude cycling between $-\hat{H}$ and $+\hat{H}$, the flux density B will be obtained from the plot, following the shown hysteresis loop. Clearly it will not have a sinusoidal shape. This creates a problem: in order to accurately model the hysteresis characteristic of a transformer, we would have to abandon the phasor circuit, which requires linearity, which is well beyond the scope of this book.

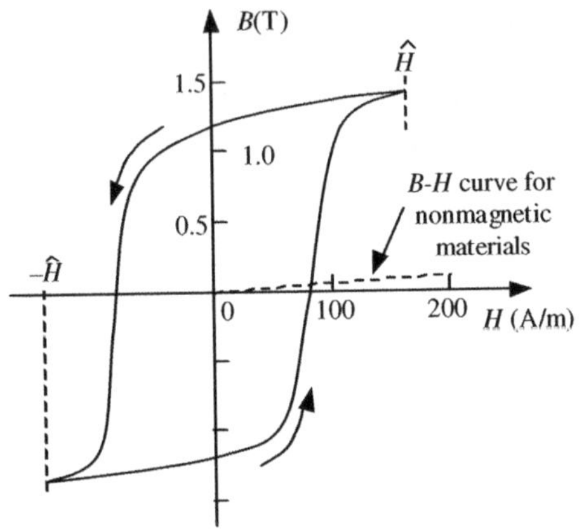

FIGURE 7.9. Typical magnetic hysteresis loop of a ferromagnetic material.

Magnetic hysteresis is important also from an energy point of view: it can be demonstrated that hysteresis phenomenon implies loss of energy in the magnetic circuit of the transformer, which reduces efficiency and heats the iron.

Another important phenomenon inside the iron core that dissipates power and contributes to heating is that of "eddy currents." The alternating field in the core generates, in accordance with Faraday's law, EMFs in the core itself; these in turn produce currents, since iron is conductive. These currents generate heat within the iron in accordance with Joule's law.

Consider a side of the toroidal structure of a transformer. The alternating flux inside creates a magnetic field in the iron itself, whose force lines belong to the plane orthogonal to the side axis. Thus currents, called *eddy currents* because of their shape, flow along these force lines and generate heat in the iron, dissipating otherwise useful energy (Figure 7.10a).

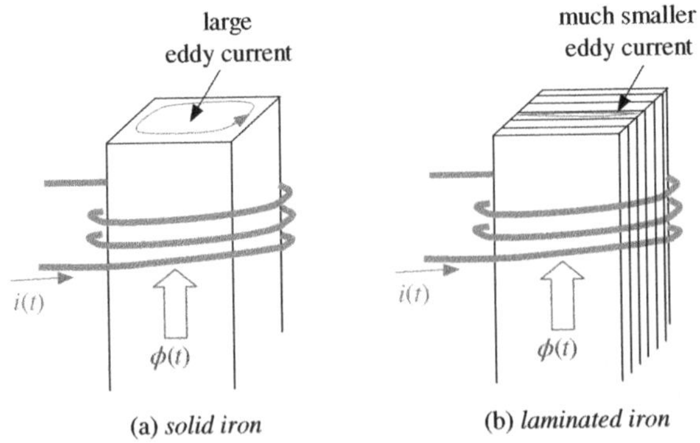

FIGURE 7.10. Eddy currents and their mitigation through lamination.

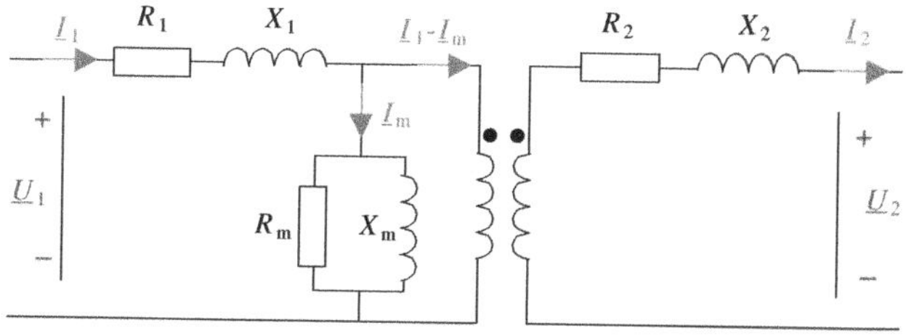

FIGURE 7.11. Full model of the single-phase transformer.

This phenomenon must be reduced. In fact, in all electric machines (i.e., transformers and rotating machines), whenever iron is traversed by a variable magnetic field, the iron is *laminated*. This means that the iron core is not solid; instead, it is composed of several thin layers covered in very thin insulating material stuck to each other. The possibilities for the current to flow are therefore reduced to very narrow paths inside the layer, and the global losses are thus dramatically reduced (Figure 7.10b).

Eddy current and hysteresis losses can be roughly taken into account by slightly modifying the equivalent circuit as shown in Figure 7.3, with the addition of a new resistor R_m (Figure 7.11).

It must be noted that the law linking the dissipated power to the circuit voltage on R_m is not the same of a ohmic resistor. Therefore this usage of R_m gives reasonable results only in small ranges of the E_1 (i.e., voltage across R_m) and therefore on small ranges of U_1 and U_2.

To conclude, consider Figure 7.12, which shows graphically (and qualitatively) the power fluxes of a transformer. It shows that in a real transformer the output power is only a fraction of the input. Losses are in the primary and secondary resistances (called copper losses), as well as in the iron, constituted by hysteresis and eddy-current losses.

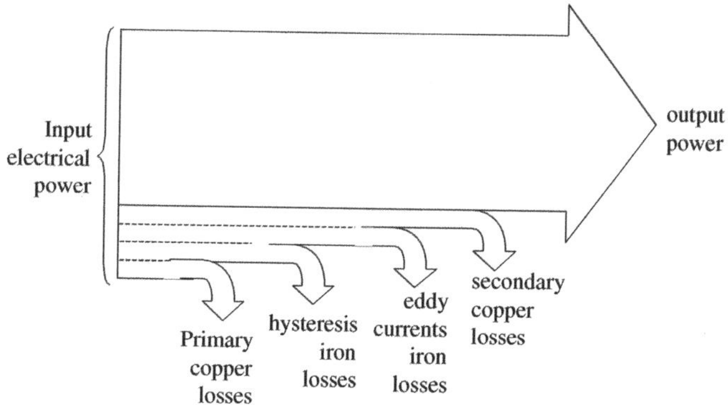

FIGURE 7.12. A qualitative chart illustrating different types of transformer losses.

7.5 OPEN-CIRCUIT AND SHORT-CIRCUIT TESTS

A method for determining the parameters of the equivalent circuit of a transformer (see Figure 7.11) consists of two tests: the "open-circuit test" and the "short-circuit test."

To perform these tests, the following hardware is needed:

- An ammeter, that is, a two-terminal device that measures current in circuits. In the case of AC ammeters, they normally directly give their rms values. Obviously an ammeter must be connected to the circuit in such a way that the current to be measured flows through it ("series" connection). The voltage between the two terminals of an ideal ammeter is null; that is, its internal impedance can be neglected with respect to the Thévenin impedance of the circuit to which it is connected.
- A voltmeter, that is, a two-terminal device that measures voltage in circuits. In the case of AC voltmeters, these normally directly give their rms values. Obviously, a voltmeter must be connected to the circuit in such a way that its terminals are applied to the points of the circuit, whose potential difference must be measured ("shunt" connection). The current absorbed by an ideal voltmeter is null; that is, its internal impedance is extremely high with respect to the Thévenin impedance of the circuit to which it is connected.
- A wattmeter, that is, a device that measures powers in circuits. AC wattmeters measure the (average) active power $P = \overline{p(t)} = \overline{u(t)i(t)}$. Since P is a function of both current and voltage, the wattmeter is provided with two current terminals (to be connected in series with the current, exactly like an ammeter) and with two voltage terminals (to be connected parallel to the voltage, exactly like a voltmeter).

In the open-circuit test (Figure 7.13) the nominal voltage is applied to the primary side of the transformer, while leaving the secondary side open. The current I_0 into the primary winding is measured, as well as the active power P_0 absorbed by the transformer, which corresponds to the sums of powers dissipated by R_1 and R_m. Since it is always $R_1 \ll R_m$, P_0 is a good estimation of iron losses, due to eddy currents and magnetic hysteresis. Such losses, which strictly depend on the voltage across R_m,

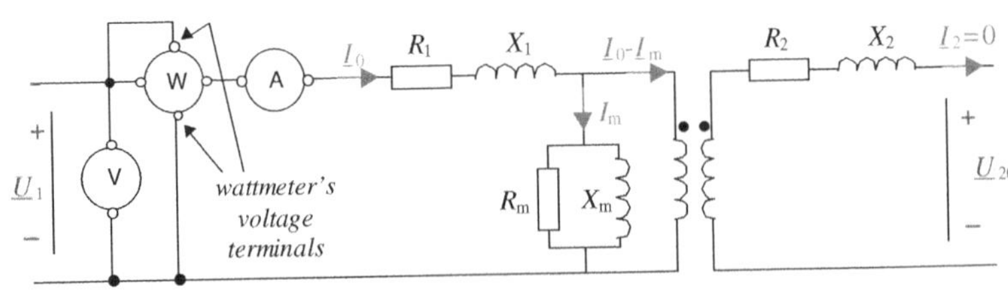

FIGURE 7.13. Open-circuit test.

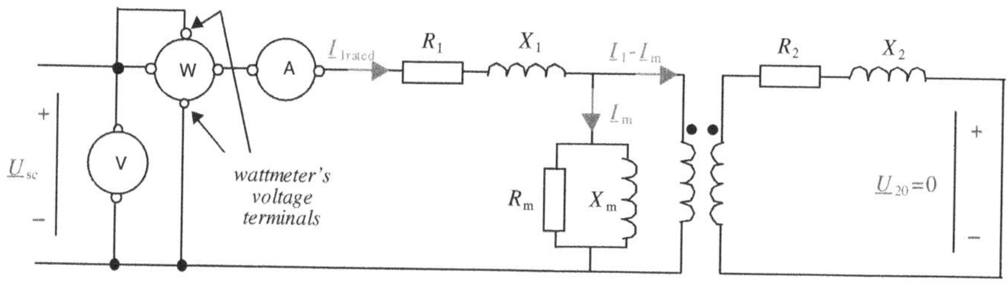

FIGURE 7.14. Short-circuit test.

remain practically constant for any loading condition, provided that the transformer is fed at its primary nominal voltage. It will therefore be

$$R_m \cong \frac{U_1^2}{P_0}, \quad X_m \cong \frac{U_1^2}{Q_0} = \frac{U_1^2}{P_0 \cdot \tan\varphi_0}, \qquad \text{where } \cos\varphi_0 = \frac{P_0}{U_1 I_0}$$

The short-circuit test (Figure 7.14) requires the secondary winding terminals to be connected to each other (Figure 7.14). This is called short-circuiting. Under these conditions, applying the nominal voltage at the primary winding would cause very high currents which, if lasting more than just a few milliseconds, would seriously damage the transformer. Therefore, a much smaller voltage U_{sc} is applied to the primary winding of the transformer in such a manner that the primary current equals the nominal primary transformer current; U_{sc} corresponds to a few cents of the nominal winding voltage.

During this test, I_m can be neglected with respect to I_1 (the core is poorly magnetised); thus the power absorbed by R_m is much lower than the one consumed by R_1 and R_2. For this reason, the power P_{sc} measured by the wattmeter is a good estimation of copper losses at rated current; in any other loading condition we have

$$P_{cu} \cong P_{sc} \cdot \left(\frac{I}{I_{rated}}\right)^2$$

The short-circuit test is unable to separate $R_1 + jX_1$ from $R_2 + jX_2$, but it allows $\underline{Z}_{sc}'$ and $\underline{Z}_{sc}''$ to be calculated (Figure 7.15; compare Section 7.2):

$$\underline{Z}_{sc}' \cong \frac{U_{sc}}{I_{1rated}} e^{j\varphi_{sc}}, \quad \text{where} \quad \cos\varphi_{sc} = \frac{P_{sc}}{U_{sc} I_{1rated}}$$

$$\underline{Z}_{sc}'' = \alpha^2 \underline{Z}_{sc}', \qquad \alpha = \frac{N_2}{N_1}$$

Remember that primary and secondary active and reactive powers are strictly related:

$$P_1 = P_2 + P_{iron} + P_{cu}$$

$$Q_1 = Q_2 + P_{iron} \cdot \tan\varphi_0 + P_{cu} \cdot \tan\varphi_{sc}$$

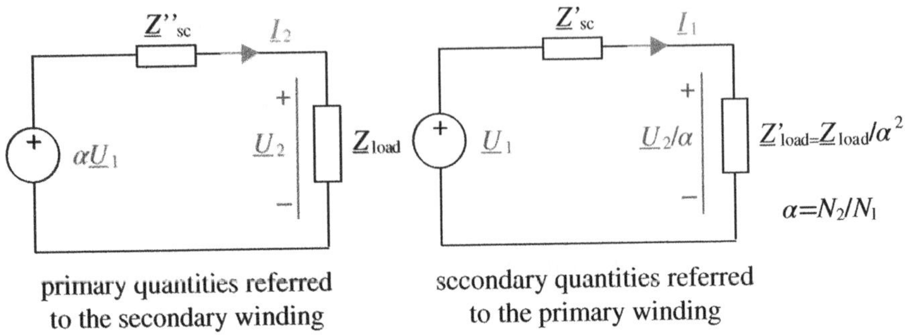

FIGURE 7.15. Use of $\underline{Z}''_{sc}$ and $\underline{Z}''_{sc}$

Example 5. A single-phase transformer rated at 50 kVA and 600/230 V has issued the following results of the open-circuit and of the short-circuit test, respectively:

$$P_0 = 2\% \text{ of rated apparent power}$$
$$I_0 = 5\text{ A}$$
$$U_{sc} = 6\% \text{ of rated primary voltage}$$
$$P_{sc} = 3\% \text{ of rated apparent power}$$

Calculate the transverse parameters (R_m and X_m), the primary and secondary short-circuit impedance.

$$\begin{aligned}
P_0 &= 2/100 \cdot S_{rated} = 2/100 \cdot 50000 = 1000\text{ W}\\
\cos\varphi_0 &= P_0/U_{1rated}/I_0 = 1000/600/5 = 0.333\\
R_m &= U_1^2/P_0 = 600^2/1000 = 360\,\Omega\\
X_m &= U_1^2/Q_0 = 600^2/(1000 \cdot \tan(\text{acos}(0.333))) = 127.3\,\Omega\\
U_{sc} &= 6/100 \cdot U_{1rated} = 6/100 \cdot 600 = 36\text{ V}\\
P_{sc} &= 3/100 \cdot S_{rated} = 3/100 \cdot 50000 = 1500\text{ W}\\
I_{1rated} &= S_{rated}/U_{1rated} = 50000/600 = 83.33\text{ A}\\
\cos\varphi_{sc} &= P_{sc}/U_{sc}/I_{1rated} = 1500/36/83.22 = 0.500, \text{thus } \varphi_{sc} = 60^\circ\\
Z'_{sc} &= U_{sc}/I_{1rated} = 36/83.33 = 0.432\,\Omega, \qquad \underline{Z}'_{sc} = 0.432\,\Omega\angle 60^\circ\\
Z''_{sc} &= Z'_{sc}/(600/230)^2 = 0.0635\,\Omega, \qquad \underline{Z}''_{sc} = 0.0635\,\Omega\angle 60^\circ
\end{aligned}$$

Please note that

- $\cos\varphi_0$ can also be calculated as $P_0\%/I_0\%$, where $P_0\% = 100P_0/S_{rated} = 2\%$ and $I_0\% = 100I_0/I_{1rated} = 6\%$.
- $\cos\varphi_{sc}$ can also be calculated as $P_{sc}\%/U_{sc}\%$, where $P_{sc}\% = 100P_{sc}/S_{rated} = 3\%$ and $U_{sc}\% = 100U_{sc}/U_{1rated} = 6\%$.

Example 6. Assuming that the transformer of the previous example is fed by its primary rated voltage, calculate the secondary voltage, the active and reactive power absorbed by the load, the primary current, and the efficiency when the load has an impedance $\underline{Z}_2 = 1.5 + j0.6\,\Omega$.

$$U_1'' = 230\text{ V}$$
$$I_2 = \underline{U}_1''/(\underline{Z}_{sc}'' + \underline{Z}_2) = 138.1\text{ A}\angle{-23.15°}$$
$\underline{Z}_2$ can be expressed as $1.615\,\Omega\angle 21.80°$
$$U_2 = Z_2 I_2 = 1.615 \cdot 138.1 = 223.1\text{ V}$$
$$P_2 = U_2 I_2 \cos\varphi_2 = 223.1 \cdot 138.1 \cdot \cos 21.80° = 28{,}592\text{ W}$$
$$Q_2 = U_2 I_2 \sin\varphi_2 = 223.1 \cdot 138.1 \cdot \sin 21.80° = 11{,}437\text{ var}$$
$$I_{2\text{rated}} = S_{\text{rated}}/U_{2\text{rated}} = 50{,}000/230 = 217.4\text{ A}$$
$$P_{\text{cu}} = P_{\text{sc}} \cdot (I_2/I_{2\text{rated}})^2 = 1500 \cdot (138.1/217.4)^2 = 605.0\text{ W}$$
$$P_{\text{iron}} = P_0 = 1000\text{ W}$$
$$P_1 = P_2 + P_{\text{iron}} + P_{\text{cu}} = 30{,}197\text{ W}$$
$$Q_1 = Q_2 + P_{\text{iron}} \cdot \tan\varphi_0 + P_{\text{cu}} \cdot \tan\varphi_{\text{sc}} = 15{,}313\text{ var}$$
$$S_1 = \sqrt{P_1^2 + Q_1^2} = 33{,}858\text{ VA}$$
$$I_1 = S_1/U_1 = 56.43\text{ A}$$
$$\eta = P_2/P_1 = 0.947$$

$\underline{I}_1$ can be also calculated as $\underline{I}_2' + \underline{I}_m$, where $\underline{I}_2' = \underline{I}_2 \cdot N_2/N_1$ and $\underline{I}_m \cong \underline{U}_1/R_m + \underline{U}_1/(jX_m)$, with $\underline{U}_1 = 600\text{ V}$.

7.6 PERMANENT MAGNETS

In Section 7.4 we introduced the magnetic hysteresis curve of ferromagnetic materials, shown again in Figure 7.16, with some additions. The curve shows that when the applied magnetic field H is gradually reduced from its maximum value $\hat{H}$ to zero, the effect (i.e., flux density B) does not go to zero, but reaches a residual value B_r.

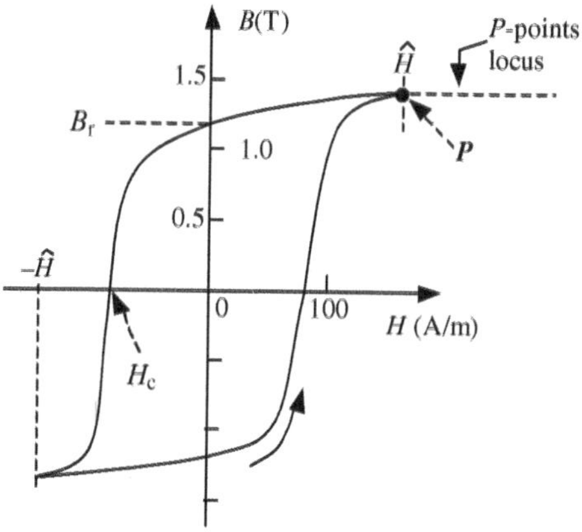

FIGURE 7.16. Typical hysteresis loop of fairly pure iron.

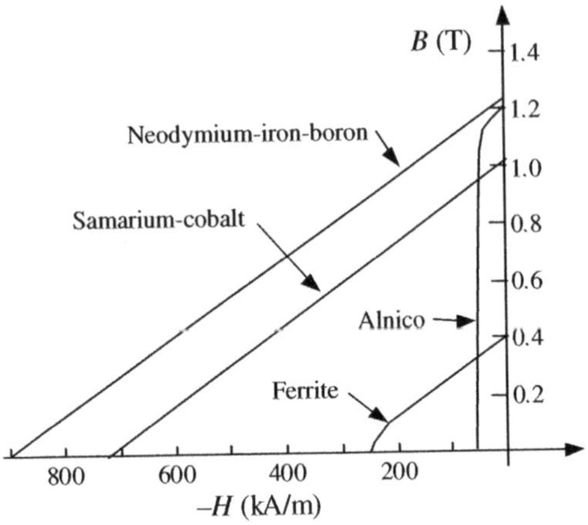

FIGURE 7.17. Demagnetisation curves for different PM magnets (at 20 °C).

This residual flux density, however, disappears when the negative field H_c is reached, called *coercive force*.

If the same material is used at a larger $\hat{H}$, the loop vertex (indicated in the figure by the point ***P***) moves along the straight line indicated in the figure as the ***P***-point locus. This line appears to be horizontal in the scale of the figure, but indeed it is not: its slope is the air permeability μ_0.

When ordinary iron is used (i.e., the one shown in Figure 7.9), the magnetic field present in the machine iron oscillates between values much higher than 100 A/m, and the B_r is just a point of the magnetic hysteresis loop experienced by the machine.

There are special materials, however, that have a coercive force several orders of magnitude higher than that of simple pure iron, and this allows them to be used in a completely different way inside electric machines. For these, the most significant part of the magnetising curve is the one in the second quadrant, called *demagnetisation curve*, and is shown for some common materials in Figure 7.17. The Alnicos are materials containing Al, Ni, Co, and Fe, which have been widely used in the past century. The demagnetisation curve on one of the alnicos is shown, having $H_c \cong -50$ kA/m, much larger than that of common iron.

Barium and strontium ferrites, invented in the 1950s, have higher $|H_c|$ than Alnico, but lower B_r. More recently, samarium–cobalt (Sm–Co) and neodymium–iron–boron (Nd–Fe–Br) were introduced, both of which have excellent characteristics in terms of B_r and H_c, but cost more than ferrites. Sm–Co and Nd–Fe–Br materials belong to the family of rare-earth permanent magnets (PMs).

Ferrites and rare-earth PMs have a slope equal to μ_0. This is very important in their usage in electrical machines: in a magnetic circuit they behave like air, into which the fixed flux density bias B_r is introduced.

Therefore, we can draw up the following rule, which will be exploited in the following chapters, when permanent-magnet DC and synchronous machines will be discussed.

Rule: *Permanent-magnet equivalent*

If ferrite or rare-earth permanent magnets are used, their effect is the production of a fixed flux density B_r (equal to the PM residual flux density), combined with the magnetic behaviour of the space occupied by the magnet, which is the same as that of air.

7.7 PROPOSED EXERCISES

Where not specifically indicated, sinusoidal quantities are expressed in rms values.

7.1. An ideal single-phase transformer is fed by a sinusoidal voltage source $U_1 = 600\,\text{V}$, thus providing 150 V at the secondary winding. If N_1 is 100 turns, $f = 50\,\text{Hz}$ and the net cross-sectional area of the core is $S = 250\,\text{cm}^2$, calculate N_2, the magnetic flux ϕ and $\hat{B}$.

7.2. An ideal single-phase transformer is fed by a sinusoidal voltage source $U_1 = 230\,\text{V}$. If the load absorbs 12 A at 46 V with $\cos\varphi_2 = 0.9$, calculate the primary current and the active and reactive power provided to the transformer by the electric grid.

7.3. An ideal single-phase transformer, fed by a sinusoidal voltage source $U_1 = 127\,\text{V}$, provides 100 W and 40 var to a load. Calculate the primary current.

7.4. An ideal single-phase transformer is rated 2 kVA and 50/300 V. A 40-Ω resistance is connected to the secondary winding, while the primary is fed by the nominal voltage. Calculate the current in the two windings and check if the transformer is overloaded.

7.5. An ideal single-phase transformer is rated 2000/200 V. A resistance, fed by the secondary winding, must have exactly 160 V and 40 A across it. Determine the additional resistance to be added to the circuit:

(a) secondary side;

(b) conversely, primary side.

7.6. The cross-sectional area of a single-phase transformer's core is $20\,\text{cm}^2$. The transformer will be fed by a sinusoidal voltage source of 150 V, 60 Hz. Calculate:

(a) the number N_1 of primary winding turns, so that $\hat{B} = 1.15\,\text{T}$;

(b) the magnetizing current, if $\hat{H} = 105$ A · turns/m and the mean magnetic length of the core is $l = 60\,\text{cm}$.

Hint: Calculate the core's reluctance or simply use Ampère's law.

7.7. An ideal single-phase transformer is fed by a distribution grid, having a Thévenin voltage source of 230 V, 50 Hz and an equivalent impedance of 1 Ω (unit power factor). The turn ratio is $N_1/N_2 = 1{:}4$. The secondary winding feeds a 32-Ω resistor. Calculate:

(a) the total resistance seen by the voltage source;

(b) the primary and secondary current;

(c) the power delivered to the load;

(d) the total efficiency of this installation;

(e) the value of the load resistance which would absorb the maximum power.

7.8. A single-phase transformer rated 250 kVA has open-circuit losses $P_0 = 1800\,\text{W}$ (at rated voltage) and short-circuit losses $P_{sc} = 3200\,\text{W}$ (at rated current). Determine the efficiency of this transformer when it provides (at rated voltage):

(a) 100%, 75%, 50%, 25% of its rated current, to a resistive load;

(b) 100%, 75%, 50%, 25% of its rated current at 0.8 power factor (lagging).

Please observe the following general result: Whatever the power factor, the efficiency is maximum when copper losses ($= P_{sc} \cdot (I/I_{rated})^2$) *equal iron losses* ($= P_0$)*. In this example, when* $I/I_{rated} = 0.75$.

7.9. A 200-kVA single-phase transformer reaches its maximum efficiency (98.4% at unit power factor) at 90% of full load. Calculate losses and the power delivered to the load in this condition. Then calculate open-circuit and short-circuit losses.
Hint: Apply the general rule stated at the end of the previous exercise.

7.10. The open-circuit current and power of a single-phase transformer are, respectively 2.5 A and 150 W when it is fed by a primary voltage of 230 V. Neglecting the resistance and the leakage flux of the primary winding, calculate:

(a) the open-circuit power factor;

(b) the equivalent transverse resistance R_m and its current I_a ("open-circuit active current");

(c) the equivalent transverse reactance X_m and its current I_μ ("open-circuit magnetizing current").

7.11. A single-phase transformer is characterized by the following parameters (see Figure 7.11): $R_1 = 2.27\,\Omega$, $R_2 = 0.0325\,\Omega$, $X_1 = 4\,\Omega$, $X_2 = 0.0573\,\Omega$. The open-circuit power is 15 W, with a power factor of 0.3.

The transformer feeds at 25 V a load of 500 VA with 0.8 power factor (lagging). Knowing that the primary current is 2.3 A, calculate the primary active and reactive power and voltage.
Hint: Use Boucherot's theorem to calculate the primary apparent power: add to the complex secondary power the active and reactive power absorbed by R_1, R_2, X_1, X_2, R_m, X_m.

7.12. A single-phase transformer rated at 3000/230 V is characterized by the following parameters (see Figure 7.11): $R_1 = 1.4\,\Omega$, $R_2 = 0.008\,\Omega$, $X_1 = 3\,\Omega$, $X_2 = 0.018\,\Omega$. Calculate the secondary short-circuit impedance and the full-voltage short-circuit current.

7.13. A single-phase transformer rated at 60 kVA and 3000/240 V has issued the following results to the open-circuit and to the short-circuit test, respectively:

$P_0 = 2.2\%$ of rated apparent power
$I_0 = 1.5$ A
$U_{sc} = 4.5\%$ of rated primary voltage
$P_{sc} = 2.2\%$ of rated apparent power

Assuming that the transformer is fed by the primary rated voltage, calculate the secondary voltage, the primary current and the efficiency, when the load impedance is $\underline{Z}_2 = 0.776 + j0.58\ \Omega$.

7.14. The primary and secondary winding of a three-phase transformer are delta- and wye-connected, respectively. The rated power is 75 kVA and the rated primary voltage is 12 kV (line-to-line). If each phase has $N_1/N_2 = 92$, calculate:

(a) the secondary rated voltage (both phase and phase-to-phase);

(b) the primary rated current (both line and phase);

(c) the secondary rated current (both line and phase).

7.15. The primary and secondary winding of a three-phase transformer rated at 400/230 kV are both wye-connected. Calculate the primary and secondary turns of each phase, knowing that the insulation between two consecutive turns can sustain 50 V. Determine also the rated primary and secondary line currents, if the rated power is 250 MVA.

7.16. An ideal three-phase transformer, fed by a balanced source of voltages ($U_1 = 400$ V line-to-line), provides 3 kW and 1.5 kvar to a balanced load. Calculate:

(a) the primary line current;

(b) the primary phase current, if the primary winding is delta-connected;

(c) $\hat{B}$ if $f = 50$ Hz, the magnetic circuit section is 300 cm^2, the primary winding is wye-connected and $N_1 = 40$ turns;

(d) N_2 if $N_1 = 40$ turns, $U_2 = 127$ V (phase-to-phase), and the primary and secondary windings are, respectively, wye- and delta-connected.

7.17. Three identical single-phase transformers are to be connected to form a three-phase bank rated at 300 MVA, 230:34.5 kV (line-to-line). For the following configurations, determine the voltage and kVA ratings of each single-phase transformer:

(a) wye–wye

(b) wye–delta

(c) delta–wye

(d) delta–delta

8

FUNDAMENTALS OF ELECTRONIC POWER CONVERSION

For the Instructor

This chapter aims to give students sufficient knowledge of how electric power is converted using electronic power converters, in order to facilitate the study of electric drives, which will be dealt with in later chapters. Furthermore, students will learn how inverters operate to allow power to be fed into grids from low-power power plants such as small photovoltaic generators. However, since it is *not essential to an understanding of subsequent chapters*, this chapter is optional.

Moreover, the teacher may choose to limit the study of this chapter to Sections 8.1–8.3 on time-domain analysis, or to approach it as a whole in order to explore frequency-domain analysis, which is used to explain filtering.

8.1 INTRODUCTION

In the next chapter we will introduce electromechanical energy conversion—that is, conversion of energy from electricity to mechanical energy and vice versa. In Chapters 10, 11, and 12 we will explore how this kind of conversion can be attained in practice—that is, how rotating electrical machines are constructed and operate.

Fundamentals of Electric Power Engineering: From Electromagnetics to Power Systems, First Edition.
Massimo Ceraolo and Davide Poli.

Nowadays, however, another kind of conversion of electric energy is carried out with increasing frequency: conversion from one electrical form into another, such as the conversion of DC power into AC and vice versa, or of a DC system into another DC system operating at a different voltage level.

This kind of electrical-to-electrical power conversion is normally effected using systems without moving parts—that is, without the interposition of mechanical energy. This is called *static power conversion* or *electronic power conversion* and is achieved using apparatuses and techniques which are analysed within the discipline of *power electronics*.

> The systems that carry out these conversions are called *electronic power converters* or *power electronics converters*, or simply *converters*. It is also worth mentioning the name used by an authoritative source [bm4] for such converters—that is, power processing units (or PPUs).

In power electronics, typical electronics components (such as diodes and transistors) are used to process not signals, but power (and energy).

8.2 POWER ELECTRONIC DEVICES

8.2.1 Diodes, Thyristors, Controllable Switches

As mentioned in the introduction, the power electronics converter is an assembly of elementary components (devices). Readers should be familiar with some of these from their study of Part II of this book—that is, resistors, capacitors, inductors, mutually coupled inductors.

Others, however, are totally different. A fairly comprehensive list is shown in Figure 8.1. In the figure, they are divided into three zones according to their behaviour. It will soon be seen that all these devices can reasonably be considered as on/off switches. Therefore, the devices are classified according to how the commutation between on and off states can occur, as follows:

- Diodes, for which the commutation between on and off states is determined by the circuit into which the diode is inserted.
- Thyristors (or silicon-controlled rectifiers, often designated by the acronym SCRs), for which commutation from off to on is determined by an electrical current pulse sent into the gate terminal ***G***, while commutation from on to off is determined by the circuit into which the thyristor is inserted (the three different symbols shown in Figure 8.1 indicate minor differences in operation that are beyond the scope of this book; only the uppermost symbol, which indicates a generic thyristor, will be used).
- Controllable switches, for which commutation between on and off is determined by electrical signals applied to the control terminal (named gate - ***G***). Since all

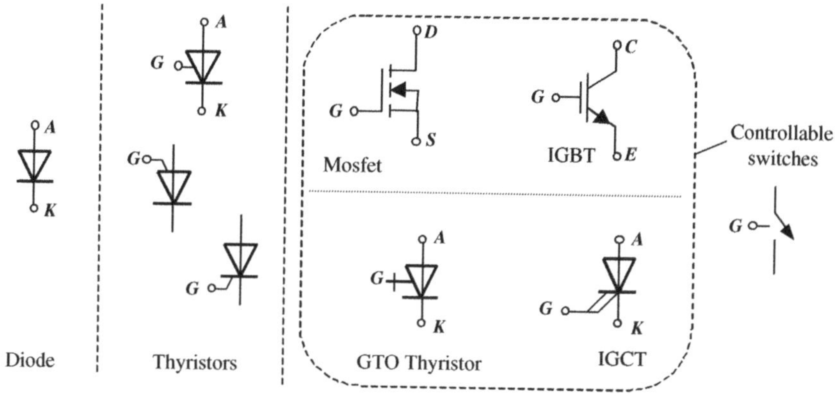

FIGURE 8.1. The power electronic devices referred to in this chapter (GTOs and IGCTs being less common than Mosfets and IGBTs).

the controllable switches show approximately the same behaviour, a single symbol is used for all of them; this symbol, though not included in the International standards, is very common in good-quality books, such as [bm2]. Since all controllable switches allow substantial amounts of current to flow only from one terminal toward another (and not vice versa), the chosen symbol is asymmetrical, and indicates preferred current flow direction by means of an arrow. The control terminal ***G*** is very often omitted in circuit representations, and it will be used very rarely in this book.

Figure 8.1 shows the commonly used letters for the device terminals. When used within circuits, these letters are often omitted. The letters are the initials of the common terminal names which, in turn, describe the inner structure of the device:

	Name	*Letter*	*Name*	*Letter*	*Name*	*Letter*
Thyristors, GTOs	Anode	A	Cathode	K	Gate	G
Mosfets	Drain	D	Source	S	Gate	G
IGBT	Collector	C	Emitter	E	Gate	G

The same figure shows that the control terminal of the controllable switch, when not omitted, is called gate (G).

The following sections include a concise description of the behaviour of these devices, as well as an introduction to approximations in such a way that they can be analysed using the linear circuit concepts. In all cases, we will analyse the behaviour of these devices as a "black box"—that is, presenting their behaviour in terms of a terminal voltage–current relationship, without analysing their inner structure or how they are produced.

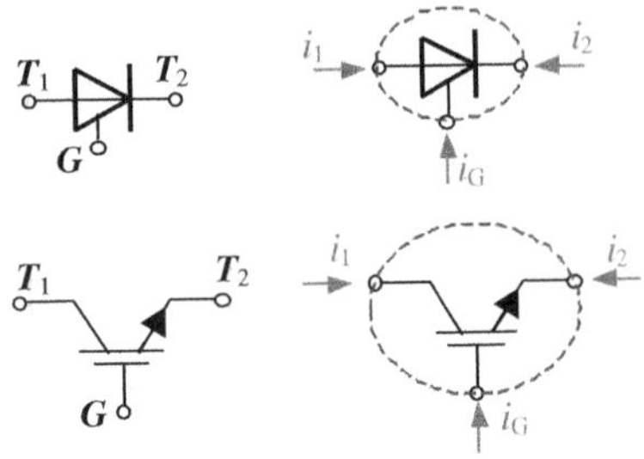

FIGURE 8.2. Application of KCL to switches.

8.2.2 The Branch Approximation of Thyristors and Controllable Switches

The reader may have noticed that all the power electronic devices listed in Figure 8.1, except for the diode, are three-terminal devices and are therefore not "branches" according to the definition given in Chapter 3.

In this same chapter, Kirchhoff's Current Law (KCL) was first used for a generic circuit zone comprised within a closed curve (see Figure 3.8); a special version was introduced for branch-based circuits. We have not yet needed to resort to the more general version of KCL.

Consider now two examples of power electronic devices, the SCR and the IGBT (Figure 8.2).

In both cases the more general KCL states

$$i_1 + i_G + i_2 = 0 \tag{8.1}$$

However, it must be noted that in the vast majority of cases, the current flowing in the control terminal G is much smaller than those flowing in the other terminals.

Therefore in these cases, (8.1) reduces to

$$i_1 + i_2 = 0 \tag{8.2}$$

which is the usual KCL for branches.

This approach is usually adopted by all textbooks and will be followed here. The following rule can now be formally stated.

Rule: *Branch approximation of power electronic devices*

In the case of three-terminal power electronic components (either thyristors or controllable switches), the current entering the control terminal is disregarded, and therefore the sum of currents entering the other terminals is assumed to be zero.

Some controllable switches (i.e., GTOs and IGCTs), draw fairly large currents from the gate for very brief periods during commutation. Gate currents are supplied by a gate driver, normally assembled with the GTO or IGCT. In this case, equation (8.2) is still valid, if i_1 and i_2 are the currents entering the device-gate driver assembly.

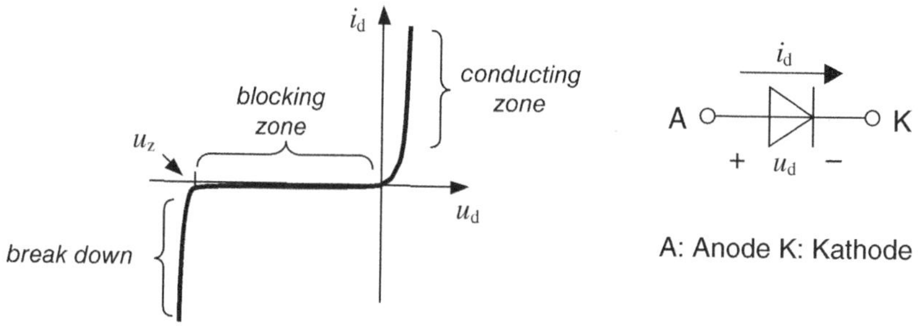

FIGURE 8.3. The diode *i-u* characteristic and its standard symbol.

Using this rule, all electronic devices can be considered as branches, in which the two main terminals can absorb currents having zero as their algebraic sum; the third terminal is represented just as a reminder that a *signal* is transmitted to send information to the devices (a command to switch on or off), but it carries no current.

8.2.3 Diodes

The diode is a two-terminal component. It can thus be connected to other components in circuits. Its specific behaviour is summarized by its *constitutive equation*, which, in general, describes the voltage across terminals, the current flowing in the component, and, possibly, their derivatives (see Chapter 3, and in particular Section 3.5.2 on "constitutive equations").

The constitutive equation of a diode is that of a nonlinear algebraic element and can be expressed in one of the following forms:

$$u_d = f_u(i_d) \qquad \text{or} \qquad i_d = f_i(u_d)$$

Function *f* has a very special form, as can be seen in Figure 8.3, in which u_d and i_d are measured according to references that follow the load sign convention.

Three zones can be singled out in the diode characteristic:

- The *conducting zone* (or *forward zone*), in which the voltage across terminals is very low (in the order of magnitude of 1 volt) and the current, it can be said, can flow freely, without significant further increases in voltage.
- The *blocking zone* (or *reverse zone*), in which the current is near zero (slightly negative), while the voltage is negative and can increase freely in absolute value, though only up to the point at which the third zone begins.
- The *breakdown zone*, in which both current and voltage are negative and not near zero.

In the blocking zone ($i_d < 0$), the diode absorbs very low power $p_d = u_d\, i_d$, because i_d is near zero. In the conducting zone ($i_d > 0$), the diode absorbs relatively low power

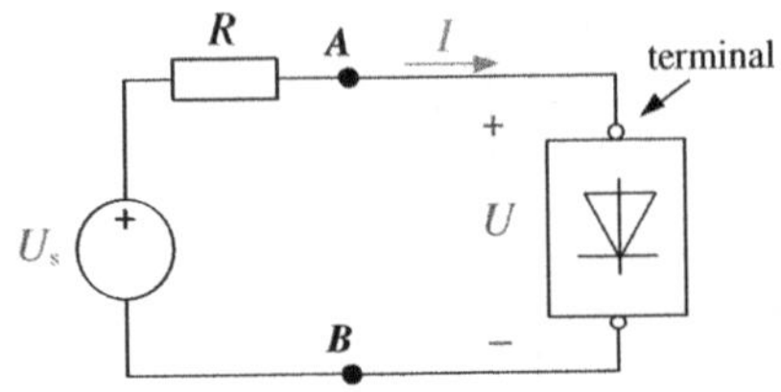

FIGURE 8.4. A simple circuit containing a diode.

because the voltage is very low. In the breakdown zone, the absorbed power is much greater than in the other two zones.

The breakdown zone is used only in special diodes called *zener diodes* (or, sometimes, *breakdown diodes*), which will not be dealt with in further detail in this book. Only the forward and reverse zones are, then, of interest in this chapter. Obviously, the systems into which the diodes are to be inserted will be such that the voltage is never lower than u_Z.

The reader might wonder how voltages and currents can be determined in a circuit containing a diode. Consider, for instance, the circuit shown in Figure 8.4. This can be seen as the composition of the part of the circuit to the left of nodes ***A*–*B*** and the one to the right. Both parts are algebraic. An algebraic two-terminal element (i.e., branch, according to this book's terminology) is characterised by a constitutive equation that contains neither derivative nor integral and is therefore an ordinary function, which can be represented by a curve in the *U–I* plane.

The operation of the circuit of Figure 8.4 in the *U–I* plane is represented in Figure 8.5. The operating point ***P*** determines the voltage and current U_P and I_P across terminals ***A* – *B***. The area of the rectangle having, as opposite vertices, the axis origin and point ***P*** is a graphical indication of the power lost in the component ($P_{lost} = U_p\, I_p$).

In the analysis of electronic power converters, a simplified version of the diode characteristic can very often be used, and this greatly facilitates circuit analysis. This is shown in Figure 8.6b. The breakdown zone is disregarded, since the converters are designed so that it is never reached. The blocking zone is idealized in such a way that the current, instead of being very low, is exactly zero. The conducting zone voltage is idealized also: the voltage across terminals, instead of being relatively low, is exactly zero.

In this way, the blocking zone of the actual diode is substituted by a nonconducting zone in the idealized diode, called *off-state*. Similarly, the forward zone is converted

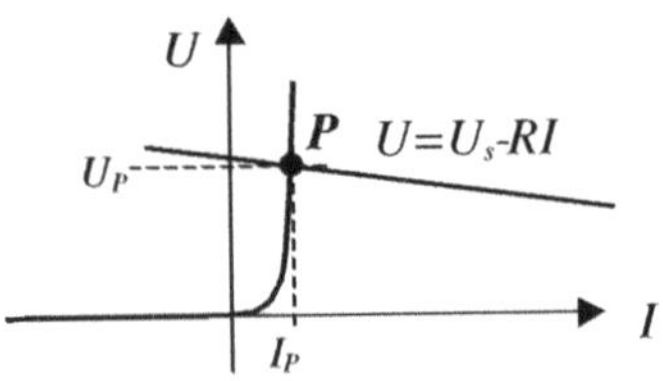

FIGURE 8.5. Determination of the operating point of the circuit of Figure 8.4.

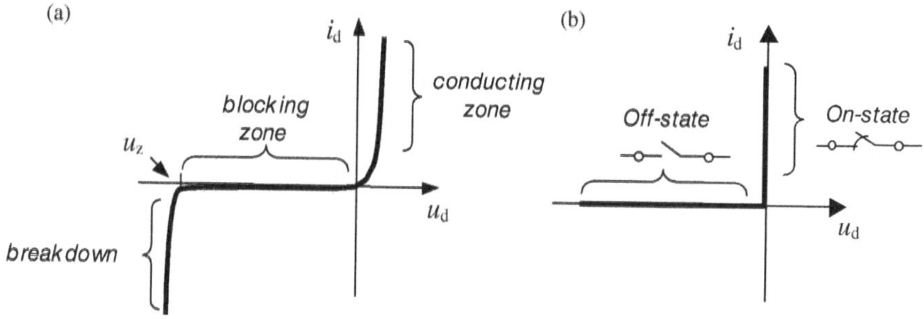

FIGURE 8.6. From real to idealized diode *i–u* characteristic.

into an ideal conducting zone, called *on-state* zone. With this idealization, the diode operates as a switch open (no current flow) in the off-state zone and closed (null voltage across terminals) in the on-state zone.

The opening/closing action is determined by the conditions in the circuit outside the diode, as is clear from the examples below.

Any change in the diode state (from on to off and vice versa) is called diode *commutation.*

In the following circuits we will need to evaluate the diode states in a couple of significant cases, which are analysed here.

Consider what happens when two diodes are connected in series, having cathodes in common (Figure 8.7a).

If $V_A > V_B$, the lower diode is off because it is reverse-biased; when $i_0 = 0$, the very small (negligible) reverse current of diode D_B will be transferred from terminal *A* to *B*. But when the load is such that some significant current i_0 flows (much greater than the diode's inverse current), this current can flow only through the diode D_A, which, therefore, becomes conductive.

If $V_A = V_B$ the situation is undetermined and the share of current between the two diodes will be determined by the details of the two real characteristics that will never be perfectly equal to each other. This condition is, however, of little interest.

Consider now three diodes with cathodes in common (Figure 8.7b). The condition in which two potentials have exactly the same value is of little interest and can be

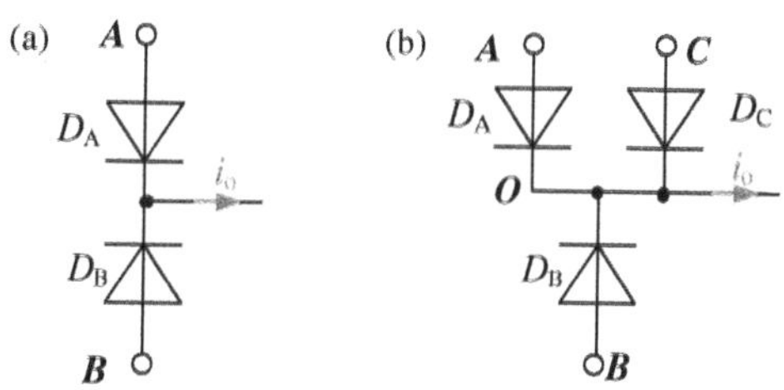

FIGURE 8.7. Two significant cases of diode connection.

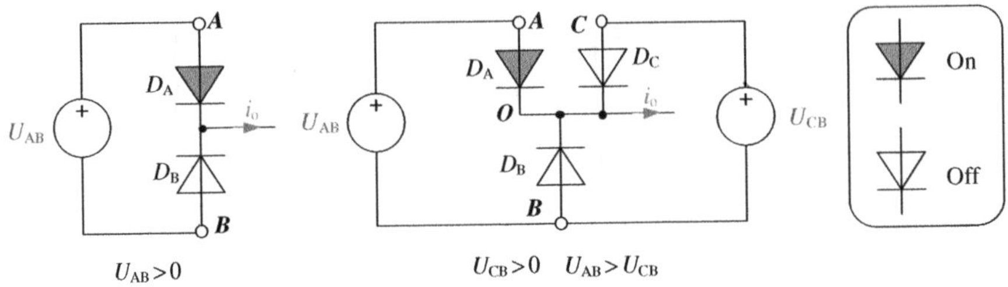

FIGURE 8.8. Conducting diodes in common-cathode configuration.

ignored. We can therefore give terminals their names in such a way that it is $V_A > V_C > V_B$. Applying what we have just learned in the case of two diodes connected by their cathodes, we infer that diode D_B is necessarily off, because both $U_{AB} > 0$ and $U_{CB} > 0$.

Applying the two-diode rule to the D_A–D_C couple, we can infer that also D_C is off. When $i_0 = 0$, the diode D_A will transfer only their very small inverse currents toward D_B and D_C. But when the load is such that a significant current i_0 flows (much greater than the inverse current of the diodes), it can flow only through the diode D_A (the only non-reverse biased), which therefore becomes conductive.

These two examples justify the following general rule:

> **Rule:** *State of diodes connected by their cathodes to a common point*
>
> When two or more diodes are connected to each other by their cathodes the diode having at its anode the highest potential will be conducting, the others will be blocked.

This is graphically expressed in Figure 8.8, in which conducting diodes are shaded.

It should now be quite easy to understand the rule for diodes with anodes in common:

> **Rule:** *State of diodes connected by their anodes to a common point*
>
> When two or more diodes are connected to each other by their anodes, the diode having at its anode the lowest potential will be conducting, the others will be blocked.

These are pictorially described in Figure 8.9.

8.2.4 Thyristors

Thyristors were widely used in the second half of the twentieth century but are now being phased out in low-power and medium-power applications because of the

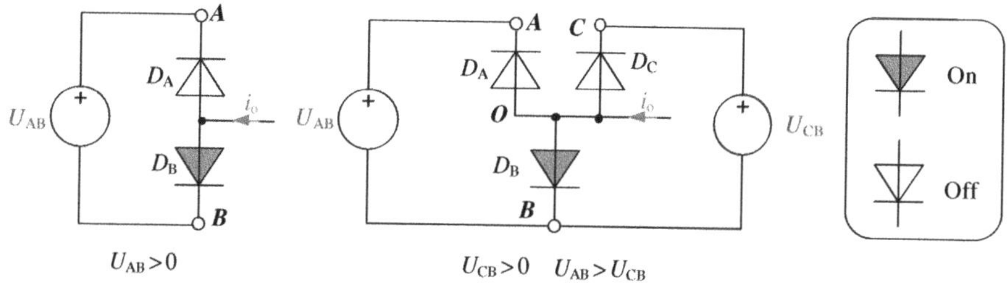

FIGURE 8.9. Conducting diodes in common-anode configuration.

diffusion of other more economical semiconductor devices, such as IGBTs, which we will study later.

Thyristors are still used in very large power applications (greater than a few tens of megawatts); but since their assessment and selection needs electric power engineers, they are beyond the scope of this book. Thyristor-based electronic converters, then, will not be dealt with in this chapter.

For the sake of completeness, the basic behaviour of thyristors is however explained below.

From Figure 8.10, it is clear that the thyristor has an on-state characteristic very similar to a diode. However, when an adequate signal is sent into the gate terminal, constituted by a pulse of very short duration (in the order of tens of microseconds) and very low value (several orders of magnitude below the maximum current that can flow in anode and cathode), the SCR is switched from off to on.

The reverse switch action (i.e., from on to off) cannot be obtained by using the gate terminal: it occurs naturally when the SCR enters the reverse zone, and it stays there for at least a characteristic time, called "turn-off time." Typical turn-off times are some tens of microseconds. As a consequence, the thyristor can be imagined as a device which operates as a switch; only the switching action from off to on is controllable, while the reverse occurs naturally.

As for the diode, any change in the SCR state (from on to off and vice versa) is called SCR *commutation*.

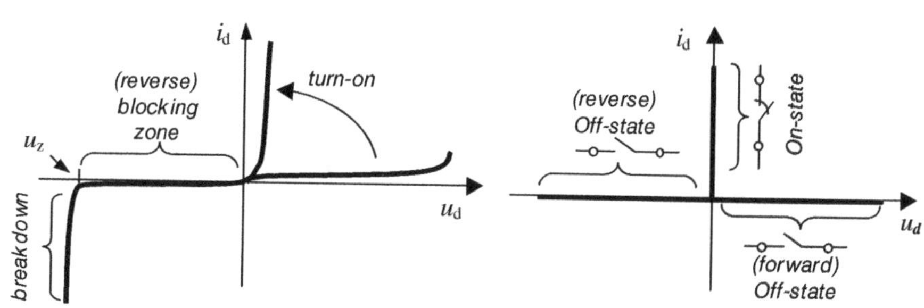

FIGURE 8.10. Real and idealized SCR characteristics.

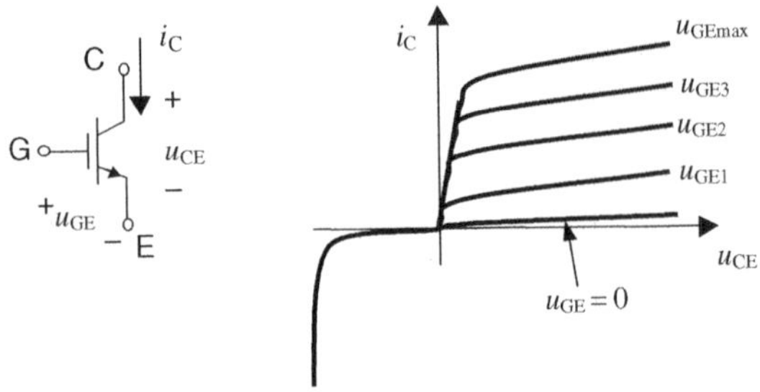

FIGURE 8.11. The IGBT symbol and characteristics.

8.2.5 Insulated-Gate Bipolar Transistors (IGBTs)

Insulated-gate bipolar transistors, like SCRs, appear from the outside to be three-terminal devices. However, they can be turned on *and* off by suitable action on gate terminal.

A qualitative representation of the IGBT's characteristics is shown in the left-hand part of Figure 8.11. The power terminals are collector (C) and emitter (E); the control terminal gate (G) always draws negligible current in comparison to the current flowing through the power terminals. The voltage u_{GE} between collector and emitter determines the behaviour of the IGBT on the C–E side.

To understand this more clearly, consider the IGBT connected to external circuits both at the GE and CE sides, as shown in Figure 8.12. The operating point ***P*** of the circuit can be determined in a very similar manner to that explained in Section 8.2.3 for diodes—that is, by finding the intersection of the *u*–*i* characteristics of the parts of the circuit to the left and right of the ***A***–***B*** pair.

The circuit to the right of the ***A***–***B*** points can be described by a constitutive equation which is equivalent to the U_s–R_i pair in series, whose intercepts to the u_{CE}–i_c axes are U_s and U_s/R_i respectively.

The circuit to the left of ***A***–***B*** is dependent on u_{GE} voltage. However, once the u_{GE} is known, the specific characteristic of the IGBT (i.e., the constitutive equation between C and E), which must intersect with the straight line representing U_s/R_i, is also immediately known: this is shown in black in the right-hand part of Figure 8.12.

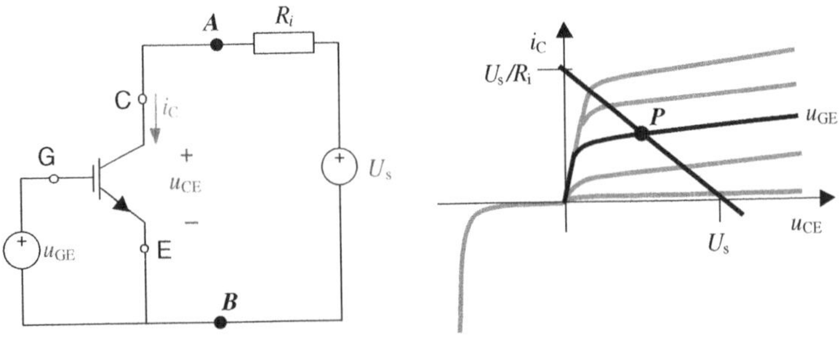

FIGURE 8.12. Operating point of an IGBT at a point ***P*** of a curve in the *i*–*u* plane.

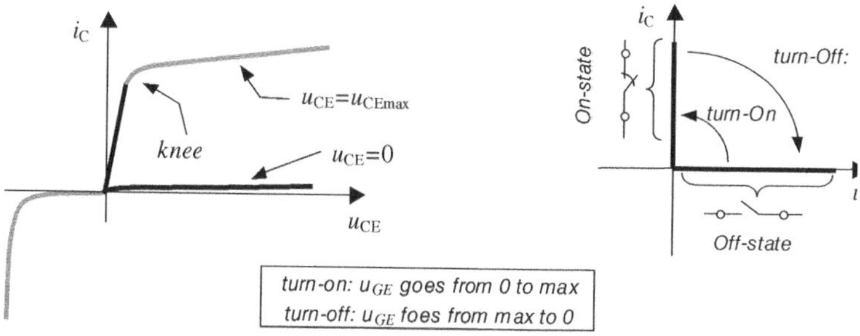

FIGURE 8.13. Real and idealized IGBT characteristics.

It must be said, however, that points like ***P*** in Figure 8.12 are never (at least as regards this book) used in power electronics. The only zone of the C–E IGBT characteristics of interest to electronic power converters is the one before the knee bend—that is, the black part as shown in Figure 8.13.

Moreover, in the vast majority of power electronics applications, only the first quadrant part of the characteristics is of interest—that is, the lines shown in black in the left-hand part of Figure 8.13. It is now natural to identify an idealized version of these characteristics, which are shown in the right part of the same figure. According to these characteristics, the IGBT operates as a switch that can be in on-state (voltage is zero) or off-state (current is zero).

The transition from on to off and vice versa is controlled by acting on the voltage applied between gate and emitter.

The idealized characteristics shown in the right-hand part of Figure 8.13 contain only positive values of voltage and current; this is acceptable in all situations in which a diode is connected in *antiparallel* with the collector-emitter terminals (Figure 8.14).

In general the word "antiparallel" can be used to indicate a special connection of two branches having a preferential direction of current flow. When these two preferential directions are opposite each other, they are said to be connected in antiparallel.

Using the connection shown in Figure 8.14, the IGBT is required to operate as a controllable switch, to be traversed by current and subject to voltage only when i_C and u_{CE} are non-negative. Whenever each of them tends to be negative, the diode switches on, and the current flows in the diode instead of in the IGBT.

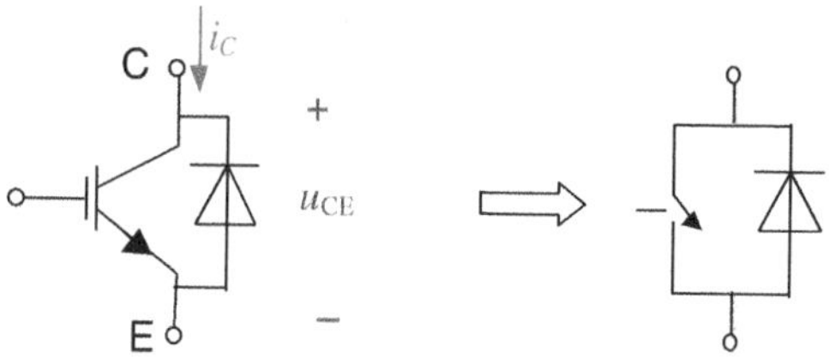

FIGURE 8.14. Connection of an IGBT and antiparallel diode.

8.2.6 Summary of Power Electronic Devices

In Section 8.2.1 we presented a list of possible controllable switches. In Section 8.2.5 we looked at the way in which an IGBT can operate as a controllable, one-directional switch.

It would be easy to repeat the analysis presented in Section 8.2.5 also for MOSs, GTOs, BJTs, and so on. However, this is not necessary for a basic understanding of the operation of power electronics conversion. What must be known is that, except for minor details, all of them operate as controllable (one-directional) switches, and indeed they are very often represented in circuits using a single, generic symbol as shown in the right part of Figure 8.1.

When a controllable switch is switched from on to off or vice versa, it is *commutated* and the switching action is called *commutation*.

Differences, however, exist:

- Not all of them have the same losses. The differences can be in terms of on-state losses (caused, as already noted, by the fact that the voltage is not exactly zero and the product voltage times current determines the losses) and switching losses—that is, the losses created each time a switch is moved from on-state to off-state and vice versa.
- Some of them are more suitable for fast commutation. For instance, Mosfets can be commutated at frequencies that can reach 100 kHz and more, while GTOs can be switched at a few hundred hertz at most.
- Some of them are able to carry huge on-state currents and/or withstand large off-state voltages, others are more limited in this respect. The high-voltage/high-current devices are more suitable as basic elements of large power converters;

In Table 8.1 the switches are compared in terms of power (product of maximum on-state direct current, and off-state reverse voltage) and commutation speeds. In general, it happens that the fastest switching devices are limited in terms of on-state current and reverse-voltage capability and vice versa. Also, power electronic devices obey the general rule of nature; that is, *big is slow, small is fast*!

The two opposite extremes in this respect are GTOs-IGCTs and MOSFETS, the first being powerful and slow, the latter fast and suitable for low powers. IGBTs are becoming more and more common because they are suitable for both average and high powers and speeds. MOSFETs are used inside converters up to a few kilowatts, while GTOs and IGCTs are required inside converters larger than a few megawatts.

TABLE 8.1. Comparison of Some Characteristics of Different Power Electronic Devices

Component	*Power*	*Commutation Speed*
MOSFET	Low	High
GTO-IGCT	High	Low
IGBT	Medium	Medium

8.3 POWER ELECTRONIC CONVERTERS

8.3.1 Rectifiers

A very important static power conversion frequently carried out in electric power systems is unidirectional AC to DC. This is performed using converters based on diodes, called rectifiers.

Rectifiers are devices that convert AC power into DC without affecting (if we disregard inner power losses) the transferred power. This concept is illustrated in Figure 8.15, showing single-phase and three-phase rectifiers. This is a generic black-box representation concealing inner details, which is useful when all that matters is their external behaviour.

In practice, output power will be slightly lower than input power. Normal efficiency (i.e., ratio of output power to input power) is around 95–98%.

It is now time to look inside the black boxes shown in Figure 8.15, showing how AC (either single-phase or three-phase) can be converted to DC using the power electronic components examined in the previous section.

In its basic form, a single-phase rectifier has an architecture called *single-phase bridge*, as shown in the left-hand part of Figure 8.16. To analyse the behaviour of

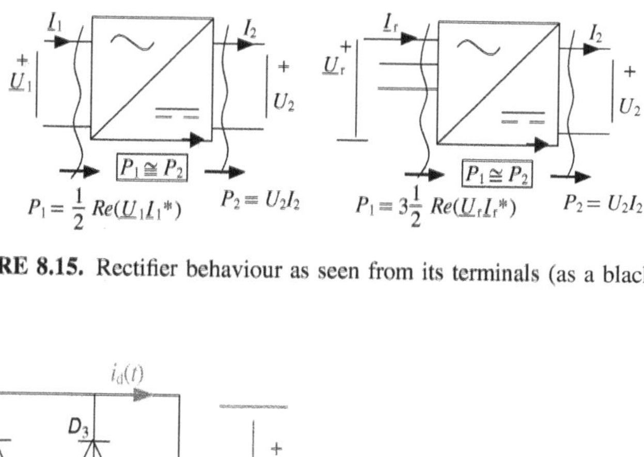

FIGURE 8.15. Rectifier behaviour as seen from its terminals (as a black box).

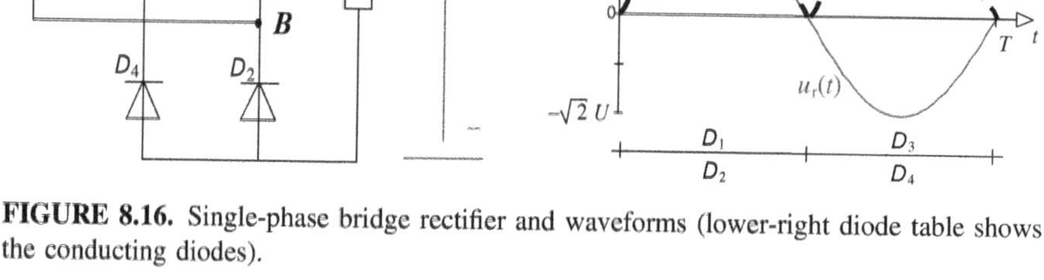

FIGURE 8.16. Single-phase bridge rectifier and waveforms (lower-right diode table shows the conducting diodes).

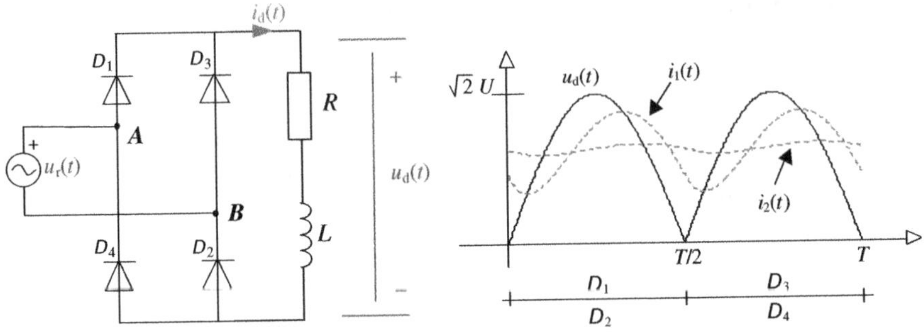

FIGURE 8.17. Single-phase bridge rectifier loaded with R–L, and waveforms (lower-right diode table shows the conducting diodes).

this bridge, we observe that the upper diodes are connected as depicted in the left-hand part of Figure 8.8, and therefore the diode corresponding to the higher potential is conducting while the other is reverse-biased (off). On the other hand, the lower diodes are connected as depicted in the left-hand part of Figure 8.9, and therefore the diode corresponding to the lower potential is conducting while the other is reverse-biased (off). Therefore, during the time interval between 0 and $T/2$, D_1 and D_2 are conducting (and the other diodes blocked), and between $T/2$ and T the opposite states occur.

As a consequence of this simple analysis, the output voltage will have the shape shown in Figure 8.16b; that is, it is unidirectional and has a nonzero mean value. It is not exactly what we would call a DC voltage, but the fact that it has an average significantly greater than zero means that, by appropriate filtering, a reasonably quasi-constant voltage can be obtained. Filtering techniques will be discussed in Section 8.4. In this figure, $U = \mathrm{rms}(u_r(t))$.[1]

There are, however, some DC loads that only require unidirectional voltages, like the one shown in Figure 8.16b; these can be fed directly from the output of the single-phase rectifier.

It is interesting to evaluate what happens in the same circuit if the load is constituted by an R–L couple, instead of a simple resistance (Figure 8.17). In this case, the current is not proportional to the voltage, but has a different shape: it is somewhat filtered $[(i_{max} - i_{min})/i_{avg}$ is lower than $(u_{max} - u_{min})/u_{avg})]$ as well as delayed. The current $i_2(t)$ refers to a larger L/R than $i_1(t)$. It is apparent that the higher this L/R ratio, the smoother the current.

The output voltage waveform is easily evaluated simply by checking the polarity of the voltage source, as was done for the example of Figure 8.16; the current wave is normally computed using one of many available simulation programmes. In this figure, $U = \mathrm{rms}(u_r(t))$.

[1]This expression indicates that U is the root mean square of the wave $u_r(t)$. The meaning of the rms value of a sine wave is discussed in Chapter 5.

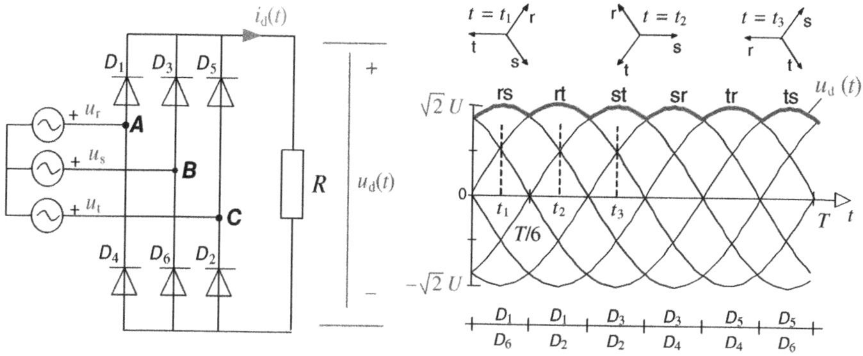

FIGURE 8.18. Three-phase bridge rectifier loaded with a resistor and waveforms (lower-right diode table shows the conducting diodes).

Current $i_d(t)$ can also be determined by means of harmonic analysis, using procedures based on the theory or Fourier series, shown in Section 8.4 and 8.5. However, because of the above-mentioned availability of computer programmes, the length of this procedure, and the introductory nature of this book, details of how to do this are not a priority here.

As a second, very significant example of rectifiers, we now introduce the three-phase rectifier, shown in Figure 8.18, as fed by ideal sinusoidal voltages and loaded with a resistor R. The conducting diodes are determined by potentials of points ***A***, ***B***, ***C***, according to the rule stated in Section 8.2.3: the upper load terminal is the common cathode of the three upper diodes, and will therefore have the highest potential from among V_A, V_B, and V_C. Similarly, the lower load terminal will have the lowest potential from among V_A, V_B, and V_C. There will thus be a voltage $u_d(t) = \max(V_A,\ V_B,\ V_C) - \min(V_A,\ V_B,\ V_C)$. This voltage will cause a current $i_d = u_d/R$ to flow.

This is applied at all times t in the plot in the right part of Figure 8.18. In this figure, $U = \text{rms}(u_r(t)) = \text{rms}(u_s(t) = \text{rms}(u_t(t))$.

The figure also shows the *commutation* times in which diodes change their state (for instance, $t = T/6$, $t = 2T/6$, etc.) Clearly, at these times actual voltages and currents are determined by the real, not ideal, diode characteristics, but this more detailed analysis will not change the results shown in Figure 8.18 except for minor details.

The reader might have noted how the diodes are numbered. This choice of subscripts is fairly standard and has a mnemonic advantage. It can indeed be seen that, using this numbering, the upper diodes, with odd numbers, enter their on-state in natural sequence: 1-3-5-1-3-5, etc. Similarly, the lower diodes, having even subscripts, enter their on-state in natural sequence as well: 2-4-6-2-4-6, etc.

It is also to be noted that while commutation takes place every sixth of the period, each diode stays in its on-state for a third of the period.

The average value of the DC voltage is easily computed using the definition of the average of a quantity:

$$U_d = \frac{1}{T/6}\int_0^{T/6} u_d(t)\,dt = \frac{6}{T}U\int_{-T/12}^{T/12}\sqrt{2}\,U\cos(\omega t)\,dt = \frac{3}{\pi}\int_{-\pi/6}^{\pi/6}\sqrt{2}\,U\cos(\omega t)\,d\omega t$$

$$= \frac{3}{\pi}\sqrt{2}\,U\left(\sin\frac{\pi}{6}+\sin\frac{\pi}{6}\right) = \frac{3}{\pi}\sqrt{2}\,U \cong 1.35U$$

This result is easily interpreted by looking at the right-hand part of Figure 8.18: the average value should be slightly lower than the maximum of $u_d(t)$ because of the ripple, and in fact it is reduced by a factor of $3/\pi = 0.955$.

To understand the behaviour of the currents delivered by the AC three-phase network, consider that at each circuit configuration (each commutation causes a configuration change) one of the three voltage sources delivers current (which is thus equal to the DC load current), another receives this current back, and no current flows through the third one.

Referring again to time t_1 of Figure 8.18, it is apparent that $i_r(t_1) = i_d(t)$, $i_s(t_1) = -i_d(t_1)$, $i_t(t_1) = 0$.

The global situation is shown in Figure 8.19 for $i_r(t)$. Obviously, $i_s(t)$ has the same shape as $i_r(t)$ but 120 degrees (i.e., $T/3$ seconds) later, and $i_t(t)$ is delayed by another $T/3$ seconds.

Finally consider a three-phase diode bridge with an *R–L* load (Figure 8.20). To show a more realistic situation on the AC side, three inductors are also shown, to simulate Thévenin's impedance of a three-phase transformer, normally present at the supply side of three phase diode rectifiers.

As for the single-phase rectifier, the load inductance makes the load current much smoother than the voltage. Moreover, the presence of inductances at the supply side has an effect on commutation; this can be verified by theoretical analysis, disregarded here for the sake of simplicity.

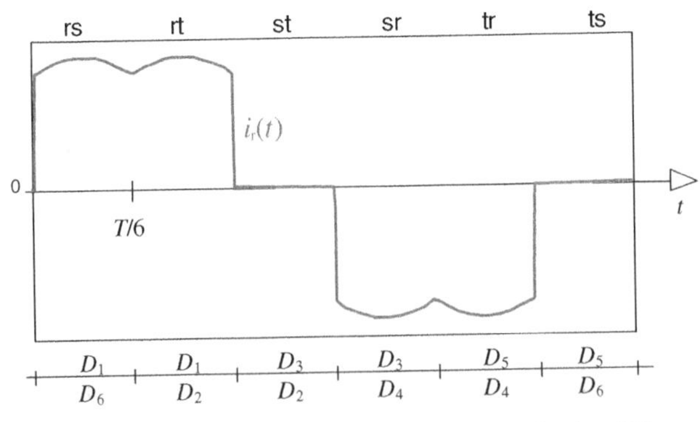

FIGURE 8.19. Current absorbed by phase r in the circuit of Figure 8.18.

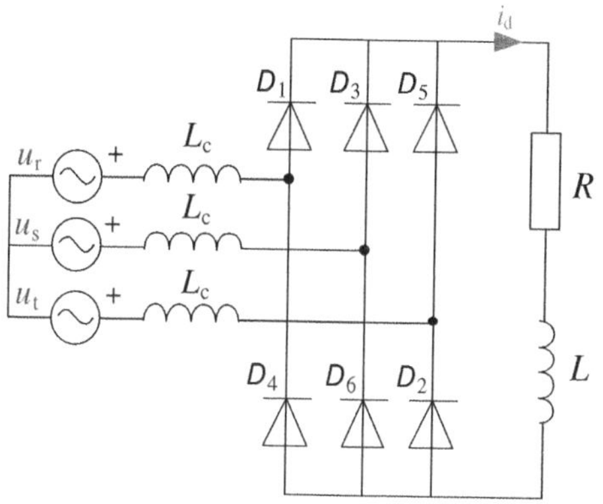

FIGURE 8.20. Three-phase rectifier loaded with R–L.

As usual, the situation can be evaluated using a simulation programme, which would produce, for instance, the results shown in Figure 8.21. In the left part of this figure, the behaviour of the bridge, shown in Figure 8.20, is viewed under the hypothesis that $L_c = 0$ and the inductance/resistance ratio in the load part is much greater than the period T of the supply voltages.

In the right-hand part, more realistic waveforms are shown, with a nonzero L_c and $L/R = T/2$.

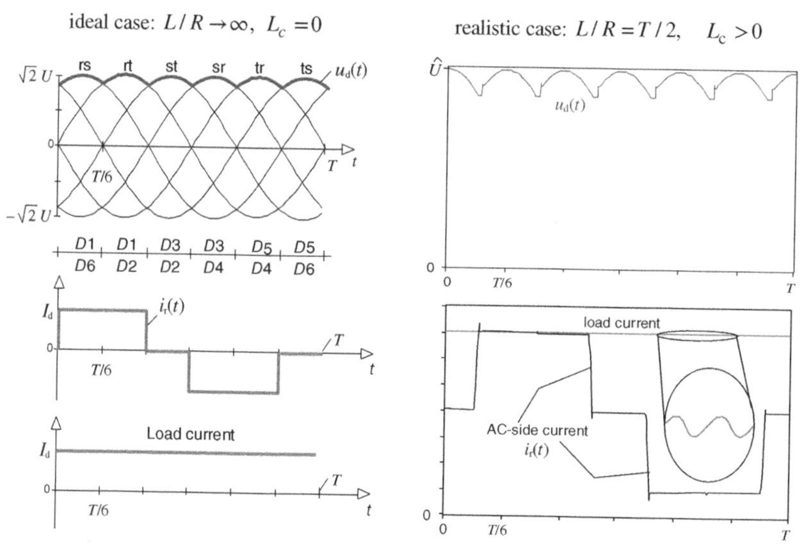

FIGURE 8.21. Voltages and currents for an ideal (*left*) and more realistic (*right*) three-phase rectifier.

Some comments can be made:

- In the ideal case, shown in the left-hand part of the figure, the DC current is flat. The corresponding AC current equals the DC one for one-third of T, is zero for $T/6$, and is equal to the DC current but with opposite sign for another $T/3$.
- In the realistic case, the DC current presents some ripple, even though in the full scale of the plot it is still not visible. The AC side current no longer has vertical segments, which are substituted by curved segments.
- In the realistic case, the DC voltage has some notches, which cause the average value to be lower than in the ideal case. It can be demonstrated that the reduction of the average DC voltage due to this phenomenon is proportional to the average DC current.

Moreover, it can be demonstrated that those curved segments which in the real case substitute the vertical segment of the AC current, are parts of sine curves. It is worth noticing that, because the constitutive equation of inductors is $u_L = L\mathrm{d}i/\mathrm{d}t$, the vertical segments are not allowed, since otherwise the voltages across L_c terminals would be infinite. This further justifies the fact that the presence of L_c causes these vertical segments to disappear.

To avoid the complexity of analysing all the details of the three-phase bridge when it is inserted into a more complex system, an equivalent DC-side circuit is shown in Figure 8.22.

In this circuit:

- U_{d0} represents the average DC voltage which is present when I_d or L_c is negligible. As already said, this is $U_{d0} = (3/\pi)\sqrt{2}U$.
- R_c represents the voltage drop due to the presence of L_C's on the AC side; as already mentioned, voltage drop is proportional to the DC current. It can be demonstrated that it is $R_c = (3/\pi)\omega L_c = (3/\pi)X_c$ (ω being $2\pi/T$) such a mathematical detail is beyond the scope of this book.
- The diode is needed to ensure that the current cannot be reversed.

Needless to say, this equivalent disregards any ripple in the quantities and takes into account only average values.

It must also be noted that R_c models the effects of L_c's, and does not imply energy losses. It is not a real resistor, but only a way to indicate that the DC voltage is reduced by an amount proportional to the delivered current.

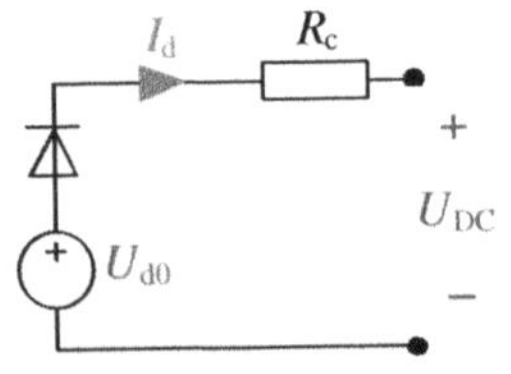

FIGURE 8.22. DC circuit equivalent to the three-phase rectifier of Figure 8.20.

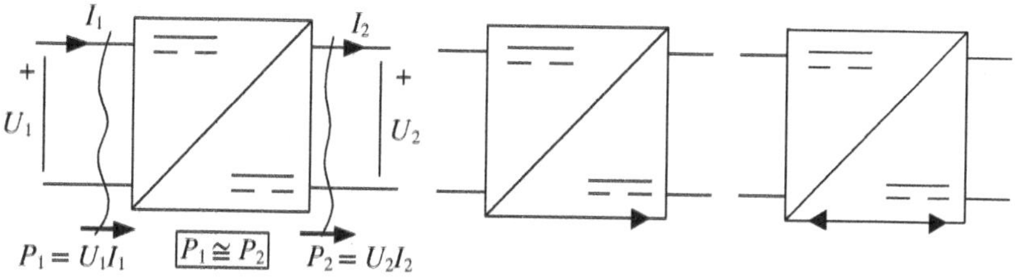

FIGURE 8.23. Black-box representation of DC–DC converters.

8.3.2 DC–DC Converters

DC–DC converters can modify the two parameters that define DC power—that is, voltage and current—while leaving their product nearly unchanged. In this respect they can be represented as black boxes as indicated in Figure 8.23.

The generic DC–DC converter shown in the left-hand part of the figure gives no indication of the possible power flows. Indeed, some DC–DC converters can transfer power only from one side to the other and not vice versa; in this case the symbol shown in the central part of Figure 8.23 is more informative. If power is able to go in either direction, the symbol shown in the right-hand part of the same figure may be used.

Ideally, a DC–DC converter operates in such a way that

$$U_1 I_1 = U_2 I_2$$

In this respect, a DC–DC converter is very similar to the ideal transformer [Chapter 5, Figure 5.14 and Chapter 7, equation (7.7)], and is similar to a "transformer operating in DC systems."

The operating principle of the transformer is based on Faraday's law and is thus unable to transform voltage or current while in DC; until the last two decades of the twentieth century, when static DC–DC converters first appeared, this limitation was a significant hindrance to the diffusion of DC as a means to carry and distribute electric power.

Today, static energy conversion is a reality and this kind of transformation is easy and cheap, even though it does not achieve the very high efficiency of the best transformers. However, using relatively high quality components and architecture, conversion efficiencies of around 98% are quite easy to obtain.

In its basic form, a DC–DC converter is simply composed of a controllable switch and a diode. Consider, as a first example, the *step-down DC–DC converter* (also known as *buck DC–DC converter*). This is able to transfer power from a system 1 to a system 2 under the condition that $U_1 > U_2$ (hence its name). The power cannot be reversed.

Two typical uses of a step-down converter are (a) to create a constant DC source whose voltage U_2 is lower than the input voltage U_1 and (b) to feed some quasi-constant current into an already-existing constant DC voltage.

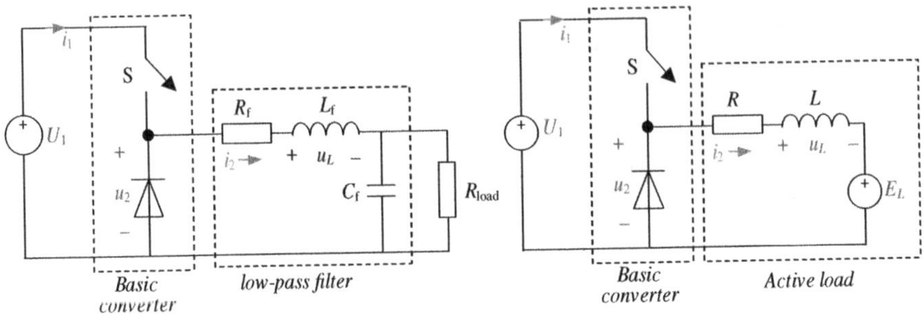

FIGURE 8.24. Two DC–DC step-down converters connected to two load types.

These two cases are shown in Figure 8.24. Remember that the controllable switch may represent any of the devices able to operate as a controllable unidirectional on-off switch (such as BJTs, GTOs, IGBTs, etc.; see Section 8.2). The output of the basic converter is far from constant and has to be improved using a filter to obtain a reasonably constant DC voltage at the load terminals; we will see in Section 8.4 how the scheme shown in the figure can effectively operate as a filter.

In the second case the load just requires some current in-feed. Both load and basic converter require the current to be constant or quasi constant; this is achieved with the help of inductor L, exploiting the inherent capability of inductances to impede current fluctuation, and no explicit filter is needed. This is the case of a DC–DC converter that feeds a DC machine: it will be seen in Chapter 10 that the machine can be effectively modelled as a voltage source E in series with an inductor and a resistor.

Imagine the filter is operating correctly; the voltage across capacitor C is, therefore, almost perfectly constant, and the behaviour of the basic converter can then be analysed in either case using the circuit shown in the right part of Figure 8.24.

In DC–DC converters the controlled switch S is continually switched on and off in a repetitive manner (Figure 8.25): it is held on—that is, in a forward conduction state—for a time T_c, and Off for a blocking time T_b. After this, it is again switched on and held in such a state for T_c, and the process continues. The process therefore repeats itself in such a way that the periodicity of all the quantities is $T = T_c + T_b$.

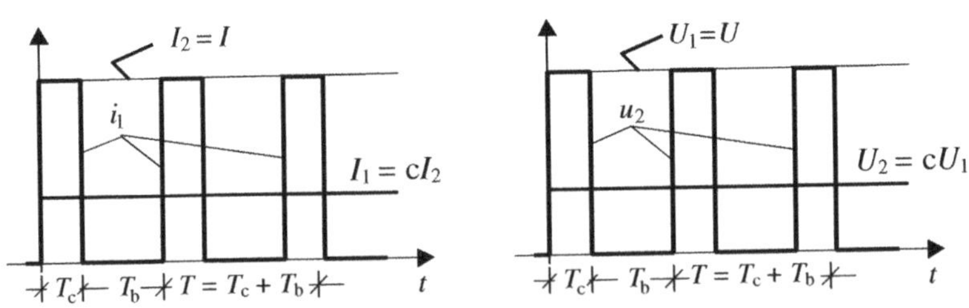

T_c = conduction time	T_b = blocking time	$T = T_c + T_b$
$c = T_c/T$ = conduction index	$b = T_b/T$ = blocking index	$c + b = 1$

FIGURE 8.25. Idealized operation of a step-down DC–DC converter.

If the inductance L is large enough to keep the converter output current $i_2(t)$ really constant,[2] it can be written $i_2(t) = I_2$ and the shapes of basic converter voltage and current are those shown in Figure 8.25.

They can be explained as follows. During T_c, S is on, and therefore the input voltage U is transferred at the output of the basic converter: $u_2 = U$. Moreover, the input current will be equal to the output current: $i_1 = I_2$.

At the end of T_c, S is switched off (by action on its gate terminal, not shown in the circuit). The inductance connected at the output terminals of the basic converter will prevent the output current from suddenly stopping, thus generating an overvoltage that counteracts any reduction of current; that is, the potential of the wire to which the diode anode is connected will rise above that of the cathode, thus causing the diode to switch on. When the diode is on, as discussed in Section 8.2.1, it can be analysed using the idealized characteristic—that is, disregarding the voltage across its terminals.

We conclude that when S is switched off, the diode is forced by the inductance action to switch on, and thus during this interval it will be $u_2 = 0$. During the same interval it is $i_1 = 0$, since S is off and therefore no path exists for the current to flow.

This analysis justifies the shape of the plots shown in Figure 8.25. The corresponding averages of the time-varying quantities—that is, $u_2(t)$ and $i_1(t)$—are easily computed:

$$U_2 = \frac{1}{T}\int_0^T u_2\,\mathrm{d}t = \frac{1}{T}T_c U_1 = cU_1 \qquad \text{and} \qquad I_1 = \frac{1}{T}\int_0^T i_1\,\mathrm{d}t = \frac{1}{T}T_c I_2 = cI_2$$

Consequently;

$$U_1 I_1 = U_2 I_2$$

as expected. The quantity

$$c = \frac{T_c}{T} \tag{8.3}$$

is called here *conduction ratio* (according to some books, *duty ratio* or *duty cycle*).

It should be apparent that the proposed circuit is only able to transfer power from the left-hand side to the right-hand side, and the left-hand side of the basic converter must have a higher voltage than the one required to supply the load—hence the name step-down converter for this circuit. The very disposition of elements in the circuit (the wire connecting U_1 to the circuit is shown higher than that leaving the diode suggests this voltage relationship.

The plots shown in Figure 8.25 consider primary secondary current as being perfectly constant. A more realistic picture of what happens is shown in Figure 8.26. Here the voltage is exactly the same as in Figure 8.25, but the secondary current is not

[2]We will soon see that this occurs when $L/R \gg T$.

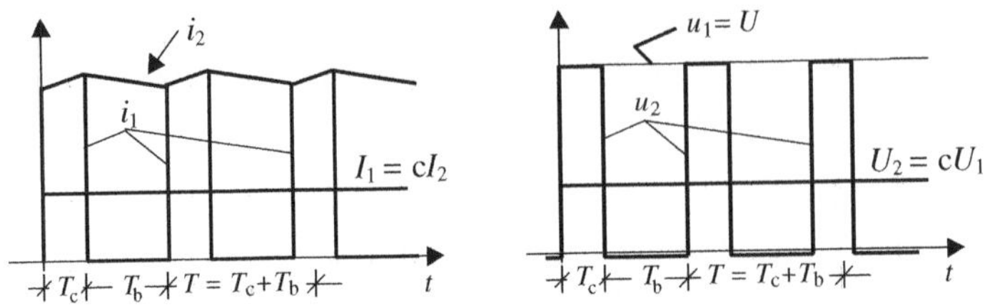

FIGURE 8.26. Realistic operation of a step-down DC–DC converter.

perfectly constant. The current i_2 rises when S is On, and falls when it is off. During these intervals it can be considered with a fair degree of accuracy to be composed of segments of a straight line.

The derivative of $i_2(t)$ is related to the voltage appearing across the inductor L according to the inductor constitutive equation (written considering the voltage and current references shown in Figure 8.24):

$$u_L = L\frac{di}{dt}$$

Indeed, during T_c the voltage u_L is positive and equal to $U_1 - U_2$ (disregarding the drop across R), while during T_b it is $u_L = -E$ and is therefore negative. In steady state the average of i_2 over T_c and T_b must be equal to each other, and this occurs when $U_2T = U_1T_c$, confirming our earlier findings.

***More in Depth:** Actual current and voltage shapes.*

On examining again the circuits that are determined when substituting devices with on-state or off-state switches students will realize that during both T_c and T_b they are composed of a loop containing just a constant voltage, a resistor, and an inductor.

Therefore, the theory of U_s–R–L circuits applies (that was dealt with in Section 5.4), and the shape of currents is exponential. However, for times which are much shorter than the time constant of U_s–R–L circuits $\tau = L/R$, the exponential function can be approximated a segment of a straight line, as seen in this section outside this *more in depth* box.

The output voltage $u_2(t)$, while having a nonzero average, is far from constant, as an ideal DC source should be.

However, a filter can be interposed between the DC–DC converter and the load. A filter is able to transfer the average value of a signal to the load, while (almost totally) blocking its fluctuation around this average.

Some information on how filters can be made in practice will be provided in Section 8.5.

More in Depth: *Harmonic spectrum of the produced voltage.*

This box can be studied at a second reading of the chapter, after reading Section 8.4.

To gain better knowledge of how to deal with this voltage in circuits, it is useful to examine its frequency spectrum. It is composed of a constant component U_0, and several harmonics. Two examples are shown in Figure 8.27.

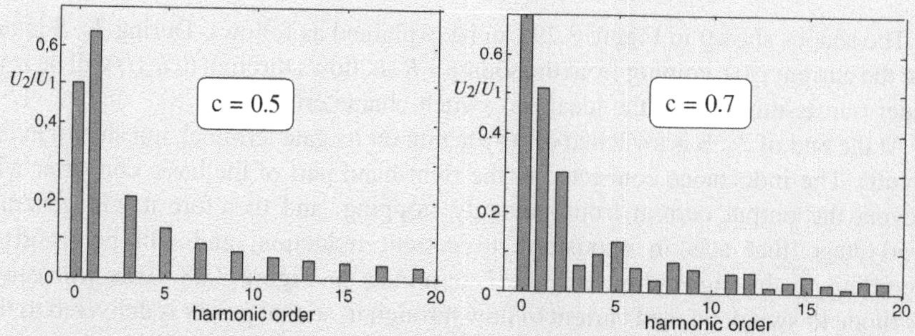

FIGURE 8.27. Frequency spectra of $u_2(t)$ of Figure 8.25 for two values of c.

Consider now the DC–DC converter shown in Figure 8.28. This circuit is designed to transfer power from the lower voltage side (represented on the right-hand side of Figure 8.28) toward the higher voltage side of the converter. It is thus called *step-up DC–DC converter* (aka *boost DC–DC converter*).

The principle of operation is very similar to that of the step-down, and it can be illustrated by referring to the plots shown in Figure 8.29. In this case also, the inductance L must be so large as to keep substantially constant the current $i_2(t)$ flowing through it.

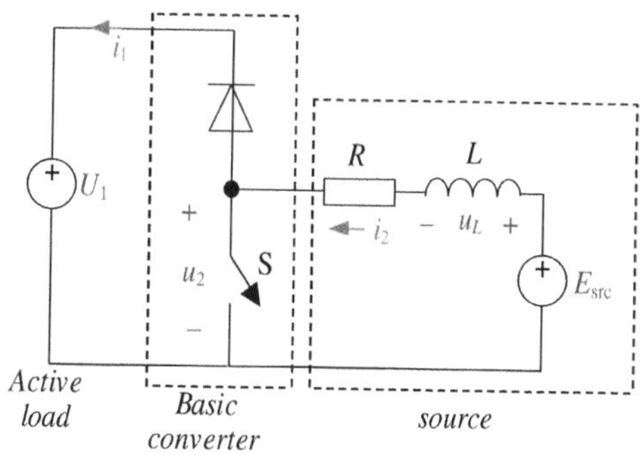

FIGURE 8.28. The DC–DC step-up converter.

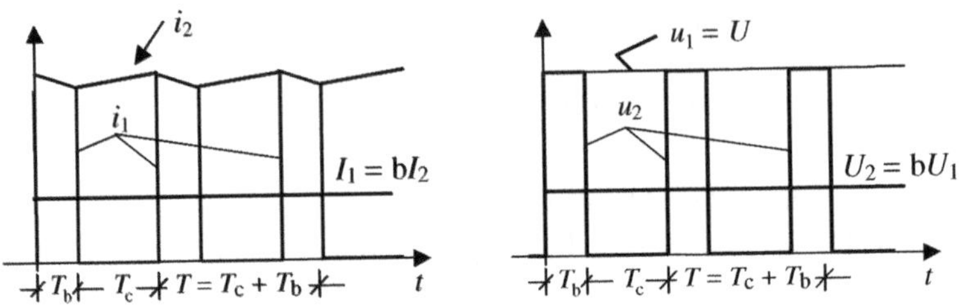

FIGURE 8.29. Operation of a step-up DC–DC converter.

The shapes shown in Figure 8.29 can be explained as follows. During T_c, S is on, and the current $i_2(t)$, coming from the source's R–L, flows through it; $u_2(t)$ will be zero under our assumption of the idealized switch characteristic.

At the end of T_c, S is switched off (by action on its gate terminal, not shown in the circuit). The inductance connected to the right-hand part of the basic converter will prevent the output current from suddenly stopping, and therefore it will generate overvoltage that acts in opposition to current reduction, and will be positive, according to the signs shown at the L terminals in Figure 8.28. This will cause the diode to switch on, and current to flow through it, so that power is delivered to the active load.

During T_c (disregarding the drop across R) it will be

$$u_L(t) = E_{src}$$

while T_b, on the contrary (again ignoring the drop across R), will be

$$u_L(t) = E_{src} - U_1 < 0$$

Thus, considering the constitutive equation of the inductance; we obtain

$$u_L(t) = L\frac{di_2}{dt}$$

Note that during T_c the current i_2 will increase while during T_b it will decrease.

From the shapes shown in Figure 8.29 it is clear that the current shape can be deemed, with a high degree of accuracy, to be formed by a sequence of linear segments. As a consequence it, will be

$$U_2 = \frac{1}{T}\int_0^T u_2\,dt = \frac{1}{T}T_bU_1 = bU_1 \quad \text{and} \quad I_1 = \frac{1}{T}\int_0^T i_1 dt = \frac{1}{T}T_bI_2 = bI_2$$

Consequently,

$$U_1I_1 = U_2I_2$$

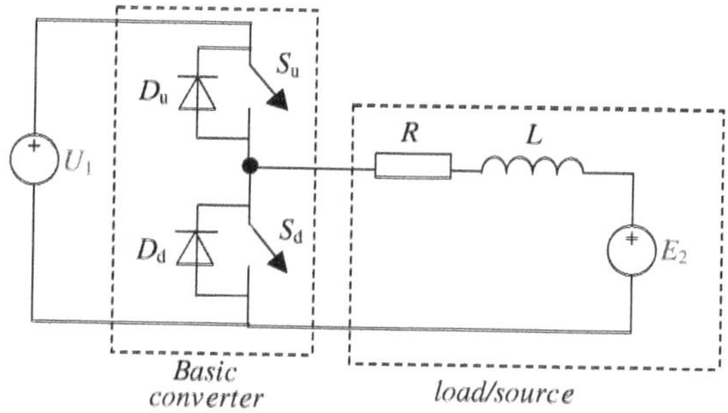

FIGURE 8.30. The step-up/step-down DC–DC converter.

The quantity

$$b = \frac{T_b}{T}$$

is called here the *blocking ratio*.

If the definition of the conduction ratio (8.3) is recalled, it is evident that

$$c + b = 1$$

As the last DC–DC converter considered here, we now show a possible construction of a step-up/step-down converter, as shown in Figure 8.30.

This is a combination of the two circuits previously described. When only switch S_u is operated, S_d is held continuously in off-state and the converter works as an ordinary step-down converter; in this case the load/source in the figure operates as a load (absorbing power).

When, on the contrary, S_u is kept off and the on–off operation takes place only on switch S_d, the converter operates as a step-up; in this case the load/source in the figure operates as source (delivering power).

Applications of this kind of converter are common. An interesting example is vehicle propulsion.

Consider a battery–electric vehicle. These vehicles get their power from a battery which is charged from the mains when the vehicle is parked, but also partially during braking. This is an advantage over conventional cars, in which kinetic energy is converted into heat in mechanical brakes and is no longer usable.

A possible way to calculate the power train of an electric vehicle is shown in Figure 8.31. Mechanical propulsion power is produced by a DC electric machine that must be fed by a variable-voltage DC source (DC machines will be discussed further in Chapter 10). The DC voltage needs to be slowly varied to accommodate different vehicular speeds.

The propulsion energy is drawn from an electrochemical battery, whose voltage varies in a way that is completely different from what is needed by the electric machine. In this case a very good solution is to connect the battery to the machine by

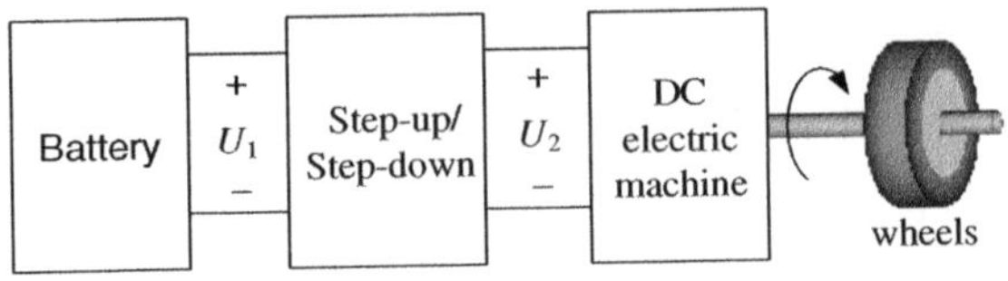

FIGURE 8.31. A simplified representation of the power train of an electric car using a DC electric machine for propulsion.

means of a step-up/step-down converter of the type shown in this section, for which $U_2 \leq U_1$.

When the battery delivers power (e.g., when the vehicle accelerates or operates at constant speed), the converter operates as a step-down and only switch S_u is operated; the higher the conduction ratio and U_2, the higher the propulsion torque. When, however, the vehicle has to brake, power is sent back into the battery by reversing the power flux: the DC machine converts mechanical energy into electricity, and this power is sent back into the battery (excluding the unavoidable inner losses) so that it can be stored inside the battery. This is achieved by operating the converter as a step-up, keeping S_u blocked off, and turning switch S_d on/off, using the technique seen above. The higher the blocking ratio, the lower the braking action.

To summarize, in this example the converter has two roles:

- It adapts the voltage at its two ends to each other; this adaptation is dynamic, since different voltage ratios correspond to different vehicular speeds.
- It transfers power from its left side to its right, and vice versa, as required for the vehicle run. This allows the propulsion power to be positive or negative; in the latter case, regenerative braking takes place.

8.3.3 Inverters

Inverters are, strictly speaking, converters designed to convert power from DC into AC. However, today's inverters have an internal structure that allows them to transfer power in both directions, and therefore they can *operate* as either inverters or rectifiers while being designated as inverters.

Thus, when the word *inverter* designates a converter, in today's terminology it indicates a bidirectional DC–AC converter; the inverter, in turn, can operate in *inverter mode* and in this mode it actually converts from DC into AC, or in *rectifier mode* and the opposite is true.

To avoid confusion, it is therefore advisable to call these "bidirectional AC–DC converters," even though the more simple term "inverter" is much more common.

An AC–DC converter can be represented as a black box using drawings similar to those already seen in the previous sections for rectifiers and DC–DC converters, as shown in Figure 8.32.

Figures 8.32a and 8.32b show single-phase and three-phase inverters respectively, without specific indications as to the allowable power flux direction. Figure 8.32c shows a three-phase inverter, explicitly indicating its bidirectional (or reversible)

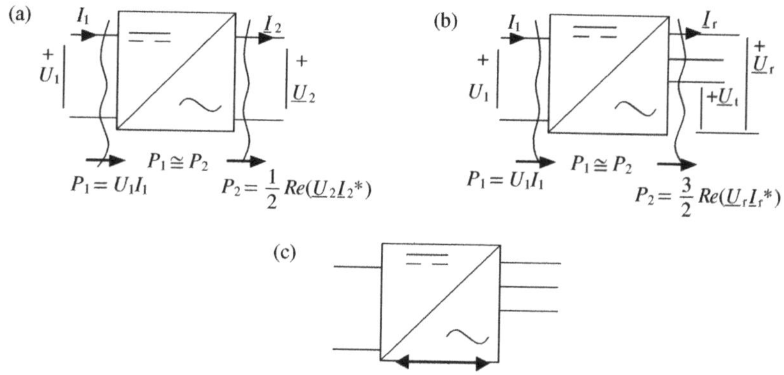

FIGURE 8.32. Black-box representation of DC–AC converters.

power flow capability. This is obtained by the bidirectional arrow shown in the lower part of its rectangular shape.

There exist different types of inverters. One much-used classification divides current source inverters (CSIs) from voltage source inverters (VSIs).

The former interface with a DC circuit that can be modelled as a *current* source (the DC current is somehow obliged to stay constant in the short term, e.g. using an inductor in series, through which flows the current entering the DC side of the inverter).

VSIs, instead, interface with a DC circuit that can be modelled as a *voltage* source (the DC voltage is somehow obliged to stay constant in the short term—e.g., using a capacitor across the DC terminals of the inverter—so that the inverter DC voltage is equal to the voltage across the capacitor).

VSIs have been increasingly used in recent years, while CSIs are used only for very large power applications and other niche markets.

The inverters studied in detail in this section are all VSIs.

The inner structure of a (VSI) single-phase inverter is shown in Figure 8.33 and is indicated as usual as a *basic inverter* by a dashed rectangle, in which a possible

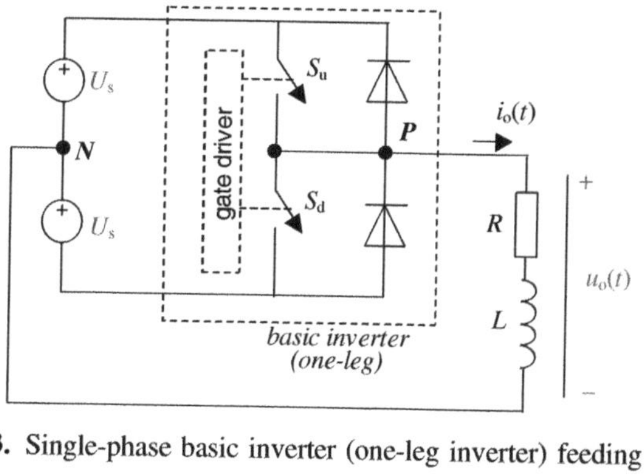

FIGURE 8.33. Single-phase basic inverter (one-leg inverter) feeding an *R–L* load.

connection to an R–L load is also shown. In the literature this structure is also called a one-leg inverter; this terminology will be clearer later when two-leg and three-leg inverters are introduced.

Remember that the controllable switch may represent any of the devices able to operate as a controllable unidirectional on-off device (such as BJTs, GTOs, IGBTs, etc., see Section 8.2).

This is a basic form, rarely used in practice but a useful starting point for our discussion.

The inverter operation is determined by controlling the on/off state of switches S_u and S_d, acting on suitable signals that are sent from an electronic board called *gate driver*. It is normal for the gate driver to be omitted in drawings so as to facilitate the representation of the power part of the inverter circuitry; this approach is also adopted here for the next figures.

The way in which S_d and S_u are commutated is called the inverter's *switching scheme*. In all main switching schemes used in practice, and in particular all those discussed here, S_u and S_d are always in opposite states: when one is on, the other is off. Therefore, whenever needed (as in the lower right-hand part of Figure 8.34), only the state of S_u is shown, since S_d is always its opposite.

The reader may object that the commutation of S_u and S_d cannot be exactly simultaneous: If commutation times are evaluated with sufficient precision, it is logical to expect that one commutates ahead of the other. This is correct: indeed, when switch commutation takes place, the switch that is going to be commutated into the on state is switched a short time after the other, in such a way as to ensure that S_u and S_d are never on at the same time. This is to avoid short-circuiting the two DC sources, which in theory would cause a theoretical infinity (in practice very high) current to flow.

Once it is stated that S_u and S_d must be operated with opposite states, it has to be decided when to switch one on and the other off. Different switching schemes have been introduced for this inverter.

In Figure 8.34 the one-leg inverter is again shown, along with the waveforms obtained when the so-called *square-wave switching scheme* is used.

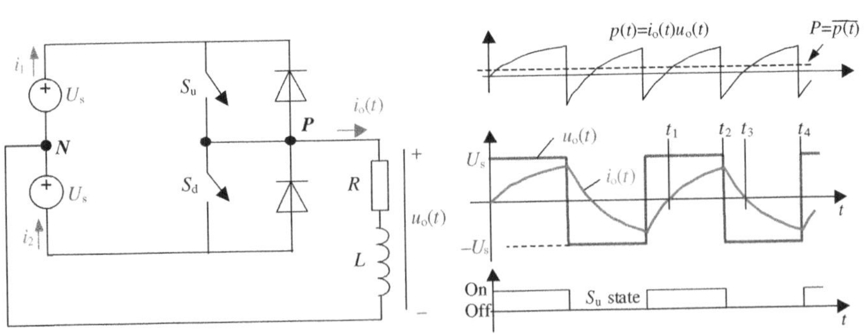

FIGURE 8.34. The square-wave switching scheme for a one-leg inverter.

We can understand how the inverter operates under this switching scheme by separately analysing the time intervals in which S_u is on and off.

When S_u is on (and thus S_d is off), the voltage that appears at the load terminals, u_o, is equal to U_s. The current could thus be evaluated simply by analysing what happens in an U_s–R–L circuit, exactly as seen in Chapter 5. The current shape is thus an exponential curve, with the continuity condition that its value at the beginning of the time interval started after a commutation must equal the value reached at the end of the previous one (i.e., before commutation).

When S_u is off, the load voltage is $u_o = -U_s$; again, the load circuit is an U_s–L–R circuit, with the new value of the source ($-U_s$ instead of U_s).

A global representation of voltages and currents is in the top right-hand part of Figure 8.34: the shape of the voltage might be deemed to be alternating with distortion,[3] and the same can be said of the current, although it has a different shape.

Note that the figure shows a transient starting from $i = 0$, but after a few periods it stabilizes at a shape which is nonzero at the beginning and end of each interval between commutations (these values are opposite numbers).

As far as power is concerned, it was seen in Chapter 5 that the instantaneous power of an AC circuit fluctuates around an average. This is true also in this case: the instantaneous and average powers relating to the circuit shown in Figure 8.34 are shown in the upper right-hand part of the figure itself. The average power is positive and corresponds to the average power converted into heat in R, while the instantaneous power has positive and negative values. If the diode and controllable switch characteristic are those indicated as idealized in Section 8.2, the power delivered by the two DC sources is exactly equal to that delivered to the AC load. In more realistic situations it will be

$$p_{DC}(t) = U_s(i_1 + i_2) \cong p_{AC}(t) = u_{o(t)} i_o(t), \quad P_{DC} = \overline{p_{DC}} \cong P_{AC} = \overline{p_{AC}}$$

The purpose of the anti parallel diodes is to allow current flow when the voltage changes its sign and the current has not yet changed its direction.

Consider, for instance, interval (t_1,t_2) in Figure 8.34. During this interval the current path is as indicated in the top left-hand part of Figure 8.35: the current flows through the upper U_s and S_u. Then S_u is switched off; the load voltage reverses, but the presence of inductance L in the load impedes instantaneous change in the current which is therefore forced to continue in the same direction as before. The current cannot get to the load by flowing through S_u (open) or D_u (the direction is opposite to the diode direction) and therefore flows through D_d. Indeed, even when a switch is on, when current must flow in the reverse direction, it will flow through its antiparallel

[3]We have seen in Chapter 5 that what is called "alternating current" and "alternating voltage" indeed indicates "sinusoidal current" and "sinusoidal voltage," respectively. The shapes shown in the top right-hand part of Figure 8.34 are not sinusoidal, but in some way are understood as "near to sinusoidal" or "sinusoidal with some distortion added." These concepts will be much clearer after Sections 8.4 and 8.5 have been studied.

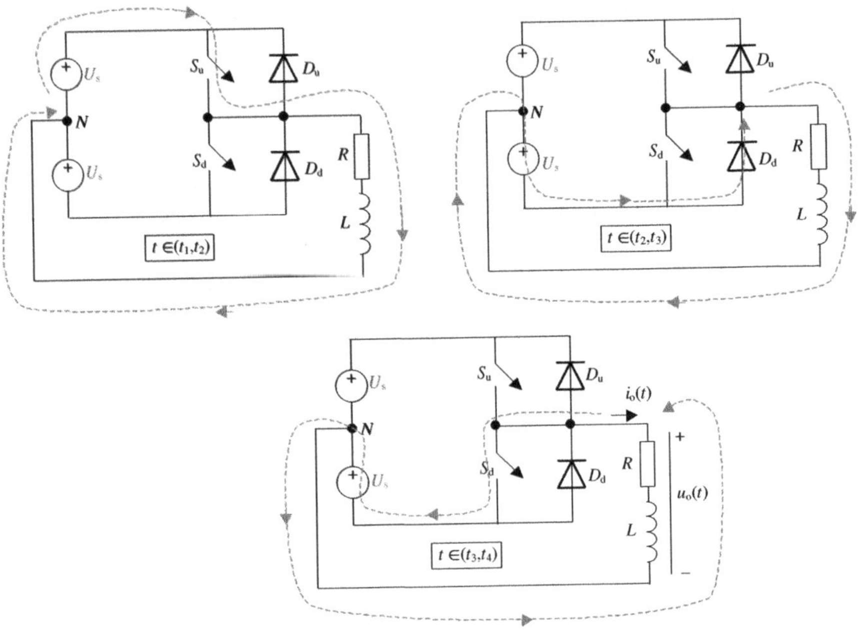

FIGURE 8.35. Current directions during some time intervals (indicated in Figure 8.34).

diode. This is shown in the top right-hand part of Figure 8.35. During this interval, however, the current path contains a voltage source U_s, oriented so that it tends to reverse the current direction. Therefore the current progressively reduces its value, and at $t = t_3$ reverses its direction.

In the interval (t_3, t_4) the situation is as depicted at the bottom of Figure 8.35. The last situation (i.e., immediately after t_4) is not examined here, but the reader is invited to reflect on it to become accustomed to the behaviour of circuits containing controllable switches with antiparallel diodes.

It can be concluded that in the inverter the current flows through an antiparallel diode whenever the delivered power is negative (i.e., actual power goes from AC to DC).

The square-wave switching scheme is not very common since in order to vary the amplitude of the output voltage, the amplitude of U_s (i.e., of the DC input voltage), must be varied, and this is not very desirable. This issue is overcome through use of the *PWM switching scheme*.

The acronym PWM stands for pulse-width-modulation, since, as we will soon see, the width (i.e., time duration) of pulses is continuously modified according to circuit needs.

The PWM switching scheme can be analysed by referring to Figure 8.36. This shows a triangular-shaped *carrier* wave $c(t)$ with frequency $f_c = 1/T_c$, a control sine signal $s(t)$ with a sinusoidal shape, and the same frequency f_s as required at the load terminals.

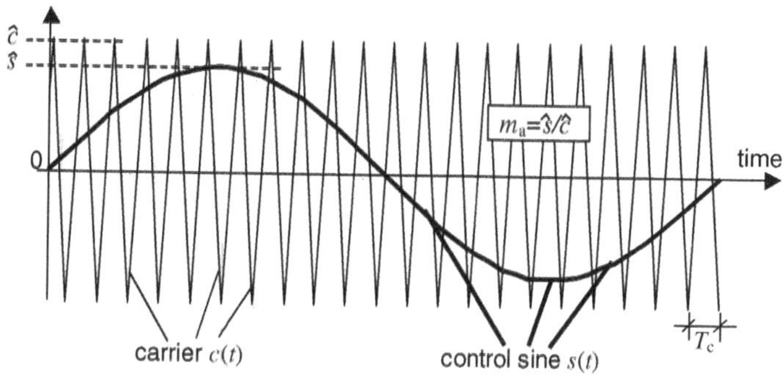

FIGURE 8.36. The carrier-sine comparison for PWM switching scheme.

The ratio of signal amplitude to that of carrier wave is called *amplitude modulation ratio* m_a; $m_a = \hat{s}/\hat{c}$, and it will prove to be very important for controlling the inverter operation.

The PWM switching scheme operates as follows:

- when $s(t) > c(t) \Rightarrow S_u$ is on
- when $s(t) < c(t) \Rightarrow S_u$ is off

The effects of this switching scheme can be seen in the curves shown in Figure 8.37. Using this scheme, the circuit can again be analysed as a sequence of U_s–R–L circuits, in which the value of the source is either $+U_s$ (at times in which S_u is on) or $-U_s$ (when S_u is off). Therefore, the current is a sequence of exponential curves which, since their duration is very short, can be approximated with straight lines.

This causes the current to have a shape of the type shown in the right-hand part of Figure 8.37—that is, similar to a sinusoid as expected.

When the load of the PWM inverter is characterized by a large inductance (i.e., $L/R \gg f_c$), and its behaviour depends just on this current (and not on the applied voltage), the PWM inverter can feed it without the need of interposing filters. It will be seen later that this is the case of three-phase machines (motors), which can be fed by

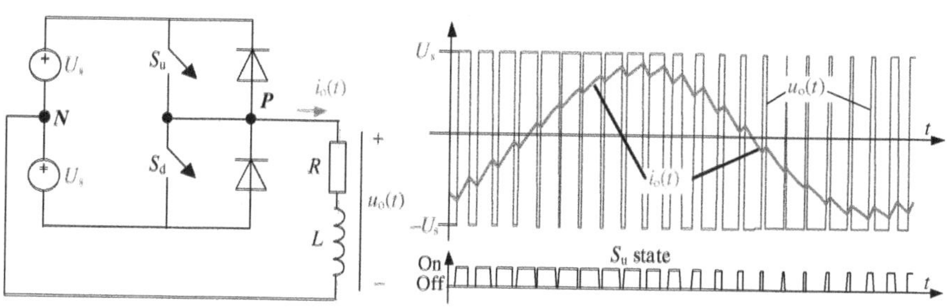

FIGURE 8.37. The PWM switching scheme for a one-leg inverter.

inverters that are a three-phase extension of the one shown in Figure 8.37 without any interposition of filters.

To feed a load that requires to be fed by a near-sinusoidal voltage shape, a filter must be interposed between the inverter and the load. This filter will have to transfer to the load the *fundamental component* of the produced voltage (i.e., the sinusoidal part) while (almost totally) blocking its fluctuation around that sine wave (see also Section 8.3.3.1).

A more precise definition of the fundamental component and an explanation of how filters can be made will be provided in Sections 8.4 and 8.5, respectively.

***More in Depth:** Harmonic spectrum of the produced voltage*

This box can be studied during a second reading of the chapter, after reading Section 8.4.

To gain better knowledge of how to deal with voltage $u_0(t)$ in circuits, it is useful to examine its frequency spectrum. This is composed of the fundamental component bar, plus a group of harmonics around the carrier-wave frequency. Two examples are shown in Figure 8.38.

The examples show a general situation: the harmonic content of the voltage produced by a PWM inverter contains a bar in correspondence with the fundamental frequency ($f_k/f = 1$), then a group of bars around the frequency of the carrier wave, here 20 times the fundamental frequency.

In between there is a region substantially devoid of harmonics.

It will be seen in Section 8.4 that this situation will greatly facilitate filtering: a filter has the purpose of allowing the fundamental component to pass through and reach the load while blocking (or greatly reducing) the other components. It can be intuited that the greater the distance separating what passes through and what is blocked, the easier it is to filter.

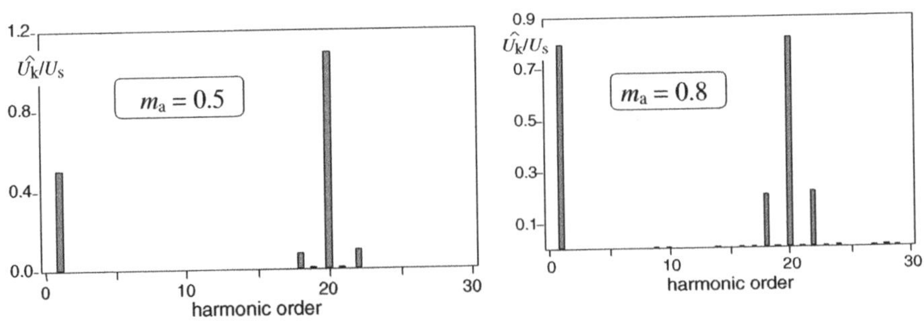

FIGURE 8.38. Frequency spectra of $u_o(t)$ of Figure 8.37 for two values of m_a.

It could be demonstrated that there is a very simple relationship between the amplitude of the fundamental AC voltage component and the DC voltage:

$$\hat{U}_1 = m_a U_s = m_a U_{DC}/2 \qquad \text{(one-leg inverter)} \tag{8.4}$$

Therefore, it is very easy to control $\hat{U}_1$, by acting on m_a, which is arbitrarily chosen by the gate driver software.

The phase of the fundamental component of the voltage produced by the PWM inverter can also be easily controlled, since it is equal to the phase of the sine signal used inside the gate driver to generate the switch commutation signals (see again Figure 8.36).

The single-phase inverter shown in Figure 8.33 requires that the DC side be split into two parts and that the middle point be made available to the load. This split is normally obtained by connecting two capacitors in series to each other and connecting this series in parallel with a unique DC source. The point where the two capacitors are connected to each other constitutes the midpoint connection (compare Figure 8.43a).

Another scheme, very often used in practice to create a single-phase inverter, is the so-called *full bridge*, or *two-leg* inverter, shown in Figure 8.39.

Switches are controlled in a similar fashion to the one leg inverter, since as it also has two states:

- In *up state*, S_{lu} and S_{rd} are on and the other two switches are off.
- In *down state*, S_{lu} and S_{rd}, are off and the other two switches are on.

When the bridge is in an up state, the upper and lower terminals of the source U_s are in connection with the upper and lower terminal of the load, respectively, and $u_0 = U_s$; when it is in the down state, the opposite happens and $u_o = -U_s$.

The shape of the voltage $u_o(t)$ does not differ from those shown in Figure 8.34 and 8.37 for square-wave and PWM switching schemes, respectively. However, the latter scheme, having a higher cost due to the higher number of power electronic devices, has the advantage of not needing user access to the central point of the DC source, and,

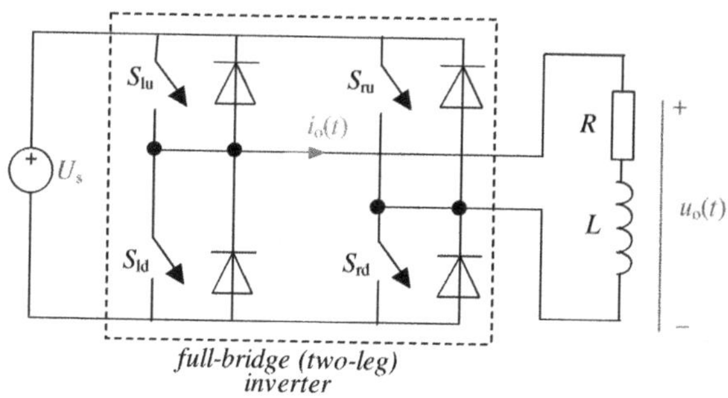

FIGURE 8.39. The full bridge (or two-leg) inverter.

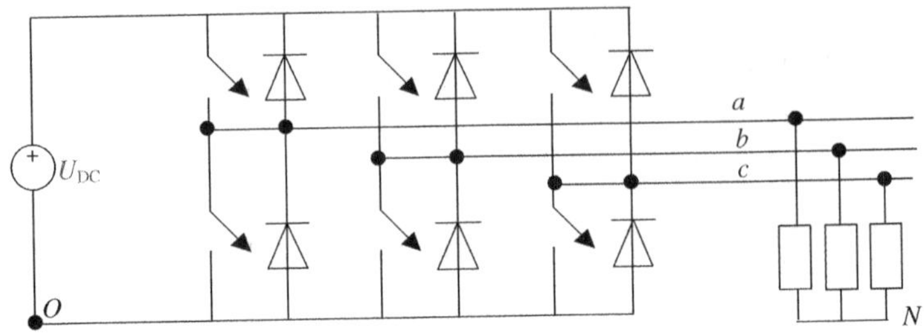

FIGURE 8.40. Three-phase inverter.

at equal DC voltage, of supplying the load with a voltage that is twice as high as the one-leg inverter:

$$\hat{U}_1 = m_a U_{DC} \qquad \text{(two – leg inverter)}$$

Finally, Figure 8.40 shows the arrangement used for three-phase inverters. The switching scheme for each of the three legs is the same as the one seen for the one-leg inverter. It can thus be either square-wave or PWM.

The three-phase bridge allows the generation of a three-phase system of voltages, without the need to access the midpoint of the DC side. This can be easily obtained by, for instance, considering the voltage $u_{ab}(t)$ as a difference between, $u_a(t)$ and $u_b(t)$. This is shown in Figure 8.41 for the square-wave switching scheme and in Figure 8.42 for PWM.

It is typical for an electric motor to be connected to the output terminals of the inverter (terminals a, b, and c of Figure 8.40).

It could easily be demonstrated that for the three-phase line-to-neutral voltage, equation (8.4) still applies:

$$\hat{U}_{1aN} = \hat{U}_{1bN} = \hat{U}_{1cN} = m_a U_{DC}/2 \qquad \text{(three – phase inverter).} \tag{8.5}$$

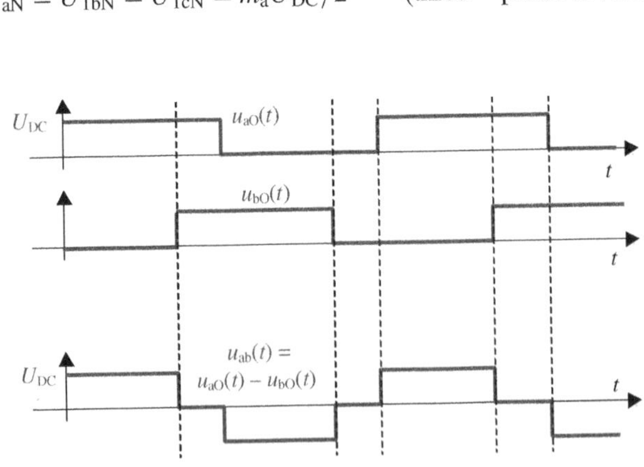

FIGURE 8.41. The line-to-line voltage when square-wave switching scheme is adopted.

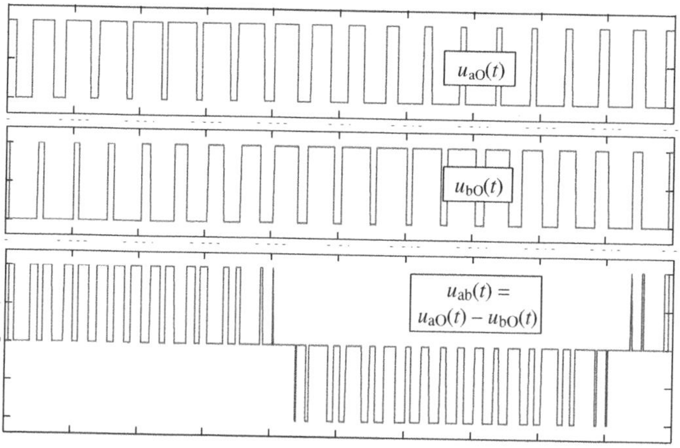

FIGURE 8.42. The line-to-line voltage when PWM switching scheme is adopted.

More in Depth

The switching scheme described here as PWM is one of several adoptable PWM techniques: every switching scheme that modulates voltages using pulses with variable widths falls into this category. The PWM switching scheme described here is more correctly called *sine-PWM*.

In recent years, for three-phase systems, the PWM technique called *space-vector PWM* (SV-PWM) has become more and more common since it has the advantage of allowing for a given DC source to obtain larger AC voltages, roughly 15% larger than the sine-PWM.

From equation (8.5) the maximum line-to-line peak voltage is immediately

$$\max(\hat{U}_{ll}) = \sqrt{3}\, U_{DC}/2 \cong 0.866 U_{DC}$$

Using SV-PWM, instead, it is max $(\hat{U}_{ll}) \cong U_{DC}$, which therefore allows better exploitation of the available DC voltage. Details on this technique can be found in reference [bm3].

8.3.3.1 ***Inverters for Sinusoidal Supply*** All the inverter switching schemes introduced in this section produce voltages that are far from sinusoidal, even though, if the load contains a sufficiently large inductance, the absorbed current tends to be much nearer to sine wave.

Inverters are normally directly connected to loads to feed motors; when a whole network requires AC supply, near-sinusoidal waveforms, thus filtering, is mandatory.

This applies to both single-phase and three-phase networks. This is summarised in Figure 8.43, where the inverter switching scheme can be square-wave or PWM.

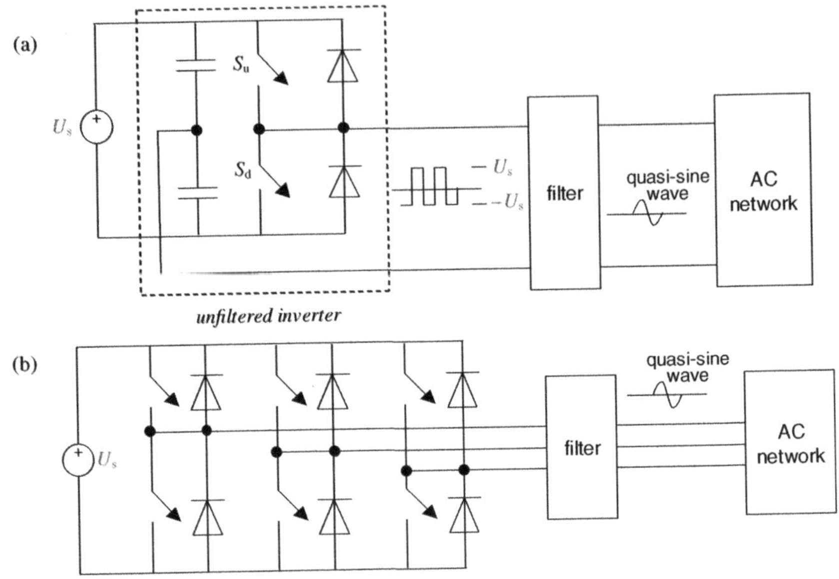

FIGURE 8.43. Inverter interfaced to AC networks with the interposition of filters.

The filtered voltage at the load side of filters is very near to the fundamental component of the voltage appearing at their inverter side. How filters can be built and how they operate will be discussed in Section 8.4.

The relationship between the amplitude of the fundamental component and DC voltage is the same as that for the one-leg inverter:

$$\hat{U}_1 = m_a \frac{U_s}{2} = m_a \frac{U_{DC}}{2}$$

Therefore, acting on m_a (compare with Figure 8.36), the amplitude of the voltage seen by the AC network can be modified at will.

It has been mentioned that the networks feeding the AC networks in Figure 8.43 apply to the AC system quasi-sinusoidal voltages. Disregarding the deviation from sine wave, they are thus AC sources, to which Thévenin's equivalents apply. For the purposes of this section the resistive part of Thévenin's impedance can be disregarded, and therefore Thévenin's equivalent of inverter plus filter is simply that represented in the left-hand part of Figure 8.44. Obviously, for three-phase inverters the equivalent represents the single-phase equivalent of the three-phase circuit.

It is very easy to express the relationship between the modulus of U_{Th} and U and their phase difference β and the active and reactive power that the inverter is able to

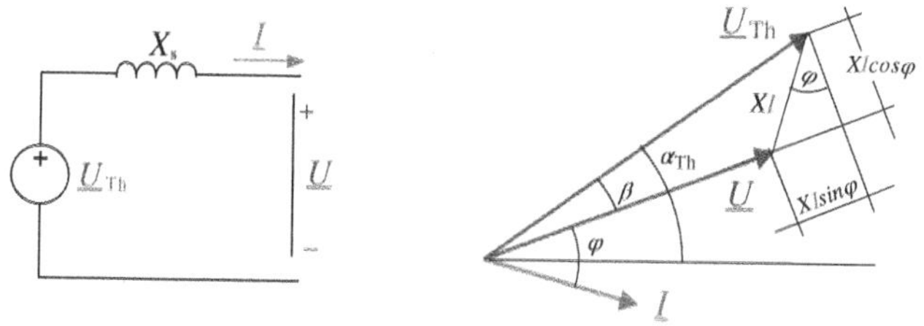

FIGURE 8.44. Simplified Thévenin equivalent of a PWM inverter with filter, and corresponding phasor diagram (for three-phase inverters it is the single-phase equivalent).

deliver to the AC network. Looking at the phasor diagram shown in the right-hand part of Figure 8.44, it can indeed be easily seen that

$$
\begin{aligned}
P &= UI\cos\varphi = U\frac{U_{\mathrm{Th}}\sin\beta}{X_s} = \frac{UU_{\mathrm{Th}}}{X_s}\sin\beta \\
Q &= UI\sin\varphi = U\frac{U_{\mathrm{Th}}\cos\beta - U}{X_s} = \frac{UU_{\mathrm{Th}}}{X_s}\cos\beta - \frac{U^2}{X_s}
\end{aligned}
\tag{8.6}
$$

Note that in these equations P and Q are real numbers, which can have positive or negative signs.

For instance, when β is negative, P is negative as well, and the inverter operates in rectifier mode; that is, it transfers net power from AC into DC.

The Thévenin electromotive force $\underline{U}_{\mathrm{Th}}$ is controlled by action on the sine voltage of the gate driver:

- The amplitude of $\underline{U}_{\mathrm{Th}}$ is proportional to m_a. Indeed, it can be seen that it is $\hat{U}_{\mathrm{Th}} \cong \hat{U}_1 = m_a U_{\mathrm{DC}}/2$.
- The phase angle of $\underline{U}_{\mathrm{Th}}$ is equal to the phase angle of the control sine wave: $a_{\mathrm{Th}} \cong angle(\underline{U}_1) = m_a U_{\mathrm{DC}}/2$ and the angle of $\underline{U}_1$ is the same as the sine-signal of Figure 8.36.

Consider an inverter connected to a prevailing network.[4] In this case (Figure 8.45), amplitude and phase of phasor $\underline{U}$ of Figure 8.44 are constant. Therefore any increase

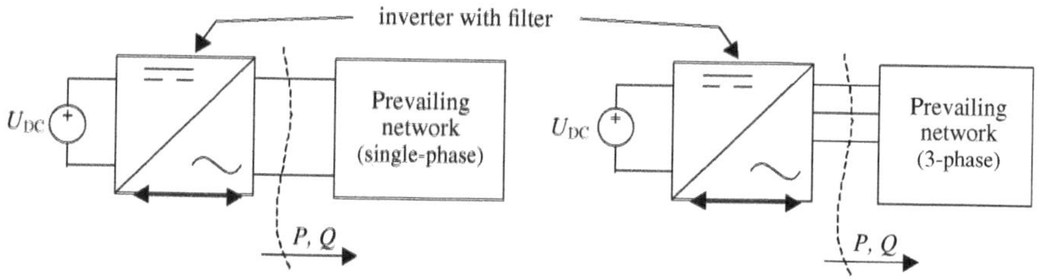

FIGURE 8.45. Inverter with filter connected to a prevailing network and reference signs for powers.

[4]Remember that this is a network whose terminal voltage phasor $\underline{U}$ has amplitude and phase that do not depend on amplitude and phase of the current injected into or drawn from it.

or decrease in the angle of the PWM sine signal will imply an equal variation of angle β; and any variation in the amplitude of that sine signal will cause a proportional variation of U_{Th}. Therefore, acting only on the PWM sine signal, the power fluxes of the inverter can be set arbitrarily, in amplitude and sign, according to equations (8.6).

This can be summarised as follows:

Result: *P and Q control of a sinusoidal voltage PWM inverter*

If a PWM inverter with filtering is connected to a prevailing network, active and reactive powers can be controlled in amplitude and sign by direct action of the modulating wave, with any combination of signs.

8.4 ANALYSIS OF PERIODIC QUANTITIES

8.4.1 Introduction

In this chapter, we introduced several power electronic converters. As already mentioned, the waveforms of voltages and currents in these circuits are usually far from sinusoidal. Therefore, in principle, none of the circuit analysis techniques described in Chapters 5 and 6 based on phasors can be used.

However, all the waveforms, when systems are in steady state, are periodic quantities; that is, they repeat themselves after a given time interval, called a *period*.

Some techniques to analyse circuits operating in periodic mode (i.e., with periodic quantities) must therefore be introduced to enable us to analyse circuits based on power electronics, and especially to understand how to filter quantities to obtain quasi-constant values (for DC circuits and systems) or quasi sinusoidal shapes (for AC circuits and systems).

This is the purpose of this section—that is, to deal with ways to analyse circuits operating in periodic mode, by expanding and reusing the knowledge gained for solving DC and AC circuits.

8.4.2 Periodic Quantities and Fourier's Series

A signal $x(t)$ that varies with time is said to be periodic if it repeats itself after a fixed time T, called *period*.

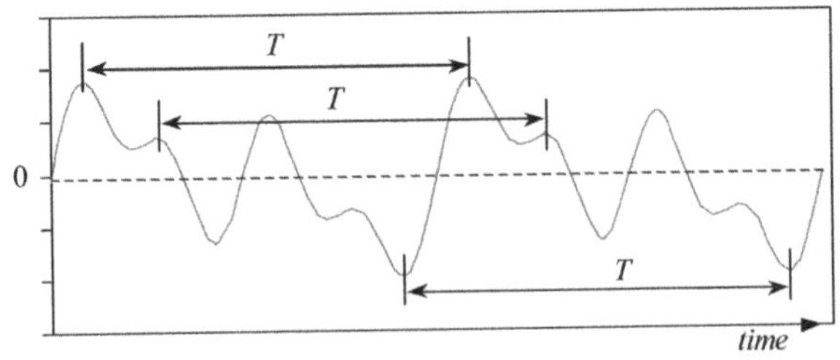

FIGURE 8.46. An example of a periodic quantity having T as period.

An example of a periodic quantity is shown in Figure 8.46, which also shows how all corresponding points share a horizontal distance equal to T.

More in Depth

A periodic quantity can be given a more formal definition; that is, a quantity $x(t)$ is periodic if a time T^* exists so that

$$x(t) = x(t - T^*) \qquad \text{for each possible value of } t$$

If such T^* exists, also $2T^*$, $3T^*$, and so on, share the same characteristic of causing $x(t)$ to repeat itself after that time:

$$x(t - T^*) = x(t) \qquad \forall t \Rightarrow x(t - 2T^*) = x[(t - T^*) - T^*] = x(t - T^*) = x(t)$$

The period T of $x(t)$ is defined as the minimum of all the possible T^*s.

The inverse of a signal period is its frequency f:

$$f = 1/T$$

Frequency is measured in hertz: one hertz is the frequency of the signal having a period of 1 s. It is also useful to define the *radian frequency* ω as follows:

$$\omega = 2\pi/T$$

The radian frequency is expressed in radians per second.

A very important result, called Fourier's theorem, applies to periodic quantities. Consider a generic periodic signal $x(t)$, whose radian frequency is ω. Fourier's theorem demonstrates that under some assumptions regarding $x(t)$, always satisfied in the case of voltages, currents, and other circuit quantities, there exist two sequences of real numbers $a_0, a_1, a_2, \ldots$ and $b_1, b_2, b_3, \ldots$ so that:

$$x(t) = \frac{a_0}{2} + \sum_{k=1}^{\infty} a_k \cos(k\omega t) + \sum_{k=1}^{\infty} b_k \sin(k\omega t) \tag{8.7}$$

Relation (8.7) is called Fourier's series of $x(t)$. Fourier's demonstration of this theorem also provides a way to compute all the terms of the two sequences, that is,

$$a_k = \frac{2}{T}\int_0^T x(t)\cos(k\omega t)\,\mathrm{d}t, \qquad b_k = \frac{2}{T}\int_0^T x(t)\sin(k\omega t)\,\mathrm{d}t$$

It can easily be seen that the term outside the two summation symbols is $\bar{x}$, that is, the average value of $x(t)$:

$$\frac{a_0}{2} = \bar{x} = \frac{1}{T}\int_0^T x(t)\,\mathrm{d}t$$

The numbers of the two sequences tend to approach 0 when their subscripts tend to infinity. This is quite reasonable, considering that the two summations in (8.7) must have a finite value—hence the practical and, for engineering purposes, necessary decision to approximate the sums using their first N terms:

$$x(t) \cong \frac{a_0}{2} + \sum_{k=1}^{N} a_k \cos(k\omega t) + \sum_{k=1}^{N} b_k \sin(k\omega t)$$

The value of N depends on the objective of Fourier's decomposition, but in practical cases is kept below 40, often even below 20.

An expression of Fourier's decomposition equivalent to that of (8.7) is

$$x(t) = A_0 + \sum_{k=1}^{\infty} A_k \sin(k\omega t + \alpha_k) \tag{8.8}$$

in which

$$A_0 = \frac{a_0}{2}, \quad A_k = \sqrt{a_k^2 + b_k^2} \;\; (k = 1, 2, \ldots), \quad \alpha_k = \tan^{-1}(a_k/b_k) + \zeta\pi \text{ (see footnote 5)}$$

A_0, as already noted, is the average value of $x(t)$, while $x_1(t) = A_1 \sin(\omega t + \alpha_k)$ is called its *fundamental component*; all the other terms of the summation in (8.8) are called *harmonic components* of $x(t)$: $x_k(t) = A_k \sin(k\omega t + \alpha_k)$ is the harmonic component at the frequency $f_k = kf$. The set of harmonic components is called harmonic contents of $x(t)$. The number k equal to the ratio of kth harmonic frequency and signal frequency is called *harmonic order* of the relative harmonic component.

The average value of signal $x(t)$ (i.e., A_0), being constant over time, is also called, when referring to circuits, the DC component of $x(t)$.

In AC circuits it is usually desirable to have quantities that are as similar as possible to sinusoids in shape. They will be so if, when decomposed into Fourier series, all coefficients of (8.8) except A_1 are null. Very often A_0 is null as a consequence of the nature of the circuits, while the other terms are different from

[5] If $b_k < 0$, then $\zeta = 1$; if $b_k > 0$, then $\zeta = 0$; if $b_k = 0$, then $\alpha_k = \mathrm{sign}(a_k) \cdot \pi/2$.

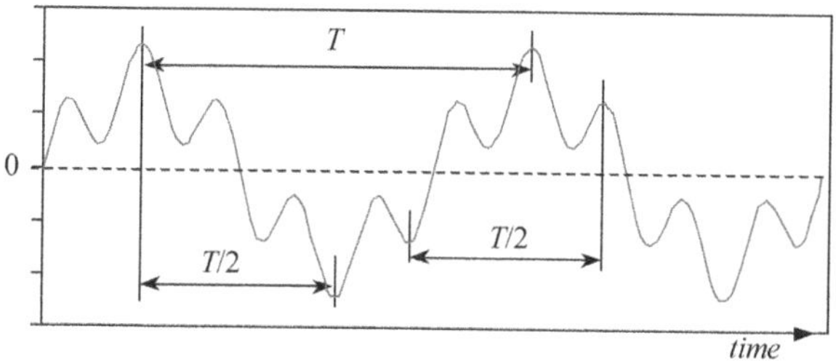

FIGURE 8.47. Example of periodic signal with half-wave symmetry.

zero. This justifies the widespread use of a normalized version of the distortion, the total harmonic distortion (THD):

$$\mathrm{THD} = \sqrt{\sum_{i=2}^{N}\left(\frac{A_k}{A_1}\right)^2} = \sqrt{\sum_{i=2}^{N} A_k^2} \bigg/ A_1$$

The higher the THD, the higher the distortion, which means deviation from the sine-wave shape. However, rigorous analysis must not also disregard A_0.

International standards state the maximum voltage distortion in terms of individual values of harmonic components, both for DC systems (for which all components A_k with $k \neq 0$ must be low in comparison with A_0) and AC systems (for which all components A_k with $k \neq 0$ must be low in comparison with A_1). In the case of AC systems, N is taken[6] as 40 or 50 and maximum acceptable voltage THD levels are 5% or 8%.[7]

8.4.3 Properties of Periodic Quantities and Examples

Some characteristics of periodic quantities $x(t)$ have immediate effects on Fourier's coefficients—for example, the coefficients in (8.7).

Here we mention only the *half-wave symmetry* because of its importance. A periodic signal $x(t)$ having a period T has half-wave symmetry if it is

$$x(t) = -x(t - T/2) \qquad \text{for each value of } t$$

An example of a $x(t)$ with half-wave symmetry is shown in Figure 8.47.

Table 8.2 shows mathematical expressions of Fourier coefficients for some typical waveforms. Table 8.3 shows approximations obtained by summing limited numbers of terms in their Fourier series. These give an idea on how the original waveforms are produced with increasing accuracy as the number of terms grows.

[6]International standards IEEE 519, IEC 61000-2-2 and 61000-2-4 require $N = 50$, while European standard EN 50160 requires $N = 40$.

[7]International standard IEEE 519 regarding voltage requires $\mathrm{THD}_{max} = 5\%$, while IEC 61000-2-2, 61000-2- and European standard EN 50160 requires $\mathrm{THD}_{max} = 8\%$.

TABLE 8.2. Some Periodic Waveforms and Their Fourier Series Coefficients

Wave	*Amplitude*	*Phase*
Square wave 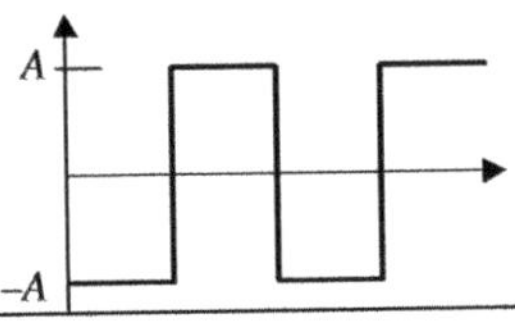	$A_k = 0$ (k even) $A_k = \frac{4A}{\pi k}$ (k odd)	$\alpha_k =$ undetermined (k even) $\alpha_k = 0$ (k odd)
Saw-tooth 	$A_k = \frac{2A}{\pi k}$	$\alpha_k = \pi$
Half sine 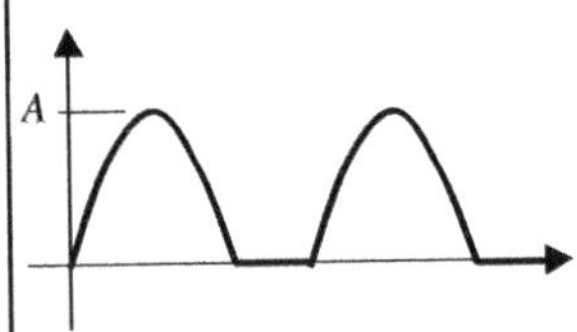	$A_0 = \frac{A}{\pi}$, $A_1 = \frac{A}{2}$ $A_k = \frac{2A}{(k^2-1)\pi}$ ($k > 0$ and even) $A_k = 0$ ($k > 1$ and odd)	$\alpha_1 = 0$ $\alpha_k = -\pi/2$ (k even) $\alpha_k =$ undetermined (k odd)

The minimum approximation is constituted only by A_0. For the half sine wave, which has a nonzero average value, this can be useful.

The DC component with the addition of the fundamental component gives a slightly better approximation of the waveforms, but still does not include much of the information content of the original waveforms. The higher the number of harmonics, the closer the reproduced form becomes to the original.

8.4.4 Frequency Spectrum of Periodic Signals

A very effective way of showing the harmonic contents of a signal $x(t)$ is to plot the amplitudes and phases of different harmonics as bar plots. This bar plot is normally called *frequency spectrum* of $x(t)$.

Since bar plots can also be made for phases of different harmonic components, the plots in Figure 8.48 are more precisely indicated as amplitude (frequency) spectra, in contrast with *phase spectra*, not shown here.

Just to give an idea, Figure 8.48 shows the frequency spectrum of amplitudes of the square and saw-tooth waves (having unity amplitudes) already presented in Table 8.2.

TABLE 8.3. Partial Fourier Series Sums of the Waveforms Considered in Table 8.2

Max k			
	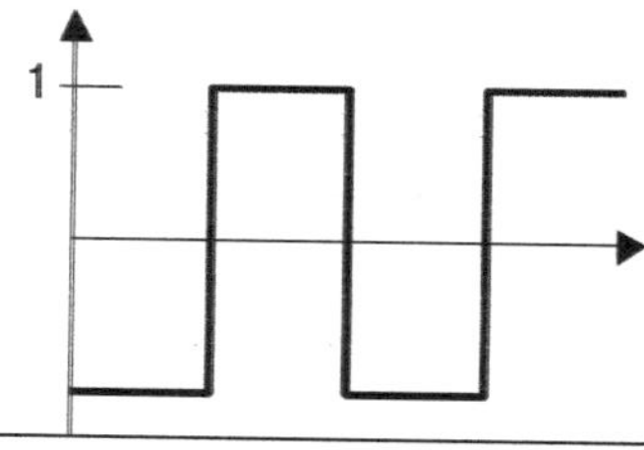	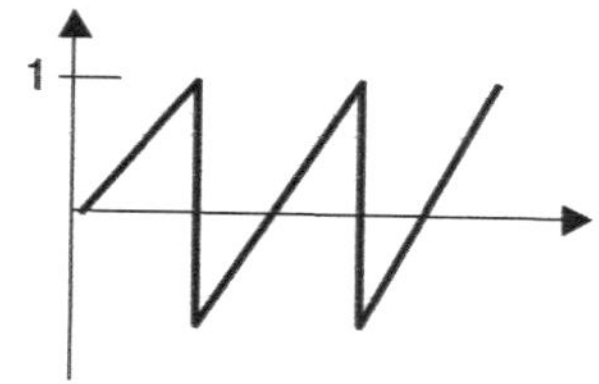	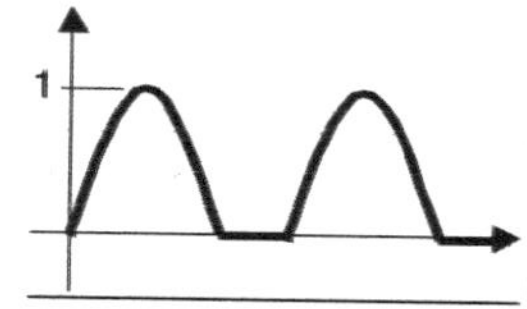
1	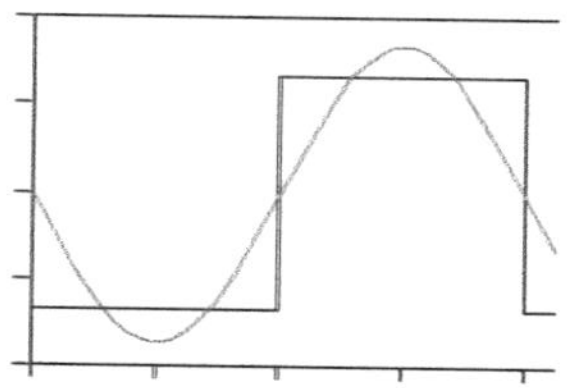	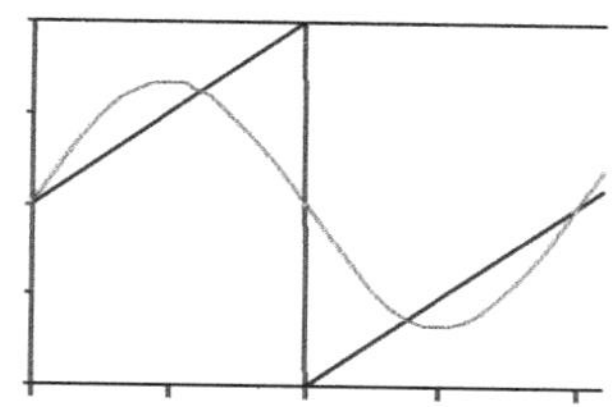	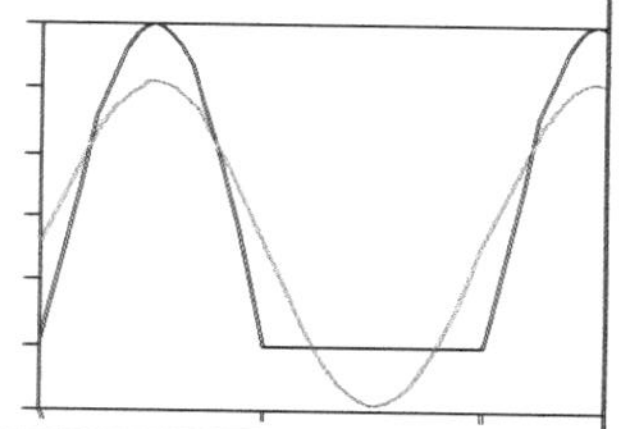
3	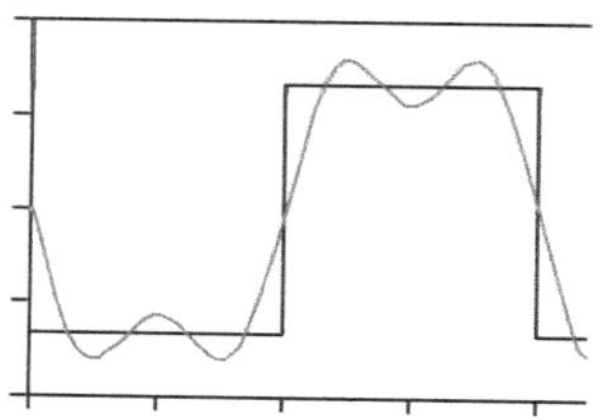	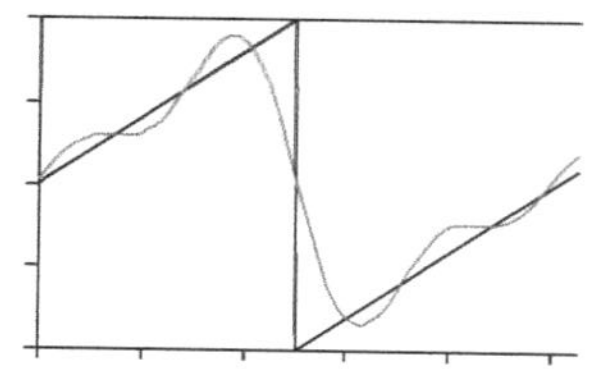	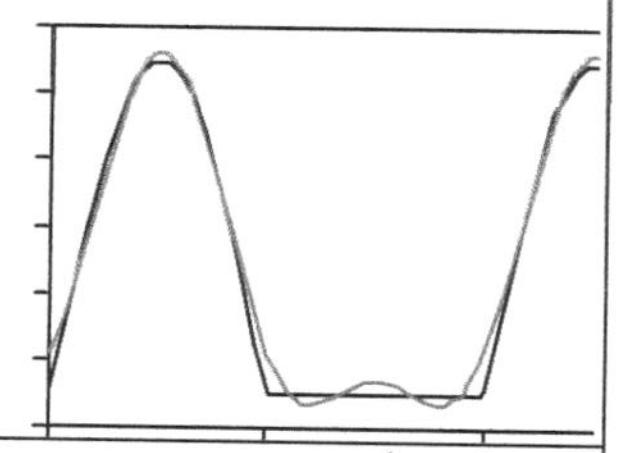

(continued)

Table 8.3 (*Continued*)

Max k			
5	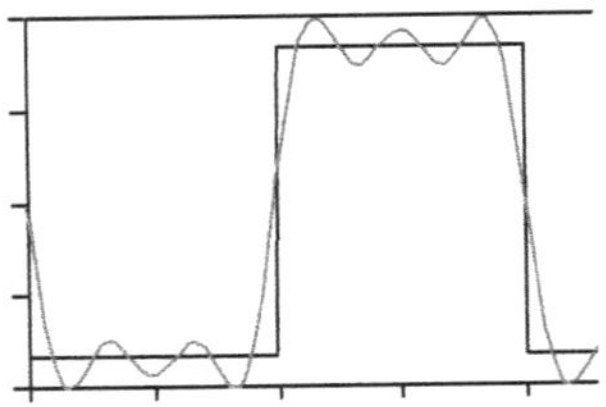	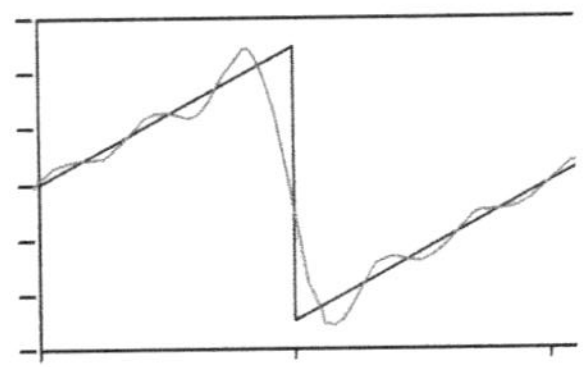	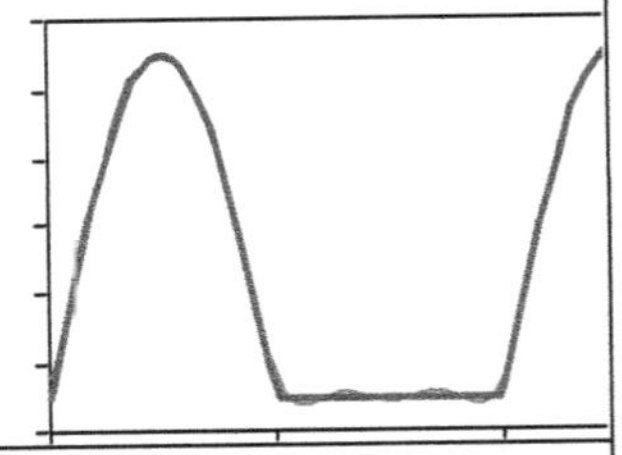
6	As above	As above	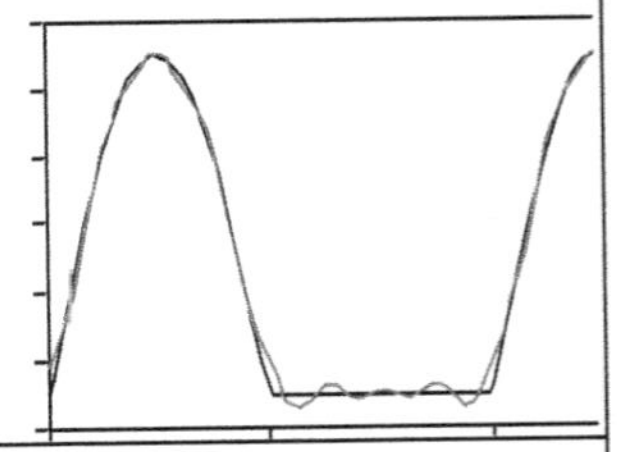
15	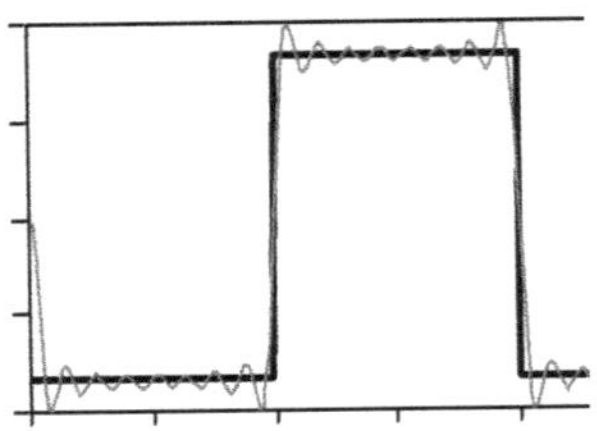	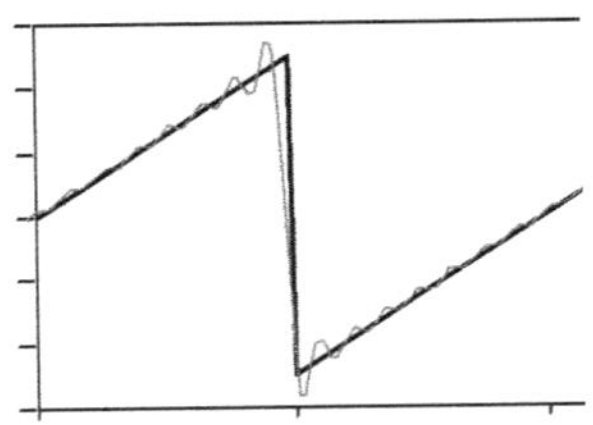	—

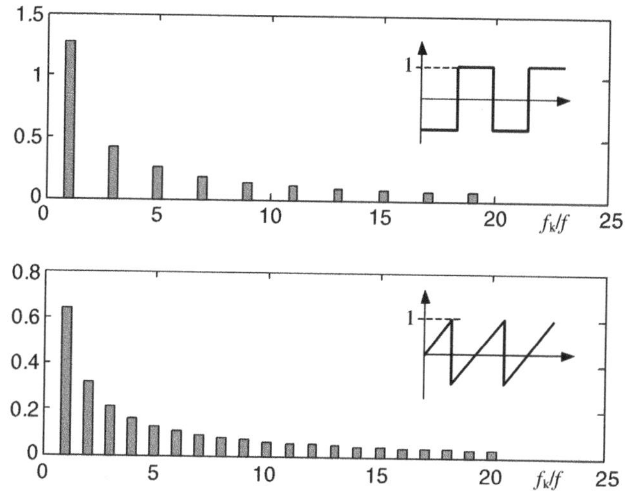

FIGURE 8.48. Frequency spectra of square-wave and saw-tooth waves.

The variable shown in the abscissa of the plots is the harmonic order of the various harmonic components of the given signals. The bar corresponding to $f_k/f = 0$ refers to the average of the signal A_0; the bar corresponding to $f_k/f = 1$ is the fundamental component. Perfect DC signals (e.g., the voltage across a DC voltage source) have a frequency spectrum constituted by only one bar for $f_k/f = 0$; ideal signals (e.g., the voltage across a AC voltage source) have a frequency spectrum constituted by only one bar, for $f_k/f = 1$.

8.5 FILTERING BASICS

8.5.1 The Basic Principle

We saw in Section 8.3 that converters are not usually able to produce output voltages as intended: converters designed to produce DC output produce voltages that have a nonzero average, but whose shape is far from constant. Similarly, inverters, which are intended to generate sinusoidal output voltages, produce voltages that have a nonzero fundamental frequency, but are far from sinusoidal.

A filter is usually used to bring the output voltage closer to what was intended. This is recalled in Figure 8.49, in which the subscript "u" stands for "unfiltered."

How can a filter obtain this result? Consider that the input voltage of the filter u_u can be expressed in a Fourier series (imagine that the summation is stopped at the N-term):

$$u_u(t) \cong U_{u0} + \sum_{k=1}^{N} U_{uk} \sin(k\omega t + \alpha_{ik})$$

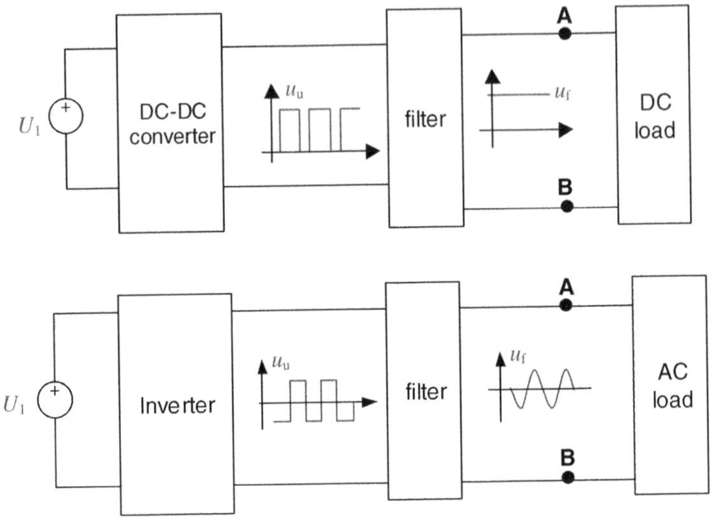

FIGURE 8.49. Need to insert filters between converters and loads.

A filter requested to filter DC voltages will have the purpose to leave unchanged

$$U_{f0} \cong U_{u0}$$

and will have to reduce the size of the other components of the Fourier series:

$$U_{fk} \ll U_{f0} \qquad \text{with } k > 0$$

Similarly, a filter required to filter AC voltages will have the purpose of leaving unchanged

$$U_{f1} \cong U_{u1}$$

and will have to reduce the size of the other components of the Fourier series

$$U_{fk} \ll U_{f0} \qquad \text{with } k \neq 1$$

In several cases, however, the component U_{u0} is already much lower than U_{u1}, and therefore a good filter could simply try to filter out U_{uk} when $k > 1$, while also leaving U_{u0} unchanged.

Filters vary in schemes and complexity. Very often filters are composed of networks of the usual circuit components: resistors, inductors, and capacitors. A very effective and widely used solution is to have an R and L in series and a C in parallel (Figure 8.50).

Now let us try to verify whether this filter is able to fulfil its purpose—that is, to transfer (almost), unchanged, the useful part of the signal, while (almost) blocking all the other components.

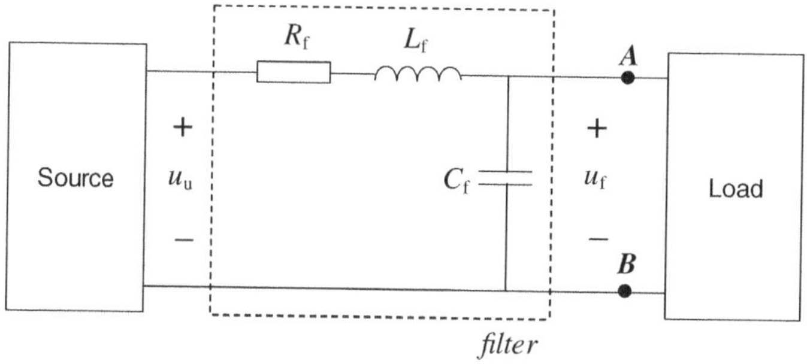

FIGURE 8.50. The square filter between source and load.

If the filter is sized well, its effectiveness will not be influenced too much by the presence of the load. This concept is used here, and the filter behaviour is analysed in a condition in which the load is disconnected (no-load operation), under the assumption that when the load is connected, filter effectiveness remains substantially unchanged.

Once the load is disconnected, the circuit is composed of the source whose input voltage u_i is supposedly known (a periodic quantity).

The circuit is as represented in the left-hand part of Figure 8.51. Since this circuit is fed by a periodic quantity, phasor analysis cannot be directly used, since it applies to sinusoidal circuits only. However, the input voltage u_u can be decomposed as a summation of sinusoids having different frequencies, according to Fourier's series (8.7):

$$u_u(t) \cong u_u + U_{u1}\sin(\omega t + \alpha_1) + U_{u2}\sin(2\omega t + \alpha_2) + \cdots + U_{uk}\sin(k\omega t + \alpha_k)$$

This allows the substitution of the circuit in the left-hand part of Figure 8.51 with the one in its right. At this point the superposition principle can be applied, and this circuit can be solved by applying sources $u_0, u_1, \ldots, u_N$, one at a time. The N circuits shown in the right-hand part of Figure 8.51 are all linear circuits with a one-frequency source.

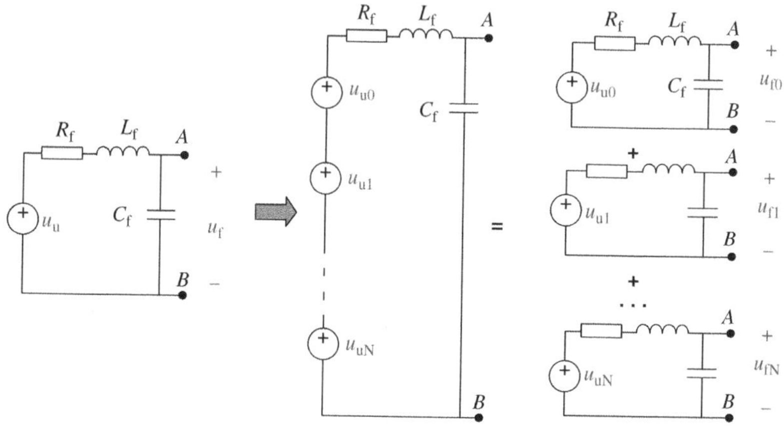

FIGURE 8.51. Application of superposition principle to solve periodic circuits using phasor analysis.

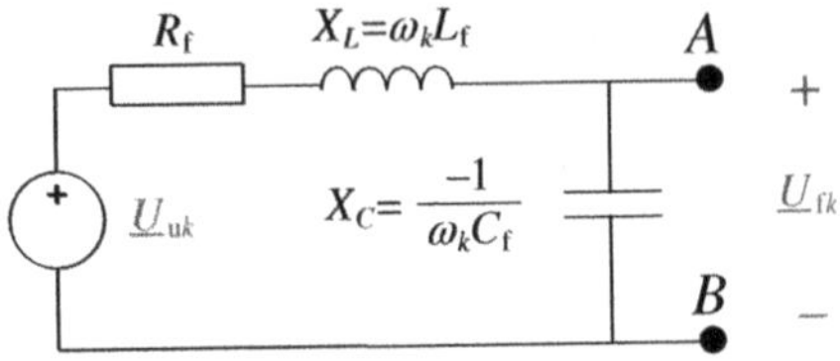

FIGURE 8.52. Generic phasor circuit for the kth harmonic component of the filter of Figure 8.51.

Therefore they can all be solved by using the usual phasor analysis. Once they are solved, the complex vectors are converted back into functions of time, to find the functions of time of the original circuit.

This process can be used to find the voltage between $\boldsymbol{A}$ and $\boldsymbol{B}$ terminals (the terminals to which the load will be connected, $u_{AB}(t)=u_f(t)$). To do this, consider the generic kth circuit shown in the right part of Figure 8.51. Since it is a single-frequency circuit, it can be converted into the phasor circuit, shown in Figure 8.52. The generic angular frequency is indicated as ω_k (obviously $\omega_k=k\omega$).

Let us first analyse this circuit, ignoring the resistance R_f. It is

$$\underline{U}_{fk}=\frac{\underline{U}_{sk}}{j(\omega_k L_f-1/\omega_k C_f)}\cdot\frac{-j}{\omega_k C_f},\qquad |\underline{U}_f/\underline{U}_s|=\frac{1}{1-\omega_k^2 L_f C_f}=\frac{1}{1-(\omega_k/\omega_0)^2}$$

where $\omega_0=1/\sqrt{L_f C_f}$. The value of angular frequency ω_k that renders null the denominator of the previous formula has a special importance and is called *cutoff angular frequency* of the filter (and $f_0=\omega_0/(2\pi)$ is called *cutoff frequency*).

When the source frequency is much larger than the cutoff frequency,

$$\omega_k \gg \omega_0 \tag{8.9}$$

it will be

$$|\underline{U}_f/\underline{U}_u|\cong\frac{1}{\omega_k^2 L_f C_f}$$

Therefore, if (8.9) is valid, the harmonic voltages will be transferred from input to output, reduced by a factor that is dependent on ω_k itself, and goes very quickly to zero as long as ω_k increases. On the opposite side, harmonic components with a frequency of

$$\omega_k \ll \omega_0$$

will be transferred, unchanged, from the filter to its right-hand side.

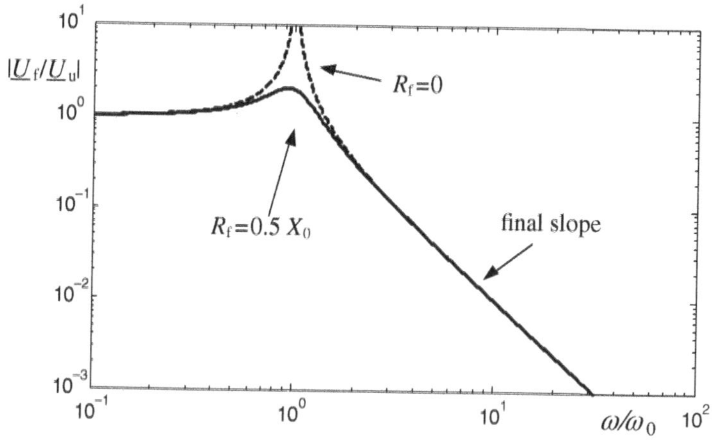

FIGURE 8.53. Effect of a square filter on voltage at different frequencies.

The values of $|\underline{U}_f/\underline{U}_u|$ as a function of ω_k/ω_0 are shown in Figure 8.53 (dashed curve).

If the computation $|\underline{U}_f/\underline{U}_u|$ is repeated with a finite value of R_f, the asymptote disappears from the diagram and is substituted by a peak (called resonance peak) whose value depends on the value of R_f.

In general, the resonance peak depends only on the ratio of R_f and the resonance reactance $X_0 = \omega_0 L_f = 1/(\omega_0 C_f)$, that is, the absolute value of the reactance of L_f and C_f (they are equal to each other) at the resonance frequency ω_0.

The values of $|\underline{U}_f/\underline{U}_u|$ in the case of $R_f = 0.5\omega_0 L$ as a function of ω_k/ω_0 are shown in Figure 8.53 (continuous curve). The plot shows several points worth noting:

- For frequencies much lower than ω_0 the output voltage is nearly equal to the input (as was already observed through the analysis carried out while disregarding R_f.
- For frequencies much greater than ω_0 the output voltage decreases fast as frequency grows: there is a fixed slope that relates the two, indicated in the plot as "final slope." This shows that for a 10-fold increase in the frequency there is a 100-fold decrease in voltage.
- The difference between the case with $R_f = 0$ and the case with $R_f = 0.5X_0$ is apparent only around $\omega_k = \omega_0$.

This suggests how to exploit this kind of filter to obtain our purposes: We must determine the size of the filter in such a way that the useful part of the signal (DC component or fundamental component) is located well to the left-hand side of $\omega/\omega_0 = 1$; on the opposite side, the components of the input voltages that we wish to be reduced as much as possible shall be located in the right-hand part of $\omega/\omega_0 = 1$.

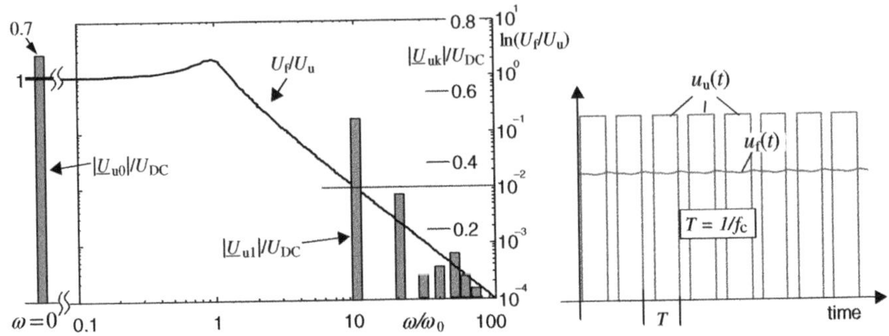

FIGURE 8.54. Filtering effectiveness for a DC–DC converter evaluated in the frequency spectrum and time domain (here the filter cutoff frequency is one-tenth of the converter frequency; $T_{ON}/T = 0.7$).

In this way, when separately computing the harmonic components according to the procedure described in Figure 8.51, the output voltage will have a very different shape from the input one: the harmonic components well below ω_0 will remain unchanged; the ones well above this limit will be much reduced in size, while no non-negligible harmonic component with frequencies around ω_0 should be present. If there were some, they would be amplified by an amount that depends on their vicinity to ω_0; moreover, the lower the R_f, value of the higher the amplification factor.

The filtering action of a filter can be evaluated by using, in the same plot, the frequency spectrum of the filter input and the ratio $|\underline{U}_f/\underline{U}_u|$.

This is shown in Figure 8.54 for a DC/DC converter (the height of the bars are the amplitudes of $u_u(t)$ harmonic components divided by DC voltage).

It can easily be seen from the figure that if the frequency at which the DC/DC converter operates is much greater than the resonance frequency of the inverter ω_0, all the harmonic components are dramatically reduced. In Figure 8.54 the ratio between these two frequencies is 10, and the first harmonic component, having an amplitude that is 71% of the DC component, is reduced by a factor of 100; therefore, in the output of the filter it will hardly be visible. The other harmonic components are reduced even further.

Now it is clear how the filter output voltage has the shape shown in the right-hand part of Figure 8.54, where the unfiltered voltage input is also shown for greater clarity.

The example shown in Figure 8.54 refers to the case of $c = 0.7$. Naturally, a similar level of effectiveness will also be obtained with other values of c, since the first harmonic will always be at a frequency 10 times ω_0 and therefore will be reduced by a factor of 100, and the others will be reduced even further.

In a similar way, the same filter structure can dramatically enhance the quality of the voltage at the AC terminals of a basic PWM inverter, as shown in Figure 8.55. Here the carrier angular frequency was set at 10 times ω_0. Again it can be seen both in the frequency spectrum and in the time-domain plots that the filter effectively improves the voltage output of the basic inverter.

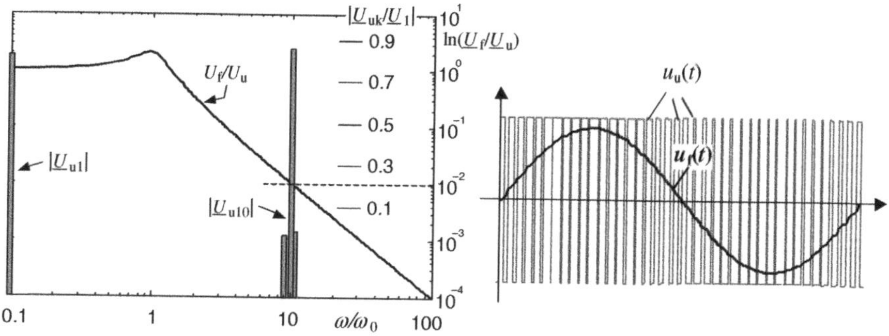

FIGURE 8.55. Filtering effectiveness for a PWM inverter evaluated in the frequency spectrum and time domain (here the filter cutoff frequency is one tenth of the converter frequency; $\hat{u}_f/\hat{u}_{tri} = 0.8$).

The example shown in Figure 8.55 refers to the case of $m_a = 0.8$; a similar level of effectiveness will also be obtained with other values of c, since the first harmonic will always be at a frequency of 10 times ω_0 and therefore will be reduced by a factor of 100, and the others will be reduced even further.

8.6 SUMMARY

In this chapter we have seen that the advent of power electronics has made the conversion of electric energy very flexible.

It is possible to convert from DC to DC, by raising or lowering the voltage at equal power (i.e., maintaining constant the product $U \cdot I$) and from AC to DC either one-way, using a diode rectifier, or bi in directionally, using an inverter which, despite its name, can normally operate both in inverter modes and in rectifier modes.

Power electronic converters operate by switching circuits; therefore, the voltages and/or currents produced when creating a DC system are never really constant, but can be rendered very nearly so by adequate filtering action.

Similarly, converters required to create AC from DC actually create nonsinusoidal (i.e., distorted) voltages and/or currents; however, particularly if the PWM technique is adopted, these distorted quantities can easily be rendered quasi-sinusoidal by proper filtering action.

The faster the basic converter (i.e., the higher the commutation frequency), the smaller and cheaper the filter. However, there is always a need to find the right compromise between speed and power: the fastest controllable switches are not able to withstand large currents and/or voltages; that is, if hundreds of amperes or thousands of volts are required, slower and stronger switches are needed (such as GTOs).

The PWM inverters can generate AC at variable frequency and amplitude, with fast and effective control also of the phase of the produced sinusoid (after filtering).

Therefore they can be used to feed variable-frequency motors, as will be seen in the (following) chapters regarding electric machines and drives. They can also be used to interface with AC networks to exchange, in either direction, either active or reactive powers. This can be exploited, for instance, to interface a photovoltaic or wind plant with the grid. This concept will be discussed with some degree of detail in Part IV of this book.

When the converters are used to feed electric motors, filters are often omitted, since the presence of large inductances inside the machines creates a natural filtering action. Thus the DC machine currents are much nearer to constant values and the AC machine currents are much more similar to sine waves than the corresponding voltages. This will also be seen in detail in the following chapters.

9

PRINCIPLES OF ELECTROMECHANICAL CONVERSION

For the Instructor

This chapter aims to provide a basic understanding of the rotating machines analysed in subsequent chapters.

In order to simplify the topic of electric machines, we take the radial component of magnetic flux density as sinusoidal.

The equation (9.1) is first derived from Lorentz's force equation (thus capitalizing on electromagnetics knowledge) and then transformed into the forms (9.8) and (9.9), which are easier for use in electric machines. The order in which the numerical factors are written in these formulas was chosen so that the vector (Lorentz's force) equation, from which they derive, is in some way reproduced. Equation (9.12) is called here (possibly for the first time) *Faraday's law for rotating coils*; we believe this terminology will simplify learning.

We have followed a similar approach for forces and torque created on machine coils which, starting from Lorentz's force expression, produces first (9.5), then (9.10), and finally (9.13).

Today, electric machines are introduced with increasing frequency through the *space phasor* concept. However, for this introductory book we did not deem this powerful approach appropriate.

Fundamentals of Electric Power Engineering: From Electromagnetics to Power Systems, First Edition.
Massimo Ceraolo and Davide Poli.

9.1 INTRODUCTION

Electricity in modern life is an *energy carrier* more than an energy source or sink. Electric energy is rarely produced in nature; one example is lightning, a kind of electricity not used by humans. Manmade electricity comes from other sources of energy, such as the potential energy of water (in hydraulic power plants), or the chemical energy of fuels like petroleum or natural gas (in thermal power plants), or nuclear energy from nuclear forces. In all these examples, energy is first converted into mechanical energy, making shafts rotate, and then this mechanical energy is converted into electricity, to be later transferred to load centres and used there.

On the other hand, electricity itself normally is not consumed directly but is instead converted into other forms such as heat (ovens), mechanical power (motors), light (lamps), and so on. One of the greatest uses of final energy from electricity is again in mechanical form, whenever electricity feeds electric motors.

There are, however, important cases in which electricity is consumed as electric energy in addition to being a carrier—for example, in electronic appliances such as TV sets or personal computers, which *use electricity as electricity*: any other possible forms of energy must first be converted into electricity before they can be used by these appliances.

These examples illustrate the great importance of electromechanical conversion—that is, conversion from mechanical energy into electricity, and vice versa.

9.2 ELECTROMECHANICAL CONVERSION IN A TRANSLATING BAR

Consider the system shown in Figure 9.1. It consists of a conducting bar b which translates (i.e., it moves without any rotation) within a space where there is a uniform magnetic flux density $\boldsymbol{B}$, at a constant speed $\boldsymbol{v}$. The bar is in electric contact with two

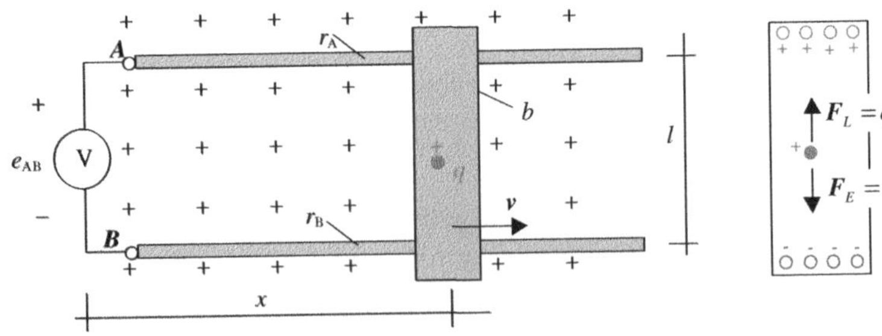

FIGURE 9.1. Effects on inner charges of a conductor sliding across a magnetic field.

rails r_A and r_B, which, in turn, are connected with the two terminals **A** and **B** of a voltmeter measuring the induced electromotive force e_{AB}.

Let the flux density be orthogonal to the plane formed by this page and directed downwards. This is indicated, as usual, by several plus sign.

The convention for indicating directions orthogonal to the page implies the use of a dot (•) for quantities directed toward the observer (i.e., upwards from the page) and a plus sign (+) for the opposite direction.

To remember this convention, the reader can visualize the situation with the help of the sketch below, which shows a 3D arrow with a cross at its tail and a dot at the very end of its tip.

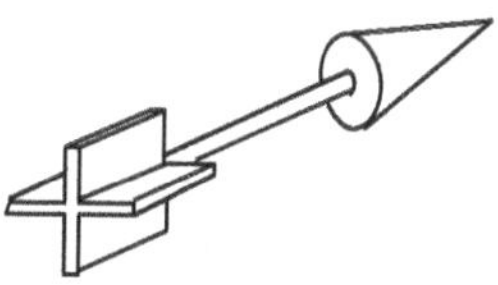

Consider a positive charge q inside the conducting bar. It shares the bar speed and because of this speed and the flux density $\boldsymbol{B}$, it will be subject to Lorentz's force:

$$\boldsymbol{F}_{\mathrm{L}} = q\boldsymbol{v} \times \boldsymbol{B}$$

and will therefore be forced to move upward in our situation (remember the right-hand rule for determining the result of a vector product). In a very short time, charge separation will occur inside the bar, with positive charges accumulated in the top part of the bar and negative ones accumulated at the bottom. The process stops when the electric field caused by the accumulated charge exactly matches Lorentz's force, as shown in the right-hand part of Figure 9.1:

$$\boldsymbol{F}_{\mathrm{E}} = -\boldsymbol{F}_{\mathrm{L}}$$

and therefore

$$\boldsymbol{E} = \boldsymbol{F}_{\mathrm{E}}/q = -\boldsymbol{F}_{\mathrm{L}}/q = -\boldsymbol{v} \times \boldsymbol{B}$$

The rearrangement of charges inside the bar is very fast because of the tiny mass of electrons; therefore, in nearly all applications of electrical engineering, this rearrangement can be taken as being instantaneous.

Consider, now, the potential of the electric field inside the bar:

$$e_{AB} = V_A - V_B = \int_A^B \boldsymbol{E} \bullet \mathrm{d}\boldsymbol{l} = \int_A^B vB\,\mathrm{d}l = vBl \qquad (9.1)$$

Note that the polarity of e_{AB} can be determined using the right-hand rule: if the thumb is in the direction of $\boldsymbol{v}$ and the index finger in the direction of $\boldsymbol{B}$, the tip of the middle finger indicates the positive polarity of the induced EMF.

It is advisable to remember equation (9.1) by leaving the three factors in the given order; this way, $\boldsymbol{v}$ and $\boldsymbol{B}$ are in the same order as in Lorentz's law, and the induced EMF will tend to push charges in the direction of Lorentz's law.

Expression (9.1) is valid only when the directions of the sliding bar $\boldsymbol{v}$, $\boldsymbol{B}$, are all orthogonal to each other. In rotating electrical machines the directions of the sliding bars are always orthogonal to $\boldsymbol{v}$, while $\boldsymbol{B}$ can have a different orientation. Therefore, the following more general expression is of the utmost importance for rotating electrical machines:

$$e_{AB} = vB_{\perp}l \tag{9.2}$$

This result, often called *Hall's effect*, is a fundamental law for electromechanical conversion:

Result: *Faraday's law for sliding conductors*

If a conductor with length l is moving at a speed $\boldsymbol{v}$ orthogonal to the conductor direction in a magnetic field having density $\boldsymbol{B}$, an electromotive force appears at either end. Its value is $e = vB_{\perp}l$, with polarity given by the right-hand rule, in which $B_{\perp}$ is the component of $\boldsymbol{B}$ orthogonal to the plane defined by $\boldsymbol{v}$ and $\boldsymbol{l}$.

It is worth mentioning that the same result (9.1) would have been found by considering that the bar, along with the rails and the connections with the voltmeter, create a mesh, in which, in accordance with Faraday's law, the following electromotive force is induced:

$$e = \frac{\mathrm{d}\phi}{\mathrm{d}t} = \frac{\mathrm{d}(l \cdot x \cdot B)}{\mathrm{d}t} = lB\frac{\mathrm{d}x}{\mathrm{d}t} = vBl \tag{9.3}$$

Let us now slightly change the system by allowing some current to circulate in the bar and the rails. The situation is depicted in Figure 9.2. The situation is pretty similar to what happens in a circuit fed by an electrochemical battery, as examined in Section 4.2.1. Lorentz's force "pumps" charges upward inside the bar, and these charges create, in the circuit outside the bar (rails, circuit containing resistor R), an electric field that causes the current i to circulate.

The very circulation of i causes the charges inside the bars to move vertically, thus causing Lorentz's force to act in an orthogonal direction to charge flow. This is again

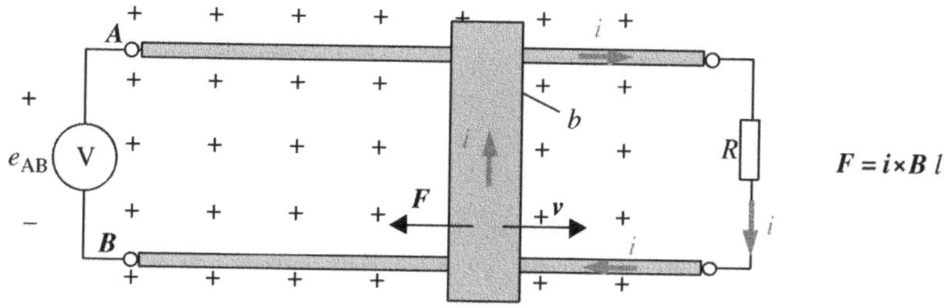

FIGURE 9.2. Effects of current flowing in the system shown in Figure 9.1 by the effect of the load R. Operating principle of an electric generator.

in accordance with Lorentz's force expression, which, in the case of currents, takes the form (see the Appendix)

$$\mathrm{d}\boldsymbol{F} = i\,\mathrm{d}\boldsymbol{l} \times \boldsymbol{B} = \boldsymbol{i} \times \boldsymbol{B} \cdot \mathrm{d}l$$

which is normally referred to as Laplace's equation of electromagnetically induced force.

Given that the current direction and magnetic field are orthogonal to each other, and considering the right-hand rule for vector products, the total force on the bar will be

$$\boldsymbol{F} = \int_A^B \mathrm{d}\boldsymbol{F} = \int_A^B \boldsymbol{i} \times \boldsymbol{B}\,\mathrm{d}l = \boldsymbol{i} \times \boldsymbol{B} \cdot l \tag{9.4}$$

It is of great interest in electromechanical conversion to compute the component of $\boldsymbol{F}$ in the direction of the sliding bar speed $\boldsymbol{v}$.

$$F_v = iB_{\perp}l \tag{9.5}$$

Equation (9.5) has the same importance in electromagnetic conversion as (9.2), thus:

> **Result:** *Force on a conductor carrying current inside a magnetic field*
>
> If a conductor with length l carrying current i crosses a magnetic field having density $\boldsymbol{B}$, a force orthogonal to $\boldsymbol{l}$ and $\boldsymbol{B}$ is generated, whose value is $F = iBl$, with polarity determined by the right-hand rule. If the conductor is moving at a speed $\boldsymbol{v}$, the component of $\boldsymbol{F}$ along $\boldsymbol{v}$ is $F_v = iB_{\perp}l$, in which $B_{\perp}$ is the component of $\boldsymbol{B}$ orthogonal to the plane defined by $\boldsymbol{v}$ and $\boldsymbol{l}$.

The two equations (9.2) and (9.5) together constitute the basis for the analysis methodology of electrical machines, or, better, electromechanical converters. By combining these two equations, for instance, we can easily find the equation of mechanical to electrical energy conversion. Indeed, to keep the bar b moving with speed $\boldsymbol{v}$, some external action is needed to produce a force that exactly counteracts the one produced by (9.4).

This external force will then deploy the power

$$P = \boldsymbol{F} \bullet \boldsymbol{v} = F_v \cdot v = iBl \cdot \frac{E}{Bl} = i \cdot e \tag{9.6}$$

where the values of F_v and v are taken from equations (9.5) and (9.1), respectively.

Equation (9.6) expresses the energy conservation law for electromechanical conversion:

> **Result:** *Energy conservation in electromechanical conversions*
>
> When a conductor constituting part of a circuit loop moves in a magnetic field, voltages, currents, and forces are generated in such a way that it is
>
> $$\boldsymbol{F} \bullet \boldsymbol{v} = e \cdot i$$
>
> Therefore, the mechanical power needed to keep the conductor moving is converted into electrical power which is generated and deployed in the circuit.

Our analysis highlights the presence of force $\boldsymbol{F}$ on the sliding bar. Naturally, the principle of action and reaction (Newton's third law) requires that another force, opposite $\boldsymbol{F}$, must be generated. Indeed, it was not shown in the analysis *how* the flux density $\boldsymbol{B}$ was produced. We would have seen that other currents produced $\boldsymbol{B}$, and applying Lorentz's force equation to these other currents would give rise to a force system equivalent to a force opposite to $\boldsymbol{F}$.

Our analysis of the system of Figure 9.2 shows a basic form of electric generator: power is converted from mechanical into electrical form.

It is now very easy to construct a similar system capable of effecting the opposite conversion. It can be structured as depicted in Figure 9.3. At the right-hand-side of the

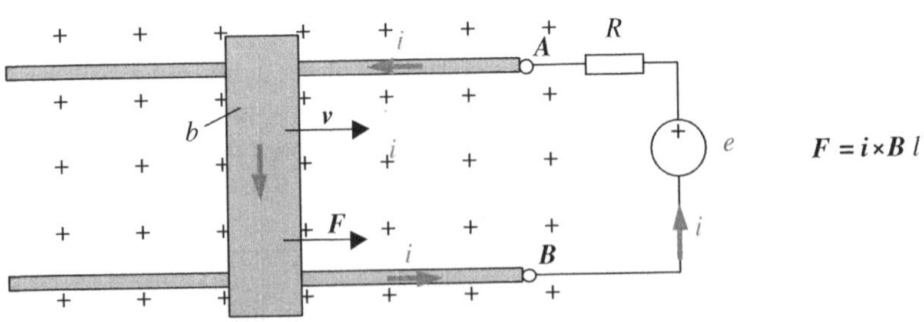

FIGURE 9.3. Principle of operation of an electric motor.

two rail terminals $\boldsymbol{A}$ and $\boldsymbol{B}$, a system capable of producing a voltage source e (e.g., a battery, or another system with a sliding bar) is connected by the interposition of an element which can be modelled as a pure resistor R. The current i will circulate in the system which, if the resistance of the rails and the bar is disregarded as before, is

$$i = e/R$$

As a consequence of (9.5), this circulation of current will cause force F whose orientation is found using the right-hand rule. This will tend to cause a movement of the bar toward the right.

If the bar starts moving to the right with speed $\boldsymbol{v}$, the mechanical power $\boldsymbol{F} \bullet \boldsymbol{v} = Fv$ is produced and transferred to a mechanical system. For instance, if a new force $-\boldsymbol{F}$ counteracts $\boldsymbol{F}$, mechanical power Fv will be transferred to the mechanical system producing this force. At the same time, the speed $\boldsymbol{v}$ will cause the EMF vBl to be produced and again it is

$$P = \boldsymbol{F} \bullet \boldsymbol{v} = F_v \cdot v = iBl \cdot \frac{E}{Bl} = i \cdot e \tag{9.7}$$

So power conservation still applies. The system of Figure 9.3 is therefore a basic form of electric motor.

9.3 BASIC ELECTROMECHANICS IN ROTATING MACHINES

9.3.1 Rotating Electrical Machines and Faraday's Law

In the previous section, the principle of electromechanical conversion was shown for a sliding bar.

A practical problem of the sliding bar is that it cannot be used for electromechanical conversion in a limited space, due to the fact that the bar needs to move further and further away. Another problem is that a uniform magnetic field is an abstraction and can only be obtained in limited regions of space.

The more natural way to exploit electromechanical conversion in practice as shown for the sliding bar is to "curl" its path—that is, to perform a closed loop. This concept is shown in Figure 9.4. Figure 9.4a shows the basic structure of a rotating machine. The main elements of the machine are:

- A stationary (i.e., nonrotating) circular structure called *stator*, which is mechanically connected to a mechanical support. The stator is made of ferromagnetic material to ease the circulation of flux density.
- A rotating part which, in principle, is cylindrical and able to rotate around its axis. The drawings in Figure 9.4 show a view of the machine in a plane which is orthogonal to the axis of rotation. The rotor is also made of ferromagnetic material to facilitate the circulation of flux density.
- The space between stator and rotor is constituted by air, to allow free rotation of the rotor, and is therefore called *air gap*.

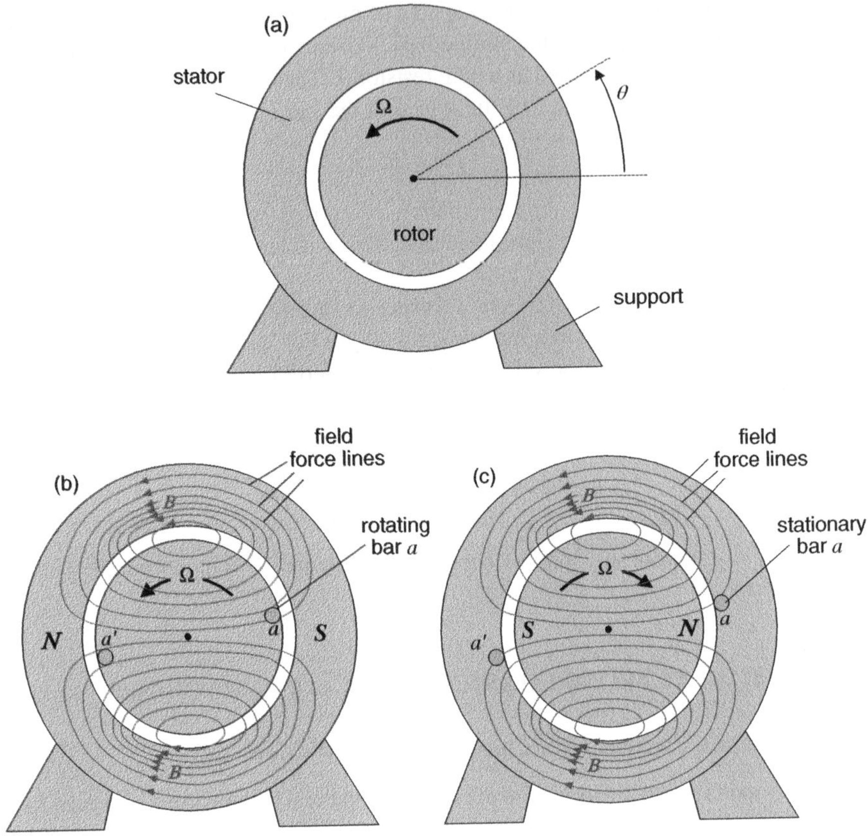

FIGURE 9.4. The basic structure of rotating electric machines. In part (b) the ***B*** map is produced by the stator and is stationary; in part (c) the ***B*** map is produced by the rotor and rotates with it.

Given this structure, relative motion between magnetic field and conducting bars can be obtained by installing the conductive bars either in the rotor (Figure 9.4b) or in the stator (Figure 9.4c). Here symbols ***N*** and ***S*** indicate "north pole" and "south pole" respectively. By convention, the flux exits from north poles and enters south poles.

In both cases, the machine can easily be analysed using one of the sliding bars, by "rectifying" the air gap; this is done by imagining to cut it in the position in which $\theta = 2\pi$ and uncurling the cylindrical surface as a rectangle in a plane. In this rectified representation, the radial component of magnetic field density ***B*** is orthogonal to the page and represented as usual by plus sign (downward) or dots (upward).

This process can be carried out for the air gap in Figures 9.4b and c, showing in particular the situation of conductor a', as shown in Figure 9.5. The position of a' is such that B_r, measured with the given reference (a plus sign indicating the direction with respect to the page), is positive.

The indicated speed $\boldsymbol{v}$ is the speed of the sliding bar, measured in a reference fixed to the map of ***B*** over space.

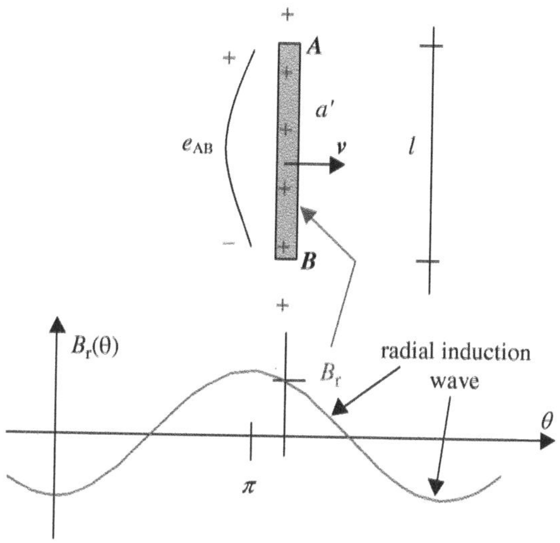

FIGURE 9.5. Movement of the conductor a' of Figure 9.4 analysed using the rectified air gap.

The direction of this is coherent with the situation depicted in Figure 9.4b—that is, when the rotor has the angular speed Ω and the map of ***B*** over space is constant, or rotates less than the rotor.

The direction of indicated speed ***v*** is also coherent with the situation depicted in Figure 9.4c, where the map of ***B*** over space rotates at an angular speed Ω in the direction shown in the figure.

In the bar position shown in Figure 9.5, B_r has just passed its maximum and is now decreasing. The voltage produced at the conductor ends is, according to (9.2), $e_{AB} = vB_r l$.

Note that the situations shown in Figures 9.4b and 9.4c define field–conductor interactions that are directly exploited in the most important electric machines. This will be analysed with some details in the following chapters:

- DC machine: situation of Figure 9.4b with ***B*** map stationary.
- Synchronous machine: situation of Figure 9.4c with rotating ***B*** map.
- Asynchronous machine: situation of Figure 9.4b with ***B*** map and rotor both rotating, but at different angular speeds.

The shape of the radial component of B_r along the air-gap periphery is normally not an exact sine wave. However, the machine is constructed in such a way that $B_r(\theta)$ is a near-sine shape.

Naturally, for all this to be useful, the voltage generated at the ends of conductors must be captured and possibly applied to a load so that some current, and then power, flows through the load at the expense of mechanical power.

Consider a view of the machine in a section that contains the rotation axis, as shown in Figure 9.6. The position of conductors a and a' is different from that in

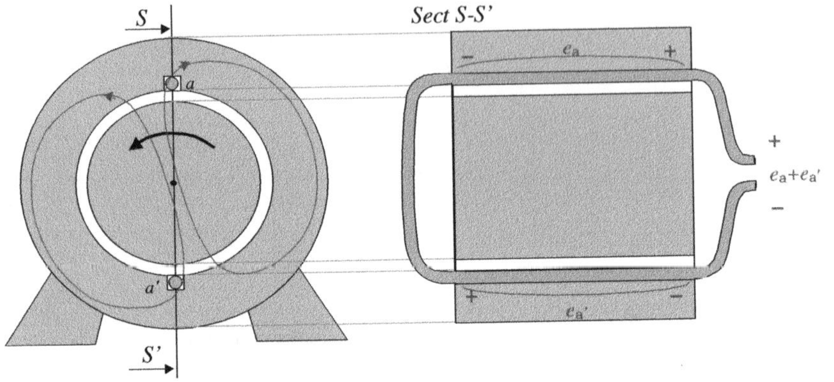

FIGURE 9.6. Connections of the conductor ends to gather the sum of voltages generated on the two wires a and a'.

Figure 9.4, so they can be visualized in the axial section. The positions of the north and south poles are also different, so that a radial component is present where the conductors reside.

Moreover, to make the drawing more similar to what is done in practice, the two conductive bars are shown inserted in slots (here square, and therefore represented by white squares), which are grooves explicitly created in the stator so that the conductors, and the insulating layer around them, can be inserted into the stator during manufacture.

Figure 9.6 shows that when the two conductive bars are connected at one end of the machine, the total EMF produced at the other end is the sum of the EMFs of the two active conductors a and a'. The connections undergo negligible electromotive forces, because the flux density in the air surrounding the machine is scarce, due to the concentration of flux occurring in the space where iron is present.

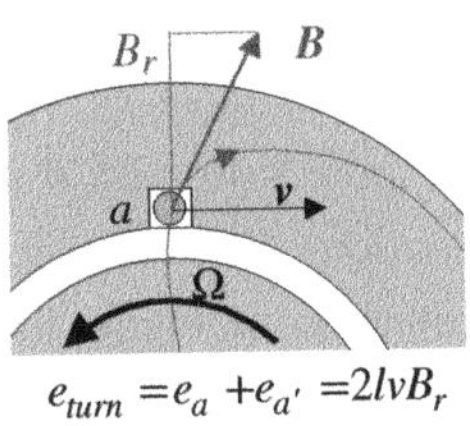

The value of the produced electromotive force $e = e_a + e_{a'}$ can be computed using the relation (9.2). The component of flux density which is orthogonal to speed now becomes the radial component (directed from centre to air gap; see sketch above):

$$e_{turn} = e_a + e_{a'} = 2lvB_r \tag{9.8}$$

The subscript "turn" indicates a single full path of wire. Similarly to normal coils and inductors, it will be seen in the next chapter that electrical machines also have coils with several turns each.

It must be noted that in order to apply the right-hand rule and to determine the polarity of induced voltages, the velocity $\boldsymbol{v}$ to be taken into account is the velocity of the conductor relative to the field, and not vice versa. This is the reason why the conductor a, a $\boldsymbol{v}$ directed to the right, should be considered, since the magnetic flux density $\boldsymbol{B}$ is imagined to be produced by the rotor and is fixed to it, thus rotating counterclockwise.

Equation (9.8) can be further processed considering that in the machine it is $v = \Omega R$, R being the radius of the coil—that is, the distance of the conductor centre from the machine centre—and $2Rl$ can be assumed to be the coil effective surface area A. Therefore it is

$$e_{turn} = 2l\Omega R B_r = \Omega B_r A \tag{9.9}$$

Note that equation (9.9) gives the EMF produced by the couple of conductors a–a', while B_r is the radial flux density corresponding to conductor a, which will therefore be called the *base* conductor of the couple a–a'.

It should be explicitly noted that equation (9.9) is valid whatever the shape of $B_r(\theta)$. However, in practical machines, as already mentioned, all efforts are made so that this shape resembles a sinusoid as closely as possible, as already shown in Figure 9.5.

Given the introductory nature of this book, a sinusoidal shape of $B_r(\theta)$, and therefore of e_{turn}, will always be assumed. For a sinusoidal shape, along with a rotor rotating at constant angular speed Ω, equation (9.9) becomes:

$$e_{turn} = \Omega BA \sin(\Omega t + \theta_0)$$

9.3.2 Generation of Torques in Rotating Machines

Let us now consider what happens in our two-wire coil when its two terminals are connected to a load. We only consider what happens in steady state—that is, when all transients are complete.

The situation is depicted in Figure 9.7, showing a physical representation to the left of terminals **A–B** and an equivalent circuit on the right. The torque created by the

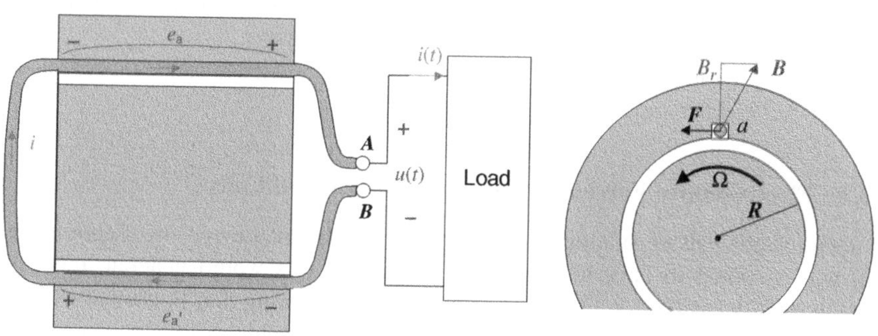

FIGURE 9.7. Principle operation of a basic electric generator.

force applied to the bar displayed in the right part of the figure, with respect to the rotational axis, will be [see equation (9.5)]:

$$T_{\mathrm{bar}} = iB_{\mathrm{r}}lR = iB_{\mathrm{r}}\frac{A}{2}$$

where, considering the turn constituted by this bar and the opposite one, $A = 2Rl$ is the area of the turn. Therefore,

$$T_{\mathrm{turn}} = 2T_{\mathrm{bar}} = iB_{\mathrm{r}}A \tag{9.10}$$

Equation (9.10) is the twin equation (9.9) and is also very important.

It should first be noted that the torque acting on the turn, as a consequence of Newton's action and reaction law, is numerically equivalent to the one acting on the machine rotor.

Thus, replicating what was seen in equation (9.7), it is immediately apparent that

$$P = T\Omega = iB_{\mathrm{r}}A \cdot \frac{e}{B_{\mathrm{r}}A} = i \cdot e \qquad (^{\circ})$$

which confirms that the electric power absorbed by the turn is perfectly equal to the one generated on the machine's rotor.

The requirement of operating in steady state to produce this result is important, since during transients the variation of current i will imply some accumulation of energy in the machine's magnetic field, and therefore the electric energy is partly stored as magnetic field energy and partly converted into mechanical form. Indeed the machine acts as an inductor and inductors accumulate energy, in accordance with the law already seen in Chapter 5: $\mathscr{E} = (1/2)Li^2$. In three-phase machines a generalization of (°) applies (as will be further discussed in the next chapters with the form $P = i_1 \cdot e_1 + i_2 \cdot e_2 + i_3 \cdot e_3$).

The behaviour of the system in Figure 9.7 is that of a basic electric motor.

If the system outside the machine generates electric power and the current flow is reversed at equal voltage polarity, the system will behave as an electric generator: Electric power will be converted into mechanical power.

Again, in the following chapters these concepts will be used to create more realistic DC and AC electric motors.

9.3.3 Electromotive Force and Torque in Distributed Coils

In the basic analysis of rotating machines performed in the previous two sections, the coil was assumed to have a single turn. In reality, this would generate too little voltage; this is easily seen using equation (9.9), considering that in normal iron the peak value of B is just over one tesla. A one pole-pair machine with a coil surface of $2\,\mathrm{dm}^2$, with a rotor rotating at 3000 rpm, would have just 6–7 V of peak voltage.

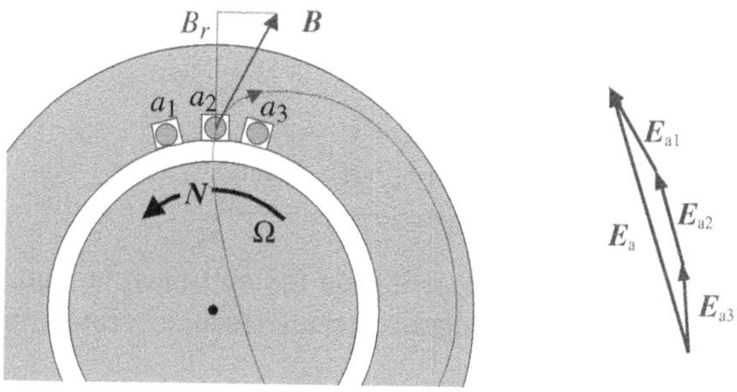

FIGURE 9.8. Drawing showing the effect of distributing coils along the air gap.

To use N turns with large values of N is not as easy as in transformers, because of the different physical structure of the machine. However, significantly larger numbers of turns per coil can be obtained by distributing the turns along the air gap, according to the scheme shown in Figure 9.8.

It is obvious that the EMFs generated in each turn are not in phase with each other, because of the angular displacement between one coil and the adjacent one. Therefore, differently from transformers, the voltages on different turns do not sum algebraically, but as vectors; the result is that the peak of the total induced electromotive force is lower than N times the peak of EMF induced on a single turn. The EMF induced on a coil with area A will therefore be

$$e_{\text{coil}} = k_{\text{w}} N \Omega B_{\text{r}} A \tag{9.11}$$

where

- N is the number of turns per pole per phase;
- k_{w}, called the *winding distribution factor*, takes into account the fact that the sum of single EMFs has an amplitude lower than N times the single EMFs[1];
- B_{r} is radial induction at the centre turn of the coil (a_2 in Figure 9.8).

We can introduce the flux $\phi = B_r A$ and the linked flux $\psi = k_{\text{w}} N B_r A = k_{\text{w}} N \phi$; therefore equation (9.11) can be written as

$$\boxed{e_{coil} = \Omega \psi} \tag{9.12}$$

Equation (9.12) constitutes a special version of equation (A.11) and will therefore sometimes called in this book *Faraday's law for rotating machines*.

[1]Other coil details might make the generated EMF different from $N\Omega B_{\text{r}} A$. Equation (9.11) is still valid, if the numerical value of k_{w} considers all these details. For more information on this, see reference [bm1].

A similar procedure can be followed for torques, and therefore the (9.10) for a distributed coil with N turns will become:

$$\boxed{T_{coil} = k_w N i B_r A = \psi i} \tag{9.13}$$

Again, considering that the torque acting on the coil must be equal (by Newton's action–reaction law) to a torque generated on the machine's rotor, mixing (9.13) and (9.11) it is immediate that

$$P = T\Omega = i \cdot e$$

that is, again the electric power associated with i and e is converted into mechanical power generated in the machine's rotor.

9.3.4 The Uniform Magnetic Field Equivalent

Although the radial component $B_r(\theta)$ of magnetic flux density as a function of the air-gap angle θ can have a large variety of shapes, it was already noted that in actual machines, very often the manufacturer tends to realise a sinusoidal shape. This hypothesis is always taken for granted from now on.[2]

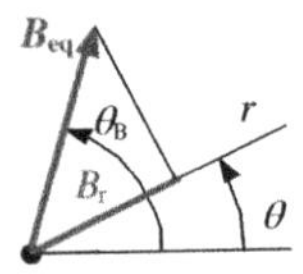

$B_r(\theta) = B\cos(\theta_B - \theta) =$
$B\sin(\theta - \theta_B + \pi/2)$

Remember that a sine wave can always be expressed as a translated cosine. Therefore the two formulas immediately above the sketch are equivalent if $\theta_{0c} = \theta_{0s} + \pi/2$ (subscripts "c" and "s" stand for "cosine" and "sine"). The same radial component of magnetic flux density can be obtained considering a uniform distribution $\boldsymbol{B}_{eq}$ having in all points of space a modulus with a maximum value of B_r; the actual flux has direction the same as that of a vector with its tail in the rotor's centre and its head in the air-gap point in which B_r is maximum.

This can be easily verified looking at the sketch above: the projection of $\boldsymbol{B}_{eq}$ on a generic ray r constitutes a sinusoidal wave:

$$B_r(\theta) = B\sin(\theta - \theta_{0s}) = B\cos(\theta - \theta_{0c})$$

[2]In reality, actual machines implement a shape that is sinusoidal with θ, but can contain more than one period in a complete rotation around the air-gap circumference: $B_r(\theta) = B\sin(p\theta - \theta_0)$, where p is an integer number fixed for a given machine. This will be explained in the next chapter, but does not cause loss of generality of the reasoning shown here.

where

$$\theta_{0s} = \theta_B - \pi/2, \qquad \theta_{0c} = \theta_B$$

Therefore in all cases in which the B_r shape is sinusoidal—that is, in all cases dealt with in this book—instead of considering the actual map of $\boldsymbol{B}$, such as those depicted using force lines in Figures 9.4b and 7.4c, just a single uniform equivalent field can be considered:

> **Result:** *The uniform equivalent magnetic field*
>
> Therefore in all cases in which the B_r shape is sinusoidal, instead of considering the actual map of $\boldsymbol{B}_{eq}$, we can consider an equivalent uniform field whose modulus is equal to $B_r(\theta)$, and whose direction has its tail in the rotor's centre and its head in the air-gap point where $B_r(\theta)$ is maximum. This field is called "uniform equivalent magnetic field".

This result will obviously ease the analysis and will be well exploited in the following chapters.

When rotating machines operate, as has already been noted, the corresponding wave of radial flux density translates in a θ–B_r Cartesian diagram (Figure 9.5), or, equivalently, the map of $\boldsymbol{B}$ force lines rotates. Using the uniform magnetic field equivalent, it is simply a matter of visualizing a unique $\boldsymbol{B}_{eq}$ rotating.

This rotation is very similar to the rotation of phasors representing sinusoidal time functions (remember the theory of phasors in Chapter 5). Therefore this rotating field equivalent is called, in modern analysis of rotating machines, *space phasor*. Space phasors are a very powerful tool to analyse electric machines, but, because of its introductory nature, it will not be used in this book.

9.4 RELUCTANCE-BASED ELECTROMECHANICAL CONVERSION

In the previous sections, electromechanical conversion was obtained using the principle of the translating bar, in which there is explicit interaction between the moving charge in the translating bar, and an external magnetic field produced by other means.

It is, however, also possible to produce or absorb mechanical power using current that interacts with the field it produces.

These cases are normally studied through use of magnetic circuit reluctance, which has been used several times already in this book; in this chapter the corresponding conversion is therefore called reluctance-based electromechanical conversion.

The principle of reluctance-based electromechanical conversion can be understood by analysing an electromagnet (Figure 9.9). In this example, the moving armature rotates around a pivot, thus causing the air-gap length x to vary.

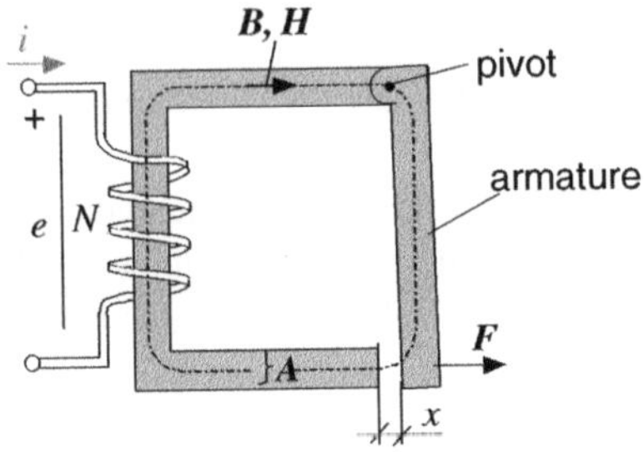

FIGURE 9.9. An electromagnet.

A simple evaluation of the force generated on the armature by the presence of a magnetic field in the magnetic circuit is carried out by ignoring any energy dissipation, and in particular the energy dissipated in the coil resistance and in the pivoting hinge friction. Under this assumption, for an infinitesimal increase dx in the air gap, the following energy balance applies:

$$\mathrm{d}\mathscr{E}_e = \mathrm{d}\mathscr{E}_\mathrm{m} + \mathrm{d}W_x \tag{9.14}$$

where

$$\begin{aligned}
\mathrm{d}\mathscr{E}_e &= ei\mathrm{d}t = i\mathrm{d}\psi = i\mathrm{d}(Li) = i^2\mathrm{d}L + Li\ \mathrm{d}i \\
\mathrm{d}\mathscr{E}_\mathrm{m} &= \mathrm{d}\left(\frac{1}{2}Li^2\right) = \frac{1}{2}i^2\mathrm{d}L + Li\ \mathrm{d}i \\
\mathrm{d}W_x &= F\,\mathrm{d}x
\end{aligned} \tag{9.15}$$

where $\mathrm{d}\mathscr{E}_e$ is the electric energy absorbed by the electromagnet and $\mathrm{d}\mathscr{E}_\mathrm{m}$ and $\mathrm{d}W_x$ are, respectively, the corresponding increase in the energy accumulated in the magnetic field and the mechanical work produced. As a consequence of these signs, F is positive when it tends to increase x. Substituting the (9.15) into (9.14) gives

$$F\mathrm{d}x = \mathrm{d}\mathscr{E}_\mathrm{e} - \mathrm{d}\mathscr{E}_\mathrm{m} = \frac{1}{2}i^2\mathrm{d}L \quad \Rightarrow \quad F = \frac{1}{2}i^2\frac{\mathrm{d}L}{\mathrm{d}x} \tag{9.16}$$

To find a practical estimate of the force based on the general equation (9.15), consider that the reluctance of the whole magnetic circuit is composed of the sum of the reluctance of its iron and air-gap parts[3]:

$$\mathscr{R}_\mathrm{tot} = \mathscr{R}_\mathrm{Fe} + \mathscr{R}_\mathrm{ag} = \frac{l_\mathrm{Fe}}{\mu_\mathrm{Fe}A} + \frac{x}{\mu_\mathrm{ag}A} \cong \mathscr{R}_\mathrm{ag} \tag{9.17}$$

[3]For a reminder of the reluctance of a magnetic circuit and its relation to inductance, see Section 5.2.3.

where l_{Fe} is the length of the iron part of the line over which the fields $\boldsymbol{B}$ and $\boldsymbol{H}$ lie (dashed in Figure 9.9) and A is the cross-sectional area of the magnetic circuit.[4] The approximate equality is justified by the fact that the iron magnetic permeability is usually thousands of times more than the permeability of air and this prevails, whenever x is not very small, over another characteristic of the circuit—that is, that $l_{\mathrm{Fe}} \gg x$.

When the approximation in (9.17) is made, equation (9.16) becomes

$$F = \frac{1}{2} i^2 \frac{\mathrm{d}}{\mathrm{d}x}\left(\frac{N^2}{\mathscr{R}}\right) = \frac{N^2 i^2}{2} \frac{\mathrm{d}}{\mathrm{d}x}\left(\frac{\mu A}{x}\right) = -\frac{N^2 i^2 \mu A}{2x^2} = -\frac{H^2 \mu A}{2} \tag{9.18}$$

Therefore the force is negative and tends to attract the armature toward the rest of the magnetic circuit.

Equation (9.18) is often shown in textbooks in other, equivalent forms, that is,

$$F = -\frac{1}{2}\frac{B^2 A}{\mu} = -\frac{1}{2}\frac{\phi^2}{\mu A}$$

These forms are practical to evaluate the force when a permanent magnet is involved, of which the magnetic flux ϕ is usually known.

It also shows that, within the limits of its applicability, the lower the area A of flux passage, the higher the force produced. This is exploited in practical electromagnets, in which the area A is kept rather small.[5]

The mechanical work $\mathrm{d}W_x$ is produced at the expense of other systems.

The balance of the three energies shown in (9.14) depends on how the electric circuit is fed. It is traditional (and useful) to consider the two special cases separately:

1. Electric circuit feeding at $i = \text{const.}$
   From (9.15) it is

$$\frac{\mathrm{d}\mathscr{E}_{\mathrm{m}}}{\mathrm{d}x} = \frac{1}{2} i^2 \frac{\mathrm{d}L}{\mathrm{d}x}, \qquad \frac{\mathrm{d}\mathscr{E}_{\mathrm{e}}}{\mathrm{d}x} = i^2 \frac{\mathrm{d}L}{\mathrm{d}x}$$

   Therefore when the current is constant, and x reduces, energy is drawn from the circuit that is equally shared between mechanical work and accumulation in the magnetic field:

$$\mathrm{d}W_x = \mathrm{d}\mathscr{E}_{\mathrm{m}}$$

[4]For a more precise evaluation, A should be slightly adjusted to take into account the fact that the magnetic field force lines are more dispersed in the air gap, and therefore the actual surface of the passage of $\boldsymbol{B}$ is a bit larger than A.

[5]Not too small, otherwise the hypotheses under which the formulas are drawn are no longer valid. For instance, small sections which are too small cause iron to saturate, thus reducing the force produced.

2. Electric circuit feeding with energy = const.
   From (9.14) it immediately follows that

$$\frac{d\mathscr{E}_m}{dx} = -\frac{dW_x}{dx} = -\frac{1}{2}i^2\frac{dL}{dx}$$

   It is clear, therefore, that when no energy is supplied by the external circuit, mechanical work is produced at the expense of the previously accumulated magnetic energy.

Principle of minimum reluctance

Let us consider again equation (9.16):

$$F = \frac{dW_x}{dx} = \frac{1}{2}i^2\frac{dL}{dx}$$

This says that in a magnetic circuit (operating at constant current), forces are generated that tend to raise the circuit inductance. Since it is also

$$L = N^2/\mathscr{R}$$

this force tends to minimize circuit reluctance. In the example of Figure 9.9 the process continues up to when $x = 0$; the reluctance is minimized since the part of it due to air gap $\mathscr{R}_a$ is no longer added.

This explains that all magnetic circuits always tend to reach a condition of minimum reluctance; this fact is sometimes called the *principle of minimum reluctance*. This principle is exploited not only in electromagnets, but also in some rotating machines, called *reluctance machines* and *switched reluctance machines*.

Although these have applications of some importance, they will not be dealt with further in this book and the interested reader is advised to refer to more specific textbooks, such as [bm1] or [bm4].

10

DC MACHINES AND DRIVES AND UNIVERSAL MOTORS

For the Instructor

It is well known that DC machines are increasingly substituted by AC machines, since their competitiveness in variable speed drives is greatly reduced by the lower cost of inverters. However, this chapter has been included in this book for the following reasons:

- DC machines have not yet abandoned the market, and they are forecast to be around for years to come.
- The operation of linear DC machines and drives is very simple and can be a useful introduction for students to more complex machines and drives.

However, if time is insufficient, the teacher may choose to omit this chapter and directly proceed to the next one.

As explained in the "For the Instructor" box of Chapter 8, the reader is not required to study power electronics before approaching this chapter. Conversely, readers who have already studied Chapter 8 will find an interesting recapitulation of electronic converters in Section 10.6, which deals with electric drives for DC motors.

Fundamentals of Electric Power Engineering: From Electromagnetics to Power Systems, First Edition.
Massimo Ceraolo and Davide Poli.

10.1 INTRODUCTION

Traditionally, the phrase "electric machine" refers to either a transformer or a rotating electric machine. This is justified by the fact that transformers share many physical similarities with rotating machines: both are based on *magnetic circuits* made of ferromagnetic materials in which variable flux is produced, so that EMFs are generated by Faraday's law action.

Rotating machines are also based on the principles of electromechanical conversion discussed in the previous chapter. The majority of machines used in practice are AC machines. However, it is common for basic textbooks to also examine DC machines. This approach is followed in this book for the following reasons:

- The reader has already studied DC circuits and may wish to see how this type of electric energy can be used to generate mechanical power.
- Machines which are structurally very similar to simple DC machines are used very frequently as single-phase AC motors (universal motors).
- Small DC machines are still used whenever a DC source is easily and cheaply available—for example, on board vehicles, where the basic energy source is an electrochemical battery.

Nevertheless, because of their reduced use nowadays, we will deal with DC machines in a particularly simplified way.

10.2 THE BASIC IDEA AND GENERATION OF QUASI-CONSTANT VOLTAGE

The general arrangement of a DC machine is shown in Figure 10.1. It is composed of a stationary part: the *stator*; a rotating part: the *rotor*; and a space of air, called *air gap*, between the two.

The stator contains a winding made by connecting in series all the conductors shown in the cross section in the figure. A simplified idea on how these connections are made can be inferred from the small sketch shown in the lower part of Figure 10.1.

This winding, called *field winding* for reasons that will be clear presently, is connected to a DC source and is therefore traversed by a DC current.

To analyse what happens in the machine, the field winding current can be imagined to be flowing according to the signs indicated in Figure 10.1; that is, it leave the page surface when a small dot is shown in the conductors, and enters it when a cross is shown (to understand the rationale beyond this graphical code, the reader is advised to revert to Section 9.2).

By effect of this current, a magnetic field is created, whose force lines flow in the stator, traverse the small air gap, enter the rotor, traverse it, and then return to the stator, as qualitatively indicated in Figure 10.1. As mentioned in Section 9.3.1, by convention magnetic flux abandons north poles (indicated with N), and enters south poles (indicated as S).

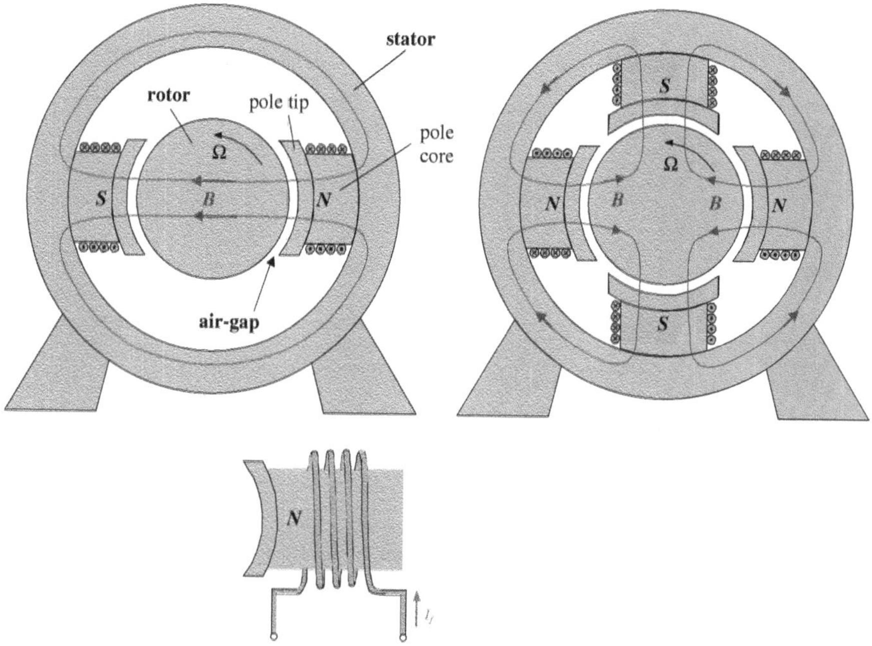

FIGURE 10.1. The basic structure of a DC machine. *Left*: Two-pole machine. *Right*: Four-pole machine. *Bottom*: Conductor connections to obtain stator windings.

In basic analysis of DC machines and in this book also, the field $\boldsymbol{H}$ produced in the air gap, and therefore the corresponding flux density $\boldsymbol{B}$, can be assumed to be uniform. $\boldsymbol{B}$ will be proportional to the current I_f circulating in the field winding.

The two machines represented in Figure 10.1 differ from each other with regard to their pole pairs: the machine on the left has one pole pair, while the one on the right has two; machines can have even more pole pairs. In all cases, all the windings are connected in series, and are thus traversed by the same current. The analysis of the machine behaviour that will be proposed presently is based on the interaction between rotor conductors and the flux density produced by the stator and is therefore applicable to either type of machine. Some differences between the two will be discussed later.

The rotor of a DC machine contains conductors at its periphery, as represented in Figure 10.2.

These conductors are inserted into *slots*, which are spaces from which iron has been removed, and are able to keep the conductors mechanically tightly connected to (but electrically insulated from) the basic iron structure of the rotor.

In the terminology of electric machines, the *armature* is the part of the machine in which an EMF is induced by Faraday's law. Therefore, in a DC machine the armature is the rotor. In other machines (such as the synchronous machine, as will be seen in Chapter 11) it is located in the stator.

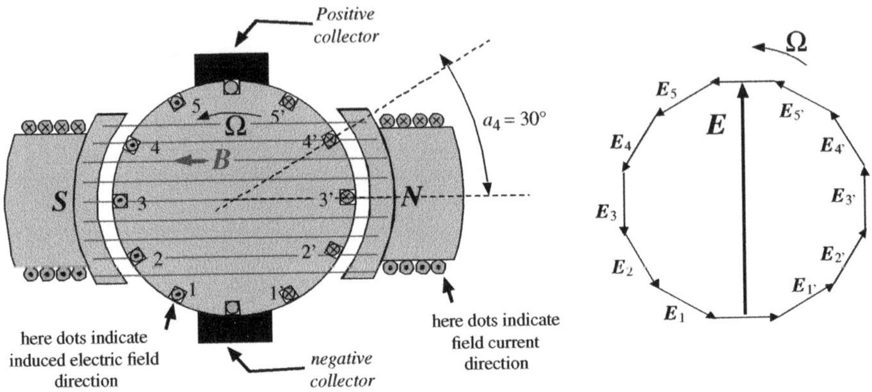

FIGURE 10.2. Principle scheme for generation of EMFs in the rotor, and total EMF collected E.

The rotor conductors are connected to each other by means of connections not shown in Figure 10.2, in such a way that the voltage across the *collectors* are the sum of all the voltages appearing in the two (left and right) parts of the rotor.

The two collectors are shown in Figure 10.2 (as black elements) in a position different from the real one, as will be commented below (when the commutator is discussed); however, the real position of the collectors is electrically equivalent to the one in the figure.

In fact even in the real position the two collectors, which must be imagined to be made of conductive material (e.g., carbon), are continuously in close, but sliding, contact with the rotor, so that from time to time it is in electric connection with the two extreme conductors (the highest and the lowest). Each rotor conductor moves in the uniform flux density distribution $\boldsymbol{B}$ and has a local speed $\boldsymbol{v}$, whose amplitude is ΩR (R being the rotor's radius) but whose direction is different from one conductor to another. On each conductor, an electric field will be generated (see Section 9.2):

$$\boldsymbol{E}_{\mathrm{c}} = -\boldsymbol{v} \times \boldsymbol{B}$$

which will produce a potential difference on the conductor:

$$e_{\mathrm{c}}(t) = vB_{\perp}l = v\hat{B}l\sin(\Omega t + \alpha_c) \tag{10.1}$$

This varies as a sine wave with time, and the angle α_{c} is specific for each conductor. For instance, in the position shown in the left-hand part of Figure 10.2, the angle of conductor 4 is $\alpha_{\mathrm{c}} = \alpha_4 = 30°$.

As long as the rotor rotates, the collectors connect to different conductors, so that the total EMF $\boldsymbol{E}$ shown in the right-hand part of Figure 10.2, and corresponding to the instantaneous sum the voltages generate on half-conductors (left or right half), always maintains roughly the same amplitude.

More precisely, the amplitude fluctuates slightly, since it is always equal to the vertical distance connecting the upper and lower points of the polygon shown in the right-hand part of Figure 10.2.

This EMF is thus is composed of an average, to which a slight fluctuation is superposed. Ignoring this small fluctuation, we can write

$$E = kN_a v\hat{B}l$$

k is a dimensionless factor that can be easily demonstrated to be equal for the armature structure illustrated in Figure 10.2 and for large numbers of conductors, to $2/\pi$. N_a is half the number of armature conductors (it is 6 in the example of Figure 10.2).

Other machine structures have different values of k, which is, however, always a fixed dimensionless number for a given machine.

Furthermore, if we consider that it is $v = \Omega R$ (R being the rotor radius), it will be

$$E = kN_a \Omega R\hat{B}l = kN_a \Omega \Phi \qquad (10.2)$$

and finally, considering the new variable $\Psi_{fa} = kN_a\Phi$, linked flux with armature:

$$E = \Omega \Psi_{fa} \qquad (10.3)$$

The procedure followed is very similar to that shown in Sections 9.3.1 and 9.3.3 for AC machines. The student is recommended to re-read this analysis and make comparisons.

The flux Φ_f, in turn, is produced by the field winding and is proportional to the field current I_f:

$$\Phi_f = (L_f/N_f)I_f \qquad (10.4)$$

The proportionality factor between Φ_f and I_f is written as L_f/N_f since, if N_f indicates the number of turns of the field winding, L_f has a precise meaning: it is the inductance of the field winding.

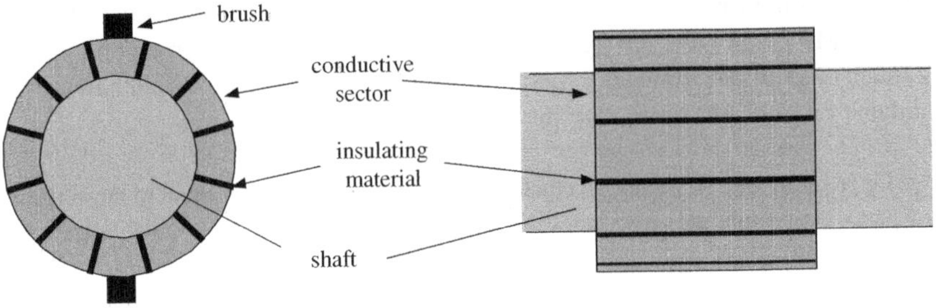

FIGURE 10.3. The basic construction of the machine commutator and its brushes.

Thus,

$$\Psi_{fa} = kN_a\Phi_f = k\frac{N_a}{N_f}L_f I_f = L_m I_f \quad \text{with} \quad L_m = k\frac{N_a}{N_f}L_f$$

This may also be written as $\tau = k\frac{N_a}{N_f}$; $L_m = \tau L_f$. The quantity τ is called *turns ratio* of the machine.

The rotor conductors are linked to a *commutator*, which contains conductive segments separated by small layers of insulating material, as depicted in Figure 10.3. The conductive elements are connected to the active conductor in the rotor periphery, creating a situation that reproduces the principle scheme of Figure 10.2.

The collector is mounted on the machine shaft, and the conductive sectors are connected to the rotor active conductors and rotate with the rotor; the brushes do not rotate and are pressed using special springs against the commutator. The brushes are made of a carbon-based material, which is much softer than copper so the sliding tends to wear out the brushes rather than the commutator. Although this means that they have to be periodically replaced, this dramatically reduces the wear and tear of the commutator, which would be much more expensive to replace.

In a reference integral with the rotor, the flux produced by the stator poles rotates, and therefore the rotor of the machine must be laminated.[1] The stator, instead, is subject (except for the tips) to constant flux and does not need to be laminated. However, it often happens that manufacturers laminate all parts of their DC machines, simply because it facilitates machine assembly or because it makes the manufacturing process easier as a whole.

If the machine is operated with a continuously varying field current, stator lamination also reduces losses.

[1]Lamination is discussed in Section 7.4.

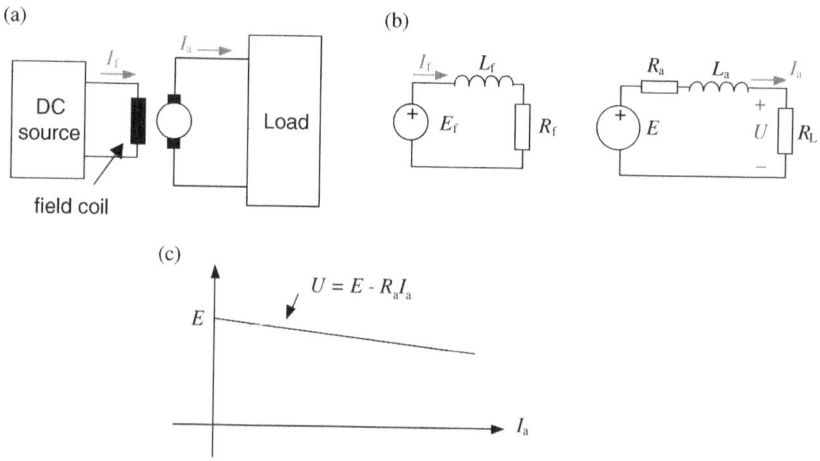

FIGURE 10.4. The DC generator with its field coil, their equivalent circuit and U–I characteristic I_a sign chosen to better describe generator operation.

10.3 OPERATION OF A DC GENERATOR UNDER LOAD

In the previous section it was seen that a DC machine is composed of a stator, which contains a coil to be fed in DC, and a rotor, to which two brushes are connected to gather the voltage produced by the interaction of the stator field and the rotor motion.

This is summarized in Figure 10.4a, in which the brushes, in turn, feed a DC load. The system shown in Figure 10.4a can be sufficiently modelled using the circuital equivalent shown in Figure 10.4b.

In fact, since the field coil is fed by an external DC source, it is constituted by a simple DC circuit, and the field current I_f is simply $I_f = E_f/R_f$, where E_f is the voltage applied to the field winding.

The armature of the DC machine can be modelled by means of an electromotive force, whose value is given by (10.2), an inner resistance, corresponding to all the conductors that constitute the armature winding (active conductors, connections between them), and an inductance that has effect only during transients, when current I_a varies with time.

In the case of I_a which is constant or slowly varying, it will be

$$I_a = E/(R_a + R_L) = kN_a\Omega/(R_a + R_L)$$

From the equivalent circuit of the machine, the U–I characteristic can also be easily determined. Ideally, this has the shape shown in Figure 10.4c. Indeed, in the case of real machines, the nonlinear behaviour of the iron that constitutes stator and rotor causes the actual characteristic to be nonlinear; this, however, is beyond the scope of

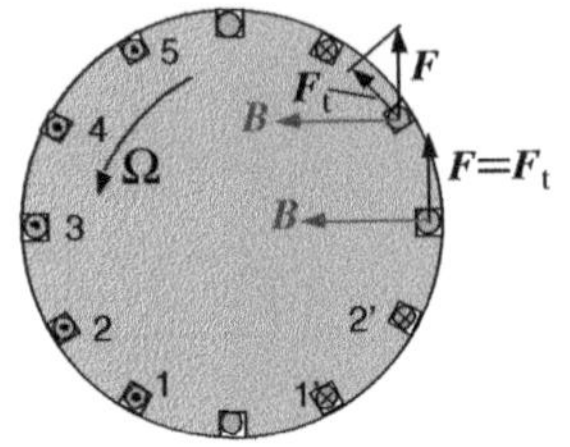

FIGURE 10.5. Forces on the rotor conductors and their contributions F_t to torque.

this book. For small armature currents, the actual machine characteristic can be assumed to be linear.

The current flowing in the machine armature (i.e., in the rotor) will flow in the active conductors in the rotor periphery, as shown in Figure 10.2. This will produce a force on each rotor conductor generated according to Lorentz's law:

$$\boldsymbol{F} = \boldsymbol{I} \times \boldsymbol{B} \cdot l$$

where $\boldsymbol{I}$ has as amplitude the current I_a and the conductor's direction, and l is the length of the conductor. These forces will have tangential components $\boldsymbol{F}_t$ whose amplitudes will depend on their position (Figure 10.5).

The set of all these forces will globally create a torque that will evidently be proportional to the current I_a, the stator flux Φ_f and the number of conductors (remember that N_a is half the number of armature conductors):

$$T = kN_a\Phi_f I_a \tag{10.5}$$

It would be easy to demonstrate that the constant k of equation (10.5) is exactly the same as the one that appears in equation (10.2). Assuming this, we obtain

$$T = I_a\Psi_{fa} \tag{10.6}$$

Considering equation (10.3), we have finally

$$T\Omega = EI_a \tag{10.7}$$

Equation (10.7) is very important, since it states the basic electromechanical conversion in the DC machine: the power inserted into the system as mechanical power $T\Omega$ is converted into an electrical form, generating the power EI_a.

To summarize, DC machine operation can be analysed using the following equations:

$$\begin{aligned} E &= \Omega\Psi_{fa} \\ T &= I_a\Psi_{fa} \end{aligned} \qquad \text{with} \quad \Psi_{fa} = kN_a\Phi_f = L_m I_f \tag{10.8}$$

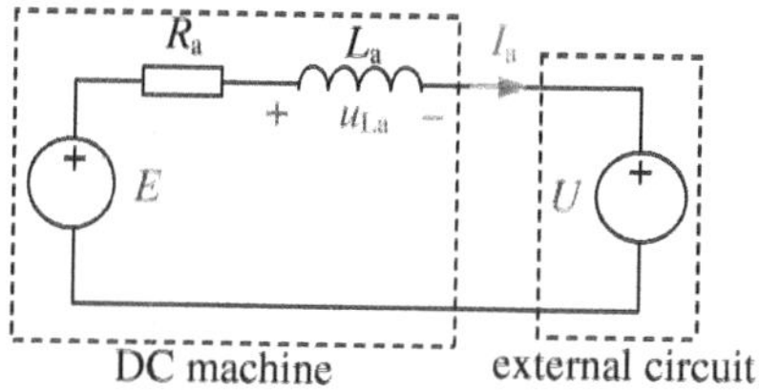

FIGURE 10.6. The equivalent circuit of a DC machine (with $I_a > 0$; it operates as a generator).

Equations (10.8) must be combined with the equation of the electrical circuit created when the machine is connected to a source with voltage U. This circuit is shown in Figure 10.6.

Between the EMF E and the terminal voltage U, there are two circuital components:

- The resistance R_a, which takes into account the resistance of the conductors making up the machine's armature.
- the armature inductance L_a. This causes a voltage drop that tends to counteract any variation in the armature current I_a: $u_{La} = L_a dI_a/dt$.[2] Its presence is justified, considering that when the current I_a flows in the machine's armature it creates a magnetic field across the machine. Associated with this field there is an armature flux $\Psi_a = L_a I_a$.

Example 1. A DC generator runs at $n = 1000$ rpm, supplying 200 V to an electric load. If the stator flux per pole Φ_f is 25 mWb, the constant kN_a equals 90 and the armature copper losses are 300 W, calculate:

(a) the electromotive force E
(b) the armature current I_a
(c) the armature resistance
(d) the electric power P_e provided to the load
(e) the mechanical power P_m and the torque T required to move the generator
(f) the efficiency, ignoring the losses in the field winding

For steady-state studies, we can use a model where the armature inductance L_a does not appear:

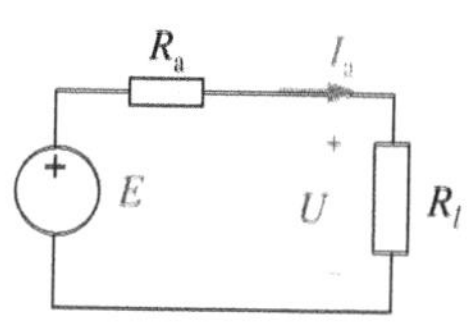

[2]Here we depart in some way from the general rule that uppercase quantities are constant and lowercase variable. Indeed, in steady state, I_a is constant and therefore is usually represented as uppercase, while $u_{La}(t)$ is lowercase since its steady state is always zero.

The angular speed Ω can be easily calculated from n:

$$\Omega = 2\pi n/60 = 2\pi \cdot 1000/60 = 104.7 \text{ rad/s}$$

Hence:

$$\begin{aligned}
&\Psi_{fa} = kN_a\Phi_f = 90 \cdot 0.025 = 2.250 \text{ Wb}\\
&E = \Omega\Psi_{fa} = 235.6 \text{ V}\\
&I_a = P_{cu}/(E - U) = 300/(235.6 - 200) = 8.424 \text{ A}\\
&R_a = P_{cu}/I_a^2 = 300/8.424^2 = 4.227\ \Omega\\
&P_e = UI_a = 200 \cdot 8.424 = 1685 \text{ W}\\
&P_m = P_e + P_{cu} = 1985 \text{ W} \qquad (\text{also } EI_a)\\
&T = P_m/\Omega = 18.95 \text{ Nm} \qquad (\text{also } I_a\Psi_{fa})\\
&\eta = P_e/P_m = 0.849
\end{aligned}$$

10.4 DIFFERENT TYPES OF DC MACHINES

In the previous description, the operation of the DC machine as a generator was taken as reference. Although this operation of DC machines was common in the past, nowadays it is obsolete. It was proposed here because it is simpler to introduce the DC generator first rather than the motor. In Section 10.4.1 we will look at how the DC machine operates as a motor.

Moreover, in the previous description, the field current I_f was generated using an independent DC source on the armature circuit. This corresponds to the *independent excitation DC machine* which is not very frequently used, since it requires that an independent DC source is available. In Section 10.4.3 other important ways to create the flux density $\boldsymbol{B}$ in the machine are introduced briefly, with reference to its operation as a motor, which, as we have just said, is *the only one* used today.

10.4.1 Generators and Motors

The DC machine is intrinsically reversible and can therefore operate as a generator or as a motor. In the upper part of Figure 10.7, where generator operation is shown, the shaft is connected to a mechanical source of power, and the electrical load is connected to the armature. Some means of generating the field must be provided, and this can be done in several ways, as will be discussed in Section 10.4.3. Here, for simplicity sake, we refer to a DC source, which is totally independent from the armature circuit.

In the lower part of the figure, instead, motor operation is depicted. The armature is to be connected to a DC source; some source of current is to be provided, as before, for the field coil. The motor generates mechanical power, which is transferred to the mechanical load through the machine shaft.

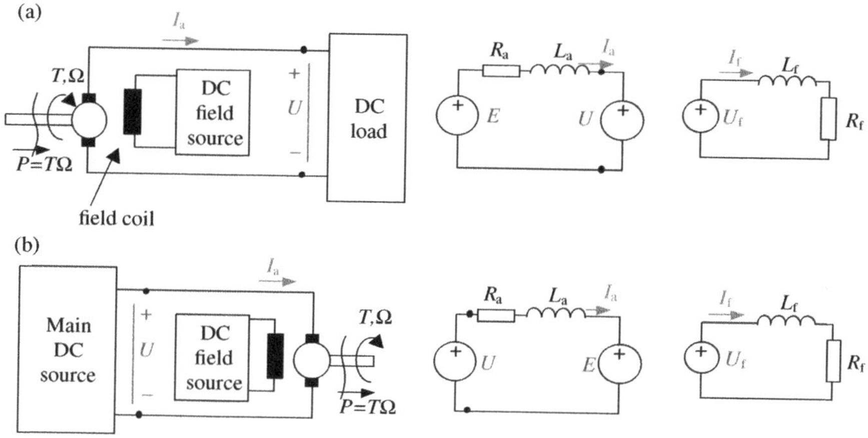

FIGURE 10.7. DC machine connections and equivalent circuits: (a) Generator operation. (b) Motor operation.

The right-hand part of the figure shows a circuit equivalent to the arrangements shown in the left part. The only difference is in the reference sign assumed for current I_a, which is combined with the voltage reference signs according to generator and motor conventions respectively.

Naturally, inductances play a role in the circuits only during transients, since when in steady state the voltage across them is null.

Understanding what happens in the machine when it is operating as a motor is simply a matter of repeating the reasoning behind its operation as generator, though in a slightly different order. This is rapidly done in this section.

Consider that the machine is initially still. Its connection to the sources causes a flux density $\boldsymbol{B}$ to be created inside the machine, and an armature current to flow in the rotor active conductors. The two interact and Lorentz's forces are generated:

$$\boldsymbol{F} = \boldsymbol{I} \times \boldsymbol{B} \cdot l$$

which globally produce a torque T that will make the rotor start rotating according to equation (10.5):

$$T = I_a \Psi_{fa} \qquad (\Psi_{fa} = L_m I_f)$$

When the rotor rotates, an EMF is produced in the active conductors according to equation (10.2):

$$E = \Omega \Psi_{fa}$$

Similarly to when it operates as a generator, the electric power EI_a is converted into the mechanical power T:

$$EI_a = T\Omega \tag{10.9}$$

The general equations of the DC machine (10.8) are therefore still valid. These are combined with the circuit equation which is normally written with generator or motor signs according to the relative operation, shown in Figure 10.7. In steady-state conditions (I_a = const), we have

$$\begin{aligned} U &= E - R_a I_a \quad \text{generator} \\ E &= U - R_a I_a \quad \text{motor} \end{aligned} \tag{10.10}$$

The field circuit equation might also be considered, that is,

$$U_f = R_f I_f + L_f \frac{dI_f}{dt}$$

This is normally considered in steady state: $U_{f,ss} = R_f I_f$

Example 2. A 240-V DC machine has an armature resistance $R_a = 0.2\,\Omega$. If the armature current is 20 A, calculate the steady-state electromotive force:

a. for generator operation
b. for motor operation.

$$E = U + R_a I_a = 240 + 0.2 \cdot 20 = 244\ \text{V}$$

$$E = U - R_a I_a = 240 - 0.2 \cdot 20 = 236\ \text{V}$$

10.4.2 Starting a DC Motor with Constant Field Current

The fundamental equations (10.8) are valid for any value of the field current I_f. To understand better how a DC motor operates, let us now study a starting transient when the machine is operated with I_f= const. This is the case in Figure 10.7, if the voltage source of the field circuit U_f is held constant. In the next section, other options of generating I_f will be examined.

Consider the transient illustrated in Figure 10.8. Initially consider that the transient is due to the closing of switch S, *while holding closed* S_1 *and* S_2.

The KVL equation of an equivalent DC motor armature circuit, which is the leftmost circuit shown in Figure 10.7b, is

$$U = E + R_a \mathrm{I}_a + L_a \frac{dI_a}{dt} = \Omega \Psi_{fa} + R_a I_a + L_a \frac{dI_a}{dt} \tag{10.11}$$

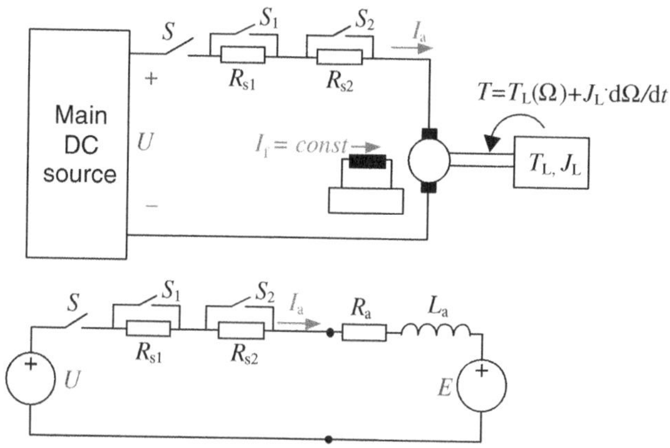

FIGURE 10.8. A start-up transient of a machine for which $I_f = \text{const}$ and the corresponding equivalent circuit.

For a given Ω, this allows us to determine the current I_a and therefore the produced machine torque:

$$T_{mach} = I_a \Psi_{fa} = I_a L_m I_f \tag{10.12}$$

The produced torque will cause the angular speed to grow according to the equation

$$T_{mach} - T_L(\Omega) = J_{eq}\dot{\Omega} \tag{10.13}$$

where $T_L(\Omega)$ is the torque produced by the load and $J_{eq} = J_{mach} + J_L$ is the moment of inertia of load and machine combined.

Equations (10.11), (10.12), and (10.13) must be solved simultaneously step by step, for which a simulation programme is required.

However, they give some qualitative indications. For instance, they show that in the beginning, when the speed Ω is very low, the EMF E will be near zero and according to (10.11) current I_a tends to grow rapidly. Were it not limited by the presence of L_a, it would reach the very high value $I_0 = U/R_a$.

This current is often too large for the machine, which would be damaged if started up directly. In these cases, use is made of the starting resistors. The number of starting resistors may vary, depending on the applications; in Figure 10.8, two of these are considered as an example.

When these resistors are used, the starting procedure is as follows:

- At the beginning, S is closed, with S_1 and S_2 kept open. Their presence will limit the initial current peak to below $U/(R_a + R_{s1} + R_{s2})$.
- When the speed grows, the term $\Omega\Psi_{fa}$ will increasingly limit the current in equation (10.11).

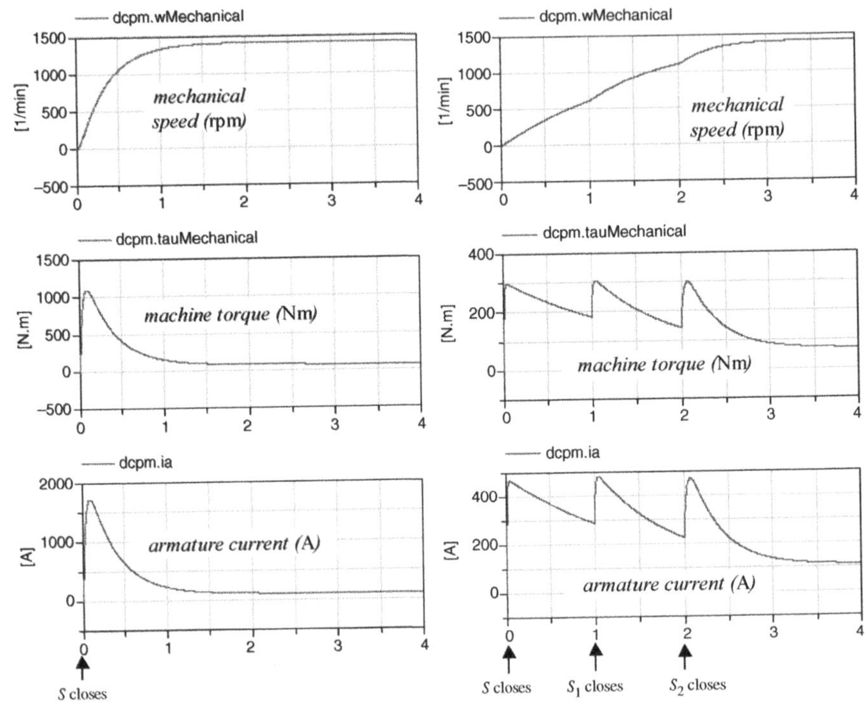

FIGURE 10.9. Simulations of a DC machine start-up. *Left*: Without starting resistors. *Right*: With two starting resistors. Horizontal variable is time in seconds (the total transient is 4 s).

- When the current is sufficiently reduced, S_1 is closed; the current initially grows and then, as Ω is increased further, will decrease again.
- Finally, S_2 is closed, and the system reaches the final steady-state speed.

This is clarified in Figure 10.9, which shows realistic values of torque, current, and speed, as obtainable by numerical simulation when no (left) and two (right) starting resistors are used.

The simulation refers to a 100-V, 100-A machine with a load requiring a torque that varies with speed: $T = T_0 + k\Omega^2$ with $T_0 = 35$ Nm and at the final speed $T = 70$ Nm. The combined moment of inertia of the machine and load is 3.0 kg m^2.

In the absence of a starting resistor, a huge current is initially generated with a peak of around 1700 A, which, depending on the machine design, could seriously damage it. In fact, even though currents a few times higher than the rated value are permissible for limited periods, 17 times this value is normally unacceptable.

In the transient shown in the right-hand part of the figure, the presence of suitable starting resistors reduces the peak current to below 500 A, which is acceptable.

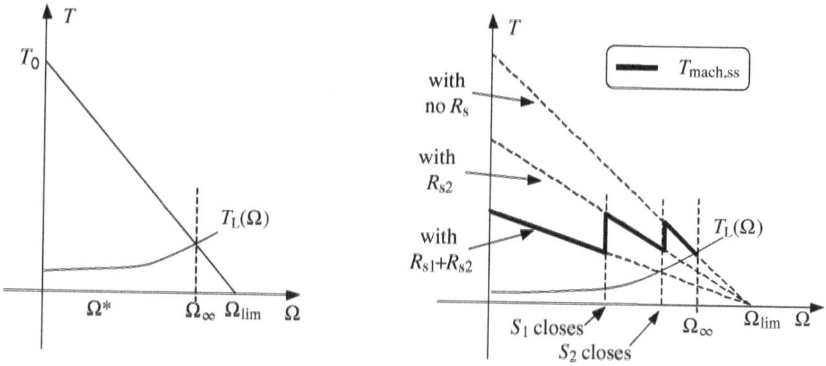

FIGURE 10.10. Analysis of machine start-up using the steady-state characteristics. *Left*: Without starting resistors. *Right*: With two starting resistors.

The torque is reduced in proportion, and the transient is slower but still lasts less than four seconds.

We can understand the two transients even better by looking at the steady-state machine and load torques shown in Figure 10.10.

The steady-state machine torque, which is roughly T_{mach} in equation (10.13), is obtained by substituting the steady-state current from equation (10.11) (i.e., the current value that can be obtained when the voltage across inductance L_a is neglected):

$$I_{a,ss} = (U - \Omega\Psi_{fa})/R_a, \qquad T_{mach,ss} = \Psi_{fa} I_{a,ss}$$

in which $\Psi_{fa} = L_m I_f = \text{const}$.

This torque, in the T–I_a plane, is a straight line; its value at zero speed is $T_0 = U/R_a \Psi_{fa}$, while it reaches 0 at a virtual speed $\Omega_{lim} = U/\Psi_{fa}$.

The left-hand and right-hand parts of Figure 10.10 show the steady-state characteristics when the motor is started both with and without two starting resistors. Differently from the values used for transients shown in Figure 10.9, for a more general picture we consider a torque load T_L as a function of the mechanical speed Ω.

In both cases, a rough estimate of the accelerating torque can be obtained from the difference $T_{mach}(\Omega) - T_L(\Omega)$. By evaluating this difference, we can interpret the plots of Figure 10.9. Equilibrium is reached when $T_{mach} = T_L$ and therefore, from equation (10.13), $\Omega = \text{const} = \Omega_\infty$. Therefore it is

$$T_{mach,\infty} = T_L(\Omega_\infty) \Rightarrow \Psi_{fa} I_{a,\infty} = \Psi_{fa}(U - \Omega_\infty \Psi_{fa})/R_a = T_L(\Omega_\infty)$$

and

$$\Omega_\infty = \Omega_{lim}(1 - T/T_0) \qquad \text{with} \quad \Omega_{lim} = U/\Psi_{fa} \text{ and } T_0 = \Psi_{fa} U/R_a \qquad (^\circ)$$

In the case of $T_L = \text{const}$, as in the transient shown in Figure 10.9, equation (°) allows us to determine Ω_∞ immediately.

Note that nowadays, with increasing frequency, large motors are started without a starting resistor and without the motor absorbing the current I_0; instead they are fed during starting with a reduced voltage. This will be discussed in Section 10.6.

Example 3. Consider the machine simulated in Figure 10.9, having the following nominal data:

$$U_n = 100\,\text{V}, I_n = 100\,\text{A}, n_n = 1425\,\text{rpm}, R_a = 0.05\,\Omega.$$

Evaluate the initial steady-state current I_0, as well as Ω_{lim}, Ω_∞, and T_0, if the load requires a torque $T = 70\,\text{Nm}$.

$$I_0 = U/R_a = 100/0.05 = 2000\,\text{A}$$

The nominal EMF is the one corresponding to the nominal current:

$$E_n = U_n - R_a I_n = 100 - 0.05 \cdot 100 = 95\,\text{V}$$

Thus

$$\begin{aligned}
\Omega_n &= 2\pi n_n/60 = 2\pi \cdot 1425/60 = 149.2\,\text{rad/s} \\
\Psi_{fa} &= E_n/\Omega_n = 0.6366\,\text{Wb} \\
T_0 &= \Psi_{fa} U/R_a = 0.6366 \cdot 100/0.05 = 1273\,\text{Nm} \\
\Omega_{lim} &= U/\Psi_{fa} = 100/0.6366 = 157.8\,\text{rad/s} = 1500\,\text{rpm} \\
\Omega_\infty &= \Omega_{lim}(1 - T/T_0) = 1500 \cdot (1 - 70/1273) = 1417\,\text{rpm}
\end{aligned}$$

The latter value is exactly the one obtained with both simulations proposed in Figure 10.9.

The peak torque obtained from the simulation without starting resistors is 849 Nm, which is below the value of T_0 calculated here. This is as it should be, because of the effect of L_a, not considered in the steady-state machine torque curve.

Similarly, the peak current from the simulation is 1300 A, compared to the higher $I_0 = 2\,\text{kA}$ calculated here.

***More in Depth:** Effects of saturation*

The analysis of the DC machine performed here was based on a linear model of the machine.

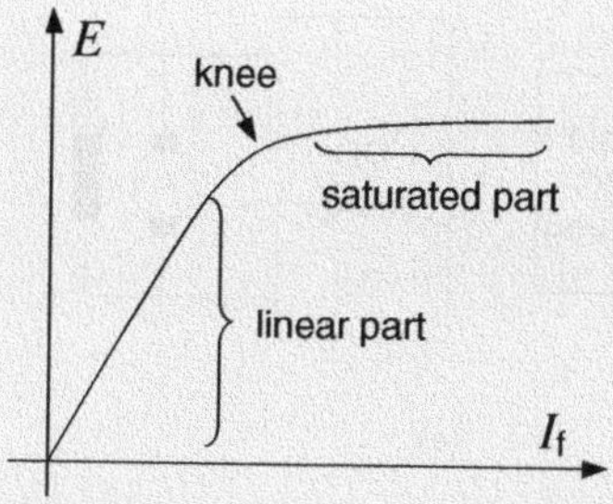

This is correct only as an approximation since, due to the nonlinear behaviour of stator and rotor iron, actual machines significantly deviate from linearity when saturation occurs. The main deviation from linearity in the relationship between the field current I_f and the generated EMF is indicated in the figure alongside.

In real machines, however, the actual operating point of excitation is held in the linear zone, or at most at the *knee*, since going further would increase excitation losses without increasing the EMF. The simulations shown disregard any saturation effect.

Example 4. A DC motor with independent excitation is supplied with 240 V, the flux per pole Φ_f and the armature resistance being 30 mWb and 0.5 Ω, respectively. If the motor develops a torque $T = 45$ Nm and the constant kN_a equals 85, calculate

(a) the armature current
(b) the electromotive force
(c) the motor speed, in rpm
(d) the mechanical power provided
(e) the efficiency

$$I_a = \frac{T}{kN_a\Phi_f} = \frac{45}{85 \cdot 0.03} = 17.65\ \text{A}$$

$$E = U - R_a I_a = 240 - 0.25 \cdot 17.65 = 231.2\ \text{V}$$

$$\Omega = \frac{E}{kN_a\Phi_f} = \frac{231.2}{85 \cdot 0.03} = 90.66\ \text{rad/s} \equiv 865.7\ \text{rpm}$$

$$P_m = T\Omega = 4080\ \text{W}$$

$$P_{el} = U \cdot I_a = 4235\ \text{W}$$

$$\eta = \frac{P_m}{P_{el}} = 0.963$$

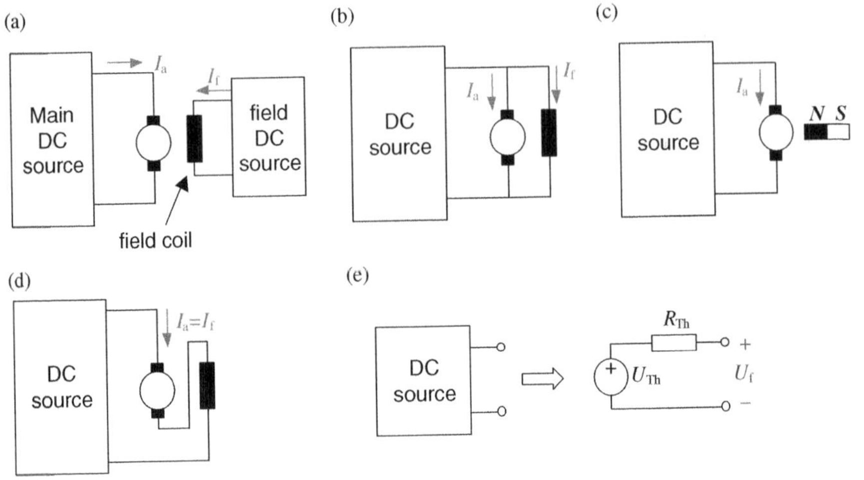

FIGURE 10.11. Some options for excitation of DC machines: (a) Independent (or separate). (b) Shunt. (c) Permanent magnet. (d) Series.

10.4.3 Independent, Shunt, PM, and Series Excitation Motors

In the previous sections the fundamental equations (10.8) were demonstrated in both generator and motor operation. They contain the field flux Φ_f.

The only way previously examined to generate this flux Φ_f was by means of an external DC source. As already discussed, this is just one option, normally referred to as *independent excitation*. Other ways are possible and are commonly used. Some of them are shown in Figure 10.11 and are briefly discussed here.

In Figure 10.11a, excitation is independent (or separate), as discussed in the previous section. In Figure 10.11b the supply to the field coil is the same source that feeds the armature. This is called *shunt excitation*. If the DC source is ideal (i.e., with zero internal resistance), this solution is perfectly equivalent to independent excitation. The electric model of a DC source is shown in Figure 10.11e.

In the case of a nonzero inner (or Thévenin) resistance of the DC, a (frequently minor) difference lies in the fact that when the machine absorbs a very high current I_a, the voltage U_f feeding the field coil is significantly lower than U_{Th}. Thus, with shunt excitation, a high armature current tends to reduce the field current.

Therefore, except for minor differences, the voltage–current and torque–current characteristics are the same as shown in the previous section for the case of independent excitation.

In Figure 10.11c, the field is created not by an actual coil, but by the insertion of permanent magnets in the stator circuit. Their effect is to create a fixed field flux Φ_f, exactly in the same way as if a fixed current flowed in an excitation coil. Therefore, PM excitation is equivalent to independent excitation; and like the latter, it is nearly equivalent to shunt excitation.

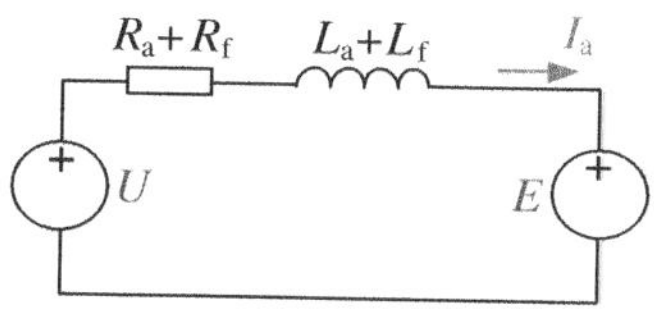

FIGURE 10.12. Equivalent circuit of series-excited DC motor.

Some important, though obvious, differences regarding these types of excitation are:

- Independent excitation requires an independent DC source; it allows easy modification of I_f, acting on the value of the voltage applied to the field coil.
- Shunt excitation is cheaper than independent excitation since it does not require an external energy source; modification of I_f is possible by inserting an additional resistance in series with the field coil, so that I_f is reduced (at equal voltage applied to the series, the total resistance is increased).
- PM excitation allows easy and robust machine construction, but does not allow flux modifications. It is usually used in small DC motors.

Finally, in Figure 10.11d, the field winding is connected in series with the machine armature. In this case the two equivalent circuits of Figure 10.7b are in series, giving rise to the combined circuit shown in Figure 10.12, whose KVL equation is (let $R_{eq}=R_a+R_f$ and $L_{eq}=L_a+L_f$):

$$U = E + R_{eq}I_a + L_{eq}\frac{dI_a}{dt} = \Omega\Psi_{fa} + R_{eq}I_a + L_{eq}\frac{dI_a}{dt} = (\Omega L_m + R_{eq})I_a + L_{eq}\frac{dI_a}{dt}$$

Thus, if I_a varies slowly (and therefore the derivative term is disregarded), we have

$$T = I_a L_m I_f = L_m I_a^2 = L_m \frac{U^2}{\left(L_m\Omega + R_{eq}\right)^2} \tag{10.14}$$

The resulting shape, drawn considering $U=\text{const}$, is summarized in Figure 10.13. Starting from the value $T_0=L_m U^2/R_a^{\,2}$ at standstill, the torque decreases continuously as Ω increases, and it is always positive. At high speeds, the term R_a in the denominator is negligible and can therefore be ignored; thus the curve decreases proportionally to $1/\Omega^2$.

The standstill current I_0 (and the standstill torque T_0) is normally too high and can cause damage to the machine. Therefore, the machine is started at a reduced voltage, and the reduction is obtained at low speeds by interposition of a starting resistor R_s, which is then bypassed when the current has become acceptable, in correspondence to the speed Ω^*. (The curve of I_0 when S_1 is open is not shown in the current plot, just to avoid too many curves on the same figure).

Series-excited machines are used when the load might have sudden peaks of torque; in this case this machine slows down and raises its torque by large amounts, thus overcoming the difficulty. These machines were widely used in the past on board trains, mainly because of their high torque capability at low speeds, even though trains were started by temporarily inserting series resistances (such as the R_s in Figure 10.13).

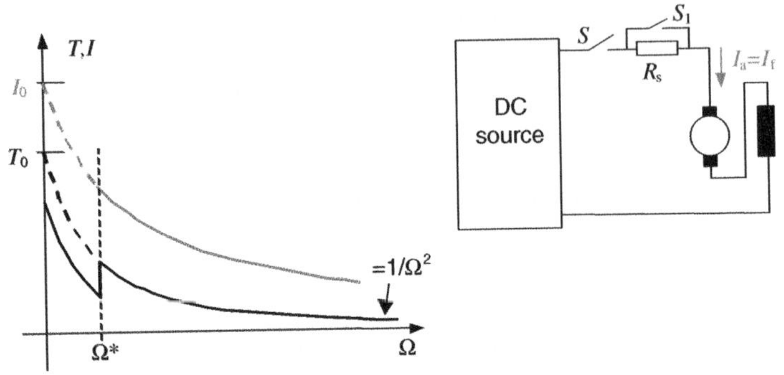

FIGURE 10.13. Current–speed and torque–speed characteristics with starting of a series-excited DC motor.

Today they are used as small motors, often sized to be able to start without R_s, and as "universal motors," which will be discussed in Section 10.5.

Example 5. A 300-V DC shunt motor absorbs 35 kW from the electric grid. The field and armature resistances are respectively 120 and 0.15 Ω. Calculate:

(a) the line, armature, and field current
(b) the electromotive force E
(c) the efficiency η
(d) the motor speed Ω, if the flux per pole Φ_f is 30 mWb and the constant kN_a equals 80
(e) the torque T developed

$$I = \frac{P_{el}}{U} = \frac{35000}{300} = 116.7\ \text{A} \quad \text{(line)}$$

$$I_f = \frac{U}{R_f} = \frac{300}{120} = 2.50\ \text{A} \quad \text{(field)}$$

$$I_a = I - I_f = 114.2\ \text{A} \quad \text{(armature)}$$

$$E = U - R_a I_a = 300 - 0.15 \cdot 114.2 = 282.9\ \text{V}$$

$$P_m = E \cdot I_a = 32{,}295\ \text{W}$$

$$\eta = \frac{P_m}{P_{el}} = \frac{32{,}295}{35{,}000} = 0.923$$

$$\Omega = \frac{E}{kN_a\Phi_f} = \frac{282.9}{0.03 \cdot 80} = 117.9\ \text{rad/s} \equiv 1125\ \text{rpm}$$

$$T = P_m/\Omega = 274\ \text{Nm} \quad (\text{also } kN_a\Phi_f I_a)$$

The reader can easily verify that losses $(P_{el} - P_m)$ equal $R_a I_a^2 + R_f I_f^2$.

Example 6. A 240-V DC series motor has an armature resistance $R_a = 0.4\,\Omega$ and a field resistance $R_f = 0.12\,\Omega$. If the motor runs at 900 rpm absorbing 30 A, calculate torque and efficiency.

$$E = U - (R_f + R_a) \cdot I = 224.4\ \text{V}$$

$$\Omega = \frac{n}{60} \cdot 2\pi = 94.25\ \text{rad/s}$$

$$P_m = E \cdot I = 6732\ \text{W}$$

$$T = \frac{P_m}{\Omega} = 71.43\ \text{rad/s}$$

$$\eta = \frac{P_m}{UI} = 0.935$$

10.5 UNIVERSAL MOTORS

Consider a series-excited DC motor and its equivalent circuit (Figure 10.14). Its steady-state torque formula of a series-excited DC motor is (10.14):

$$T = L_m I_a^2 = L_m \frac{U^2}{\left(L_m \Omega + R_{eq}\right)^2}$$

This formula shows that the torque generated does not depend on the voltage sign. If, therefore, the source polarity is reversed, torque T and speed Ω do not change their signs.

Consequently, if a series-excited "DC machine" is connected to an alternating source that continuously reverses its polarity, it will operate as a series-excited DC machine, with a torque–speed relationship similar to that discussed in the previous section.

A motor with commutator, designed to be fed by an AC source, is normally called *universal motor*.[3] Under AC operation, the motor will draw a current that can be assumed to be sinusoidal:

$$i(t) = \hat{I} \sin \omega t, \quad \text{where } \omega \text{ is the angular frequency of the AC source.}$$

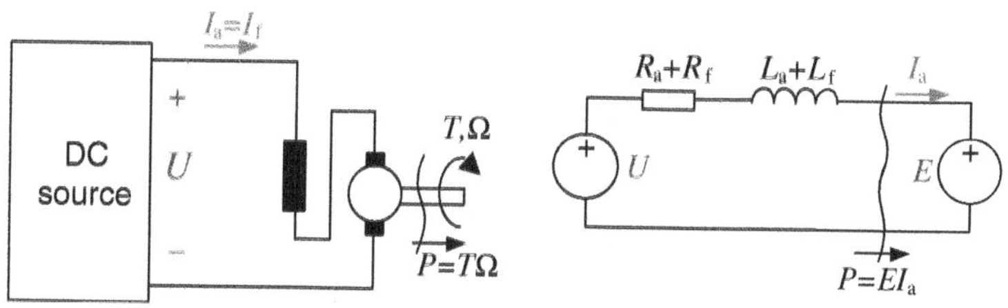

FIGURE 10.14. Series-excited DC motor with indication of electric and mechanical reference signs.

[3]Sometimes they are also called AC commutator motors.

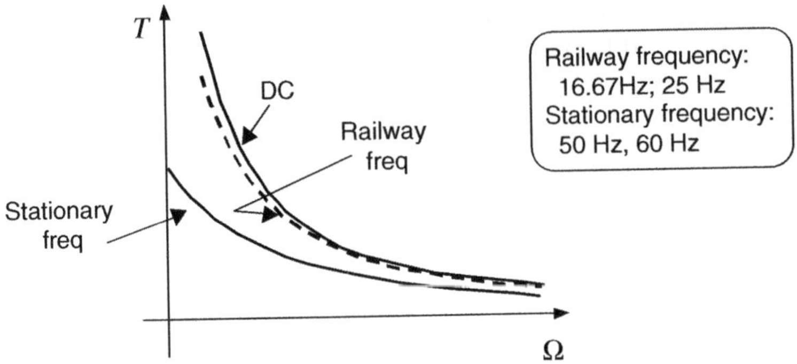

FIGURE 10.15. Universal motor torque behaviour at different frequencies.

This will generate a torque:

$$T(t) = L_m I^2 (1 - \cos 2\omega t)$$

which has an average value and has a fluctuation around that value at twice the frequency of the supply voltage.

The presence of the large inductors $L_a + L_f$, however, tends to oppose the current variation, and therefore the current flowing in the circuit will fall as long as the circuit frequency rises.

The higher the frequency applied to the universal motor, the more its operation will differ from DC characteristics. For instance, the torque–speed relationship of the same machine operating at different frequency is qualitatively represented in Figure 10.15.

In the recent past, when electric trains were usually operated using commutator motors, very large AC commutator motors—up to hundreds of kilowatts—were used. To keep the advantage of AC (allowing the use of transformers to feed the contact lines) and obtain a torque behaviour similar to the case of a DC supply, many railway contact lines were operated at a frequency lower than the one typically used for stationary applications: 25 Hz in some American states, 50/3 = 16.67 Hz in some European countries. In Europe, thousands of kilometres are still operated at 16.67 Hz, even though modern trains first internally convert electricity into DC and then supply power trains that are usually based on synchronous or induction motor drives.

Today, small universal motors are used very frequently in small domestic appliances, where simple construction and high torque at small speeds are required. Therefore, typical applications are in drills, food mixers, saws, and vacuum cleaners. Characteristic of these motors is a loud noise and small arcs in the commutator; both are mainly due to the continuous switching of current between the armature conductors, commanded by the switching of brushes from one commutator sector to another.

Needless to say, universal motors must always be laminated in both stator and rotor.

10.6 DC ELECTRIC DRIVES

The characteristic of a standard supply does not always meet the requirements of an electric machine for optimal behaviour.

We have already encountered one such situation in this chapter, when it was seen that a constant voltage is not normally adequate for starting a DC machine, since when its mechanical speed is null or very low, its electromotive force E is null or very low, and the machine draws a very large current which can damage it.

The characteristics of the supply system used to be matched to the machine by electromechanical means. As regards the DC motors, it was seen in Section 10.4.2 that one or more starting resistors R_s placed in series with the armature, can carry out this matching. These electromechanical means, however, have distinct disadvantages. For instance, the resistor R_s dissipates significant quantities of energy during starting and does not easily allow continuous variation of the DC voltage during start-up or during operation at low speeds.

In recent years, new apparatuses called *electronic converters* have appeared which allow very flexible and effective modification of the supply characteristic and which allow machines to be more easily and more effectively matched to their supply system.

Electronic converters are sometimes also called power processing units (PPUs), to evidence their characteristic of changing the terms (voltage and current) of the power transferred to the load.[4]

Therefore we can insert an electronic converter between the source and the machine, so that voltage current at the machine's terminals are different from those at the DC source (Figure 10.16a). This is normally done when a cheap DC source is available to feed the converter—for example, in a system in which the main source uses an electrochemical battery.[5] In the majority of cases, however, an AC source is easily available. For powers larger than just a kilowatt, this is typically a three-phase source, as shown in Figure 10.16b.

The general purpose of the electronic converter is to modify the parameters of the electricity supply, ideally leaving the power unchanged. Should the main source be DC, the following equation will ideally apply:

$$P = U_s I_s = U_m I_m \tag{10.15}$$

In cases in which the source is three-phase AC, indicating with the symbols U_l and I_l the rms value of the line-to-line voltages and line currents, respectively, and

[4]The acronym PPU is used throughout the recent book [bm3]. We dealt with electronic converters in Chapter 8. However, it is not mandatory to study Chapter 8 to be able to understand this analysis of electric drives.

[5]This is, for instance, the case of some Italian submarines. Submarines are indeed often fed by electrochemical sources, because when submerged they cannot consume the oxygen required by combustion-based engines.

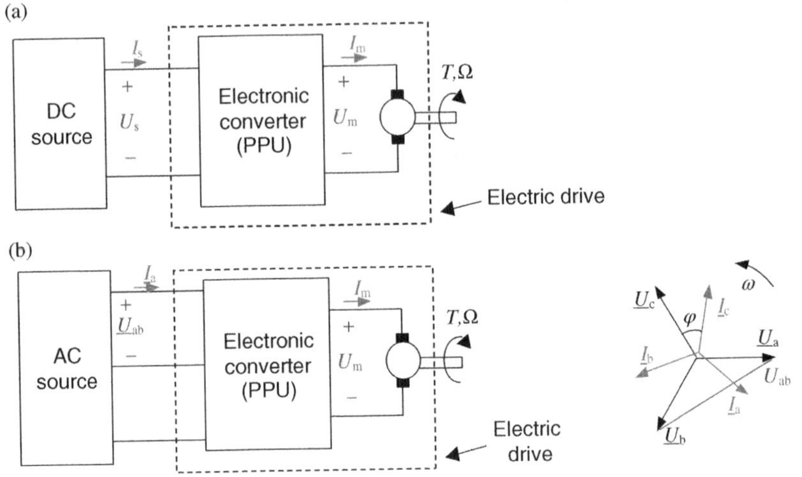

FIGURE 10.16. An electric drive is composed of a machine and an electronic converter that allows the matching of electric source and machine.

with $\cos\varphi$ the power factor (phase displacement between phase voltage and phase current), it will be

$$P = \sqrt{3}U_l I_l \cos\varphi = U_m I_m \tag{10.16}$$

It is fairly obvious that it is not possible to achieve the exact equivalence of input and output power, because, to effect voltage and current conversion, some power loss would unavoidably occur. For the purpose of basic analysis of electric drives, these losses can be initially disregarded: they are considered only at a later stage to evaluate global efficiencies. Therefore in this section (10.15) or (10.16) can be assumed to be true.

Actual electronic converters can have very high efficiencies. It is not uncommon to have efficiencies[6] of around 97–98%.

A converter corresponding to the basic equation (10.15) is called *DC–DC converter* or *chopper*. It performs in much the same way as an ideal transformer: It is like the transformer, but operates in DC.

The ratio between output and input voltage and between input and output current, namely,

$$\alpha = U_m/U_s = I_s/I_m$$

[6]The efficiency η of a converter will be computed as the ratio of output and input power: $\eta = U_m I_m/(U_s I_s)$ or $\eta = U_m I_m/(\sqrt{3}U_l I_l \cos\varphi)$, respectively, for choppers and three-phase rectifiers.

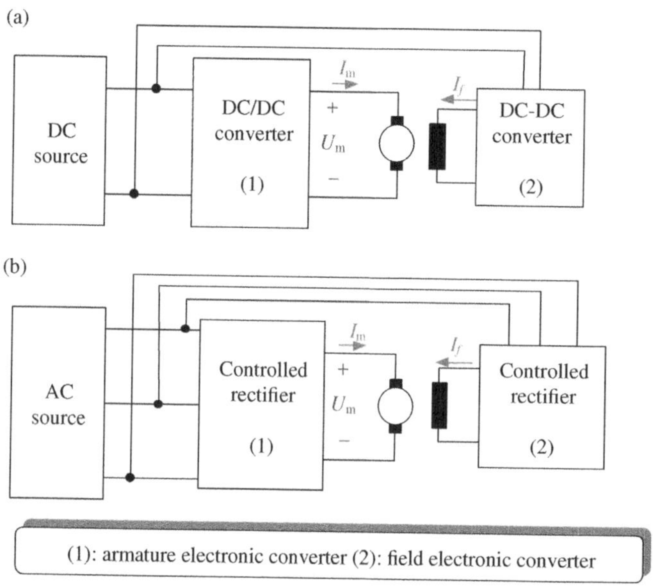

FIGURE 10.17. Electric drives with controlled armature and field voltages. (a) Starting from a DC source. (b) Starting from an AC source.

can be set by a control action on the inner operation of the converter, differently from the transformer, in which this ratio is fixed because it equals the turns ratio.

A converter corresponding to the basic equation (10.16) is called *AC–DC converter* or *rectifier*. Rectifiers can be created in different ways in practice. To obtain an effective drive, a *controlled rectifier* is needed—for example, a system able to transfer power from an AC source to a DC load in a controlled way—so that at given AC voltages, the produced DC voltage can be modified by control action.

A second electronic converter can be used to vary the voltage applied to the field coil, so that the field current $I_f = U_f/R_f$ is modified according to requirements. The arrangements in this case can be of the type shown in Figure 10.17.

It is important to evaluate the maximum torque and power performance of an electric drive of the type shown in Figure 10.17. This can be easily done by referring to the fundamental equations of the DC motor (10.8):

$$E = \Omega \Psi_{fa}$$
$$T = I_a \Psi_{fa}$$

in conjunction with the circuit equation, considered for slow-varying currents:

$$U = E + R_a I_a$$

and considering that the machine will have maximum values of armature voltage and current U_{max} and I_{max} (which cannot be overcome) and that in normal operation $U \approx E$ can be assumed.

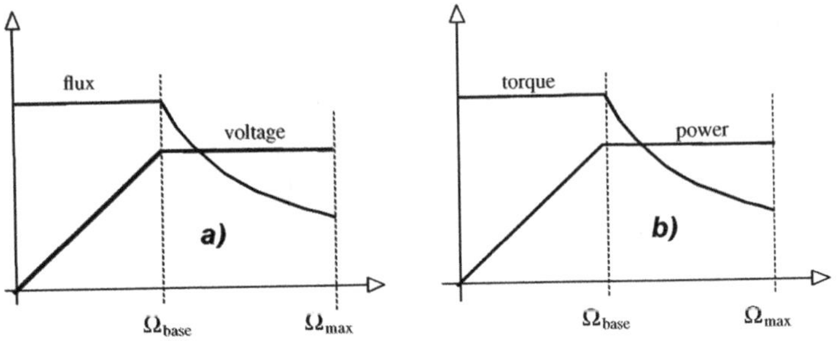

FIGURE 10.18. Flux, voltage, torque, power trends at the maximum current I_{max} of a DC drive controlled in armature and field voltages.

Consider the machine operating at steady state at a speed Ω. For low, increasing speeds, in the beginning the linked flux Ψ_{fa} is held equal to its maximum value. The armature voltage will be approximately linearly increasing with Ω, because $U \cong E = \Omega\Psi_{fa}$ (left-hand side of Figure 10.18a). Correspondingly, the maximum torque available will stay at the fixed value $T = \Psi_{fa}I_{max}$ (left-hand side of Figure 10.18b). When the maximum voltage has reached its maximum value U_{max} at speed Ω_{base}, it cannot increase further.

The machine speed can further increase over Ω_{base}, while maintaining the armature voltage constant. This can be done by reducing the linked flux Ψ_{fa}; this is the *field weakening* region, which can continue up to the point in which the maximum machine speed Ω_{max} is reached (right-hand part of Figures 10.18a and 10.18b).

Note that this behaviour allows exploitation of the maximum torque the machine is able to deliver, up to the point at which the machine has reached its maximum power; starting from that point, the machine is able to deliver its maximum power up to the maximum allowed speed.

Clearly, the drive is able to make the machine deliver torque and power that are lower than the maximum admissible values shown in Figure 10.18; it is just a matter of making the machine absorb currents that are lower than I_{max}. Controlling the drawn current is easy done. Consider the equation

$$U = E + R_a I_a = \Omega\Psi_{fa} + R_a I_a$$

where, at given flux Ψ_{fa}, E varies slowly since its variation is connected with the mechanical inertia of the machine rotor and its mechanical load.

Any variation in the applied armature voltage U will therefore result in a corresponding variation of the drawn current:

$$I_a = \frac{U - E}{R_a}$$

and therefore of the torque produced: $T = \Psi_{fa} I_a$.

The usual rule for controlling the DC drive at partial load is to keep the flux Φ_f at its maximum value—that is, that shown as a function of Ω in Figure 10.18. In this way, for a given torque request the armature current I_a and the corresponding Joule losses are at their minimum (remember that $T = \Psi_{fa} I_a$).

For further information on DC electric drives, the reader is advised to consult the book [bm1].

10.7 PROPOSED EXERCISES

10.1. Calculate the voltage induced in the armature of a DC machine running at 1500 rpm, if the flux per pole Φ_f is 30 mWb and the constant kN_a equals 120.

10.2. The DC generator of the previous exercise has an armature resistance of 0.4 Ω and supplies a load of 40 Ω. Calculate:

(a) the armature current

(b) the voltage applied to the load

(c) the electric power provided to the load

(d) the torque and the mechanical power required to move the generator

10.3. A separately-excited DC generator runs at 1500 rpm, supplying 240 V to a certain load. The flux per pole Φ_f is 20 mWb and the constant kN_a equals 85. The armature copper losses are 600 W and the total brush-contact drop is 2 V. Calculate:

(a) the electromotive force

(b) the armature current

(c) the armature resistance

(d) the electric power provided to the load

(e) the torque required to move the generator

10.4. A 200-V separately-excited DC machine has an armature resistance of 0.3 Ω. If the armature current is 24 A, calculate the electromotive force:

(a) for generator operation

(b) for motor operation.

10.5. A 250-V shunt generator provides 5 kW to a certain load. The armature and field resistances are respectively 0.5 and 200 Ω. Calculate the armature and field currents, the electromotive force, and the torque required to move the generator, if it runs at 1200 rpm.

10.6. A separately-excited DC motor is supplied with 200 V. The flux per pole Φ_f is 25 mWb and the constant kN_a equals 70. If the motor develops a torque of 40 Nm and the armature resistance is 0.3 Ω, calculate:

(a) the armature current

(b) the electromotive force
(c) the motor speed, in rpm
(d) the mechanical power provided
(e) the efficiency

10.7. Repeat the previous exercise, assuming that the mechanical load requires a torque $T_{\text{load}} = 5 + 0.4\,\Omega$ Nm, where Ω is the angular speed of the motor.
Hint: Using equations (10.8) and (10.10), express T as a function of Ω, U being fixed. Then impose $T = T_{\text{load}}$ in order to find the angular speed of the motor.

10.8. A 400-V DC shunt motor has field and armature resistances of respectively 100 and 0.1 Ω. If the motor absorbs 40 kW from the electric grid, calculate:

(a) the line, armature, and field current
(b) the electromotive force
(c) the efficiency

10.9. The 400-V DC motor of the previous exercise is running at 1100 rpm and absorbing a line current of 80 A. Calculate the torque developed and the efficiency.

10.10. A 230-V DC shunt motor, with an armature resistance of 0.4 Ω and a field resistance of 200 Ω, absorbs 2.5 A in no-load operation. In another loading condition, the same motor runs at 300 rpm while absorbing 30 A. Calculate the no-load speed.
Hint: Use the second loading condition to calculate $kN_a\Phi_f$. Then use it to analyse the no-load operation.

10.11. A 400-V DC series motor, with an armature resistance of 0.3 Ω and a field resistance of 0.15 Ω, absorbs 40 A. If the motor runs at 800 rpm, calculate the torque and the efficiency.

10.12. A 230-V DC series motor, with a total circuit resistance of 0.45 Ω, absorbs 40 A while running at 600 rpm. Calculate the speed when the motor absorbs 30 A and the torque in both conditions. Compare Ω_1/Ω_2, I_1/I_2, T_1/T_2.
Hint: Use equations (10.2) and (10.4) to demonstrate that in a series motor $E = k'\Omega I$, where $I = I_a = I_f$. Use the first loading condition to calculate k', and then apply this value in the second loading condition.

10.13. A 230-V DC series motor, with a total circuit resistance of 0.5 Ω, absorbs 20 A while running at 800 rpm. Calculate the torque. Then calculate the motor speed when the torque is reduced by 30%.
Hint: Same hint of the previous exercise.

11

SYNCHRONOUS MACHINES AND DRIVES

For the Instructor

Synchronous machines are very important as generators, as they are the main way of producing electricity in power stations. Today, however, they are frequently used as motors when, usually with permanent magnets on the rotor, they are coupled to inverters to form brushless drives.

Therefore a significant part of this chapter is devoted to the presentation of a simplified analysis of synchronous drives.

Naturally, we had to find a way to analyse synchronous drives while avoiding the complexity Park's equations. We thus decided to take as a reference the isotropic construction of the permanent-magnet synchronous machine, and we analysed the drive by making use of phasors, a technique used throughout the book. Although it is probably new (we had not it seen previously in books or papers), we hope that this simplified way of analysing synchronous drives will be useful for instructors and students.

Fundamentals of Electric Power Engineering: From Electromagnetics to Power Systems, First Edition.
Massimo Ceraolo and Davide Poli.

11.1 THE BASIC IDEA AND GENERATION OF EMF

The synchronous machine is a three-phase machine, able to exchange power with a three-phase network. Although it can also operate as a motor (absorbing electric energy) in the majority of applications, it is used as a generator (delivering electric energy).

The vast majority of power stations generate electricity in three-phase form, using synchronous machines.

To explain the basic operation of a synchronous machine, we start from what we learned in Chapter 9 regarding rotating machines. The reader will recall that the basic components of a rotating machine are a stator and a rotor.

Figure 11.1 is a simplified representation of a three-phase machine stator. This is similar to the one already shown in Chapter 9, but here there are three coils, constituted by the six conductors a–a', b–b', c–c' and distributed evenly along the air-gap circumference.

Each of the coils is composed of the active conductors and the connections at their ends, as discussed in Chapter 9 (see Figure 9.6) and recalled in the right-hand part of Figure 11.1 for coil a (constituted by conductors a and a') only. The coil terminals are named like the corresponding conductors (a_+ and a_- for coil a–a'). When the machine operates under load, it will obviously deliver current through the coil terminals; the reference signs for these currents are shown in Figure 11.1 by the dot–cross convention (introduced in Section 9.2) in the left-hand sketch and with the usual arrow used also in circuits in the right-hand one.

These reference directions also allow oriented axes to be defined for the three coils: the ones represented by the bold symbols ***a*, *b*,** and ***c*** in Figure 11.1; their orientations are chosen in conjunction with the reference signs for currents, and they are related to each other by the right-hand rule for coils.

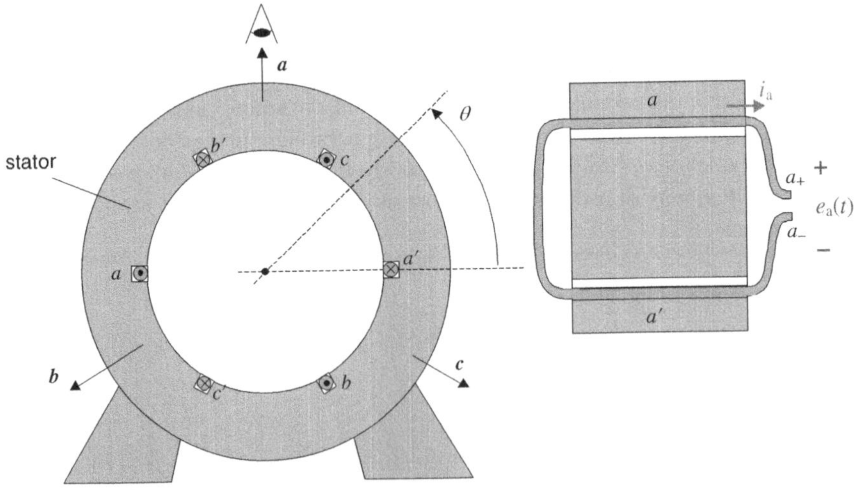

FIGURE 11.1. A simplified representation of a three-phase machine stator.

The right-hand arrow rule for coils can be expressed in two ways:

1. If the hand is positioned with the longer fingers curved to simulate a coil, and their tips are thought to be arrow heads of the current, the thumb tip will indicate the arrowhead of the oriented axis.
2. If an eye looks at the tip of the oriented axis, it will see the current, imagined to be flowing according to the assumed direction, moving counterclockwise. An eye is schematically shown in Figure 11.1 for axis $\boldsymbol{a}$.

Figure 11.1 also shows an angular abscissa θ which is able to represent any air-gap position in terms of an angle between 0 and 2π radians, as seen in Chapter 9. The origin of this abscissa is obviously arbitrary, but should be clearly defined. In the figure, an angle $\theta = 0$ is chosen in such a way that the oriented axis $\boldsymbol{a}$ of coil a has an abscissa of $\pi/2$ radians.

Finally, it is also important to represent the whole set of coils with axis $\boldsymbol{a}$ (= oriented axis of the first coil) in vertical position and oriented upward; this will facilitate analysis, as will be seen later.

As far as the rotor is concerned, two versions of them exist: *cylindrical*, Figure 11.2a, or *salient-pole* (or simply salient), Figure 11.2b; that is, a synchronous machine can be built with cylindrical or salient pole rotor.

The salient-pole version allows savings in iron and therefore in cost, but is not suitable for machines that are operated at very fast speeds because of the higher centrifugal forces. Therefore, salient-pole rotors are used for lower speeds, while cylindrical ones are used for machines that rotate faster (say above 1000–1500 rpm).

There are significant differences in the electromechanical behaviour of the two kinds of machines, stemming from the fact that cylindrical machines have radial symmetry, since the path of magnetic forces has the same reluctance whatever the angle; conversely, salient-pole machines have a lower reluctance in the axis joining ***N-S*** poles of the rotor—that is, the parts of the rotor where the flux abandons the rotor

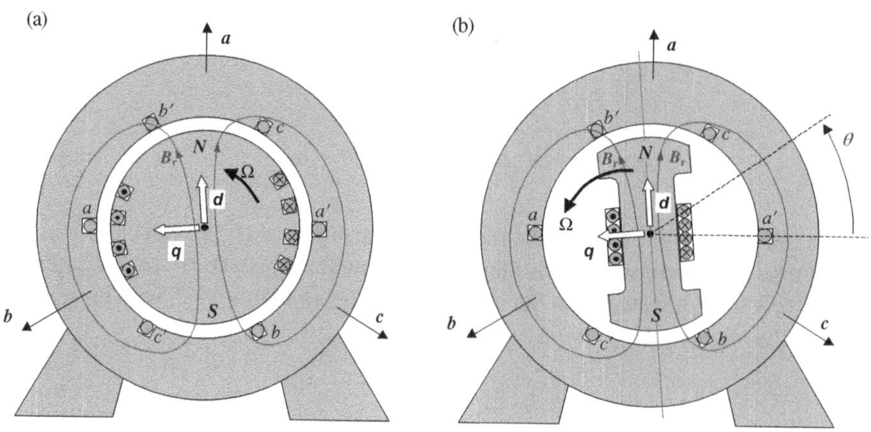

FIGURE 11.2. Three-phase synchronous machines: (a) With cylindrical rotor. (b) With salient rotor.

itself or enters it, respectively. This axis is the *direct axis*. It has larger inductance than in the direction orthogonal to it, the *quadrature axis*.

Since cylindrical rotor machines have isotropic behaviour (reluctance does not depend on the radial position of the ray considered), they are often called *isotropic*; conversely, salient-pole are also called *anisotropic*.

These differences, however, are not so great as to merit analysis in an introductory course such as the one offered by this book. Therefore we state the following:

Assumption: *Synchronous machine isotropy*

The mathematical model of the synchronous machine proposed in this book does not take into account reluctance differences between the direct and quadrature axes. It is, therefore, better suited to the analysis of cylindrical-rotor machines, although it is also sufficient, for basic purposes, for salient-pole machines.

Following the practice of most textbooks, drawings will, however, usually be made using salient-pole machines because they allow more immediate visual analysis. Consider the drawing in Figure 11.2a, referring to a cylindrical rotor. The rotor contains conductor bars parallel to the rotation axis, through which flows a DC current, commonly called *field current* (and therefore indicated as I_f). Later we will explain how this current can be transferred to a moving part such as a rotor.

This current creates a distribution of flux density $\boldsymbol{B}_r$ (here the subscript "r" does not mean radial but, instead, "rotor"), two force lines of which are represented. When the rotor is rotated at a speed Ω, by means of mechanical connection to an external supplier of mechanical energy such as a diesel engine or a turbine, all the $\boldsymbol{B}_r$ map rotates along with the rotor.

Figure 11.2 also shows angle θ discussed earlier.

The radial component B_{rr} of $\boldsymbol{B}_r$ as a function of angle θ can be approximated by a sinusoidal function, as discussed in Chapter 9, each time. Let it be represented as a cosine[1]:

$$B_{rr}(t,\theta) = B\cos(\theta - \tilde{\theta}), \qquad \tilde{\theta} = \omega(t - t_0) \tag{11.1}$$

where, for the machine depicted in the figure,[2] the ω equals Ω. We can choose as the initial time t_0 the time when $\boldsymbol{B}_r$ is aligned to the ray (half-line) $\theta = 0$. In this case it is simply $\tilde{\theta} = \omega t$. For instance, in the situation depicted in Figure 11.2, with this choice of t_0, ωt is slightly above 90°.

[1]It should be totally clear to the reader that any sine wave can be represented either as a sine or as a cosine, since one of these functions can be converted into the other by horizontal translation.

[2]The reader will soon learn that in general $\omega = p\Omega$, where p is the number of pole pairs of the machine.

Using (11.1) and Faraday's law for rotating coils, as seen in Chapter 9, we obtain

$$e = \omega B_{rr} A$$

where e is the induced electromotive force and A is the coil surface area; it is easy to find expressions of electromotive forces of the three machine coils.

Remember that B_{rr} is the component of the flux density $\boldsymbol{B}_r$ produced by the rotor, measured in radial direction in the position where the conductor with the quotation mark (a', for the couple a–a', etc.) is positioned. So the value of B_{rr} to be considered at a given time is different for the three coils and is the one corresponding to angle θ describing the position of the base conductor of the three coils. With the disposition of coils shown in the figures of this section, and in particular in Figure 11.2b, it will be $\theta_a = 0$, $\theta_b = (2/3)\pi$, $\theta_c = (4/3)\pi$: therefore,

$$\begin{aligned} B_{rra} &= B_{rr}(t, 0) = B\cos(-\omega t) = B\cos(\omega t) \\ B_{rrb} &= B_{rr}(t, 2\pi/3) = B\cos[2\pi/3 - \omega t] = B\cos[\omega t - 2\pi/3)] \\ B_{rrc} &= B_{rr}(t, 4\pi/3) = B\cos[4\pi/3 - \omega t] = B\cos[\omega t - 4\pi/3)] \end{aligned}$$

and

$$\begin{aligned} e_a &= \omega A B_{rra} = \omega B A\cos(\omega t) \\ e_b &= \omega A B_{rrb} = \omega B A\cos[\omega t - (2/3)\pi] \\ e_c &= \omega A B_{rrc} = \omega B A\cos[\omega t - (2/3)\pi] \end{aligned}$$

which shows that the machine produces a three-phase balanced system of sinusoidal electromotive forces that can be represented by their phasors, as shown in the right-hand part of Figure 11.3. This indicates that a three-phase synchronous machine can operate as a three-phase voltage source.

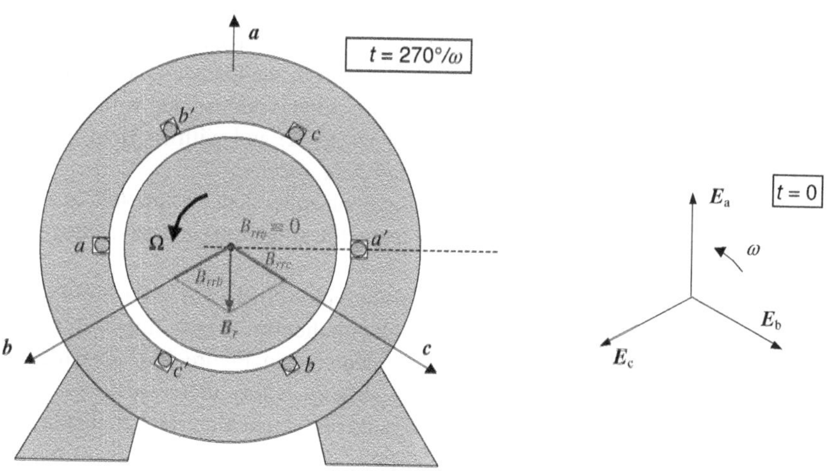

FIGURE 11.3. Radial components of $\boldsymbol{B}_r$ at the different coils at a given time (*left*), and the produced set of EMF phasors (*right*) by a three-phase synchronous machine.

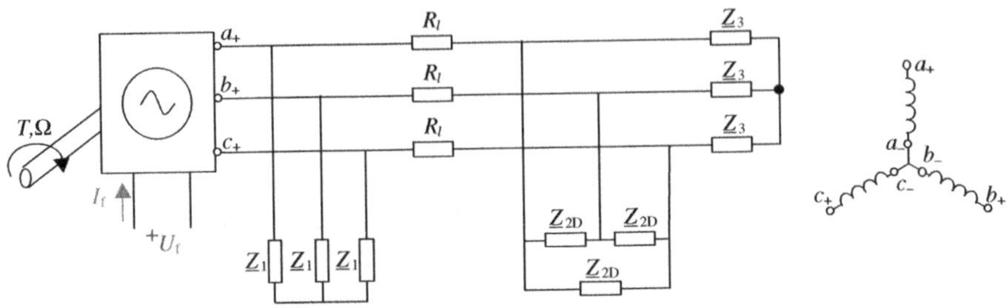

FIGURE 11.4. Star connection of coils to generate a three-terminal three-phase generator.

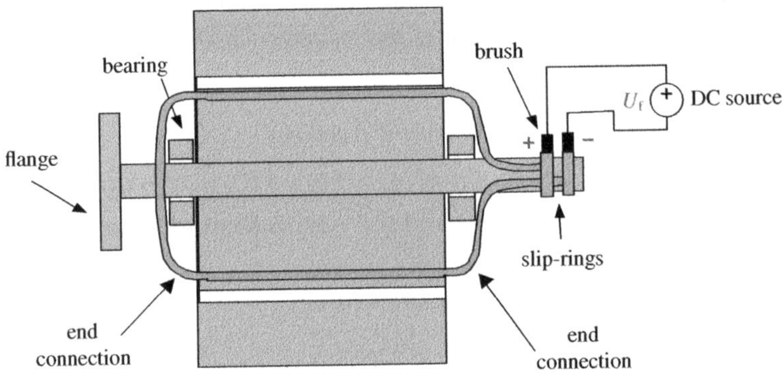

FIGURE 11.5. Connection of the field coil to a stationary DC source by means of slip rings and brushes.

To provide the three-phase source required by loads, the three coils are normally connected to each other in a star configuration: conductor terminals a_-, b_-, and c_- are connected to each other and ends a_+, b_+, and c_+ constitute the three machine terminals, to be connected to phases *a*, *b*, and *c* of loads.[3] This way, the three-phase balanced system of EMFs is applied to the line and to the loads connected to it. An example is shown in Figure 11.4, in which the machine is represented with its rotating shaft (carrying torque T and having angular speed Ω) and with its field current I_f supplied by the DC voltage source U_f, called *field voltage*.

In the following sections there is a drawing of an equivalent circuit, able to analyse systems such as that shown in Figure 11.4 using the single-phase equivalent concept discussed in Chapter 6.

The DC current can be transferred to the rotor with a brushed or a brushless system. For simplicity's sake, this book will only deal with the former and simpler of the two.

The brushed system can be arranged as depicted in Figure 11.5 in which, for simplicity, the rotor has a single-turn coil, while in reality there are several turns, connected in series to each other at the two ends of the machine rotor as displayed in the figure.

[3]The naming convention of coil terminals is shown in Figure 11.1.

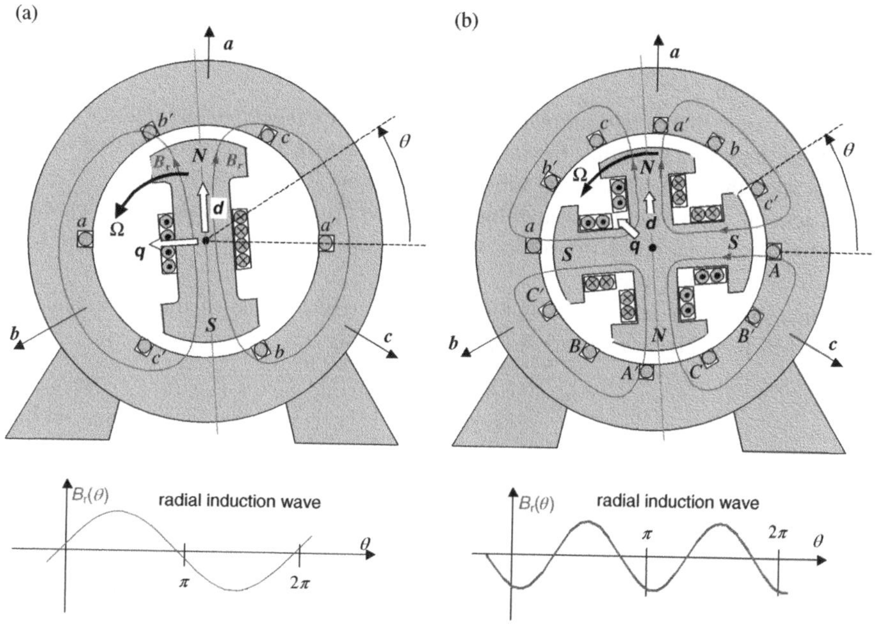

FIGURE 11.6. Comparison of machines with one and two pole pairs.

A stationary DC voltage source is electrically connected to two brushes, which are conductive elements of carbon-like material and pressed, using springs, to two slip rings that are mechanically connected to the rotor and therefore rotate along with it. When the rotor rotates, electricity is transferred from the stationary brushes to the rings, even though there is relative movement between the two parts.

It can be said that this mechanism is similar to the trolley-catenary contact so widely used to feed trains by means of overhead lines: there the carbon-like elements are in the trolley and the contact line is in copper.

These sliding couples allow DC current to be transferred to the rotor. The brushes are made of a material which is much softer than copper, so that the brushes have to be periodically replaced; the slip rings, which would probably be much more difficult to replace, last much longer.

Ignoring any iron saturation phenomenon, it can be said that the field produced is proportional to the producing current I_f:

$$B \equiv I_f, \qquad \hat{E} = \omega AB \Rightarrow \hat{E} = \omega L_f I_f \tag{11.2}$$

in which L_f, the proportionality coefficient between ωI_f and $\hat{E}$ (that is, the peak of any of the electromotive forces induced in the stator), has the dimensions of an inductor. The machine structure shown in Figure 11.2, and reproduced in Figure 11.6a only as a salient-pole configuration, has for both stator and rotor one pole pair (one ***N-S*** couple). The corresponding wave of radial flux density is of the type shown in the same figure below the machine drawing. It is also possible, and indeed very frequent, to produce machines with more than one pole pair. To clarify the concept, a machine with two pole pairs and the corresponding wave of radial induction density is shown in Figure 11.6b.

It is apparent that every complete rotation of the rotor is seen by the stator windings as if there were two rotations: the radial induction wave contains two full sine waves, and as such they are seen by all the coil conductors.

In general, for a machine containing p pole pairs (p can reach values much greater than 10), each rotation of the rotor causes the coil conductors to be subjected to p periods of the radial wave.

Consequently, if $\Omega = \mathrm{d}\theta/\mathrm{d}t$ is the physical angular speed of the rotor, and the induced EMF is $e(t) = \hat{E}\sin(\omega t + \alpha)$, it is obviously

$$\boxed{\omega = p\Omega} \tag{11.3}$$

Despite its simplicity, (11.3) is a fundamental law for synchronous machines, and a very similar relation will play a very important role also for induction machines, to be dealt with in the next chapter.

Quantities ω and Ω are also called *electrical* and *mechanical angular speed*, respectively. The same equation is also often written using different units of measure:

$$n = \frac{60f}{p}$$

where

- n is the number of revolutions per minute (rpm),
- f is the electrical frequency (Hz), and
- p is the number of pole pairs.

The great importance of (11.3) is immediately apparent, when we consider that all the interconnected circuits and networks share the same frequency. For instance the whole of Europe operates at 50 Hz and European nations, along with their houses, factories, offices, etc. are all connected to each other. The result is that all the one-pole-pair synchronous machines of Europe rotate at 3000 rpm, and all the two-pair machines at 1500 rpm, and so on.

The rotational speeds of all the synchronous machines of an interconnected system are integer submultiples of $60f$—that is, of 3000 rpm for 50-Hz systems (e.g., Europe, some American countries) and of 3600 for the 60-Hz systems (e.g., the United States, some American countries, some Japanese islands).

In a multiple-pole-pair machine there are multiple conductors per coil. For instance, in Figure 11.6a, while the coil for phase a is constituted only by conductor a and a', Figure 11.6b will be constituted by a–a'–A–A'; the voltages induced in a and A and in a' and A' will be exactly the same and therefore they can be connected in series, according to the sketch shown in Figure 11.7.

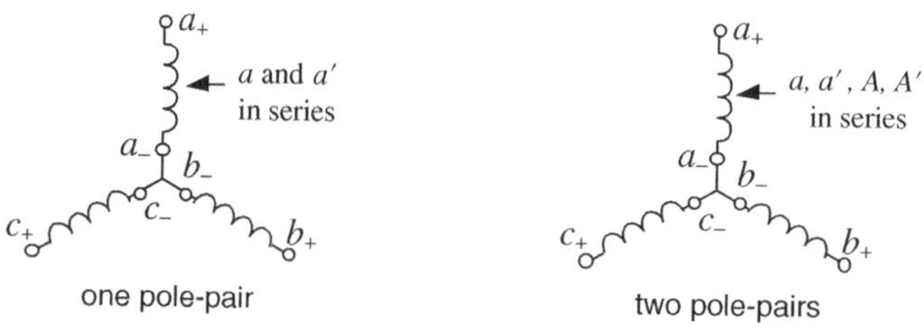

FIGURE 11.7. Connections of coil conductors in the case of star connection of windings and two different values for the number of pole pairs.

Example 1. Calculate the speed of a 60-Hz synchronous machine, if the pole pairs p are respectively 1, 2, 3, or 4.

Using $n = 60f/p$, it is easy to complete the following table:

$p = 1$	$p = 2$	$p = 3$	$p = 4$
3600 rpm	1800 rpm	1200 rpm	900 rpm

11.2 OPERATION UNDER LOAD

11.2.1 The Rotating Magnetic Field

Consider a three-phase machine connected to a balanced load—for example, constituted by three impedances $\underline{Z}$ equal to each other (Figure 11.8). The centre of the coils star is shown for completeness (conductor n), but this is not connected to the load. It was seen in the previous section that the synchronous machine generates, by effect of the field current and physical rotation, a three-phase balanced set of EMFs. Because of the symmetry of the machine and of the load, it is unquestionable that when the machine is connected to a balanced load, these EMFs will cause a three-phase balanced set of currents to circulate.[4]

Now it is time to see what this symmetrical set of currents do inside the machine.

The situation can be analysed with reference to Figure 11.9. It shows, just for simplicity and as a general example, a two-pole machine. The time at which the currents are considered is the one shown by the phasor diagram to the right of the figure—that is, a time in which the current in phase $\boldsymbol{a}$ is at its maximum. As discussed in Section 11.1, the directions of coil axes $\boldsymbol{a}$, $\boldsymbol{b}$, and $\boldsymbol{c}$ also set the reference signs for coil currents, which are also shown for greater clarity using the cross–dot convention in Figure 11.1.

Therefore, the component of the displayed vector $\boldsymbol{B}_{\mathrm{a}}$ on axis $\boldsymbol{a}$, also has a maximum; the value of its radial component B_{ar}, as a function of both time and angle θ, is

$$B_{\mathrm{ar}}(\theta, t) = \hat{B} \sin\theta \cos\omega t$$

[4]Remember that in this book the analysis of salient-pole machines ignores the effects of saliency. It can, however, be demonstrated, and will be discussed later, that even for salient-pole machines, this conclusion holds true.

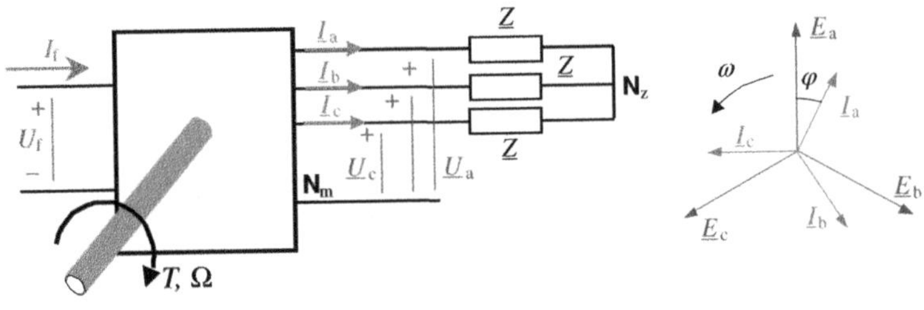

FIGURE 11.8. A three-phase synchronous machine connected to a passive load (three impedances).

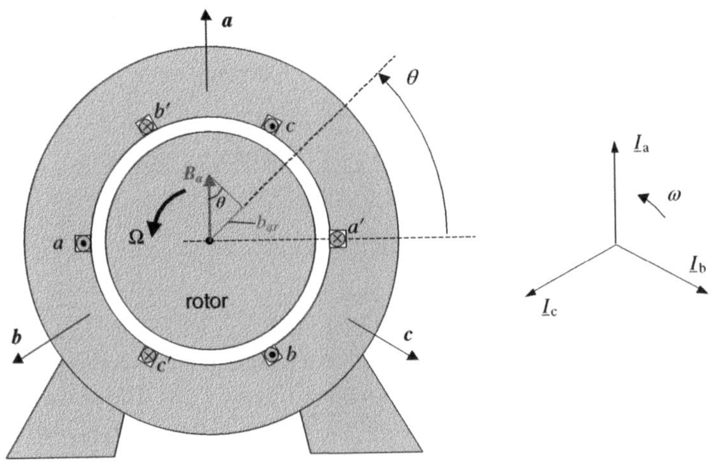

FIGURE 11.9. Basic three-phase machine arrangement of stator coils.

Since all the three currents circulate in the three stator coils, not only is $\boldsymbol{B}_a$ generated, but so also are the corresponding flux densities of phases b and c, that is, $\boldsymbol{B}_b$ and $\boldsymbol{B}_c$.

Given the physical position of the three coils and the time displacement of the three currents, this will be

$$B_{br}(\theta, t) = \hat{B}\sin\left(\theta - \frac{2}{3}\pi\right)\cdot\cos\left(\theta - \frac{2}{3}\pi\right), \qquad B_{cr}(\theta, t) = \hat{B}\sin\left(\theta - \frac{4}{3}\pi\right)\cdot\cos\left(\theta - \frac{4}{3}\pi\right)$$

The total radial induction wave will be

$$B_r(\theta, t) = B_{ra}(\theta, t) + B_{rb}(\theta, t) + B_{rc}(\theta, t) \tag{11.4}$$

Before analysing (11.4), let us first recall the following identities:

$$\sin x \cos y = \frac{1}{2}[\sin(x+y) + \sin(x-y)] \qquad \text{(from goniometry)} \qquad (^{\circ})$$

$$\sin x + \sin\left(x - \frac{2}{3}\pi\right) + \sin\left(x - \frac{4}{3}\pi\right) = 0 \qquad \text{(can easily be demonstrated)} \qquad (^{\circ\circ})$$

We can use a very fast method to demonstrate the latter identity. If we write $x = \omega t$, its left member is the sum of three phasors that constitute a balanced star of phasors. It is therefore, as we saw in Chapter 6, identically null.

Using (°), equation (11.4) becomes

$$B_r(\theta, t) = \frac{\hat{B}}{2}\left[3\sin(\theta - \omega t) + \sin(\theta + \omega t) + \sin\left(\theta + \omega t - \tfrac{4}{3}\pi\right) + \sin\left(\theta + \omega t - \tfrac{2}{3}\pi\right)\right]$$

and by applying (°°) we obtain

$$B_r(\theta, t) = \tfrac{3}{2}\hat{B}\sin(\theta - \omega t)$$

More in general, for a p pole pair machine the following expression applies:

$$B_r(\theta, t) = \tfrac{3}{2}\hat{B}\sin(p\theta - \omega t) = \tfrac{3}{2}\hat{B}\sin(\theta_e - \omega t) \qquad (11.5)$$

where $\theta_e = p\theta$ is also called electrical angle, while θ is called mechanical angle.

Equation (11.5) represents a translating wave in the θ_e axis, with speed ω (Figure 11.10). If we stick to a given abscissa θ_e we see a sine-wave radial B_r; the same B_r is a sinusoid also at a given time t, throughout the θ_e span.

Equation (11.5) shows that at a time in which the instantaneous current in phase a is at its maximum (e.g., $\omega t = 0$), the maximum value of $B_r(\theta)$ is when $\theta_e = \pi/2$—that is, along the direction of the oriented axis of coil $\boldsymbol{a}$ (cf. Figure 11.9). In a similar way, the maximum of the B_r wave is along $\boldsymbol{b}$ or $\boldsymbol{c}$ when the respective currents are at their maxima, respectively.

We have seen in Section 9.3.4 that a sinusoidal translating radial wave can be imagined as being obtained from a uniform field distribution, rotating at the angular electrical speed ω (and mechanical $\Omega = \omega/p$). Therefore, the result (11.5) is often referred to as the "rotating field theorem":

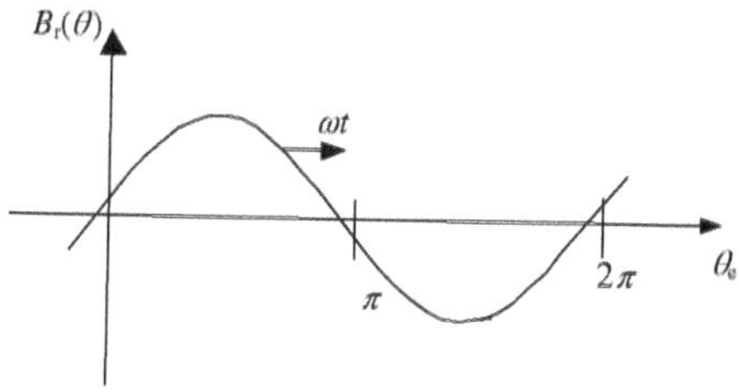

FIGURE 11.10. Translating flux density wave due to the stator currents.

Result: *The rotating field theorem*

If a balanced set of currents with angular frequency ω is introduced in windings that are separated from each other by 120° (electrical), this produces a field that is equivalent to a uniform field $\boldsymbol{B}_{\mathbf{eq}}$ rotating at a constant electrical speed ω (and mechanical $\Omega = \omega/p$), called rotating field.

The rotating field has the direction of the oriented axis of a coil when the instantaneous value of the coil current is at its maximum.

It is commonly agreed that the rotating field theorem was independently discovered by Nicholas Tesla and Galileo Ferraris, in the years 1882–1889.

11.2.2 Stator–Rotor Interaction

The rotating field theorem indicates that the stator currents generate a field similar to the one generated by the rotor. Both are rotating fields: the one produced by stator currents rotates by effect of the amplitudes and phases of stator currents (and of the arrangement of conductors around the air gap), while the one produced by the rotor rotates because of mechanical rotor rotation.

The two fields are synchronous, in the sense that they rotate at exactly the same speed.

The effects of the interaction of the two rotating fields can easily be qualitatively evaluated, bearing in mind the "permanent magnet equivalent" as shown in Figure 11.11, which shows the two fields. When a field abandons the iron part in which it is produced, a north pole (*N*) is shown with an opposite, a south pole (*S*). The interaction of the two fields can be qualitatively evaluated through our understanding of permanent magnets: similar poles repel each other, whereas opposite poles attract.

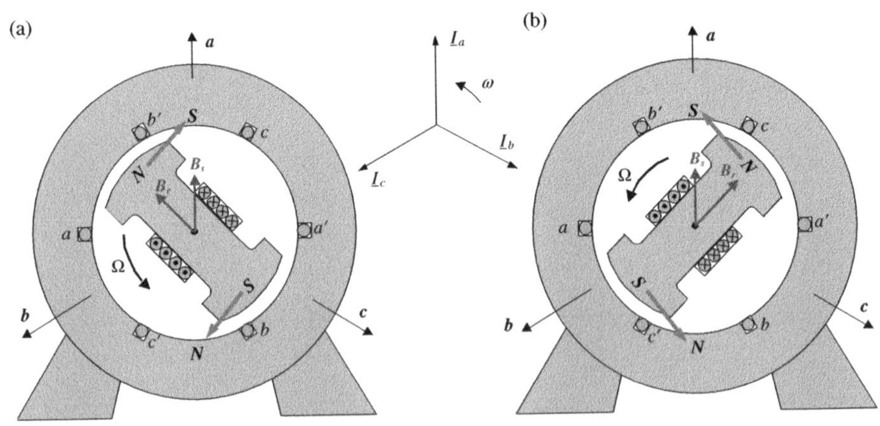

FIGURE 11.11. Forces acting on the rotor, due to the interaction of stator and rotor magnetic fields: (a) Generator. (b) Motor.

In the position represented in Figure 11.11, the field caused by the three stator currents is directed upward, because in that instant the current flowing in phase a is at its maximum (the corresponding phasor is oriented upward). As shown in the previous paragraph, the rotating field is aligned with the oriented axis of a phase when the current in that phase is at its maximum.

This means that the rotating field is, in all points of space, oriented upward: say it is the vector $\boldsymbol{B}_s$ shown in Figure 11.11. The magnetic flux thus abandons the stator in the bottom position, flows through the rotor, and enters the stator again in the stator **S** pole.

Correspondingly, **N** and **S** poles can be determined on the rotor. Two different rotor positions are depicted in Figures 11.11a and 11.11b. In the case of Figure 11.11a, an attractive force between opposite poles generates a torque that counteracts the rotor's movement; mechanical power is thus delivered by the rotor and converted into electrical power. In this case the machine is operating as an electric generator.

In the case indicated in Figure 11.11b, the relative positions of magnetic fields, and therefore of poles, are different; the net result is the production on the rotor of a torque that draws it in the direction of positive angular speeds, and power is converted from electricity into mechanical power. In this case the machine is operating as an electric motor.

The switch from motor and generator operations is very smooth and can occur at any time while the machine is working. Therefore the expressions "synchronous motor" and "synchronous generator" should be understood as shortened versions of "synchronous machine operating as a motor" and "synchronous machine operating as a generator," respectively.

More in Depth

An explanation of torque generation can also be derived using Lorentz's force law (in particular, $\boldsymbol{F}_\mathrm{L}$ in (2.7)) and the procedure indicated in Sections 9.3.2 and 9.3.3.

In the case of anisotropic machines, there is an additional mechanism of torque generation, based on the reluctance differences existing on the machine's **d** and **q** directions (see Figure 11.2a). This mechanism is very similar to that occurring in electromagnets, as discussed in Section 9.4. In practice, the generated torque tends to force the **d** direction (i.e., the lower reluctance direction) of the rotor to be aligned with the stator field direction.

This can also be visualized using the North–South pole attraction rule as shown in Figure 11.11: even in the absence of current in the rotor coils, a south pole is induced (by the so-called *magnetic induction* effect) in the rotor's iron where the stator flux enters, and a north pole on the opposite side, thus creating the reluctance torque.

This detailed evaluation of where forces are actually produced is often not required for global machine operation evaluation.

11.2.3 The Phasor Diagram and the Single-Phase Equivalent Circuit

Now that we have a fair knowledge of the physical phenomena occurring inside the machine, it is time to try to find a single-phase equivalent circuit for it. Such a circuit could allow steady-state analysis of the machine, even when it is connected to other systems, using the highly effective phasor circuit analysis and the single-phase equivalent circuit concept, with which the reader is now familiar from Chapters 5 and 6, respectively. To do this, it is convenient to introduce the linked flux phasor $\underline{\Psi}$.

As seen in the previous sections, the flux linked with any stator phase varies, as a sine wave, with time. Consider, for instance, the flux ψ_a linked with phase coil a of the machine and produced by a rotating field $\boldsymbol{B}$. It was observed in Section 11.2.1 that for analysis, this field can be considered as being uniform. Therefore ψ_a is equal to the component of $\boldsymbol{B}$ along the oriented axis $\boldsymbol{a}$ of the coil times the surface area A of the coil:

$$\psi_a(t) = B_r(\theta_{ea}, t) \cdot A = \hat{B}\cos(\theta_{ea} - \omega t) = \Psi\cos(\omega t - \theta_{ea})$$

As far as the phase of this flux is considered, it is recalled that, because of the rotating field theorem, $\psi_a(t)$ has its maximum when the current in phase a has its maximum. Therefore, the phasor representing it has to be parallel to $\underline{I}_a$. Therefore it will be

$$\underline{\Psi}_{sa} = L_{sa}\underline{I}_{sa} \qquad \text{or, simply,} \quad \underline{\Psi}_s = L_s\underline{I}_s$$

where:

- the subscript "a" may be omitted when writing the single-phase equivalent of the machine since, by convention, it always refers to phase a;
- the proportionality coefficient between current and flux is dimensionally equivalent to an inductance and has therefore been indicated with the usual symbol of inductances (L_{sa} or L_s);
- the subscript "s" has, traditionally, both the meaning of "stator" and "synchronous," since it refers to the synchronous machine and recalls that the field produced by the stator currents is synchronous (rotates at the same angular speed) with the rotor.

When the machine operates under load, two different fields rotate and generate radial waves on the air gap: stator and rotor flux density waves.

The voltage generated by these two fields can be found using Faraday's law in their phasor form:

$$\underline{E}_{tot} = \underline{E}_r + \underline{E}_s = -\frac{d\underline{\Psi}_r}{dt} - \frac{d\underline{\Psi}_s}{dt} = -j\omega\underline{\Psi}_r - j\omega\underline{\Psi}_s = \underline{E}_r - jX_s\underline{I} \qquad (11.6)$$

where, obviously, $X_s = j\omega L_s$.

Equation (11.6) can be written when there exists a linear relationship between magnetic field $\boldsymbol{H}$ and flux density field $\boldsymbol{B}$. This is valid only as an approximation in electric machines, as already noted in the study of DC machines.

The electromotive force $\underline{E}_{tot} = \underline{E}_r$ can be measured at the coil terminals, as $\underline{U}_a$ shown in Figure 11.8 before connecting the three impedances.

When the load impedances are connected, the currents flowing in the stator's coils cause $\underline{E}_s$ to be produced, as well as an ohmic drop:

$$\underline{U} = \underline{E}_{tot} - R_s\underline{I} = \underline{E}_r - jX_s\underline{I} - R_s\underline{I} = \underline{E}_r - \underline{Z}_s\underline{I}$$

where, obviously, $\underline{Z}_s = R_s + jX_s$.

This can be interpreted in a very easy and effective way using the phasor diagram and the corresponding single-phase equivalent circuit shown in Figure 11.12.

In the phasor diagram it is noted that:

- all the electromotive forces are lagging by 90° the corresponding flux phasors, since they are obtained by multiplying the latter by a positive constant and $-j$ (which rotates counterclockwise by 90°);
- the stator flux phasor is aligned with the current; this is a consequence of the already noted peculiarity of the rotating field theorem—that is, that the rotating field is aligned with the axis of a coil when the current in that coil is at its maximum. Therefore, the flux linked with that coil reaches its maximum when the current is at its maximum.

The synchronous machine model shown in Figure 11.12 and the corresponding equivalent circuit define a model often called Behn-Eschemburg's model, named after the scientist that first introduced it.

More complex models, such as Park's model, are beyond the scope of this book.

To have an idea of the usage of the single-phase equivalent circuit of the synchronous machine, take, for instance, the system shown in Figure 11.13. It can be effectively analysed using its single-phase equivalent, as shown in the right-hand part of the figure.

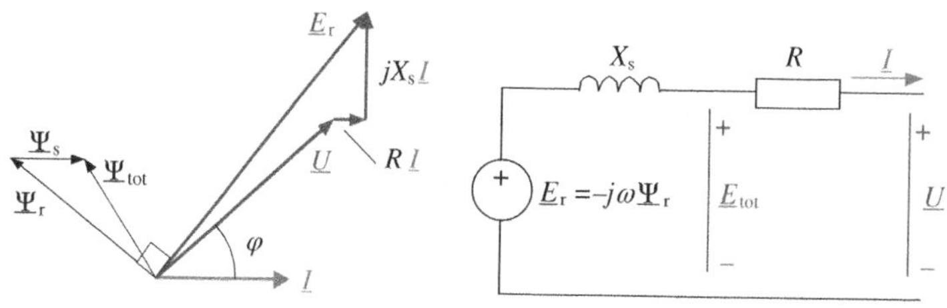

FIGURE 11.12. Phasor diagram and single-phase circuit equivalent of a synchronous machine.

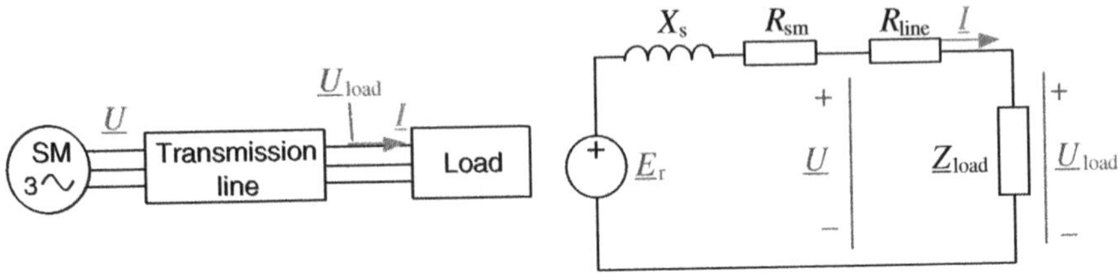

FIGURE 11.13. A simple system containing a synchronous machine and its single-phase equivalent (transmission line modelled using only its resistance).

More in Depth

Should we need to take into account the saliency of a salient-pole synchronous machine, the phasor diagram and circuit shown in Figure 11.12 can still be valid, but the value of X_s is a function of the angle between $\underline{U}$ and $\underline{I}$—that is, what happens outside the machine. As a consequence, a phasor circuit cannot be used in the way shown in this book.

It should also be added that both cylinder and salient-pole machines are not totally linear, since iron at high values of flux densities saturates, as discussed in Section 7.4, and saturation is a nonlinear phenomenon. The analysis of saturated machines is well beyond the scope of this book.

Example 2. Calculate the electromotive force of a wye-connected 5000-kVA 6.6-kV synchronous generator, delivering its rated current to a load having 0.85 power factor (lagging). The synchronous reactance and resistance per phase are 12 Ω and 0.07 Ω, respectively. Determine also the angle between $\underline{E}_r$ and $\underline{U}$, as well as the active and reactive power delivered to the load.

Let's consider Figure 11.12.

$$\underline{I} = \frac{\sqrt{2}\cdot 5\cdot 10^6}{\sqrt{3}\cdot 6.6\cdot 10^3}\angle 0° = \sqrt{2}\cdot 437.4\angle 0°\ \text{A}$$

Since $\cos\varphi = 0.85$, $\varphi = 31.79°$ and the voltage applied to the load is

$$\underline{U} = \sqrt{2}\cdot 6.6\cdot 10^3/\sqrt{3}\angle 31.79° = \sqrt{2}\cdot 3810\angle 31.79°\text{V}$$

$$\underline{E}_r = \underline{U} + (R + jX_s)\underline{I} = \sqrt{2}\cdot 3810\angle 31.79° + (0.07 + j12)\cdot \sqrt{2}\cdot 437.4 = \sqrt{2}\cdot 7959\angle 65.74°$$

then the electromotive force is 7959 V (rms) per phase (13,785 V line-to-line) and the angle between $\underline{E}_r$ and $\underline{U}$ is 33.95°.

Since $E_r > U$, the machine is said to be in "overexcitation."

It will be soon seen that when a machine is in overexcitation it delivers reactive power, while when it is in underexcitation ($E_r < U$) it absorbs reactive power.

$$S = (3\underline{UI}^*)/2 = 4.25 \cdot 10^6 + j2.634 \cdot 10^6$$

The generator delivers 4.25 MW and 2.634 Mvar to the grid.

Example 3. Repeat the previous exercise, assuming that this time the generator delivers one-third of its rated current to a load having 0.75 power factor (leading).

This time,

$$\underline{I} = \sqrt{2} \cdot \frac{437.4}{3} \angle 0° = \sqrt{2} \cdot 145.8 \angle 0° \text{A}$$

Since $\cos \varphi = 0.75$ (leading), $\varphi = -41.41°$ and the voltage applied to the load is

$$\underline{U} = \sqrt{2} \cdot 6.6 \cdot 10^3 / \sqrt{3} \angle -41.41° = \sqrt{2} \cdot 3810 \angle -41.41° \text{V}$$

$$\underline{E}_r = \underline{U} + (R + jX_s)\underline{I} = \sqrt{2} \cdot 3810 \angle -41.41° + (0.07 + j12) \cdot \sqrt{2} \cdot 145.8$$

$$= \sqrt{2} \cdot 2970 \angle -15.04° \text{ V}$$

The electromotive force is then 2970 V per phase (5144 V line-to-line) and the angle between $\underline{E}_r$ and $\underline{U}$ is 26.37°.

Since $E_r < U$, the machine is said to be in "underexcitation."

$$S = (3\underline{UI}^*)/2 = 1.25 \cdot 10^6 \text{ W} - j1.102 \cdot 10^6 \text{ var}$$

The generator delivers 1.25 MW while absorbing 1.102 Mvar from the grid.

11.3 PRACTICAL CONSIDERATIONS

11.3.1 Power Exchanges

It is very important to evaluate the relationship between the quantities shown in Figure 11.12 and the exchanges of active and reactive powers P and Q with the external network.

It would be especially useful to have very simple formulas which can give immediate information of what happens, for instance, if the field current is raised or if the mechanical power entering the machine is increased. To pursue this simple task, analysis is carried out while disregarding the stator coil resistance from the single-phase circuit. Therefore, the circuit and phasor diagram shown in Figure 11.14 can be used.

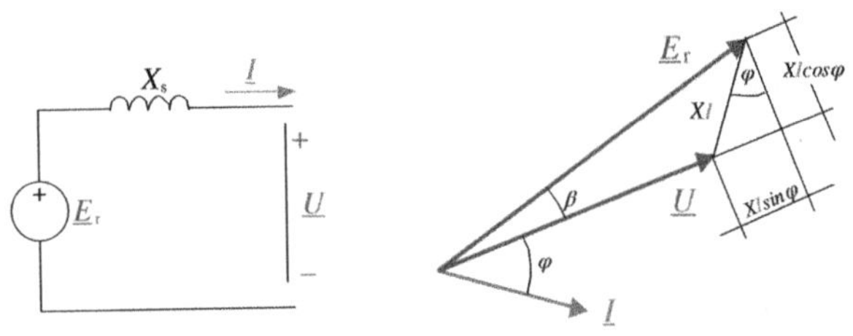

FIGURE 11.14. Machine circuit and diagram to determine the delivered active and reactive powers.

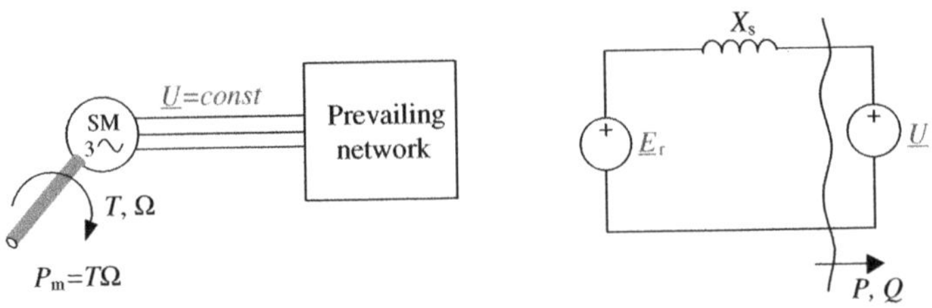

FIGURE 11.15. Operation of a synchronous machine connected to a prevailing network.

If the symbol β indicates the phase displacement between $\boldsymbol{E}$ and $\boldsymbol{U}$, from the construction shown in the right-hand part of Figure 11.14, we immediately have

$$
\begin{aligned}
P &= UI\cos\varphi = U\frac{E_r\sin\beta}{X_s} = \frac{UE_r}{X_s}\sin\beta \\
Q &= UI\sin\varphi = U\frac{E_r\cos\beta - U}{X_s} = \frac{UE_r}{X_s}\cos\beta - \frac{U^2}{X_s}
\end{aligned}
\tag{11.7}
$$

which obviously refer to the powers flowing (going out of the machine) in the single-phase equivalent circuit—that is, one-third of the whole power leaving the machine.

Equations (11.7) are able to offer an immediate interpretation of a machine connected with a so-called *prevailing network*—that is, a network so powerful that its voltage phasors remain the same, in amplitude and angle, whatever the machine does (Figure 11.15).[5]

In terms of circuit analysis, a prevailing network is characterized by a single-phase equivalent whose Thévenin equivalent has null impedance. Therefore, the interaction of the machine and the network, disregarding the coil resistance, can be evaluated using the single-phase equivalent shown in the right-hand part of Figure 11.15.

One consequence of the constancy of $\underline{U}$ is the constancy of its phase, and therefore of its frequency. Bearing in mind the fundamental law of the synchronous machine $\omega = p\Omega$, in steady state, the machine mechanical rotational speed Ω is also constant.

[5]Other names for such a component frequently found in the literature are: *prevailing power system* or *infinite power system*.

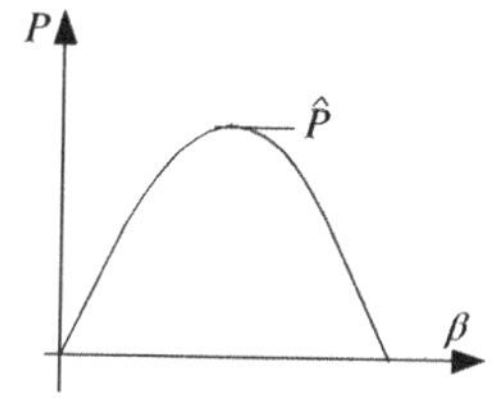

FIGURE 11.16. The power–angle curve of a synchronous machine.

Using (11.7) and ignoring all the active power losses within the machine (hence $P_m = P$), several pieces of information can be inferred regarding the situation depicted in Figure 11.15.

First of all, it should be recalled, from (11.2), that E_r and I_f are proportional. At I_f constant, therefore, the mechanical power P_m entering the machine, assumed to be equal to the active power delivered to the prevailing network, is proportional to the sinus of the so-called *torque angle* β. Under these conditions the maximum power that the machine can generate is

$$\hat{P} = \frac{UE_r}{X_s}$$

This is an electrical limit to the power that can be transferred, even when there exists more than this power available at the mechanical shaft and even if the stator and rotor conductors have not reached their thermal limit (maximum current).

The situation can be visualized in the plotted power-angle relationship shown in Figure 11.16. It is possible to demonstrate that the power exchange between the machine and the network is usually stable only for $\beta \leq 90°$; however, the machine does not work safely in proximity to $\beta = 90°$ because any further increase of power would cause the system to collapse. In a typical power system, where synchronous machines are used to generate power to be transferred to the network (see Part IV of this book), when the system operates in steady state, angle β is often maintained below 45°, to keep within adequate operational limits.

Under the assumptions made, mechanical and electrical powers are equal to each other. The reactive power flow, indeed, is governed by modification in the field current I_f.

In fact, the second equation of (11.7) shows that any increase in I_f causes a corresponding increase in Q. The effects of I_f can be made even clearer if it is assumed that $\cos\beta \cong 1$. In this case we have

$$Q \cong U\frac{E_r - U}{X_s} \tag{11.8}$$

which indicates that the machine absorbs or delivers reactive power when E_r is lower or higher than U, respectively.

Synchronous machines used in power stations, therefore, not only generate active power for the loads, but also contribute to supplying the loads with the needed reactive power. While the active power production is regulated by acting on the source of mechanical power connected to the shaft (turbine, internal combustion engines, etc.), the reactive power production is controlled acting on the field current I_f, which in turn controls the EMF E_r.

Example 4. A wye-connected synchronous machine is connected to a three-phase 3000-V (phase-to-phase) prevailing network. The synchronous reactance is $2\,\Omega$, while the stator resistance can be disregarded. Calculate the electromotive force, the torque angle, the stator current, and the reactive power delivered to the network:

(a) just after connection
(b) if the machine delivers 1.2 MW, keeping the excitation current constant to the value required for case a).

(a) $U = 3{\cdot}10^3/\sqrt{3} = 1732\ \mathrm{V}$

$$E_r = U, \quad \beta = P = Q = I = 0$$

(b) From equations (11.7), multiplied by 3 in order to consider three-phase powers, we obtain

$$\beta = \sin^{-1}\left(\frac{P\cdot X}{3E_r U}\right) = \sin^{-1}\left(\frac{1.2{\cdot}10^6{\cdot}2}{3{\cdot}1732^2}\right) = 15.46°$$

$$Q = \frac{3E_r U}{X}\cos\beta - \frac{3U^2}{X} = \frac{3{\cdot}1732^2}{2}\cos\ 15.46° - \frac{3{\cdot}1732^2}{2} = -163.9\ \mathrm{kvar}$$

$$S = \sqrt{P^2+Q^2} = 1.211\ \mathrm{MVA}$$

$$I = \frac{S}{3U} = 233.1\ \mathrm{A}$$

Example 5. With reference to the machine in the previous example, calculate the excitation voltage, the angle β, and the stator current if the machine delivers 1.8 MW and 0.8 Mvar to the network.

$$\underline{U} = \sqrt{2}\cdot 3\cdot 10^3/\sqrt{3}\angle 0° = \sqrt{2}\cdot 1732\angle 0°\ \mathrm{V}$$

$$S = \sqrt{P^2+Q^2} = 1.97\ \mathrm{MVA}$$

$$\varphi = \arg\ \cos\ (P/S) = 23.96°$$

$$\underline{I} = \sqrt{2}\cdot\frac{S}{3U}\angle-\varphi = \sqrt{2}\cdot\frac{1.97\cdot 10^6}{3\cdot 1732}\angle-23.96° = \sqrt{2}\cdot 379.1\angle-23.96°\ \mathrm{A}$$

$$E_r = \underline{U} + jX_s\underline{I} = \sqrt{2}\cdot 2154\angle 18.76°\ \mathrm{V}$$

The electromotive force is then 2154 V per phase (3731 V line-to-line) and the angle β is 18.76°.

The same result can be obtained without using phasors from equations (11.7):

$$\begin{cases} P = \dfrac{3E_r U}{X} \sin\beta \\ Q = \dfrac{3E_r U}{X} \cos\beta - \dfrac{3U^2}{X} \end{cases} \rightarrow \begin{cases} 1.8 \cdot 10^6 = \dfrac{3E_r 1732}{2} \sin\beta \\ 0.8 \cdot 10^6 = \dfrac{3E_r 1732}{2} \cos\beta - \dfrac{3 \cdot 1732^2}{2} \end{cases}$$

$$\tan\beta = \frac{1.8 \cdot 10^6}{0.8 \cdot 10^6 + \dfrac{3 \cdot 1732^2}{2}} = 0.3396, \qquad \text{hence } \beta = 18.76°$$

$$1.8 \cdot 10^6 = \frac{3E_r 1732}{2} \sin\beta \quad \rightarrow \quad E_r = 2154\ \text{V}$$

11.3.2 Generators and Motors

Synchronous machines were discussed previously in their more usual mode of operation—that is, as generators. Indeed, they can operate equally as generators or motors. The equivalent circuit of Figure 11.12 is still valid, as is the phasor diagram, but with different angles.

Indeed, in the motor operation $\underline{U}$ leads $\underline{E}$, since the value of P in equation (11.7) must reverse, and therefore angle β has to change its sign.

The phasor diagram in the motor operation has the appearance shown in Figure 11.17, which was drawn with the current $\underline{I}$ oriented as shown in the single-phase circuit alongside.

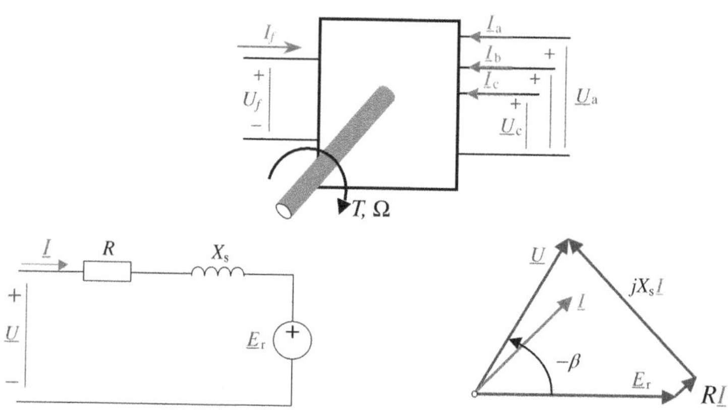

FIGURE 11.17. Motor operation of a synchronous machine.

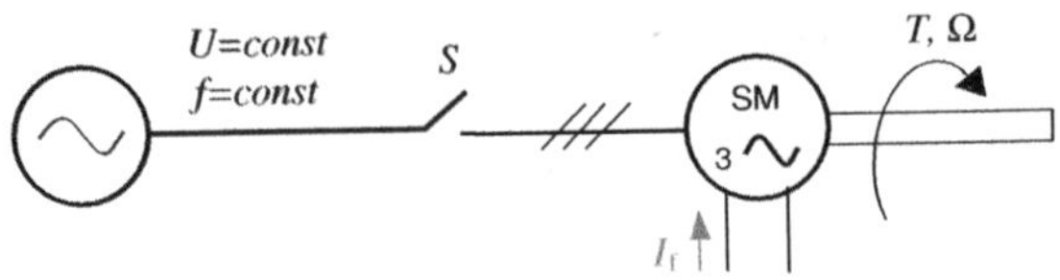

FIGURE 11.18. The attempt of a synchronous machine to start up from a network with constant voltage and frequency (prevailing network).

Motor operation of a synchronous machine is somewhat tricky. Indeed in common use, motors have to be frequently started up—that is, moved from $\Omega = 0$ up to their steady-state angular operation. Unfortunately, starting a synchronous motor is not easy, unless the motor is connected to a special device called an inverter (see Section 11.5).

If, for instance, a machine which is already fed with the field voltage U_f (making the field current I_f circulate in the rotor coils) is abruptly connected to a three-phase network (Figure 11.18), it will absorb currents that generate forces incapable of accelerating the machine. It is easier to understand what happens if we remember the rotating field theory. If we imagine that after the switch S is closed, a three-phase balanced system of currents flows in the stator, it generates rotating poles in the air gap, which interact with the (fixed) rotor poles (see Figure 11.11). The interaction of these pole pairs will generate an alternating torque, with a null average, and therefore the rotor will continue to oscillate without picking up useful speed.

One way to start up a synchronous machine is to feed it using a variable-frequency source: its frequency will start from zero and will be progressively raised: for instance, the torque angle β can be maintained constant during this acceleration transient. Details on how to do this in a special case are discussed in Section 11.5.2.

Another way is to move the machine rotor mechanically by an external method (another electric machine, a diesel engine, etc.), connecting the stator with the external three-phase network and/or starting to excite the rotor through the field current I_f only when the actual rotor speed is very near to that of the rotating field, that is, $\Omega = \omega/p$.

Example 6. A 400-V three-phase wye-connected synchronous motor has synchronous impedance of $0.3 + j3\ \Omega$ per phase. Calculate the electromotive force, the angle β, and the active and reactive power absorbed from the grid, if the stator current is 25 A:

(a) at 0.85 lagging power factor
(b) at 0.85 leading power factor

The single-phase voltage phasor is:

$$\underline{U} = \sqrt{2} \cdot 400/\sqrt{3}\angle 0° = \sqrt{2} \cdot 230.9\angle 0°\ \text{V}$$

(a) Since $\cos\varphi = 0.85$, we obtain $\varphi = 31.79°$ and

$$\underline{I} = \sqrt{2}\cdot 25\ \mathrm{A}\angle -31.79°$$

$$\underline{E}_{\mathrm{r}} = \underline{U} - (R + jX_{\mathrm{s}})\underline{I} = \sqrt{2}\cdot 230.9 - (0.3 + \mathrm{j}3)\cdot \sqrt{2}\cdot 25\angle -31.79°$$

$$= \sqrt{2}\cdot 194.5\ \mathrm{V}\angle -17.91°$$

The electromotive force is 194.5 V per phase (336.8 V line-to-line) and the torque angle is $-17.91°$.

$$S = (3\underline{UI}^{*})/2 = 14,720 + j9124\ \mathrm{VA}$$

The motor absorbs 14.72 kW and 9.12 kvar from the grid.

(b) This time, $\varphi = -31.79°$; thus we have

$$\underline{I} = \sqrt{2}\cdot 25\angle 31.79°\mathrm{A}$$

$$\underline{E}_{\mathrm{r}} = \underline{U} - (R + jX_{\mathrm{s}})\underline{I} = \sqrt{2}\cdot 230.9 - (0.3 + \mathrm{j}3)\cdot \sqrt{2}\cdot 25\angle 31.79°$$

$$= \sqrt{2}\cdot 272.6\ \mathrm{V}\angle -14.38°$$

The electromotive force is 272.6 V per phase (472.2 V line-to-line) and the torque angle is $-14.38°$.

$$S = (3\underline{UI}^{*})/2 = 14,720\text{-}j9124\ \mathrm{VA}$$

The motor absorbs 14.72 kW while delivering 9.12 kvar to the grid.

11.4 PERMANENT-MAGNET SYNCHRONOUS MACHINES

In Chapter 7 it was shown that a fixed value of flux density B_{r} can be obtained by the introduction into a magnetic circuit of a permanent magnet (PM) having B_{r} as residual flux. In a synchronous machine rotor, therefore, the insertion of PMs can have the same effect as a DC-fed coil with a fixed current flowing in it. To evaluate the circuit reluctance, the space occupied by the PMs must be considered as pure air—that is, having the same magnetic permeability of empty space.

Practical arrangements PM synchronous machine rotors are shown in Figure 11.19. In all cases, the permanent magnets generate a field $\boldsymbol{B}$ with arrows indicating the direction.

Two of the three arrangements cause the machine to behave in a particular way by effect of their anisotropic iron structure. Consider, for instance, the central drawing: the reluctance along direct axis is lower than the one along quadrature axis, since the

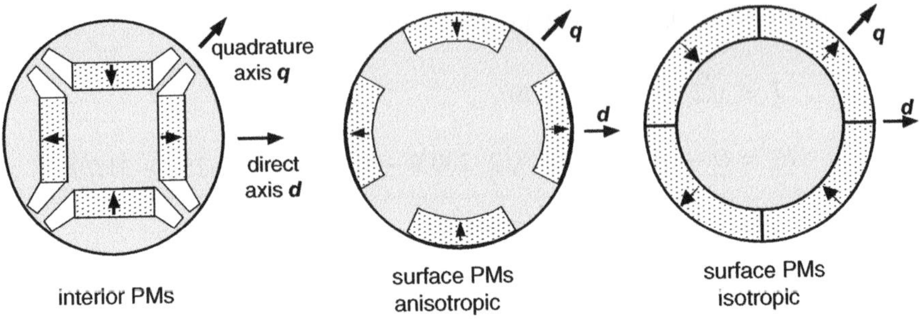

FIGURE 11.19. Different installation options of PMs in synchronous machine rotors.

space occupied by the PMs is to be considered as air. However, in this book, according to the "synchronous machine" isotropy assumption, the analyses are made disregarding any anisotropy effects.

The presence of the magnets is equivalent to conductors traversed by a DC current and producing the same flux density as the magnets.

The advantages of PM machines over conventional machines with field coils are as follows:

- Slip rings are avoided, with advantages in terms of machine cost, complexity and maintenance.
- The rotor copper losses are avoided, which in turn brings advantages in term of efficiency and heat dissipation needs.

An obvious disadvantage of this solution is that these machines have no ability to regulate the produced field, and, therefore, they cannot be used for reactive power regulation (see Section 11.3.1). Thus they tend to be used mainly as motors, within synchronous motor drives (Section 11.5).

Permanent magnet synchronous machines can also be seen as synchronous machines without brushes. They are therefore very often called *brushless machines*.

11.5 SYNCHRONOUS ELECTRIC DRIVES

11.5.1 Introduction

In Section 11.1 the fundamental equation (11.3) was obtained, telling us that a machine fed by a constant-frequency source rotates at a constant speed. It is therefore impossible for a synchronous machine to work at variable speeds if it is fed by a constant-frequency source.

In recent years, electronic devices have appeared that are able to convert constant-frequency sources into variable frequency, thus allowing modification of voltage and frequency characteristics of a three-phase source. The set consisting of a suitable electronic converter (also called power processing unit or PPU) and a motor is

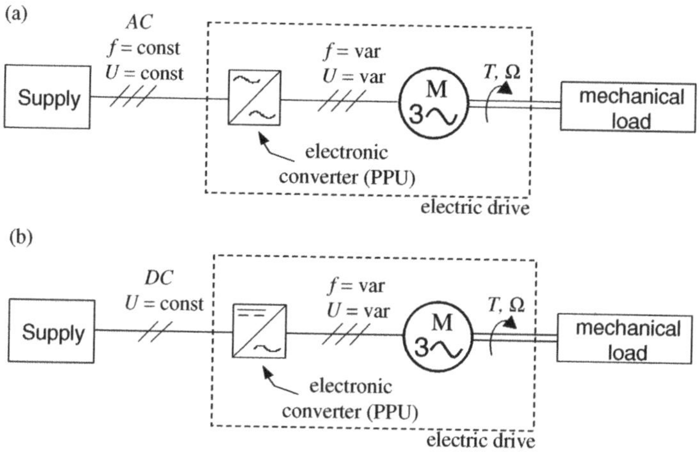

FIGURE 11.20. Combination of a PPU and an AC machine to obtain an AC electric drive.

generally called an electric drive; if the motor is synchronous, the drive is called a synchronous electric drive. Two different kinds of synchronous motor drives are shown in Figure 11.20: part (a) refers to a case in which the drive supply is an AC source, while Part b to the case of a DC supply.

Details of the converters able to perform the functions shown in Figure 11.20 are given in Chapter 8. The present chapter was deliberately written in such a way that it can be understood without the need to read Chapter 8. The latter can thus be considered as a more in-depth analysis of converters and drives.

11.5.2 PM, Inverter-Fed, Synchronous Motor Drives

In Section 11.3.2 we examined the operation of synchronous machines as motors, and saw that for starting and variable speed operation it is mandatory to feed the synchronous machine from a variable frequency source.

Information on how this can be done in practice is given here. Consider the circuit and phasor diagrams shown in Figure 11.17. For the evaluation of the machine at a variable speed, to make things even simpler the machine losses are disregarded, and the simpler circuit and diagram shown in Figure 11.21 are taken as reference.

The shaft mechanical power, when losses are disregarded, equals the one absorbed from the electric supply and is equal to[6]

$$P_{mecc} = 3UI\cos\varphi = 3E_rI\cos\gamma \tag{11.9}$$

[6]Remember that the circuit is the single-phase equivalent of a three-phase machine, in which one-third of the full machine power circulates. Moreover, Φ, Ψ, I, and U are all rms values of the corresponding sine waves.

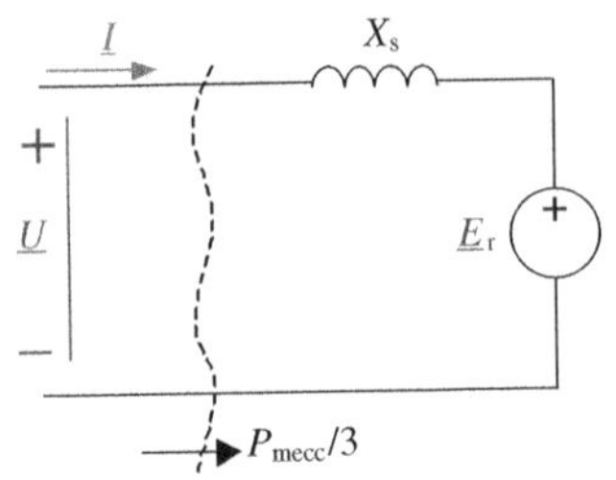

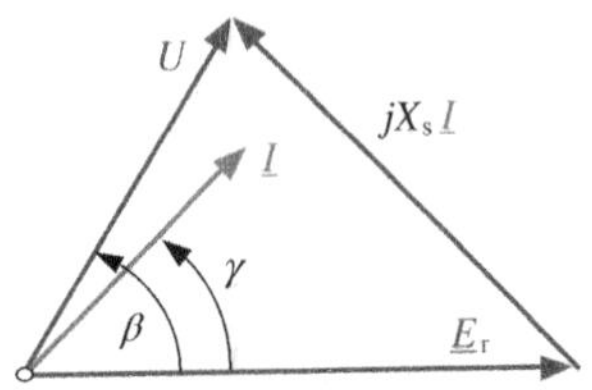

FIGURE 11.21. Motor operation of a synchronous machine disregarding losses.

where [see equation (11.6)]

$$E_r = |-j\omega k \underline{\Phi}_{PM}||-j\omega \underline{\Psi}_{PM}| = \omega \Psi_{PM}$$

$\underline{\Phi}_{PM}$ and $\underline{\Psi}_{PM}$ are respectively the flux and the linked flux produced by the permanent magnets, and k is a proportionality factor that takes into account the number of turns and the number of pole pairs. As a consequence, it is also

$$T_{mecc} = P_{mecc}/\Omega \cong 3\omega\Psi_{PM} I \cos\gamma/\Omega = 3p\Psi_{PM} I \cos\gamma \tag{11.10}$$

Remember that Ψ_{PM} and I are rms values of corresponding sinusoidally varying quantities. Equation (11.10) is often found in the literature as $T_{mecc} = (3/2)p\Psi_{PM} I \cos\gamma$; in this case, Ψ_{PM} and I indicate peak values of the corresponding sinusoidal quantities.

Also, the synchronous reactance X_s is proportional to ω, and therefore can be written as

$$X_s = \omega L_s$$

where L_s is the proportionality factor and has the dimensions of an inductance.

Since the copper losses are related to the current flowing in the machine's stator, it is convenient, whenever possible, to operate the machine so that γ is zero; this way, for a given power, there will be minimum current (and thus copper losses).

When γ is zero, the phasor diagram of the machine will be a specialized version of the diagram of Figure 11.21, and will appear as indicated in Figure 11.22 (from now on, Ψ_{PM} is simply indicated as Ψ).

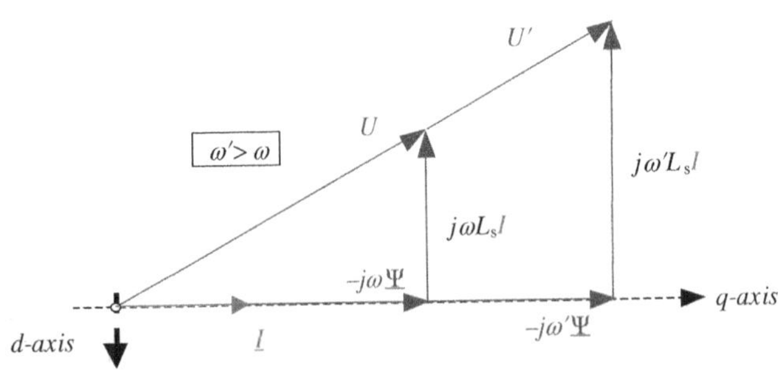

FIGURE 11.22. Phasor diagram of the PM machine when γ is zero (at low speeds).

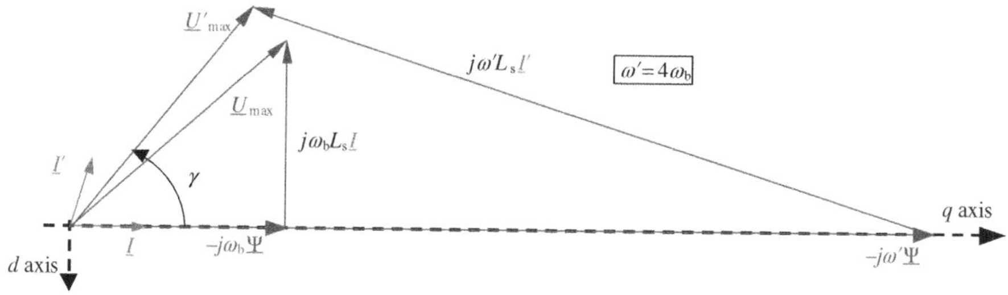

FIGURE 11.23. Phasor diagram of the PM machine at base speed ω_b and $\omega' = 4\omega_b$. (with the Ψ/L_s value of Figure 11.22 $\underline{I}'$ would be in phase with $\underline{U}$ when $\omega \cong 2.5\omega_b$.)

During operation at zero γ, if the machine current is held constant when speed varies, equation (11.10) says that the torque also stays constant. If this constant value of the current is the maximum current (also called *nominal* current,) the torque will stay constant at its maximum value.

As long as the machine's angular speed Ω and the angular frequency ($\omega = p\Omega$ increase, the machine voltage increases proportionally. Certainly, the process cannot continue endlessly: when the maximum voltage is reached, control of the machine must be changed to avoid exceeding this value. The maximum allowed voltage may be due to the need not to exceed machine insulation or because of the needs of the electronic converter connected to the machine (remember Figure 11.20).

Once the maximum value U_{max} for the machine phase voltage has been reached, to avoid further voltage increases, angle γ must become greater than 0, and the vector diagram takes the shape already seen in Figure 11.21.

The maximum speed compatible with the condition $U \leq U_{max}$ and with $\gamma = 0$ at the maximum continuous current I_n is called *base speed* of the machine, similar to what we have already seen for DC and asynchronous AC drives.

By way of an example, at a speed four times the base speed, the diagram has a shape of the type shown in Figure 11.23.

The condition to be satisfied at all speeds, namely,

$$U \leq U_{max}$$

can be written in terms of the various quantities applying Carnot's theorem at the triangle shown in Figures 11.21 or Figure 11.23 (written for the maximum continuous current I_n)[7]:

$$\omega^2(\Psi^2 + L_s^2 I_n^2 - 2\Psi L_s I_n^2 \sin\gamma) \leq U_{max}^2 \tag{11.11}$$

[7]Since U and I are rms values corresponding to $\underline{U}$ and $\underline{I}$, to obtain equation (11.11), Carnot's theorem is applied to a triangle similar to the one shown in Figure 11.23, but whose sides are all reduced by a factor of $\sqrt{2}$.

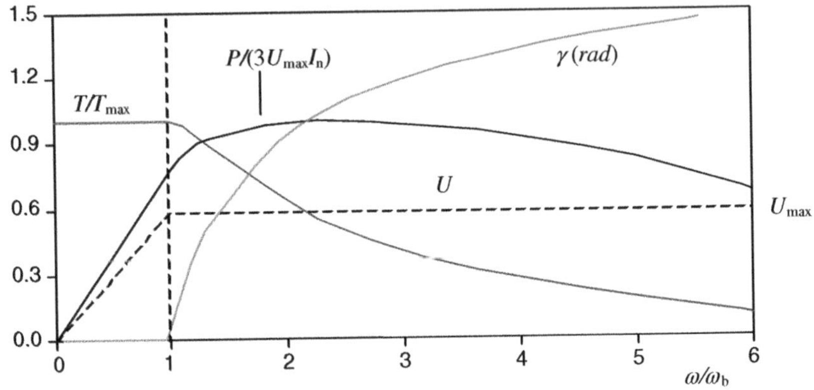

FIGURE 11.24. A sample set of PM machine characteristics at $I=I_n$, $U \leq U_{max}$ ($\omega_b\Psi = 0.70U_{max}$, $\omega_b L_s = 0.58U_{max}/I_n$).

By substituting $\gamma=0$ and using the equality sign, equation (11.11) yields the base speed:

$$\omega_b^2(\Psi^2 + L_s^2 I_n^2) = U_{max}^2 \Rightarrow \omega_b = \frac{U_{max}}{\sqrt{\Psi^2 + L_s^2 I_n^2}}$$

At angular speeds higher than $\Omega_b = \omega_b/p$, the condition (11.11) determines the value of angle γ.

As an example, Figure 11.24 shows, for a wide speed range, the trends of torque, power voltage, and γ angle for a given couple of Ψ and L_s. To make the numerical values of both quantities more useful, they are expressed as ratios to significant machine quantities, as directly shown in Figure 11.24 and in its caption.

Figure 11.24 can be commented upon as follows:

- Below the base speed the machine voltage increases proportionally with speed, and the torque stays at its maximum value.
- Once the base speed has been surpassed, γ must become larger than 0 and the torque reduces; however, according to equation (11.9), at the beginning the decrease in power due to the increase of γ does not compensate for the increase due to the ω growth.
- However, around $\omega = 2.5\omega_b$, the power reaches a maximum, after which the reduction effect prevails. When at maximum, the current absorbed by the machine is in phase with the voltage, and therefore the power is $3U_{max}\,I_n$.
- Above $\omega = 2.5\omega_b$ the current leads the terminal voltage, and the phasor diagram could be the one represented in Figure 11.23.

During the Whole range of speeds displayed, from base to maximum, power is sustained; therefore, due to this, this diagram is suitable for installation in a vehicle: within a speed ratio of 6 (ratio between maximum and base speed) the available power always remains above 50% of its maximum value.

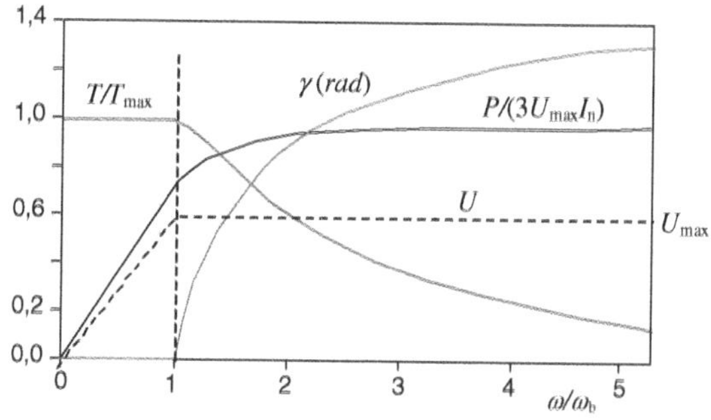

FIGURE 11.25. A sample set of PM machine characteristics at $I = I_n$, $U \le U_{max}$ under the special condition $\Psi = L_s I_n$.

There is a condition for which the power never reaches its maximum, at least with our simplified model. In fact, if the triangle of Figure 11.22 is equilateral, that is, if it is

$$\Psi = L_s I_n$$

the relation (11.11) becomes

$$2\omega^2 L_s^2 I_n^2 (1 - \sin\gamma) \le U_{max}^2 \tag{11.12}$$

At very large values of ω, this relation tends to be satisfied with values of γ approaching 90° that implies (remember Figure 11.23) that the terminal voltage U is in phase with the current I.

The curves of Figure 11.24, under condition (11.12), thus take the shape of those shown in Figure 11.25.

Our analysis of the PM machine disregards magnetic anisotropy (see again Figure 11.19). Anisotropic machines can be demonstrated as having much better overloading capabilities than isotropic ones (the topic is discussed in [p2]).

Some details on how to determine the currents to be fed into the PM synchronous machine to obtain the desired value of torque at the various speeds will be discussed in the next section.

The diagrams shown in Figures 11.24 and 11.25 show only one of the possible quadrants in the speed–torque plane. Indeed, in many cases more than a single quadrant is exploited. For instance, for vehicular propulsion, even only for forward driving, the electrical machine can give positive propulsion force (e.g., when accelerating) or negative propulsion force (when regenerative braking) when mechanical power is converted into electricity. In this case the drive operates in the two quadrants in which speed is positive and the torque can have either sign.

In other cases all the four quadrants are needed.

This is possible and easy with all electric drives: DC drives, asynchronous machine-based drives, and synchronous drives. Therefore, in some cases, the operating region is often shown as two, or even four, quadrants—that is, all four regions, as in the two diagrams shown in Figure 11.26.

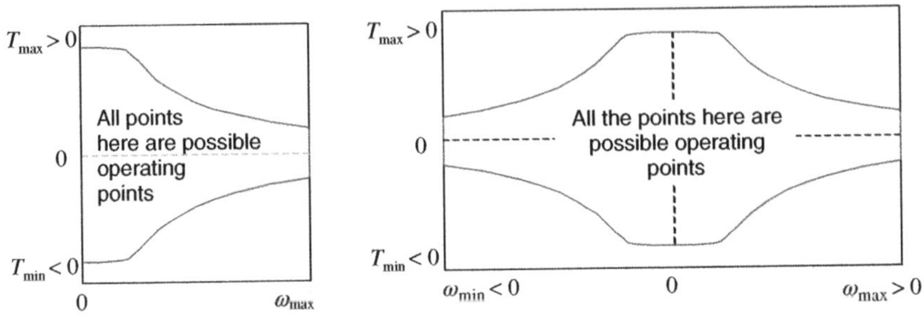

FIGURE 11.26. Possible operating regions of electric drives: (a) Two-quadrant electric drive. (b) Four-quadrant electric drive.

11.5.3 Control Implementation

The control of PM synchronous machines, when analysed in detail, is very complex. The basic operation of the control algorithms, however, can be explained rather simply, since it is a direct consequence of what was said in the previous section regarding the case of the isotropic machine.

Similarly to the control of the DC machine in Chapter 10, the purpose of the control system is to deliver a given torque while preventing the machine from exceeding the maximum current and voltage allowed.

Consider again (11.10):

$$T = \frac{P}{\Omega} = 3p\Psi_{PM} I \cos \gamma \tag{11.10}$$

This is the fundamental equation governing our control actions.

As we said in the previous section, at low speeds the control strategy will first be to try to maintain the gamma angle equal to zero and also control the torque acting on the current entering the machine; at a higher speed, if the maximum current is requested, γ cannot be maintained at 0, but instead must be raised according to equation (11.11) and the corresponding diagrams, such as the one shown in Figure 11.24.

Since the angle γ is so important for the machine control, let us first recall, before discussing the control diagrams, what the physical meaning of the γ angle is.

Consider Figure 11.27. In the left-hand part an isotropic rotor of the type shown in Figure 11.19 is inserted into a synchronous machine stator (whose stator coils are not represented) while the right-hand one is a slightly modified version of Figure 11.21.

Since the machine represented has a number of pole pairs $p = 2$, the physical angles are half the corresponding electrical angles.

Once the physical angles are multiplied by two, the situation is represented as the phasor diagram shown in the right-hand part of the figure. The direction of the phasor current $\underline{I}$ (which, as discussed in the first part of this chapter, is the current flowing in the coil a of the machine) represents the direction of the rotating field produced by the set of stator currents. Therefore, in the phasor diagram, $\underline{\Psi}_a$ is aligned with $\underline{I}$.

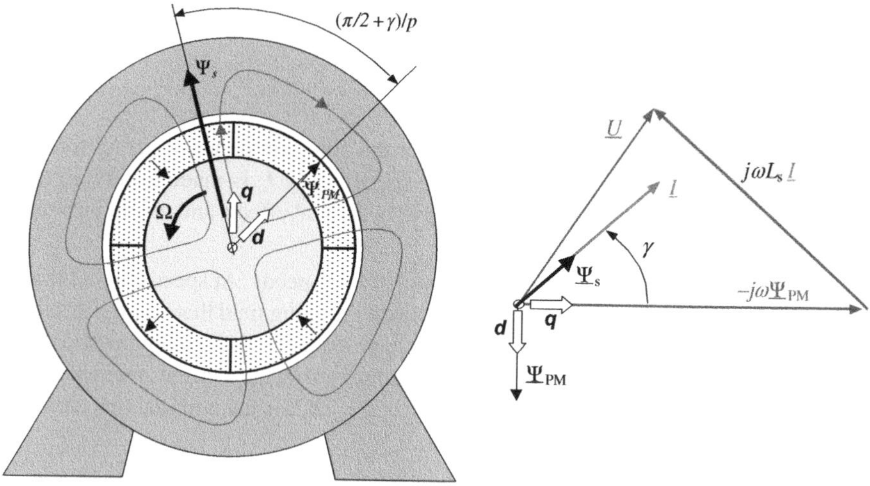

FIGURE 11.27. Correspondence between machine fields and phasor diagram.

Consequently, when $\gamma = 0$ there always exists an angle of 90 degrees between the angle of the rotor and stator fields, the latter lagging with respect to the former.

To keep $\gamma = 0$ the angular position of rotor is measured by a specific sensor, since $\underline{\Psi}_{PM}$ is constant (in amplitude and phase) in a rotor-fixed frame, the rotor position gives the $\underline{\Psi}_{PM}$ position; therefore it is possible to operate the converter feeding our motor so that there is a 90° angle between $\underline{\Psi}_{PM}$ and $\underline{I}$.

Usually, the controller of PM synchronous machines indicates the phasor $\underline{I}$, instead of with its modulus and angle, with its two components along $\boldsymbol{d}$ and $\boldsymbol{q}$ axes, which are the same as those used in the first part of this chapter and also shown in Figure 11.27: $\boldsymbol{d}$ axis is oriented with the direction of the north pole of the permanent magnets, and $\boldsymbol{q}$ axis is leading $\boldsymbol{d}$ axis by 90 electrical degrees.

Once $\underline{I}$ has been decomposed in its d and q components I_d and I_q,[8] equations (11.9) and (11.10) can be written as follows:

$$P = \frac{3}{2}\omega\hat{\Psi}_{PM}I_q, \qquad T = \frac{P}{\Omega} = \frac{3}{2}p\hat{\Psi}_{PM}I_q \tag{11.13}$$

This leads to a very effective interpretation of the role of I_d and I_q in the machine's operation: the torque is produced only by effect of I_q, while I_d has the purpose of weakening the total flux in the machine at higher speeds, and indeed, when $\gamma > 0$, it creates a magnetic field component that is opposite to the one created by the rotor's permanent magnets (check again Figure 11.27).

[8]Remember that we indicate with I the rms value of phasor $\underline{I}$, which has as amplitude $\sqrt{2}I$. Therefore it will be $\sqrt{2}I = \sqrt{I_d^2 + I_q^2}$; Moreover, $\hat{\Psi} = \sqrt{2}\Psi$.

Result: *Role of I_d and I_q in an isotropic PMSM*

In a permanent-magnet isotropic synchronous machine:

- The q-axis component I_q of a current is the sole component causing torque generation. At low speeds (below base speed), only I_q is made to flow in the machine, since nonzero I_d would not have beneficial effects and would increase copper losses.
- The d-axis component I_d of current is null at low speeds. At speeds higher than base speed, it is negative in value and creates a rotating field that counteracts the PM field, to avoid that the machine voltage exceeds the maximum allowed value U_{max}. This component causes a reduction in the current available for torque generation, since it must be $I_d^2 + I_q^2 \leq 2I_n^2$ (if I_n is the nominal rms current value).

Now that we have examined the control strategy (in Section 11.5.2) and recalled the correspondence between the phasor diagram and physical quantities, the simplified control diagram shown in Figure 11.28 can be introduced and examined.

The diagram operates as follows:

- Block 1 converts the torque request (T^*) into current amplitude (I^*) and phase (γ^*) requests: below ω_{base} angle, γ is set to zero and current $I = I_q$ is determined by equation (11.12); above ω_{base}, at a first attempt $I = I_q$ is assumed and it is verified whether *or not* a value of γ exists to satisfy equation (11.11); if this is not possible, $I = \sqrt{2}I_n = \hat{I}_n$ is assumed, γ is determined from equation (11.11) in which the equality sign is placed between the two members, $I_q = \sqrt{\hat{I}_n^2 - I_d^2}$, and the delivered torque will be lower than T^*; the voltage limit U_{max} is an input, since it can depend on quantities external to the machine itself—for example, the DC source voltage.
- Block 2 makes a coordinate transformation from modulus-and-phase into d–q components of the current request.
- Block 3 transforms the $I_d - I_q$ requests in terms of instantaneous current requests for the three machine phases; to do this, as discussed earlier in this section, a measure of actual rotor angle θ_{meas} must be received.

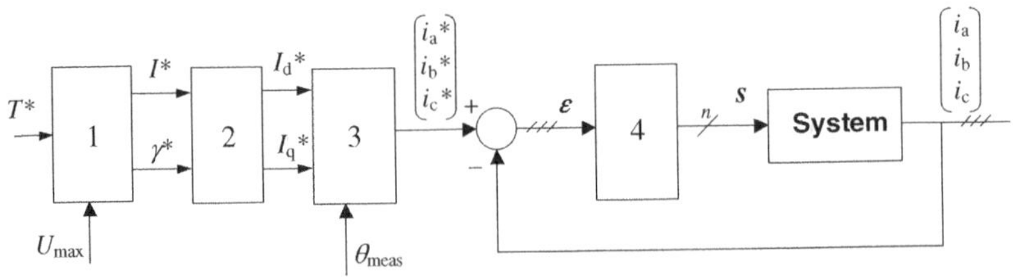

FIGURE 11.28. Simplified representation of a PMSM control diagram.

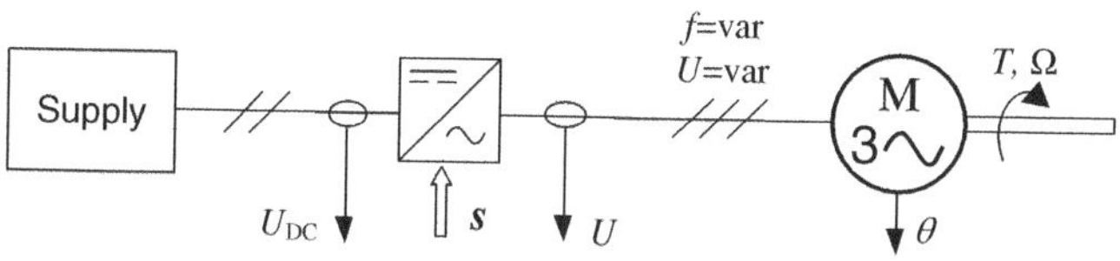

FIGURE 11.29. Representation of the system to be controlled using the control diagram of Figure 11.28, and representation of the input and output signals.

- Once the three current references are computed, they are compared with the corresponding values measured on the machine, and the corresponding errors are derived (one per phase).
- The errors are entered into block 4, the regulator, which computes the actions to be performed on the electronic converter feeding the machine to offset the errors, acting on the *n* signals entering the converter (*n* is normally equal to six).
- The *system* is the physical system, depicted in Figure 11.29: it contains the DC source, the converter, the machine, and the measures of U, U_{DC}, and θ. The measures U_{DC} and U are usually integrated into the converter, since they must already be available internally for reasonable converter control.

It must be stressed that actual PMSM controllers are more complicated than those represented in Figure 11.29, and may also have significant differences in architecture. However, the principle of operation is very often very similar to the one shown in Figure 11.28 and is therefore useful as a first reference.

More in Depth: *U_d–U_q control*

The diagram shown in Figure 11.28, where phase currents are directly regulated by block 4, can be implemented in a PPU control which is called "hysteresis control"; this has not been dealt with in this book and is not frequently used.

A more common control in important drives works in the ***d–q*** plane (thus avoiding hysteresis control), but this requires an intermediate stage in which components U_d and U_q of voltages (which have never been introduced in this book) are computed.

It can be demonstrated that a conversion can be effected between phase and ***d–q*** quantities by multiplication by a matrix, whose coefficients are a function of angle θ (under the hypothesis that $A_a + A_b + A_c = 0$):

$$\begin{pmatrix} 0 \\ A_d \\ A_q \end{pmatrix} = T(\theta) \begin{pmatrix} A_a \\ A_b \\ A_c \end{pmatrix}, \qquad \begin{pmatrix} A_a \\ A_b \\ A_c \end{pmatrix} = T^{-1}(\theta) \begin{pmatrix} 0 \\ A_d \\ A_q \end{pmatrix}$$

and this conversion can be made not only on currents but also on voltages.

Using this conversion, it is possible to introduce another control diagram, more commonly used than that of Figure 11.28:

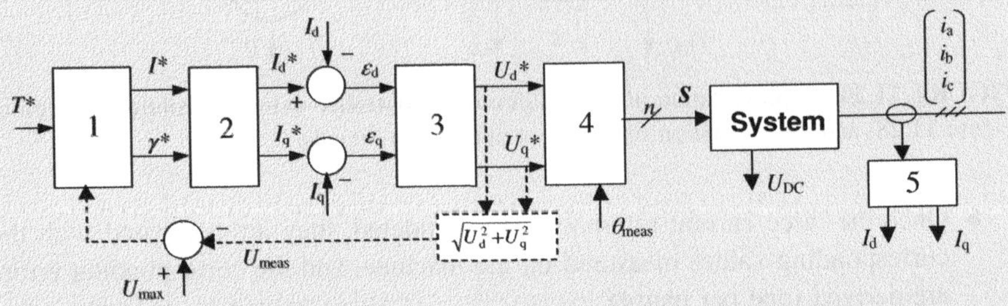

In this case the errors on I_d and I_q, (ϵ_d and ϵ_q, feed block 3, the controller, which determines the reference values of the transformed voltages U_d and U_q*; these are transformed back into phase quantities in block 4, which also determines the control signals for the inverter of the system, again the ones shown in Figure 11.28. Block 5 makes a ***d–q*** transformation on the current quantities.

The use of the *T* transformation (called Park's transformation after the scientist who first introduced it) has a lot of advantages for the study of synchronous machines and drives, since it allows analysis during transients and also allows the analysis of anisotropic machines.

It must be noted that $\sqrt{2}U = \sqrt{U_d^2 + U_q^2}$ and therefore the condition $U \le U_{max}$ can be verified by closed loop control using the measure of U from U_d and U_q, instead of open loop from (11.11), as indicated in the above diagram, with the dashed box and lines. When U tends to overcome U_{max}, a reduction in I_d is imposed, that limits the availability of I_q, and therefore of torque, according to the plots of figure (11.24).

It is finally to be noted that U_{max} is not only due to the need to keep within the maximum machine voltage, but also to the need to allow the PPU to generate the wanted voltage: in fact, as was seen in Chapter 7, the maximum voltage the PPU can generate is limited by the available DC-side voltage. Therefore, in applications in which the DC voltage can vary significantly, U_{max} is proportional to the measured U_{DC} voltage.

11.6 HISTORICAL NOTES

11.6.1 Short Biography of Ferraris and Behn-Eschemburg

Galileo Ferraris (Livorno Vercellese, 1847, Turin 1897) was an electrical engineer. He conducted research on AC circuits.

Galileo Ferraris is credited as having discovered the *rotating magnetic field*, discussed in this chapter, independently of Tesla (whose short biography is included at the end of Chapter 6), who made the same discovery in the same period.

Hans Behn-Eschemburg (Zurich 1864–1938) was a scientist and engineer. He lived in the period when the first large-scale electrification of stationary and railway systems took place. He contributed to both applications of electricity. His name is especially related to his mathematical model of synchronous machines.

11.7 PROPOSED EXERCISES

Where not specifically indicated, electromagnetic quantities are expressed in rms values.

11.1. Calculate the speed of a 50-Hz synchronous machine, if the pairs of poles p are respectively 1, 2, 3, or 4. Express the speed in revolutions per minute (rpm).

11.2. Calculate the electromotive force and the torque angle of a wye-connected 3000-kVA 6-kV synchronous generator, delivering its rated current to a load having 0.8 power factor (lagging). The per-phase synchronous reactance and resistance are 10 Ω and 0.05 Ω, respectively. Determine also the active and reactive power delivered to the load.

11.3. Calculate the per-phase electromotive force and the torque angle of a delta-connected 50-kVA, 400-V synchronous generator, delivering its rated current to a load having 0.85 power factor (lagging). The per-phase synchronous reactance and resistance are 9 Ω and 0.1 Ω, respectively. Determine also the active and reactive power delivered to the load.

11.4. Repeat exercise 11.2 for a load with 0.9 power factor (leading).

11.5. An eight-pole wye-connected synchronous machine is connected to a three-phase 3500-V (phase-to-phase) prevailing network. The synchronous reactance is 2.1 Ω, while the stator resistance can be disregarded. Calculate the electromotive force, the torque angle, the stator current, and the reactive power delivered to the network:

(a) just after paralleling

(b) if the machine delivers 1.5 MW, keeping the excitation current constant to the value required for paralleling. Determine the power factor and compare φ with β; justify this result by drawing a phasor diagram.

11.6. Consider the machine of exercise 11.5 and calculate the excitation voltage, the torque angle, and the stator current if the machine delivers 1.2 MW and 0.5 Mvar to the network:

(a) calculating φ, $\underline{I}$, and then $\underline{E}$, using phasor equations

(b) using equations (11.7)

11.7. A 900-kVA 6-kV three-phase wye-connected synchronous generator supplies a 400-kW 0.85-leading-power-factor load. The synchronous reactance is 62.5% of the "rated impedance" U^2/S_n, where U is the phase-to-phase rated voltage and S_n is the rated apparent power. The stator resistance is negligible. Calculate the synchronous reactance per phase, the excitation voltage, the torque angle, and the stator current.

11.8. A 1250-kVA 10-kV three-phase wye-connected synchronous generator has a synchronous reactance of $6\,\Omega$ and an armature resistance of $0.4\,\Omega$.

(a) If the generator delivers its rated current to a 10-kV 0.95-lagging-power-factor load, calculate the electromotive force.

(b) If the excitation current is unaltered, calculate the terminal voltage and the active and reactive power delivered to a 0.95-leading-power-factor load, at rated current.

11.9. A 6-kV three-phase wye-connected synchronous generator has a synchronous reactance of $5\,\Omega$ per phase and negligible stator resistance. If the open-circuit voltage is 7.5 kV, calculate the maximum power that can be delivered by the generator and the correspondent stator current.

11.10. A 400-V three-phase wye-connected synchronous generator has synchronous impedance of $0.2 + j4\,\Omega$ per phase. It feeds a 0.85 leading power factor load; that is, the generator generates active power but absorbs reactive power (underexcitation). Calculate the electromotive force, the torque angle, and the active and reactive power delivered to the grid, if the phase current is 20 A.

11.11. Repeat exercise 10 with 0.85 lagging power factor load (overexcitation).

12

INDUCTION MACHINES AND DRIVES

For the Instructor

This chapter first introduces the induction machine, starting with a physical analysis in a natural progression from the basic analysis described in Chapter 9, and also drawing on the reader's knowledge of electromagnetism.

This is followed by an explanation of the commonly used single-phase steady-state model of induction machines. This model, in addition to steady-state operation, is able to reproduce slow transients (in which electrical dynamics are disregarded and the only dynamic taken into account relates to the acceleration of rotating inertias). Structurally, this is the same as the power transformer-like model found in most books, already encountered in Chapter 7. All our mathematical analysis is based on this model, which allows understanding many of the induction machine features in a very simple way.

As regards induction-motor drives, however, we present a simulation using a more detailed model. This will help toward a clearer understanding of the limits of applicability of the steady-state single-phase model.

Finally, we examine single-phase induction motors to see what the most common household AC machines look like inside and to determine how they work.

Fundamentals of Electric Power Engineering: From Electromagnetics to Power Systems, First Edition.
Massimo Ceraolo and Davide Poli.

12.1 INDUCTION MACHINE BASICS

The *induction machine* is also called an *asynchronous machine*. While, in our opinion, the latter name is more correct (since induction plays a fundamental role both in synchronous and induction machines), it has the drawback of being very similar to synchronous machines, which may be confusing.

Like the synchronous machine, the induction machine is a bidirectional electromechanical converter—that is, a system able to convert electrical energy into mechanical form (when it operates as a motor) and vice versa (when it operates as a generator). It is, however, much more frequently used as a motor than as a generator, for reasons that will be discussed later.

As regards its construction, the machine stator is substantially identical to the one used for the synchronous machine, and discussed in Chapters 9 and 11 (see Figure 12.1a). For the induction machine also, the windings are normally distributed along the air gap, although for simplicity's sake in the figures of this chapter they are represented as being concentrated in a single turn. Moreover, also for this machine, the windings can have a number of pole pairs p, different from one.

The rotor is constituted by a three-phase system of windings, such as the one shown in Figure 12.1b, but with conductor bars distributed along the air gap and not concentrated in single turns. These three windings are connected to each other, or, to use a common expression, they are *short-circuited*, as in Figure 12.1b, which refers to the winding star connection.

To be more accurate, we could say that the induction machine rotors can have two possible constructions: *wound* and *squirrel cage*. The wound rotor is constituted by three coils that are star- or delta-connected, whose free ends are short-circuited. A wound rotor with a single turn per phase has the appearance of Figure 12.1b. Normally, and similarly to the synchronous machine, both stator and rotor coils are distributed along the air gap, spanned through several slots; the different turns of the same coil are connected in series using connections at the two machine ends.

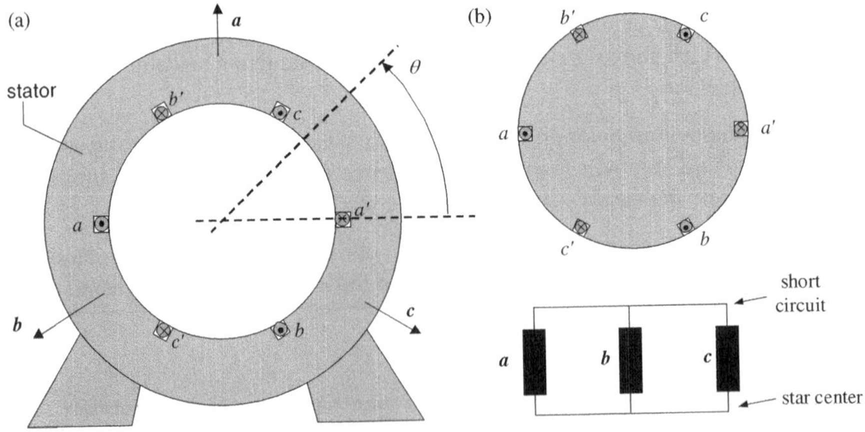

FIGURE 12.1. The basic structure of an induction machine stator (*left*) and rotor (*right*).

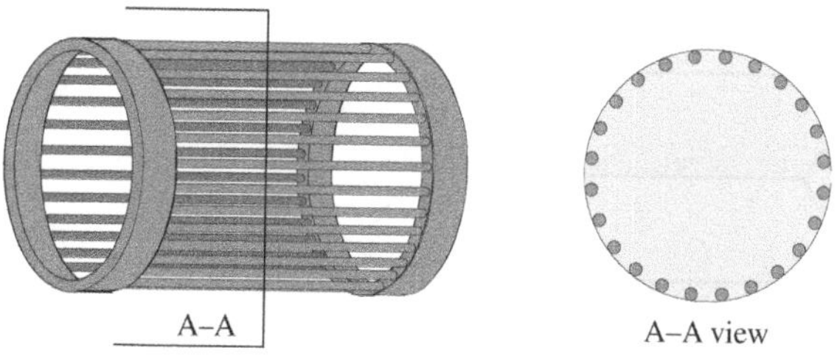

FIGURE 12.2. The structure of a squirrel-cage rotor: (a) Cross section A–A showing iron and bars. (b) Perspective view of only the squirrel cage.

Induction machines with wound rotors have the advantage of giving access, when needed, to the rotor windings, mainly for starting the machine, as discussed in the *more in depth* text of Section 12.2. The more frequent rotor construction, however, is the so-called squirrel-cage rotor, depicted in Figure 12.2.

The right-hand part of the figure shows a view orthogonal to the rotor axis, while the left-hand part is a perspective view of the conductors only. The longitudinal conductor bars are connected to the two front rings. This construction makes the rotor equivalent to a multiphase system of windings: each phase is constituted by a single bar; one of the two rings is the centre of the star connection of the bars, white the other is the short-circuit connection.

The squirrel-cage construction gives no access to the rotor conductors, but is much simpler, more rugged, and cheaper than the wound rotor. In fact, the wound construction nowadays tends to be phased out.

In cases in which frequent starting operations or even a continuous variation of speed is required (e.g., in the operation of a lift[1] or an electric car) a much more flexible operation of an induction machine can be obtained by varying stator feeding frequency, as discussed in Section 12.4. This, however, requires the addition of a complex power electronic converter between the machine and the power supply.

To understand the operating principle of an induction machine, imagine its stator winding, whose rotor is initially at standstill, connected to an external three-phase source of power, which, for our purposes, is constituted by an ideal three-phase set of ideal sinusoidal sources (Figure 12.3).

We analyse the machine under the assumption that the electric parts of the machine are in steady state; that is, all electrical transients can be ignored, and all the AC quantities are sinusoidal. Under this assumption, we understand, from what we know from the study of Chapters 5 and 6, that all voltages and currents in the machine are sine waves. In particular, we can state that the three currents drawn by the machine from the source are sinusoids. We can also assume that the machine has full three-phase symmetry: there is no reason to think that what happens in phase a should be different from what happens in phase b or c, except for the time displacement induced by one of the three sources.

[1]All lifts are subject to very frequent start-ups. Lifts of very high buildings, in addition, require gradual acceleration/deceleration phases, which require variable-speed operation of the machine.

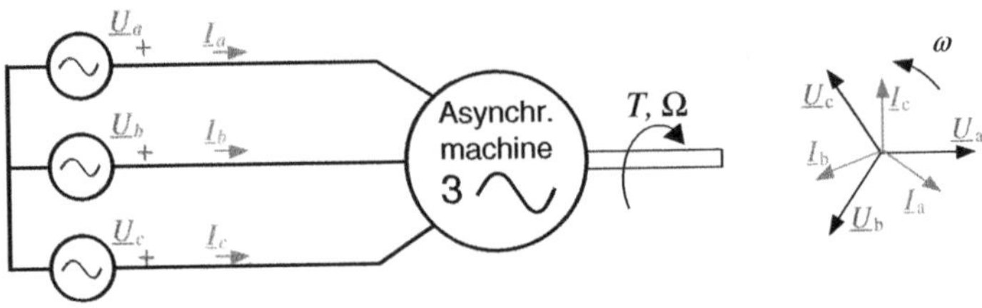

FIGURE 12.3. An induction machine, in steady state, absorbs a three-phase set of currents, because of the symmetry of the feeding voltages and of its inner construction.

Therefore, after this simple qualitative reasoning, we can conclude that the three currents constitute a three-phase balanced set, as shown also in Figure 12.3.

This set of currents, because of the rotating field theorem (Section 11.2.1), will create a rotating field $\boldsymbol{B}_s$ that produces—in any of the rotor windings—sine flux linkages which, in turn, induce a three-phase set of EMFs, by Faraday's law.

These EMFs induced in the rotor, since the rotor windings are short-circuited, create a three-phase set of currents. These, in turn, generate a new rotating field which combines with the one produced by the stator currents.

So the global operation of the machine is determined by the total flux density field $\boldsymbol{B}$—that is, the combination of stator and rotor ones:

$$\boldsymbol{B} = \boldsymbol{B}_{\text{stat}} + \boldsymbol{B}_{\text{rot}}$$

The EMF created on any of the rotor conductors can be evaluated using equation (9.2):

$$e_{PQ} = |\boldsymbol{v}_{\text{rel}} \times \boldsymbol{B}| l = v_{\text{rel}} B_{\perp} l \tag{9.2}$$

Now the speed $\boldsymbol{v}_{\text{rel}}$ is the speed of the rotor conductor as measured in a frame that is rotating along with $\boldsymbol{B}$, or a B-fixed frame,[2] and $B_{\perp}$ is the component orthogonal to $\boldsymbol{v}_{\text{rel}}$, and therefore is the radial component B_r.

The reference polarity of $\boldsymbol{e}_{PQ}$ should be set in such a way that $\boldsymbol{v}$, $\boldsymbol{B}$, and a vector going from $\boldsymbol{Q}$ to $\boldsymbol{P}$ have a right-hand orientation; therefore the "+" marking on e must be set using the right-hand rule, as shown in Figure 12.4, for which $\boldsymbol{B}$ is assumed to be rotating (in a stationary frame of reference) counter clockwise.

The stator windings can be created using just two poles, or several pole pairs. The construction of multi-pole-pair windings is the same as that of synchronous machines and some details can thus be found in Chapter 12. Moreover, the stator coils are distributed along the air gap, so the voltage produced by N turns is $k_w N$ times the voltage of a single turn, where k_w is the dimensionless factor discussed in Section 9.3.3.

[2]The rotating field theorem states that a balanced set of currents will generate an induction wave on the air gap, which is sinusoidal and rotates at a speed that for a p pole-pair machine is $\Omega_0 = \omega/p$, ω being the radian frequency of the currents. When we state here "a B-fixed frame," therefore, we refer to a frame of reference that is rotating at the same angular speed Ω_0 and the same direction as the rotating field. It will be soon seen that both fields $\boldsymbol{B}_{\text{stat}}$ and $\boldsymbol{B}_{\text{rot}}$ rotate at the same speed Ω_0.

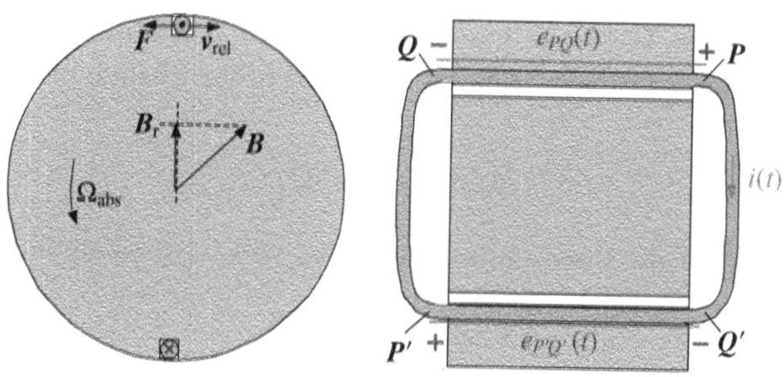

FIGURE 12.4. Induction and force generation on a rotor of an induction machine (rotor rotating counterclockwise, v_{rel} rotor speed in a frame rotating along with $\boldsymbol{B}_s$).

Therefore, the total voltage amplitude of a coil containing N distributed turns and p pole pairs can be expressed using equation (9.11), that is:

$$\hat{E}_{coil} = kN\omega BA = kNp\Omega BA$$

Any of the rotor conductors is thus subject to field $\boldsymbol{B}$, which belongs to the plane orthogonal to the conductors itself. Being traversed by currents, each rotor conductor is therefore subject to Lorentz's force (9.5):

$$\boldsymbol{F} = \boldsymbol{I} \times \boldsymbol{B} \cdot l, \qquad F = I \cdot B_{\perp} \cdot l \tag{9.5}$$

in which the direction of $\boldsymbol{F}$ is determined by using the right-hand rule with vectors $\boldsymbol{I}$ (having the same direction as the conductor and arrowhead at the same end as the arrowhead of I) and $\boldsymbol{B}$. The force has the direction shown in Figure 12.4. It tends to move the rotor in a direction which opposes the flux variation—that is, in the example given, counterclockwise.

Now that the rotor is rotating, what happens inside the machine? The reasoning above can be repeated: there still exist three stator currents which create a stator rotating field that generates three EMFs on the rotor. These, in turn, generate three currents, and forces, on the rotor conductors.

The field rotational speed is the same as before, since it depends only on the angular frequency ω of the supply voltages and on the number of the machine's pole pairs p:

$$\Omega_0 = \omega/p$$

If r is the rotor radius,[3] the speed of the rotor conductors is now

$$v_{rel} = (\Omega_0 - \Omega)r = (\omega/p - \Omega)r$$

[3]More precisely, r is the distance of the conductor's cross section center from the rotor's cross section center.

By effect of this speed, EMFs are induced in the rotor, constituting a three-phase set. The angular frequency of these voltages is related to the relative movement of rotor and stator field:

$$\omega_{rot} = p(\Omega_0 - \Omega)$$

and therefore each of the three induced voltages will have as amplitude

$$\hat{E}_{coil} = kN\omega_{rot}BA$$

In the rotor coils, these voltages will induce currents that constitute a three-phase balanced set, which in turn will generate a rotating field. This, in a frame reference rotating with the rotor, will rotate at the speed

$$\omega_{rot}/p = \Omega_0 - \Omega \tag{12.1}$$

The field produced by the rotor currents, evaluated in a stationary frame, will then rotate at a speed equal to Ω_0—that is, the same speed as the stator field.

This is an important conclusion:

Result: *Rotational speed of the magnetic field produced by the rotor*

In an induction machine, the rotating field generated by the rotor's currents rotates at the same speed as the field produced by the stator's currents. This speed is called *synchronous speed* and is equal to the angular frequency of the source, divided by the number of machine pole pairs.

As long as the rotor speed Ω remains lower than the synchronous speed Ω_0, forces will be generated on the rotor conductors which will globally produce a positive torque that will tend to accelerate the rotor. But when the actual speed approaches Ω_0, the reason for the torque to be generated (i.e. a relative motion between rotor and rotating fields) will tend to vanish. When the speed is exactly synchronous, the torque will be zero, because in a rotor-fixed frame of reference, the machine field will be seen to be stationary, and therefore no EMF is generated.

In the next section a very simple and effective mathematical model will be presented that allows the electrical and mechanical quantities of the machine to be quantitatively evaluated at the various speeds.

12.2 MACHINE MODEL AND ANALYSIS

In the previous section, a description of the physical phenomena occurring within the induction machine was shown under the hypothesis that the machine operates in steady state, it is physically symmetric, and it is connected to a three-phase balanced system of voltages.

Under these assumptions, the currents absorbed by the machine also constitute a balanced three-phase system of currents, and the behaviour of the machine can be analysed using the *single-phase equivalent* concept (see the relevant section of Chapter 6 for a reminder of this concept).

We will now examine this equivalent circuit and its usage. It is not derived from electromagnetic equations so as to reduce the mathematical burden for the reader.

Before introducing the single-phase equivalent of an asynchronous machine, a very important quantity is to be defined, the so-called *slip s* of the machine:

$$s = \frac{\Omega_0 - \Omega}{\Omega_0} \tag{12.2}$$

When a machine is at a standstill, its speed Ω is zero and the slip is unity. When the rotor rotates at the same speed as the rotating fields Ω_0—that is, when no current is induced in the rotor and therefore no torque is produced—slip is zero. In all the other cases, some power is converted between the electrical circuit and the mechanical shaft of the machine.

Using equation (12.1) and (12.2), it can be immediately demonstrated that the frequency of rotor currents and EMFs is

$$f_r = \omega_{\mathrm{rot}}/2\pi = sf \tag{12.3}$$

where f is the frequency of electrical (stator) supply. Equation (12.2) can also be written expressing speeds in terms of revolutions per minute (rpm):

$$s = \frac{n_0 - n}{n_0} \tag{12.4}$$

where

$$n = \frac{60\Omega}{2\pi} \text{ is the rotor speed in rpm}$$

$$n_0 = \frac{60\Omega_0}{2\pi} = \frac{60\omega}{2\pi p} = \frac{60f}{p} \text{ is the synchronous speed in rpm}$$

Example 1. A six-pole 60-Hz induction motor runs with a slip of 4%. Determine the synchronous speed n_0, the rotor speed n, the frequency f_r of rotor currents, and the speed of the rotor rotating field with respect to the rotor (n_{rr}) and with respect to the ground (n_r).

Expressing speeds in revolutions per minute (rpm), we obtain

$$n_0 = \frac{60f}{p} = \frac{60 \cdot 60}{3} = 1200\,\text{rpm}$$

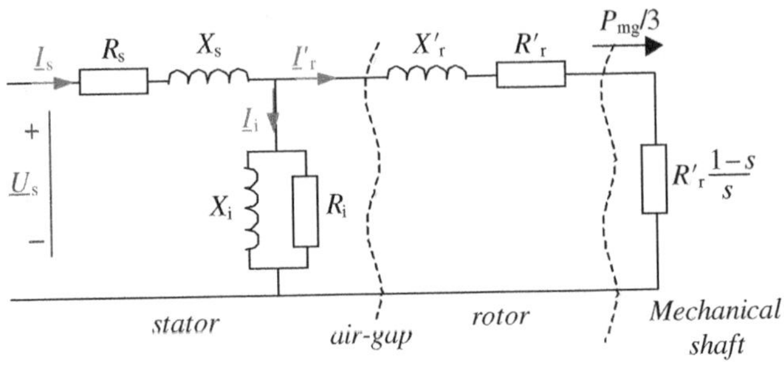

FIGURE 12.5. Single-phase equivalent circuit of an induction machine.

From equation (12.4), $n = (1 - s)n_0 = (1 - 0.04) \cdot 1200 = 1152$ rpm.

$$f_r = sf = 0.04 \cdot 60 = 2.4 \text{ Hz}$$

$$n_{rr} = \frac{60f_r}{p} = \frac{60 \cdot 2.4}{3} = 48 \text{ rpm}$$

$$n_r = n_{rr} + n = 48 + 1152 = 1200 \text{ rpm} = n_0$$

Hence stator and rotor fields are synchronous.

$$\Omega_0 = \frac{2\pi f}{p} = \frac{2\pi n_0}{60} = 125.7 \text{ rad/s}$$

$$\Omega = (1 - s)\Omega_0 = \frac{2\pi n}{60} = 120.6 \text{ rad/s}$$

Using the slip, a single-phase equivalent of a synchronous machine can be created as shown in Figure 12.5.

Readers will have certainly noted a strong similarity to the transformer circuit. Indeed the very mechanism of power transfer from the stator to the rotor has a lot in common with the power transfer from the primary to secondary windings of a transformer. Readers are invited to find similarities between the two machines, after reading and studying this chapter.

The two curved dashed lines indicate important physical transformations: The leftmost part of the circuit models the stator quantities, the air-gap dashed curve is then crossed, and the circuit section describes what happens in the rotor. The right-most section of the circuit represents the mechanical shaft.

Let's first analyse the circuit in terms of power. The powers shown in the circuit are, obviously, always one-third of those circulating in the actual machine. Therefore, the power $U_s\, I_s \cos\varphi$ entering the leftmost terminals is one-third of the total power P_{el}.

Part of the power entering the machine is dissipated in the resistance of the stator coils, and this power, usually referred to as *stator copper losses*, is

$$P_{ls} = 3R_s I_s^2$$

Further dissipation occurs in the stator and rotor iron because of parasitic currents and hysteresis. This is simulated by the resistance R_i, and the power dissipated as a result constitutes the so-called *(stator) iron losses*[4]:

$$P_{li} = 3R_i I_i^2$$

The power then traverses the air gap; and the power entering the rotor, the so-called air-gap power, is

$$P_{ag} = P - P_{ls} - P_{li}$$

Part of the air-gap power is dissipated in the rotor coils, and the power lost (*copper rotor losses*) is

$$P_{lr} = 3R_r' I_r'^2$$

It must be noted that the resistance R_r' and the current I_r' are not exactly the resistance and current of the rotor coils, but are related to them. It can be shown that

$$R_r' = \alpha^2 R_r, \quad I_r' = \frac{1}{\alpha} I_r, \qquad \text{where } \alpha = \frac{k_s N_s}{k_r N_r} \tag{12.5}$$

in which N_s and N_r are the turns of stator and rotor coils respectively, k_s and k_r are the corresponding coil distribution factors, and α is the *equivalent turns ratio*.

These expressions imply that the power dissipated in the three resistors R_r' by effect of the three currents I_r' is exactly the same power which is dissipated in the rotor coils:

$$P_{lr} = 3R_r' I_r'^2 = 3R_r I_r^2$$

The rightmost part of the circuit models the power behaviour of the shaft. Indeed, the component indicates that a resistor with resistance

$$R_{m,eq} = R_r' \frac{1 - s}{s} \tag{12.6}$$

is a fictitious resistor, since the "dissipated" electrical power is actually converted into mechanical form. The rightmost dashed curve in the circuit is the borderline between

[4]The subscript i stands for iron.

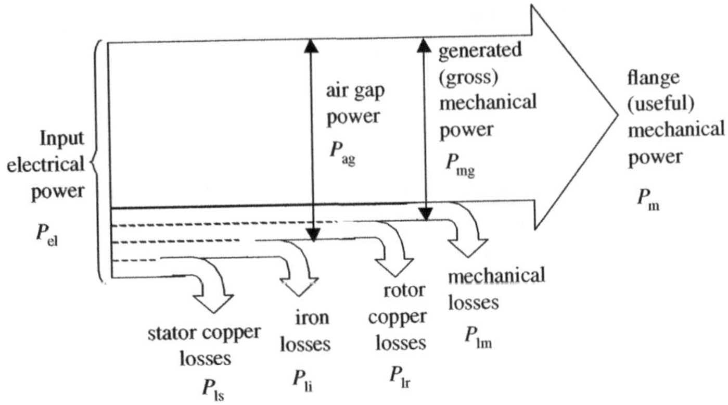

FIGURE 12.6. A qualitative chart illustrating different types of machine losses.

the electric domain of the machine and its mechanical domain. If the mechanical power generated in the machine is called P_{mg}, it can thus be written as

$$P_{mg} = 3R_{m,eq}I_r'^2 \tag{12.7}$$

From this gross mechanical power, with the obvious deduction of mechanical losses in the bearings and in the resistance encountered by the rotor by effect of the air surrounding it, we obtain the useful (net) mechanical power P_m, available at the machine flange.

The power flows in the machine are summarized in Figure 12.6.

By substituting (12.6) into (12.7) and observing[5] that $P_{mg} = P_{ag} - P_{lr}$, it can easily be demonstrated that

$$P_{ag} = P_{lr}/s \tag{12.8}$$

In Section 5.4 it was demonstrated that the instantaneous power flowing in three-phase balanced systems is perfectly constant over time. This is valid for all powers discussed in this section, including the mechanical power delivered to the load; $p_{mg}(t)$ being constant, if the mechanical load and the machine speed are constant, also the delivered torque is not fluctuating. This is one of the main advantages of three-phase induction machines with respect to its single-phase counterpart, which will be briefly discussed in Section 12.5.

Example 2. A 50-Hz three-phase induction motor has a nominal voltage[6] of 400 V. When operating at full load, it develops its nominal mechanical power of 18 kW at

[5]Rotor iron losses are usually negligible, due to the very low frequency of rotor EMFs [see equation (12.3) and consider that s is rarely higher than 0.07–0.08].

[6]Readers should become accustomed to the fact that whenever a three-phase voltage value is referred to, if nothing else is specified, this value indicates, the rms value of the phase-to-phase voltage.

705 rpm, absorbing a line current of 35 A and an electrical power of 21 kW from the electric grid. Calculate:

(a) the synchronous speed n_0
(b) the slip s
(c) the power factor $\cos\varphi$
(d) the torque T
(e) the efficiency η

With $p=1$, the synchronous speed n_0 would be $60f=3000$ rpm.
Since $n=705$ rpm, we have $3000/n=4.25$, and then $p=4$.

$$n_0 = \frac{60f}{p} = \frac{60 \cdot 50}{4} = 750 \text{ rpm}$$

$$s = \frac{n_0 - n}{n_0} = \frac{750 - 705}{750} = 0.06$$

$$\cos\varphi = \frac{P_{\mathrm{el}}}{\sqrt{3}UI} = \frac{21,000}{\sqrt{3} \cdot 400 \cdot 35} = 0.866$$

$$\Omega = \frac{2\pi n}{60} = 73.82 \text{ rad/s}$$

$$T = \frac{P_{\mathrm{m}}}{\Omega} = \frac{18,000}{73.82} = 243.8 \text{ Nm}$$

$$\eta = \frac{P_m}{P_{\mathrm{el}}} = \frac{18,000}{21,000} = 85.71\%$$

Example 3. In a four-pole 50-Hz induction motor, the power crossing the air gap P_{ag} and the gross mechanical power generated by the machine P_{mg} are respectively 22 kW and 20.8 kW. Calculate the copper rotor losses and the slip. If the rotational losses P_{lm} are 450 W, determine the net output torque.

Disregarding iron rotor losses (see footnote 5), copper rotor losses are the difference between the power crossing the air gap and the gross mechanical power:

$$P_{\mathrm{lr}} = P_{\mathrm{ag}} - P_{\mathrm{mg}} = 22 - 20.8 = 1.2 \text{ kW}$$

From (12.8), $s = P_{\mathrm{lr}}/P_{\mathrm{ag}} = 1.2/22 = 0.0545$.

The net mechanical power can be calculated by subtracting rotational losses from P_{mg}:

$$P_{\mathrm{m}} = P_{\mathrm{mg}} - P_{\mathrm{lm}} = 20.8 - 0.45 = 20.35 \text{ kW}$$

$$\Omega = \frac{2\pi f}{p}(1 - s) = 148.5 \text{ rad/s}$$

$$T = \frac{P_\text{m}}{\Omega} = \frac{20{,}350}{148.5} = 137.0 \text{ Nm}$$

Example 4. A 400-V 50-Hz three-phase wye-connected induction motor has a $0.7 + j1.4\,\Omega$ per phase stator impedance. The rotor impedance referred to stator is $0.6 + j1.5\,\Omega$ per phase. The magnetizing reactance X_i is $40\,\Omega$ and the transverse resistance R_i is $150\,\Omega$.

At 4% slip, calculate the input (stator) current, the power factor, the power crossing the air gap, the mechanical power, and the efficiency.

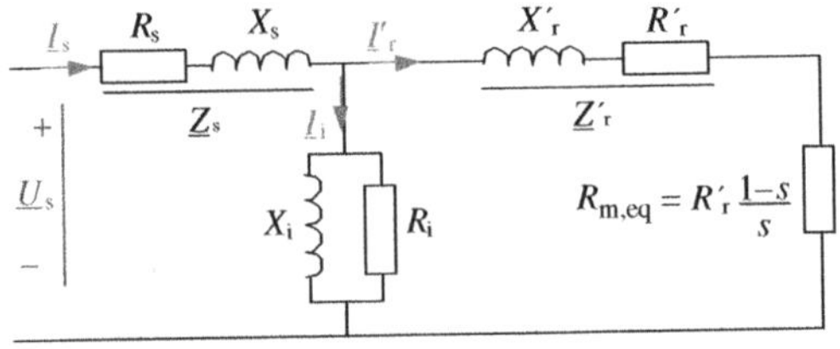

$$\underline{Z}_\text{s} = 0.7 + j1.4\,\Omega$$

$$\underline{Z}'_\text{r} = 0.6 + j1.5\,\Omega$$

$$R_\text{m,eq} = R'_\text{r}(1 - s)/s = 0.6 \cdot (1 - 0.04)/0.04 = 14.4\,\Omega$$

$$\underline{Z}'_\text{r-tot} = \underline{Z}'_r + R_l = 15 + j1.5\,\Omega \qquad (\text{also } R'_\text{r}/s + jX'_\text{r})$$

The total impedance seen by the stator is

$$\underline{Z}_\text{tot} = \underline{Z}_s + jX_i \| R_i \| \underline{Z}'_\text{r-tot} = 12.27 + j6.432\,\Omega = 13.86\angle 27.66^\circ$$

Hence:

$$\cos\varphi = \cos(27.66^\circ) = 0.886$$

$$|\underline{U}_\text{s}| = \sqrt{2} \cdot 400/\sqrt{3} = \sqrt{2} \cdot 230.9 \text{ V}$$

$$\underline{I}_\text{s} = \underline{U}_\text{s}/\underline{Z}_\text{tot} = 20.88 - j10.94 \text{ A}$$

$$\underline{I}'_\text{r} = (\underline{U}_\text{s} - \underline{I}_\text{s} \cdot \underline{Z}_\text{s})/\underline{Z}'_\text{r-tot} = 19.44 - j3.382 = \sqrt{2} \cdot 13.95 \text{ A}\angle -9.87^\circ$$

$$P_{mg} = 3 \cdot R_l \cdot I_r'^2 = 3 \cdot 14.4 \cdot 13.95^2 = 8409 \text{ W}$$

$$P_{ag} = P_{mg}/(1 - s) = 8760 \text{ W}$$

$$P_{el} = 3 \cdot \text{Re}\,(\underline{U}_s \cdot \underline{I}_s^*)/2 = 10,228 \text{ W}$$

$$\eta = P_{mg}/P_{el} = 8409/10,228 = 82.22\%$$

Let us now briefly consider the meaning of the reactances present in the machine equivalent circuit:

- reactance $X_s = \omega L_s$, where L_s is the proportionality coefficient between the stator current and the stator leakage linked flux—that is, the quote of the linked flux created by the stator currents that does not cross the air gap, and therefore does not interact with the rotor;
- reactance $X_r' = \alpha^2 \omega L_r$, where L_r is the proportionality coefficient between the rotor current and the rotor leakage linked flux—that is, the quote of the linked flux created by the rotor currents that does not cross the air gap, and therefore does not interact with the stator;
- $X_i = \omega L_i$ where L_i is the proportionality coefficient between the stator currents and the linked flux that crosses the air gap.

The equivalent circuit has longitudinal components R_s, X_s, R_r', and X_r' and transverse components X_i and R_i. In normal machine operation, all the longitudinal impedances are much smaller than the transverse one. Only when s is near to one is the situation different, since the load resistor $R_{m,eq}$ is null. In this situation, which occurs at start-up, very large currents are drawn from the mains and flow through the machine, a situation which is similar to short-circuit conditions in power systems. For this reason, the longitudinal impedance $(R_s + R_r') + j(X_s + X_r')$ is also called "short-circuit impedance" ($\underline{Z}_{sc}$). In fact, dividing the supply voltage by $\underline{Z}_{sc}$, we obtain the "short-circuit current," which corresponds (neglecting $\underline{I}_i$) both to the starting current and to the current absorbed by the machine when it is kept blocked and is fed from the mains. As discussed below, this test is used by engineers to evaluate the numerical values of the equivalent circuit parameters; obviously the test is operated at reduced voltage, in order to avoid damages due to the high currents that would otherwise circulate.

This gives us a first indication that the start-up currents of this kind of machine are much larger than the currents occurring in steady state, and therefore either frequent start-ups must be avoided or the machine must be sized to withstand these larger currents. A typical application in which the machine operates with frequent start-ups is in lifts.

Since the single-phase equivalent circuit contains only reactors and resistors, it is obvious that in all its operating conditions the machine will absorb active and reactive power, and therefore its power factor is always lagging.

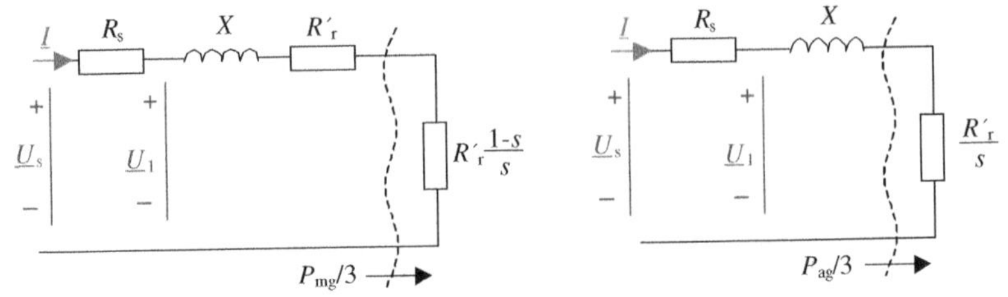

FIGURE 12.7. Simplified circuits of an induction machine.

Indeed, the statement that the machine "always absorbs active power" needs to be corrected since the load resistance is a fictitious one and, for negative slips, it has a negative value. Under these conditions—that is, when the machine rotor rotates at a speed Ω larger than Ω_0—the slip s becomes negative, the resistor $R_{m,eq}$ (equation 12.6) actually delivers power, and the machine operates as a generator.

Even when the machine generates active power, however, it still absorbs reactive power, since the reactive components of the equivalent circuit remain positive.

The single-phase equivalent enables important parameters to be determined as a function of the rotational speed Ω of the machine rotor, like the current absorbed from the supply network and the generated mechanical torque.

Here, for simplicity's sake, this is computed disregarding the effects of the transversal components R_i and X_i. This will introduce some error, but the results obtained are qualitatively correct. The simplified circuit can be expressed in one of the two forms shown in Figure 12.7, in which, obviously, $X = X_s + X_r'$. Note that if the two resistances R_r' and $R_{m,eq}$ are summed, the resulting resistor will absorb P_{ag}, which is equal to the sum of the generated mechanical power and the rotor's copper losses.

We want now to derive the absorbed current and produced torque under the condition $U_s = \text{const}$. Indeed, as a further approximation, we consider here that the voltage U_1 instead of U_s is constant.

Using the right-hand circuit in Figure 12.7, the following can be written:

$$I^2 = \frac{U_1^2}{X^2 + R_r'^2/s^2}, \qquad P_{ag} = 3\frac{R_r'}{s}I^2 = 3\frac{R_r'}{s}\frac{U_1^2}{X^2 + R_r'^2/s^2} = 3\frac{U_1^2 R_r' s}{s^2X^2 + R_r'^2} \tag{12.9}$$

and

$$T_{mg} = \frac{P_{ag}}{\Omega_0} = \frac{P_{mg}}{\Omega} = \frac{3}{\Omega_0}\frac{U_1^2 R_r' s}{s^2X^2 + R_r'^2} \tag{12.10}$$

in which the symbol T_{mg} indicates the "mechanical, generated" torque. The useful torque $T_m = P_m/\Omega$ will be lower than this, by effect of mechanical losses P_{lm}.

If we imagine the rotor speed Ω to go from zero to Ω_0, the corresponding slip will go from one to zero; the denominator of the fraction describing I^2 will steadily grow, and therefore the current I will decrease monotonically.

To analyse the shape of $T_{mg}(\Omega)$, consider first that two important points of the curve are those corresponding to $s=0$ and $s=1$. It is obviously

$$T_{mg}(s=0)=0, \qquad T_{mg0}=T_{mg}(\Omega=0)=\frac{3}{\Omega_0}\frac{U_1^2R_r'}{X^2+R_r'^2}=\frac{3}{\Omega_0}\frac{U_1^2R_r'}{Z^2}$$

T_{mg0} is normally called *starting torque* of the machine.

It is also of interest to evaluate whether between these two points there is a peak. This is done equating to zero the first derivative of $T_{mg}(s)$:

$$\frac{\partial T_{mg}}{\partial s}=0 \Rightarrow \hat{s}=R_r'/X, \qquad \hat{T}_{mg}=T_{mg}(\hat{s})=\frac{3U_1^2}{2\Omega_0 X} \tag{12.11}$$

It can be easily verified that it is always, whenever $s>0$, $\hat{T}_{mg}\geq T_{mg0}$ and therefore the curve has a maximum for $s=\hat{s}$. The corresponding angular speed will be called $\hat{\Omega}$. Note that while $\hat{s}$ depends on the rotor's resistance, $\hat{T}_{mg}$ does not.

$\hat{T}_{mg}$ is normally called the *pull-out torque* of the machine.[7]

Example 5. For the motor of Example 4, operating at $U_1=U_{nom}$, calculate the maximum torque $\hat{T}_{mg}$, the corresponding slip ($\hat{s}$), and mechanical power (P_{mg}), the supply frequency, being 50 Hz and the pole pairs two.

$$X=X_s+X_r'=1.4+1.5=2.9\,\Omega$$

$$\hat{s}=R_r'/X=0.6/2.9=0.207$$

$$\Omega_0=2\pi f/p=157.1\text{ rad/s}$$

$$\hat{T}_{mg}=\frac{3U_1^2}{2\Omega_0 X}=\frac{2\cdot 230.9^2}{2\cdot 157.1\cdot 2.9}=175.6\text{ Nm}$$

$$P_{mg}=\Omega_0(1-\hat{s})\hat{T}=21879\text{ W}$$

Please note that the equation used to calculate $\hat{T}_{mg}$ does not consider R_s. More in general, it is possible to demonstrate that

$$T_{mg}=\frac{3}{\Omega_0}\frac{U_1^2R_r's}{s^2X^2+(sR_s+R_r')^2}$$

$$\hat{s}=\frac{R_r'}{\sqrt{R_s^2+X^2}}, \qquad \hat{T}_{mg}=\frac{3U_1^2}{2\Omega_0(R_s+\sqrt{R_s^2+X^2})}$$

Typical shapes of the absorbed current (rms value) and the generated mechanical torque, as a function of the rotational speed, are shown in Figure 12.8.

[7]In reality the derivative of T_{mg} is zero also for $\tilde{s}=R_r'/X$. This corresponds to the torque minimum shown in figure 12.8 occurring at speed equal to $\tilde{\Omega}$

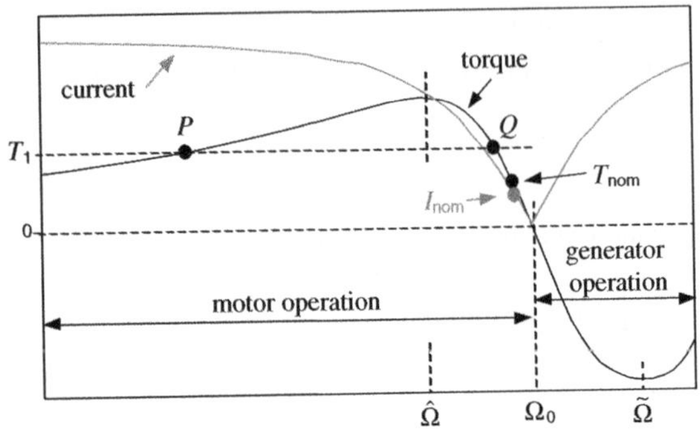

FIGURE 12.8. A typical torque–speed curve of an induction machine.

These curves were derived using realistic numerical parameters of a 50-kW squirrel-cage machine, which has a pull-out torque that is 2.35 times the starting torque. The starting torque can be higher than the value depicted in the figure, but this is obtained using a larger rotor resistance, thus reducing machine efficiency. Conversely, in order to have high efficiency, the copper losses and the rotor resistance must be small, but this reduces the starting torque (if pull-out torque is left unchanged).

The torque, as expected, becomes zero when the synchronism speed Ω_0 is reached, then it reverses, s becomes negative, and the machine operates as a generator. Obviously, for the generator region to be actually reached, a mechanical motor (i.e., a device able to supply mechanical power) should attached to the machine shaft.

Normally, however, as already noted, induction machines are operated as motors. The curves shown in Figure 12.8 clearly show that the normal operation zone of the machine is between the pull-out torque (either positive or negative) and the synchronous speed. For example, the torque T_1 can be delivered at the speed corresponding to points ***P*** and ***Q***; but point ***P***, while corresponding to a much lower mechanical power, implies the absorption of a much higher current and losses! In the next section it will be shown that by using a motor drive it is possible to deliver a given torque at different speeds while absorbing roughly the same current from the supply.

The nominal torque T_{nom} normally indicates the torque that can be delivered continuously, without damage; its product by the corresponding speed will give the *nominal* machine power. T_{nom} is much lower than the maximum torque $\hat{T}_{mg}$; for example, it could be $\hat{T}_{mg}/3$. The speed at which nominal torque and powers are delivered ("full load" condition) is the machine's nominal speed; the current absorbed when the nominal power is delivered is the nominal current I_{nom} and is typically 5–7 times lower than the starting current.

Comparing equations (12.9) and (12.10), it is possible to observe that

$$\frac{sT}{I^2} = \frac{3R_r'}{\Omega_0} = \text{constant} \qquad (°)$$

for *any working condition*. This observation is very useful to correlate T_{mg0}, $\hat{T}_{mg}$, and T_{nom} with their correspondent currents.

Example 6. At full load, an induction motor has a nominal slip of 1.5%. The starting current is six times the nominal one. Calculate the ratio between the starting and the nominal torque.

From equation (°) we obtain

$$\frac{T_{mg0}}{T_{nom}} = \left(\frac{I_{starting}}{I_{nom}}\right)^2 \frac{s_{nom}}{s_{starting}} = 6^2 \frac{0.015}{1} = 0.54$$

The zone between $\Omega = 0$ and $\Omega = \hat{\Omega}$ is normally referred to as an unstable region, while the region between $\hat{\Omega}$ and Ω_0 is considered stable. This is not completely correct since stability depends on the dynamic behaviour of a system, while the torque curves we are discussing are stationary. However, there is some truth in it: In the more common cases, indeed, equilibrium points between the torque characteristic of the machine and of the load are stable only (in the machine operating as a motor) between $\hat{\Omega}$ and Ω_0.

The operation of an induction machine as a motor with its mechanical load is depicted in Figure 12.9, which shows both the motive torque T_m delivered by the machine and the resistive one absorbed by the load T_{load} (a fan is shown as an example).

Just to give a qualitative explanation of "stable and unstable regions" of the torque curve of the machine, consider that the system operates at the equilibrium point ***P*** between the machine torque and a possible load torque T_{unst}. Should there be a slight increase in Ω, the machine torque becomes larger than the load, and the speed tends to increase further. Also, should there be a speed decrease, the system tends to move far from the equilibrium point. This does not happen in the case of point ***P*** and load curve T_{st}, nor in the case of point ***Q*** of intersection with curve T_{load}, which shows stable behaviour.

From Figure 12.9 it is possible to form an idea of a starting-up transient of the machine, using the mechanical equation

$$T_m - T_{load} = J\dot{\Omega}$$

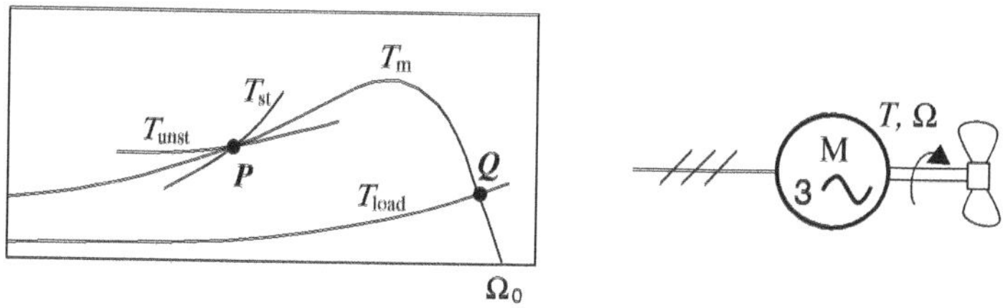

FIGURE 12.9. Machine and load torques of an induction machine with a mechanical load.

where J is obviously the combined moment of inertia of machine and mechanical load. The difference between the motive and the load torque is called *accelerating torque*, since it is proportional to the rotational acceleration $\dot{\Omega}$.

It must, however, be clarified that the shapes shown in Figures 12.8 and 12.9 are valid only as a first approximation, since they are computed using the equivalent circuit that was drawn assuming *steady-state operation* of the machine, while during the *transient*, especially the first part, their actual behaviour is rather different.

The shape of the torque of an induction machine has the inconvenience that its start-up value is smaller—often much smaller—than the peak torque and that during start-up the currents absorbed by the machine are very high. This can be mitigated by the use of wound-rotor machines or can be solved, but at greater cost, by adopting electric drives instead of simple machines.

As seen earlier, wound–rotor machines allow access to the rotor windings, to connect them electrically to a stationary circuit, usually for limited durations. Consider now that according to equation (12.11) the slip at which the machine torque has its maximum is $\hat{s} = R_r'/X = \alpha^2 R_r/X$ and is therefore proportional to the rotor resistance, while the corresponding maximum torque $\hat{T}$does not depend on this resistance. Wound-rotor machines can thus be started by inserting, in series with the rotor windings at slow speed, additional resistors $R_1, R_2, \ldots$ which are progressively bypassed, as far as the speed increases (Figure 12.10). This is done by collecting currents to the rotor coils through slip-ring/brush coupling, of the same type as those present in synchronous machines, using three rings to obtain the three-phase set of currents.

When the system is near to its steady-state speed, all the external resistors are bypassed and, by means of a special mechanical arrangement, the rotor windings are short-circuited; the brushes are then lifted from the rings to avoid friction and the wear and tear of brushes.

This system is now becoming obsolete, since it has several disadvantages compared to the more modern solution of using induction motor drives, which are discussed in the next section: the external resistors produce extra losses, and the ring–brush connection is mechanically complicated and expensive and requires maintenance; furthermore, the change between the different torque curves is discrete and not continuous.

To enhance the machine's starting torque without reducing its efficiency too much, some means must be provided to enlarge rotor resistance at starting time, without

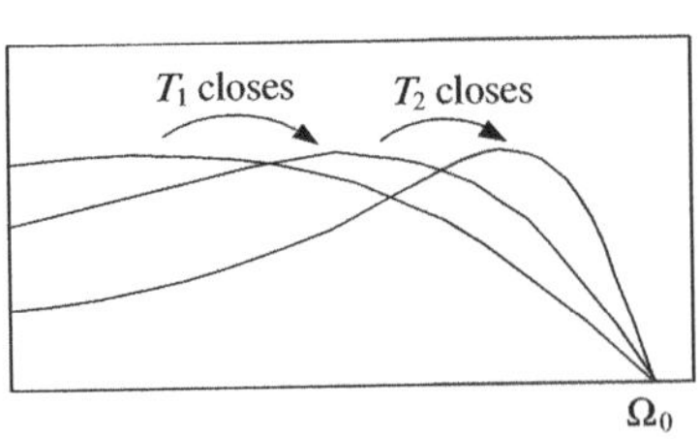

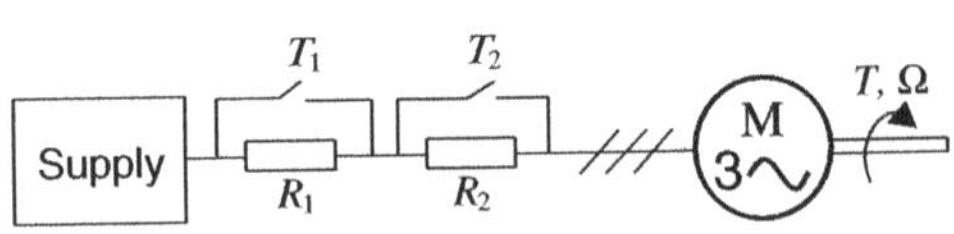

FIGURE 12.10. Motor starting using additional rotor resistors.

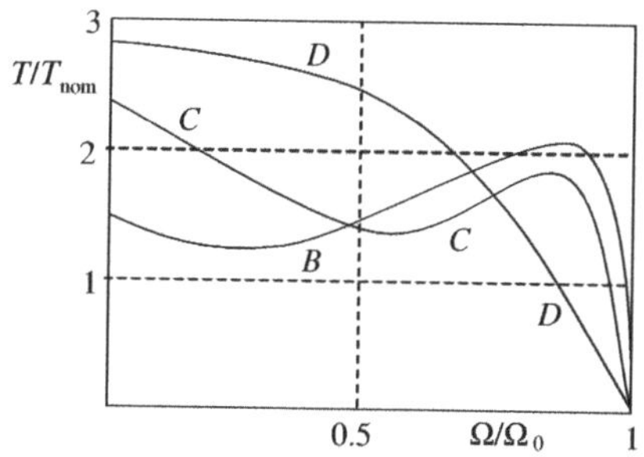

FIGURE 12.11. NEMA torque–speed curves for different designs of induction motors.

significantly affecting it near synchronous speed. For example, some dependence of R_r from slip s can be introduced. This is done by using techniques that cannot be described here. The interested user is advised to refer to [bm1] or [bm5].

The final result is a torque curve that has a similar trend to the one shown in Figures 12.8 and 12.9 near synchronous speed, while experiencing higher values at larger slips.

The induction machine's torque–speed curves are classified by NEMA (National Electric Manufacturer's Association) of the United States. The most common standard design types, named B, C, and D, have curves that are qualitatively shown in Figure 12.11. On the vertical axes the torques are shown as a ratio to the nominal torque. As previously stated, usually the higher the ratio pull-out/starting torque, the higher the efficiency. This is true also for the NEMA curves. For instance, design D is very good for starting but has a low efficiency; design B has reasonable starting and pull-out torques and therefore is the most widely used design for general-purpose applications, while design C has a high starting torque (but lower than design D) but intermediate efficiency and pull-out torque.

When an induction machine is required to operate at continuously variable speed, such as in electric trains or cars, the mechanical characteristic of the machine fed at constant voltage and frequency, such as that shown in Figure 12.8 and 12.9, is totally inadequate.

In these cases, induction machines can be used and are indeed used frequently today, using special devices to feed them, which are able to generate voltages with the desired frequency and amplitude. The machine and the feeding system constitute a subsystem that is called "motor drive." Some basic information about induction machine-based motor drives is given in Section 12.4.

12.3 NO-LOAD AND BLOCKED-ROTOR TESTS

A method for determining the parameters of the equivalent circuit of an induction machine consists of two tests: the no-load test and the blocked-rotor test.

In the former, the nominal voltage is applied to the machine while leaving the rotor free to rotate (no mechanical load is applied, thus $\Omega \cong \Omega_0$ and $I_r \cong 0$). The current I_0

absorbed by the motor is measured, as well as the correspondent active power P_0 (dissipated by R_s and R_i; see Figure 12.5). Since the stator current and the corresponding copper losses are small, P_0 is a good estimation of stator iron losses[8] (at low slip, rotor losses are always negligible in an induction machine, since the rotor frequency is low). Such losses, which strictly depend on the voltage applied to R_i, remain practically constant for any loading condition, provided that the machine is supplied around its nominal voltage.

$$R_i \cong \frac{U^2}{P_0}, \quad X_i \cong \frac{U^2}{Q_0} = \frac{U^2}{P_0 \cdot \tan\varphi_0}, \qquad \text{where } \cos\varphi_0 = \frac{P_0}{\sqrt{3}UI_0}$$

In these formulas, U is the line-to-line voltage P_0, and Q_0 is the three-phase power. Note that $U^2_{line-line}/P_{three-phase} = U^2_{phase}/P_{phase}$.

The second test is performed by increasing the supply voltage until the nominal current I_{nom} is absorbed by the motor, the rotor being mechanically blocked. The corresponding voltage U_{sc} is then measured, as well as the active power P_{sc}. Since U_{sc} is usually only a low percentage of the nominal voltage, during this test I_i it can be disregarded because it is much smaller than $I_s = I_{nom}$; thus the power absorbed by R_i is much lower than the one consumed by R_s and R_r'. For this reason, P_{sc} is a good estimation of copper losses at rated current; in any other loading condition, total copper losses can be estimated as follows:

$$P_{cu} \cong P_{sc} \cdot (I/I_{nom})^2$$

The test is unable to separate $R_s + jX_s$ from $R_r' + jX_r'$, but it allows the calculation of $\underline{Z}_{sc}$:

$$\underline{Z}_{sc} \cong \frac{U_{sc}}{\sqrt{3}I_{nom}} e^{j\varphi_{sc}}, \qquad \text{where } \cos\varphi_{sc} = \frac{P_{sc}}{\sqrt{3}U_{sc}I_{nom}}$$

Finally, electrical and mechanical powers are strictly related, their difference being iron and copper losses:

$$P_{el} = P_m + P_0 + P_{cu}$$

The reader should note the strict analogy with open-circuit and short-circuit tests of a transformer (see Section 6.5, also for measuring issues).

[8]If R_s is known, for example, thanks to a previous DC measure, to estimate the iron losses P_0 can be corrected by subtracting $3R_sI^2$.

Example 7. A 400-V 8-kW three-phase 4-poles induction motor has issued the following results to the no-load and blocked-rotor test:

$$I_0 = 5\%, \qquad P_0 = 1.2\%$$
$$U_{sc} = 12\%, \qquad P_{sc} = 7\%$$

The stator resistance is 0.12 Ω (each phase), the nominal line current is 15 A, and the stator windings are delta-connected.

Calculate R_i, X_i, and $\underline{Z}_{sc}$:

$$I_0 = 0.05 \cdot 15 = 0.75\ \text{A}$$

$$P_0 = 0.012 \cdot 8000 = 96\ \text{W}$$

$$\cos\varphi_0 = \frac{P_0}{\sqrt{3}UI_0} = 0.1848, \qquad \tan\varphi_0 = 5.319$$

$$U_{sc} = 0.12 \cdot 400 = 48.0\ \text{V} \qquad \text{(line-to-line)}$$

$$P_{sc} = 0.07 \cdot 8000 = 560\ \text{W}$$

$$\cos\varphi_{sc} = \frac{P_{sc}}{\sqrt{3}U_{sc}I_{nom}} = \frac{560}{\sqrt{3} \cdot 48 \cdot 15} = 0.4490\ \Omega, \qquad \varphi_{sc} = 63.32°,$$

Since the three windings are delta-connected, the star-connected equivalent stator resistance must be computed: $R_{seq} = 0.12/3 = 0.04\ \Omega$.

The iron losses can be calculated by subtracting the no-load stator copper losses from P_0:

$$P_i = 96 - 3 \cdot 0.04 \cdot 0.75^2 = 95.93\ \text{W (no-load stator copper losses are negligible)}$$

$$R_i = U_i^2/P_i \cong U_{nom}^2/P_i = 400^2/95.93 = 1669\ \Omega$$

Disregarding X_s, we obtain

$$X_i = U_i^2/(P_0 \cdot \tan\varphi_0) \cong U_{nom}^2/(P_0 \cdot \tan\varphi_0) = 400^2/(96 \cdot 5.319) = 313.3\ \Omega$$

$$Z_{sc} = \frac{U_{sc}}{\sqrt{3}I_{nom}} = \frac{48}{\sqrt{3} \cdot 15} = 1.848\ \Omega$$

$$\underline{Z}_{sc} = Z_{sc}\angle\varphi_{sc} = 1.848\angle 63.32° = 0.8296 + j1.651\ \Omega = (R_s + R_r') + j(X_s + X_r')$$

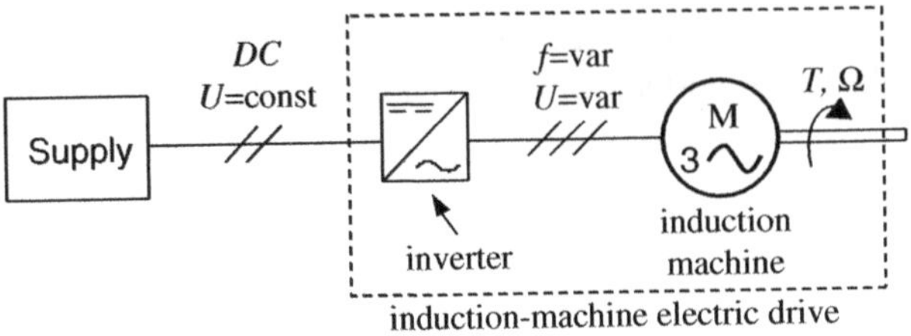

FIGURE 12.12. Combination of inverter and induction machine to obtain an induction electric drive.

12.4 INDUCTION MACHINE MOTOR DRIVES

The concept of an electric drive was introduced in Chapters 10 and 11. In particular, AC electric drives were examined in Section 11.5.1, to which the reader is referred. Here only the figure describing a DC-fed electric drive is recalled (Figure 12.12).

Details of the converters able to perform the functions shown in Figure 12.12 are given in Chapter 8. The present chapter was deliberately written in such a way that it can be understood without the need to read Chapter 8, which should thus be considered a more-in-depth analysis of converters and drives.

In this section we will learn how to operate the inverter in order to obtain a good performance in variable-speed operation from the machine.

Consider the expressions of the machine current and torque, derived in Section 12.2:

$$I^2 = \frac{U_1^2}{X^2 + R_r'^2/s^2}, \qquad T_{mg} = \frac{P_{ag}}{\Omega_0} = \frac{P_{mg}}{\Omega} = \frac{3}{\Omega_0}\frac{U_1^2 R_r' s}{s^2X^2 + R_r'^2} \tag{12.10}$$

Let us evaluate what happens to these expressions when the machine is fed in such a way that it is

$$U_1 = K_f f = K_\Omega \Omega_0 \quad (K_\Omega = pK_f/(2\pi)) \tag{12.12}$$

that is, the machine is fed in such a way that the supply frequency and voltage always remain proportional to each other (remember that $\Omega_0 = \omega/p = 2\pi f/p$).

Because of its importance, a supply compliant with (12.12) has a name of its own: It is normally called *constant voltage/frequency* supply.

Using constant voltage/frequency supply and considering that it is

$$s = \frac{\Omega_0 - \Omega}{\Omega_0} = \frac{\Delta\Omega}{\Omega_0}$$

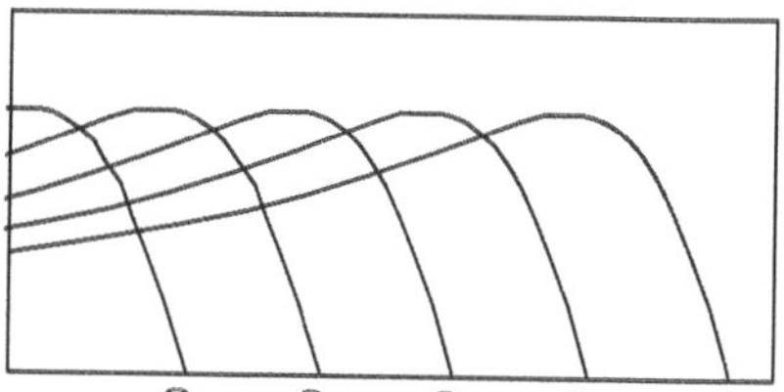

FIGURE 12.13. Translation of torques when feeding the machine with the rule $U_1 = K_{f}f$.

the squared current becomes

$$I^2 = \frac{s^2 U_1^2}{s^2 X^2 + R_r'^2} = \frac{\Delta^2\Omega \cdot K_\Omega^2}{(p\Delta\Omega L)^2 + R_r'^2} \tag{12.13}$$

In the latter equality it was considered that X is proportional to the angular frequency ω:

$$X = X_s + \alpha^2 X_r = \omega(L_s + \alpha^2 L_r) = p\Omega_0 L = p\frac{\Delta\Omega}{s}L$$

The torque is

$$T_{mg} = \frac{3}{\Omega_0}\frac{R_r'}{s}I^2 = 3\frac{R_r'}{\Delta\Omega}\frac{\Delta^2\Omega \cdot K_\Omega^2}{(p\Delta\Omega L)^2 + R_r'^2} = 3R_r'\frac{\Delta\Omega \cdot K_\Omega^2}{(p\Delta\Omega L)^2 + R_r'^2} \tag{12.14}$$

Equations (12.13) and (12.14) indicate that when the machine is fed according to (12.12) the current absorbed and the torque produced depend only on $\Delta\Omega = \Omega_0 - \Omega$ and not on Ω_0 alone.

Since the shape of T_{mg} as a function of $\Delta\Omega$ does not depend on Ω_0, it can be derived as seen in the previous section. Evidently, the function $T(\Omega)$ is the same one as in previous figures, shown again in Figure 12.13. Changing Ω_0 leaves the shape intact; only the curve is translated in such a way that the torque becomes zero for the given value of Ω_0.

This technique can be used so that the machine always operates in the most efficient region of the curve $T(\Omega)$, i.e. between $\hat{\Omega}$ and Ω_0. In this region, as was seen earlier, the torque is delivered at the lowest current and highest efficiency.

As an example, Figure 12.14 shows the same machine starting, as simulated, with sufficiently detailed PPU machine models, under two conditions: using a constant terminal voltage and using the technique (12.12). In the latter case, during start-up, Ω_0 is continuously changed in such a way that $\Delta\Omega = \Omega_0 - \Omega =$ constant, and therefore, the torque generated according to the theory developed is constant. The value of $\Delta\Omega$ is such that the machine-generated torque is larger than the load torque. When the radian frequency $\omega_{base} = \Omega_{base}/p$ is reached,[9] however, the electric supply system works at that frequency and the system reaches its equilibrium point. The figure is obtained using a simulation programme that models the machine with a higher level of detail.

[9]The variable p, as usual, indicates the number of the machine's pole pairs.

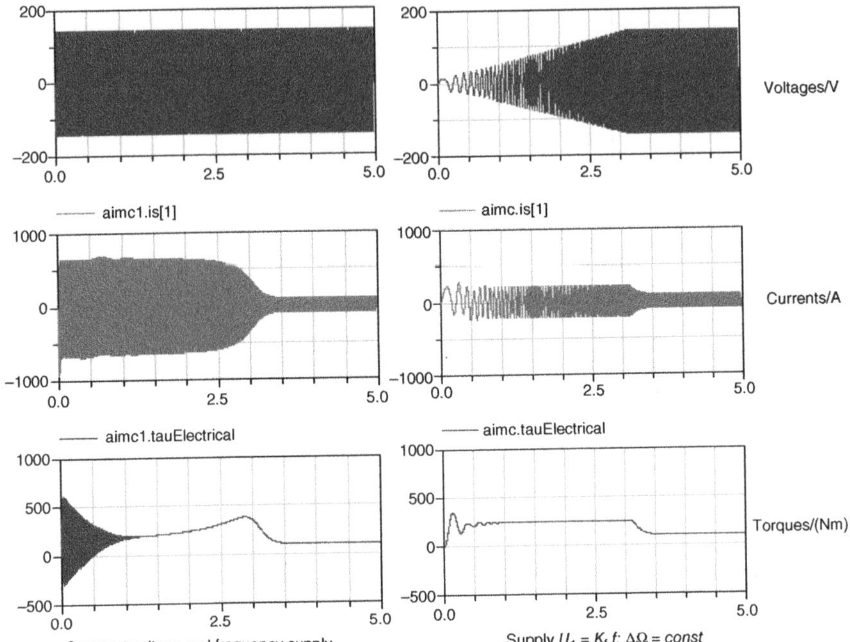

FIGURE 12.14. Simulation of the starting of an induction machine either fed directly from the mains (*left*) or with variable voltage and frequency (*right*).

More in Depth

Readers will probably already know that to obtain detailed data on the behaviour of complex systems today a manual 'ink on paper' analysis should be complemented with *simulations*. Simulators are able to tell us the behaviour of dynamic systems over time, which is nearly impossible with manual computation. The results shown in Figure 12.14 can be obtained with any simulation programme, provided that the correct numerical data of the system are given. Since the purpose of Figure 12.14 is to show just the qualitative behaviour of the voltage/frequency starting of an induction machine, the list of numerical parameters is not provided here.

It is, however, worth mentioning here that the programme used for this simulation was written in a new language that is specific for simulating systems of any complexity, mixing mechanical, electrical, electronic, and fluidic (and other systems) together. This language is called Modelica; it is widely used and students are quite likely to come across it in their future careers. (More information about Modelica can be found at the Internet site https://modelica.org.)

To understand the shown plots, it must first be noted that the time span is five seconds; therefore sinusoids operating at 50 Hz are seen as a thick band, since they evolve too rapidly to be seen well in the shown scale. This is particularly true for the top left-hand plot, in which the uniform band represents a constant-amplitude, constant-frequency 50-Hz sinusoid.

To facilitate comparisons, the vertical range of left- and right-hand plots are equal to each other for corresponding variables.

We want to make the following remarks:

- At the beginning of the transient, in the case of constant-frequency supply, the torque shows a high-frequency fluctuation around a steadily increasing value. This fluctuation is basically due to the fact that initially the three stator currents are not balanced, and the corresponding rotating field has a variable amplitude. This actually occurs in real machines but cannot be highlighted here by the equivalent circuit-based model used, since all electrical transients are disregarded. In the variable-frequency supply the torque fluctuation is reduced to a minimum.
- In the simulation proposed here, the variable-frequency supply allows the machine to be started up in the same time, but at a much lower current and torque fluctuations.

The supply voltages and frequencies required to obtain the curves shown in the right-hand part of Figure 12.14 are obtained using a control logic of the type described in Figure 12.15.

The reference torque during the start-up T^* is first converted into a $\Delta\Omega$ signal (taking into account the inner parameters of the induction machine), saturated at the

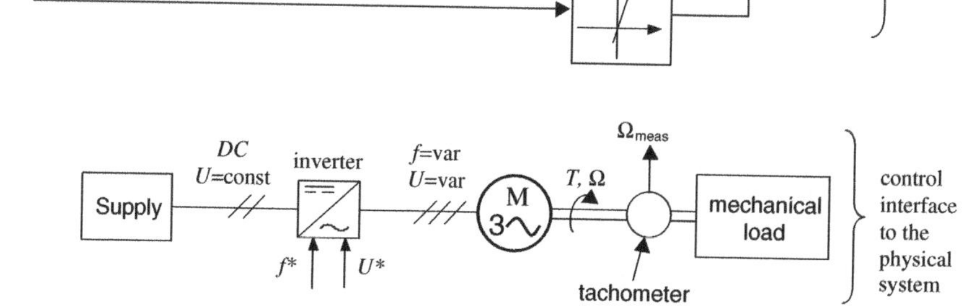

FIGURE 12.15. A possible torque control logic (used to obtain the plots shown in Figure 12.14. Here “var” stands for “variable”).

value $\Delta\hat{\Omega}$ to impose operation of the machine in the more efficient zone of its torque. Adding the measured value Ω_{meas} of Ω determines the synchronous speed Ω_0, which is converted into frequency using the pole pairs number p. Before this, the Ω_0 signal is saturated to the steady-state desired value Ω_{0ss}, since the machine is to be started up at this value of synchronous speed. Once the frequency is known, the voltage $U_1 = kf^*$ is known and from it, using the estimate I^* of the current computed from T^*, we can determine the voltage U^* desired at the machine terminals.

As already mentioned, a start-up which is faster than that shown in the right-hand part of Figure 12.15 is simply obtained by raising T^*, which in turn will raise $\Delta\Omega$ in proportion.

The values of the voltage U^* and frequency f^* to be applied to the machine are then transferred to the supply system, which creates a three-phase voltage supply with this amplitude and frequency.

It is apparent that variable-frequency control allows the machine to stay stable at any speed, because the currents are kept under control. The value of $\Delta\Omega$ determines the torque and the current, the latter being the value corresponding to the more efficient region of the torque-speed curve of the machine—that is, the one between $\hat{\Omega}$ and Ω_0.

The operation of the induction machine under (12.12) is also called *constant flux operation*. This is because the voltage U_1 corresponds in the machine to the EMF produced by the flux linked with the stator $\underline{\Psi}$. For each phase we have

$$\underline{U}_1 = j\omega\underline{\Psi}$$

and therefore $U_1/f = \text{const}$ implies $\Psi = \text{const}$.

As speed increases, the voltage to be applied to the machine terminals increases. When the maximum continuous voltage of the machine is reached (normally called nominal voltage), this quantity cannot be increased further, while it may happen that a further increase in speed is required. The speed that corresponds to the nominal voltage is normally called the *base* speed of the machine. Beyond the base speed, the voltage is kept constant while Ω_0 is raised, increasing the frequency of the supply voltages. In these conditions, the curve of the torque is progressively reduced, being proportional to $1/\Omega_0^2$:

$$T_{mg} = \frac{3}{\Omega_0}\frac{U_1^2 R_r' s}{s^2X^2 + R_r'^2} = \frac{3}{\Omega_0^2}\frac{U_1^2 R_r' \Delta\Omega}{\Delta^2\Omega L^2 p^2 + R_r'^2}$$

The full picture of the machine, showing the *constant flux* and *flux weakening* parts, is shown in Figure 12.16. It shows the zone below Ω_{base}, in which the torque characteristics translate, and also shows the zone above this value where the torque reduces in proportion to $1/\Omega_0^2$. A possible shape of the load torque T_{load} and a possible locus of the machine operating points $T_m = T_{load} + J\dot{\Omega}$ are also shown.

Using this T_m, the mechanical power increases linearly in the constant flux region; when Ω_{base} is overcome, the T_m locus shown in the figure is such that the power P is

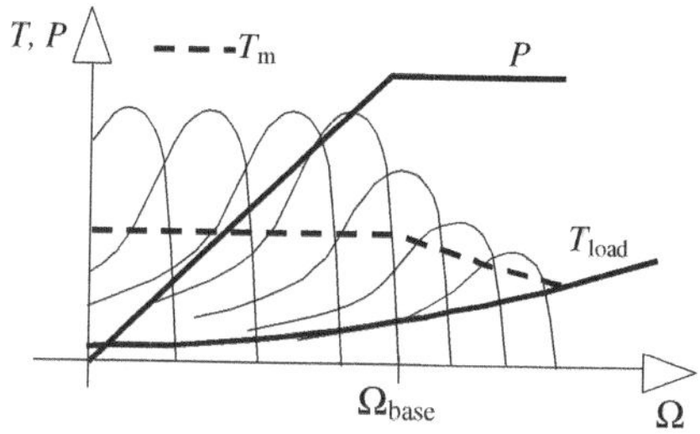

FIGURE 12.16. Induction machine drive composed of the constant flux ($\Omega < \Omega_{base}$) and flux weakening ($\Omega > \Omega_{base}$) regions.

maintained constant and therefore the actual torque reduces in proportion to Ω_0. Since the machine torque shape reduces in proportion to $1/\Omega_0^2$, the machine operating point becomes increasingly near to the peak torque.

It is fairly obvious that the zone with the flux reduction cannot be very large, because of the corresponding torque reduction. A typical ratio of $\Omega_{max}/\Omega_{base}$ does not exceed 2 in a normal induction drive.

A few final words about induction motor drives.

The constant voltage/frequency control enables the machine to be used smoothly and efficiently at all speeds. However, in fast-changing conditions it is not optimal because its theory comes from a steady-state model of the machine.

More advanced control techniques, normally called *vector control*, which keep the inner machine magnetic field under control, are used today in advanced drives and perform better in dynamic conditions.

These techniques, however, are well beyond the scope of this book and will not be dealt with here.

12.5 SINGLE-PHASE INDUCTION MOTORS

12.5.1 Introduction

In this book we will use the term *single-phase induction motors* for small motors with an operating principle similar to the three-phase motors dealt with in this chapter, but fed by a single phase sinusoidal supply. In some books, such as reference [bm6], they are also called *two-phase motors* since they have two stator windings.

While three-phase induction motors are prevalent in medium and large-power applications, for domestic appliances, where distribution is single phase and power is very low, it is usual to use single-phase induction motors. They are common in washing machines, fans, refrigerators, air conditioners, and so on.

In this section, single-phase induction motors are qualitatively evaluated. A more analytical treatment of the topic can be found in reference [bm6].

Consider at first an asynchronous (aka induction) machine containing a single winding on the stator and fed by a single-phase, constant-frequency AC source.

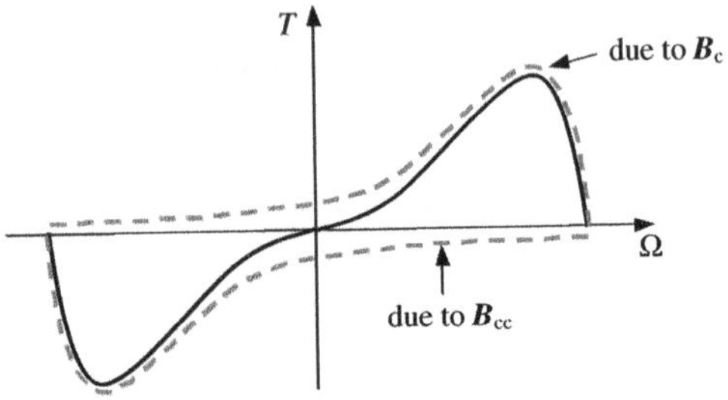

FIGURE 12.17. Torque–speed curve of an AC machine with a unique winding.

It could be experimentally verified that it would have a torque–speed curve with the shape, as qualitatively represented in Figure 12.17, with a solid line. In the region near to synchronous speed, it is very similar to the shape already seen for three-phase machines, but the behaviour is very different at low speeds: the single-phase machine develops zero torque at zero speed! This impedes the natural starting of this motor, since there always exists some—large or small—load torque at zero speed.

This characteristic can be explained by considering what happens when a machine contains a single winding. An alternating current flowing across it generates, along the winding axis (and at the air-gap position), an induction $\boldsymbol{B}$ whose values do vary as a sine wave over time but do not constitute a wave moving along the air gap.

This induction can be assumed to be composed of two different $\boldsymbol{B}$'s, $\boldsymbol{B}_c$ and $\boldsymbol{B}_{cc}$, which rotate clockwise and counterclockwise, respectively:

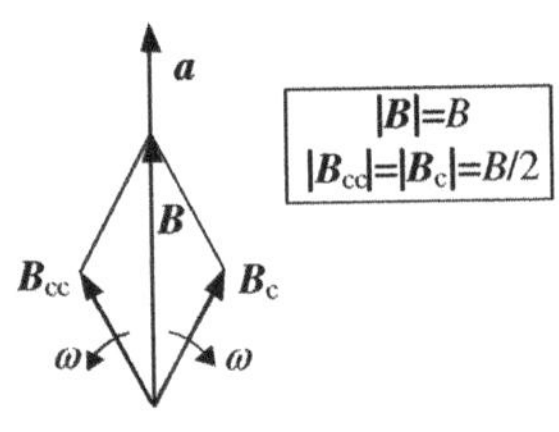

If the rotational speed is ω, the component of $\boldsymbol{B}$ along axis $\boldsymbol{a}$ (the axis of the unique winding of the machine) is

$$B_a = B_{c,a} + B_{cc,a} = \frac{B}{2}\cos \omega t + \frac{B}{2}\cos \omega t = B\cos \omega t$$

The machine's behaviour is, according to this simple explanation, determined by the presence of both $\boldsymbol{B}_c$ and $\boldsymbol{B}_{cc}$. Therefore, the torque generated can be imagined to be the sum of two different curves: one due to $\boldsymbol{B}_c$, the other to $\boldsymbol{B}_{cc}$. The machine's rotor, however, rotates in one of two possible directions: either $\boldsymbol{B}_c$'s or $\boldsymbol{B}_{cc}$'s (see again Figure 12.17, in which it is supposed that $\boldsymbol{B}_c$ is rotating in the same direction as the machine's rotor).

This is a simplified explanation of the machine's behaviour which, however, is sufficient for our purposes.

The explanation of the machine's behaviour based on the two counter-rotating fields also explains torque oscillation which is present in single-phase induction motors. This is caused by interaction between the two fields. If we analyse the machine from a system rotating at the (positive) synchronous speed, the field $\boldsymbol{B}_c$ is stationary within this reference, as discussed at the end of Section 12.1 and shown in a "Result" box. Therefore, the torque produced is constant. However, as seen from this same reference, $\boldsymbol{B}_{cc}$ is rotating at 2ω. The interaction between $\boldsymbol{B}_c$ and $\boldsymbol{B}_{cc}$ produces an oscillating torque with an angular frequency of 2ω.

The fluctuation of the machine's torque is related also to the fluctuation of the instantaneous power of any single-phase AC circuit (see Chapter 5): if the instantaneous power at the electric terminals were converted into mechanical power, because of the rotor's inertia, the fluctuation would immediately be transferred onto the electromechanical torque.

In fact it is

$$P_m = T_m \Omega \qquad (°)$$

P_m fluctuates, as mentioned, at twice the synchronous speeds:

$$\Omega_{fluct} = 2\Omega_0 = 2p\omega$$

where Ω_0 is the synchronous speed, ω is the angular frequency, and p is the number of the machine's pole pairs. The rotor's inertia causes the Ω variation to be much lower than the P_m variation, and therefore speed fluctuation is negligible; thus the fluctuation of power is nearly totally transferred into the fluctuation of the torque [consider again equation (°)].

A reduction of the mechanical power fluctuation can be obtained by introducing some fluctuating storage of energy between the machine's terminals and air gap. This is exploited in the capacitor-start/capacitor-run motor, discussed in the next section.

To make a single-phase AC machine perform more like a three-phase machine, the stator is made using *two* windings instead of one, physically displaced in the machine's air-gap periphery by 90° (electrical).

These two windings, if fed with currents with a phase difference of 90°, are able to create a rotating field just like the one created by a three-phase machine. The demonstration is omitted here, but it is very similar to the demonstration of three-windings physically displaced by 120°, shown in Section 10.2.1:

Result: *The rotating field on two-phase machines*

If two sine currents with equal amplitudes, 90° relative phase displacement and angular frequency ω, are introduced into windings that are physically separated by 90° (electrical), a field is produced that is equivalent to a uniform distribution field rotating at a constant electrical speed ω.

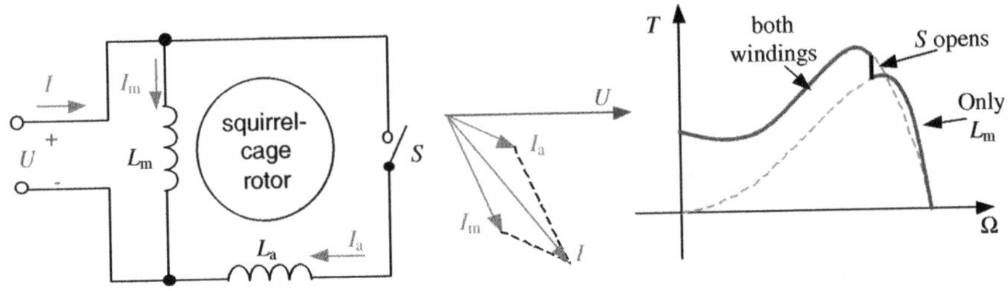

FIGURE 12.18. Single phase capacitor-less induction motor with starting phasor diagram and typical torque curve.

The vast majority of single-phase AC machines are fed from a single-phase source containing these two windings. The difference between one machine type and another resides in the physical structure (e.g., the conductor's cross-sectional area) of these windings and depends on how these are fed, exploiting the unique single-phase AC source available.

12.5.2 Different Motor Types

It was mentioned at the end of the previous section that the single-phase machines considered here have two windings. Since in general they are not identical to each other, normally they are indicated as main (subscript "m") and auxiliary (subscript "a") coils.

A possible single-phase AC machine is the so-called *capacitor-less motor* (Figure 12.18). In this machine, the auxiliary winding uses conductors with smaller cross-sectional area A than those of the main. This will imply a larger resistance according to the resistance law (Section 4.2.2):

$$R = \rho \frac{l}{A} = \frac{l}{\sigma A}$$

Its structure is sketched in the left-hand part of Figure 12.18.

As a consequence, currents $\underline{I}_m$ and $\underline{I}_a$ will have a qualitative relative position like those shown in the phasor diagram in Figure 12.18: since the displacement is not 90° and the amplitudes are not equal, the field will be composed of a rotating and a stationary part; the rotating component will cause enough torque to be generated at starting.

When the machine has started, near 75% of the synchronous speed, the switch S is opened, and only the main coil operates. The opening of S is automatic, usually by means of a centrifugal device.

Another solution, *called permanent-capacitor motor*, is described in Figure 12.19.

In this case, the presence of a capacitor in series with the auxiliary winding causes it to absorb a current that leads voltage U and has large phase displacement with $\underline{I}_m$ (which is lagging); this allows the generation of some starting torque.

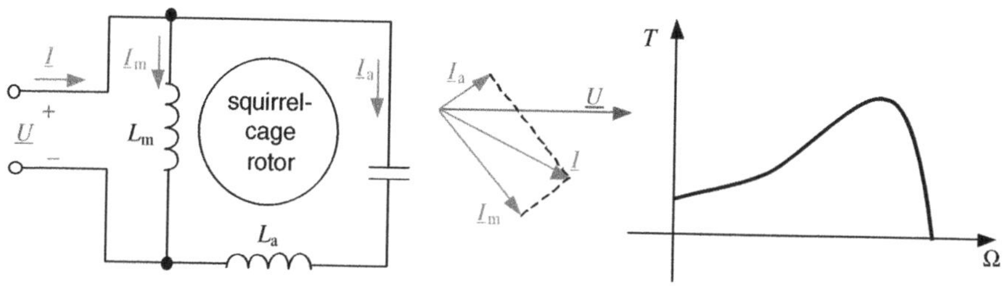

FIGURE 12.19. Single-phase permanent-capacitor induction motor with starting phasor diagram and typical torque curve.

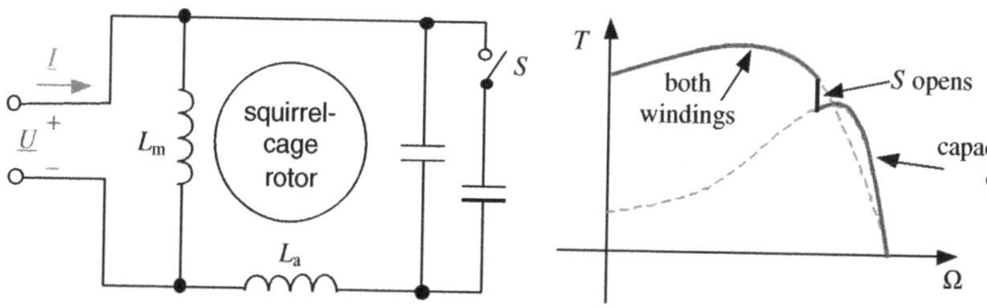

FIGURE 12.20. Single-phase capacitor-start/capacitor-run induction motor.

The capacitor value is optimized for the rated load, for which it can create a near-ideal condition, with $\underline{I}_a$ and $\underline{I}_m$ equal in amplitude and displaced by 90°, which could in principle create a pure rotating field and no fluctuating torque.

To enhance the starting torque, while keeping the optimization of the two currents in nominal conditions, a common solution is the *capacitor-start/capacitor-run motor*, shown in Figure 12.20.

The circuit with *S* closed is optimized for start-up, while the circuit with *S* open for a normal run. The commutation (opening of *S*) normally occurs at around 75% of the synchronous speed and is automatic, usually by means of a centrifugal device.

Another possibility is the *capacitor-start motor* (Figure 12.21). Here the capacitor is used only for starting; it is sized for intermittent operation and is therefore cheap, while near synchronous speed the machine operates with a single active winding. This gives enough torque, albeit significant fluctuating torque which creates vibration and noise.

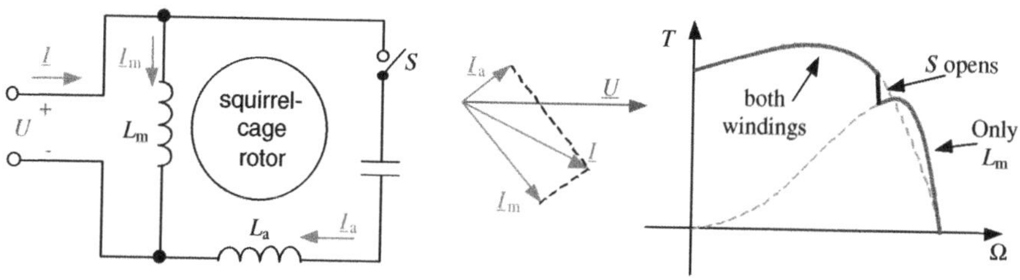

FIGURE 12.21. Single-phase capacitor-start induction motor with starting phasor diagram and typical torque curve.

The four single-phase induction motors are in a kind of order: the capacitor-less motor capacitor has the lowest cost and worst performance, next comes the permanent capacitor motor, and then the capacitor-start motor. The capacitor-start/capacitor-run is the most effective and expensive solution.

The choice between the different options depends on cost, quality (quietness), and starting torque required.

12.6 PROPOSED EXERCISES

Where not specifically indicated, electromagnetic quantities are expressed in rms values.

12.1. An eight-pole 50-Hz induction motor runs with a slip $s = 5\%$. Determine the synchronous speed n_0, the rotor speed n, the frequency f_r of rotor currents, the speed of the rotor rotating field with respect to the rotor (n_{rr}) and with respect to the ground (n_r). Express all speeds in rpm (revolutions per minute).

12.2. A four-pole 60-Hz induction motor runs at 1750 rpm. Determine the synchronous speed n_0, the slip s and the frequency f_r of rotor currents.

12.3. A six-pole 50-Hz induction motor has a full-load (nominal) slip of 3%. Determine:

(a) the synchronous speed n_0

(b) the rotor speed n at full load

(c) the frequency of rotor currents at the instant of starting

(d) the frequency of rotor currents at full load

12.4. The nameplate speed of a 25-Hz induction motor is 720 rpm. If the no-load speed is 745 rpm, estimate the number of poles and calculate the synchronous speed and the slip at full load.

12.5. A 50-Hz three-phase induction motor has a rated phase-to-phase voltage of 400 V. When operating at full load, delivering its rated mechanical power of 30 HP, the line current is 42 A, the rotor speed is 710 rpm, and the motor absorbs 25.7 kW from the electric grid. Calculate:

(a) the synchronous speed n_0

(b) the slip s

(c) the power factor $\cos\varphi$

(d) the torque T

(e) the efficiency η

Hint: 1 HP = 736 W

12.6. An eight-pole 50-Hz three-phase induction motor has an efficiency of 86% and absorbs 48.8 kW from the electric grid. If the slip is 5%, calculate the shaft torque.

12.7. Induction motors are often braked, reversing the phase sequence of the voltage supplying the motor (*plugging*). A motor with six poles is operating at 1150 rpm while supplied at 60 Hz. Two of the stator supply leads are suddenly interchanged. Calculate the new slip and the new rotor current frequency. *Hint:* n_0 changes its sign.

12.8. In a four-pole 50-Hz wound-rotor induction motor, the stator and the rotor have the same effective numbers of turns per phase. If the motor has 230 V per phase across its stator and the voltage induced in the rotor is 6.6 V per phase, calculate the slip and the motor speed.

12.9. In a two-pole 60-Hz induction motor, the power crossing the air gap and the developed power are respectively 21.6 kW and 20.1 kW. Calculate the slip s. If the rotational losses are 400 W, determine the net output torque.

12.10. A three-phase six-pole 50-Hz induction motor develops its maximum torque of 612 Nm at 600 rpm. The rotor resistance to stator, R_r', is 0.5 Ω. Calculate the supply voltage (line-to-line) and the torque developed by the motor at 750 rpm.

12.11. An induction motor develops a maximum torque which is twice the starting torque. Calculate X/R_r'. If the full-load torque equals the starting torque, determine the full-load slip.
Hint: Use equations (12.10) and (12.11).

12.12. An induction motor has a slip of 4% at full load. The starting current is eight times the full-load current. Calculate the ratio between the starting torque and the full-load torque. If the motor is started at reduced voltage, so that the starting current is only four times the full-load currents, what is the new ratio of the starting torque to the full-load torque? Comment on this result.

12.13. An induction motor has a slip of 5% at full load. At rated voltage, the starting current is six times the full-load current. Calculate the ratio between the starting torque and the full-load torque.

If the motor employs a wye-delta starter, which connects the motor phases in wye for starting and in delta when already running, what is the new ratio between the starting torque and the full-load torque? Comment on this result.

12.14. The resistance measured between two stator terminals of an induction machine is 0.14 Ω; the stator is delta-connected. The motor delivers a mechanical power of 15 kW while supplied at 230 V (line-to-line), absorbing a line current of 54 A at 0.82 power factor (lagging). The no-load test has issued the following results: $I_0 = 20$ A, $\cos \varphi_0 = 0.12$. Calculate:

(a) the resistance of each stator phase

(b) the efficiency

(c) the no-load power

(d) the sum of iron and mechanical losses

(e) the copper losses at the current load condition

(f) the slip

12.15. A three-phase induction motor has a full-load slip of 4%. The wounded rotor is star-connected, with a phase resistance of 0.12 Ω. Calculate the additional rotor resistance required to obtain the full-load torque at starting conditions.

12.16. An eight-pole 230-V 42-Hz induction motor absorbs a line current of 60 A with a 0.82 power factor (lagging) and a slip of 3.5%. The resistance measured between two stator terminals is 0.162 Ω, the stator being delta-connected. The no-load test has issued the following results: $I_0 = 18$ A, $\cos \varphi_0 = 0.12$. Calculate:

(a) the resistance of each stator phase

(b) the sum of iron and mechanical losses

(c) the stator copper losses

(d) the power crossing the air gap

(e) the power and torque delivered

(f) the efficiency

12.17. A six-pole induction motor delivers 20 kW with an efficiency of 91%, a power factor of 0.9, and a slip of 3%. The motor is supplied by a 400-V 50-Hz phase-to-phase voltage. Calculate the line current and the delivered torque.

12.18. A six-pole 230-V 50-Hz induction motor absorbs 22 kW with a power factor of 0.68 and a slip of 3.4%. The resistance measured between two stator terminals is 0.023 Ω, the stator being delta-connected. The iron losses are 820 W, and the mechanical losses are 440 W. Calculate:

(a) the stator copper losses

(b) the net mechanical power and torque

(c) the efficiency

(d) the capacity of delta-connected capacitors, required to compensate the motor to a unit power factor ($\cos \varphi = 1$)

12.19. A four-pole induction motor with a wye-connected stator, having a phase resistance of 0.4 Ω, is supplied at 400 V (line-to-line), 50 Hz. Iron and mechanical losses equal 350 W and 300 W, respectively. The motor moves a DC generator, which delivers 70 A at 125 V. The motor absorbs 12.25 kW with a $\cos \varphi = 0.9$, rotating at 1446 rpm. Calculate:

(a) the overall efficiency

(b) the motor efficiency

(c) the DC generator efficiency

12.20. A four-pole three-phase induction motor is supplied by a 400-V 50-Hz voltage. The rotor resistance shown to stator is 0.15 Ω; the overall leakage reactance shown to the stator is 1 Ω. Calculate:

(a) the maximum torque and the correspondent slip and line current

(b) the starting torque and line current

12.21. The motor of the previous exercise is applied to a mechanical load requiring a torque of $30 + 0.005\,\Omega^2$ Nm. Using equation (12.10) and a numerical technique, calculate at steady state-conditions:

(a) the slip

(b) the line current

(c) the rotor speed in rpm

(d) the mechanical power and torque delivered

12.22. A 50-Hz 15-kV (phase-to-phase) network supplies, by means of a MV/LV transformer, a 400-V three-phase four-pole induction motor, with a delta-connected stator. The transformer has the following parameters:

15,000/400 V (phase-to-phase)
delta/wye connection
18 kVA, 50 Hz
$I_0 = 2\%$, $P_0 = 0.4\%$
$U_{sc} = 7\%$, $P_{sc} = 3\%$

The motor has the following parameters:

$P_{m\ nom} = 8$ kW
$I_{nom} = 16$ A (line)
$I_0 = 6\%$, $P_0 = 1.2\%$
$U_{sc} = 14\%$, $P_{sc} = 8\%$
$R_s = 0.12\,\Omega$ (phase)

Between the transformer and the motor, there is an 80-m LV cable, with $r = 0.9\,\Omega$/km and $x = 0.15\,\Omega$/km. Thévenin's impedance of the MV network can be disregarded.

Draw the single-phase wye-equivalent of this system. Assuming that the motor is running with a 3% slip, calculate:

(a) the motor speed in rpm

(b) the mechanical power and torque delivered to the load

(c) the phase-to-phase voltage at motor mains

(d) the stator phase current of the motor

(e) the primary phase current of the transformer

PART IV

POWER SYSTEMS BASICS

Fundamentals of Electric Power Engineering: From Electromagnetics to Power Systems, First Edition.
Massimo Ceraolo and Davide Poli.

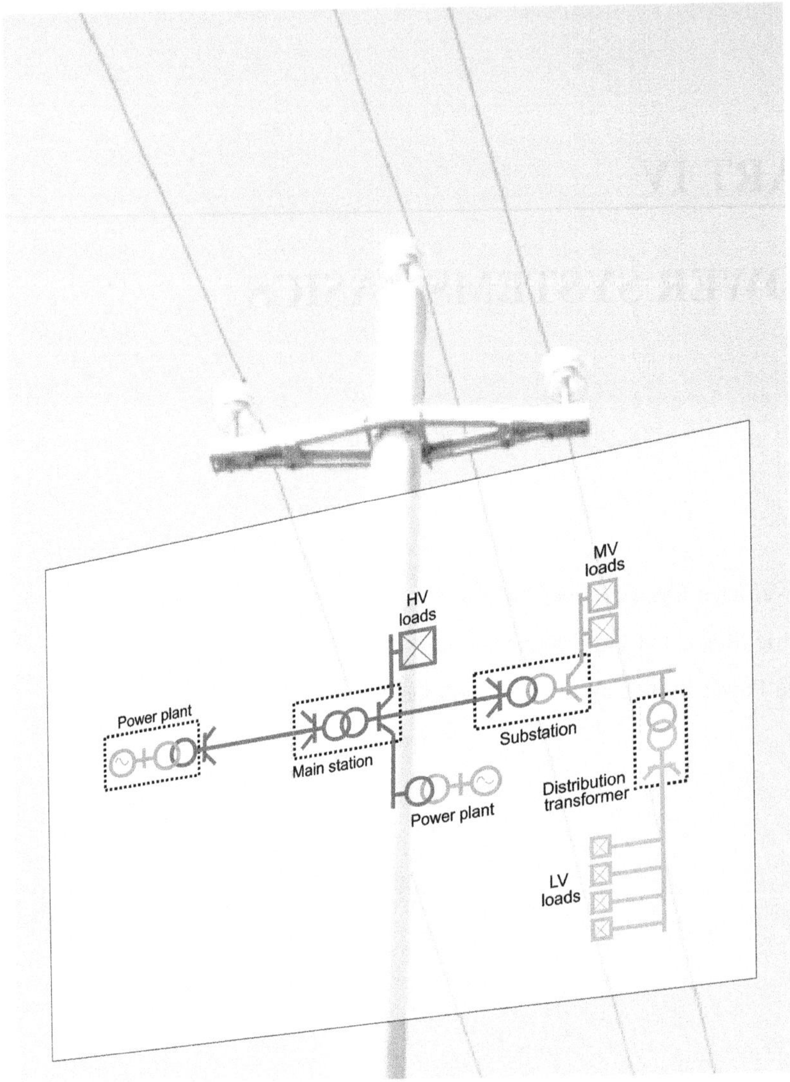
MV
loads
HV
loads
Power plant
Main station
Substation
Power plant
Distribution
transformer
LV
loads

13

LOW-VOLTAGE ELECTRICAL INSTALLATIONS

For the Instructor

The main purpose of this chapter is to give students an understanding of how low-voltage electrical installations are structured and how to determine the required sizes of cables and circuit breakers. We will develop a theory and supply numerical tables to facilitate the explanation of simple examples. However, we are unable, within the limits of this book, to show enough numerical tables to allow readers to determine the cables and breakers of electrical installations in a sufficiently large number of cases.

A good guide on how to design electrical installations, covering the topics of this chapter and much more, is the [bp2].

13.1 ANOTHER LOOK AT THE CONCEPT OF THE ELECTRIC POWER SYSTEM

Electric power systems were introduced in Chapter 1. We said there that power systems are either *full* power systems, which manage electric power by generating it and supplying it to loads, or *partial* power systems that need energy in-feed from an external electricity source and/or send the electricity produced to an external grid.

Fundamentals of Electric Power Engineering: From Electromagnetics to Power Systems, First Edition.
Massimo Ceraolo and Davide Poli.

A full power system can be as big as a continent, such as Europe or North America; in this case it constitutes the general reference for electricity of that region and is called *the* Power System (with uppercase initials), for example, the European Power System. A full power system can also be much smaller. Examples of small full power systems are:

- the electric system of a ship or a car (which involves the generation, distribution and utilization of energy);
- the electric system of a hospital which, in the event of a mains failure, disconnects from the external grid and obtains its electricity from a diesel generator;
- a toy or an electric torch, which are even smaller examples.

In Parts I and II of this book, we bridged the gap between electromagnetics and circuits. Circuits, as we have noted several times, are mathematical/graphical models that very effectively model many physical systems (called "circuital systems") and can be analysed with fast and fairly simple tools.

In Part III, we presented some very important components of power systems—that is, rotating and static converters, which are able to convert energy from one form to another (from mechanical to electrical and vice versa) and between electrical forms with different characteristics. We introduced these components both on their own and combined in *electric drives*.

An electric drive is a partial power system, since it gets energy from a source and manipulates it, first by changing the characteristic of the electric energy, then by converting it into mechanical form, and finally by using it for final purposes. An example of an electric drive as a partial power system was already provided in Chapter 1 and is repeated here for convenience (Figure 13.1). In Figure 13.1 a fan is operated by an electric drive (instead of the simple electric motor) to be able to operate at variable speed.

Two examples in which variable-speed fans may be required are:

1. *In a computer.* Mother-board fans are usually capable of operating at up to several thousand rpms (e.g., 5000 rpm), but this is required in a worst-case scenario when the room is hot and the computer is heavily loaded. In most cases, much lower speeds, even as low as 1000 rpm, are acceptable, thus dramatically reducing both fan noise and power consumption. The best PC manufacturers install variable speed drives for CPU (Central Processing Unit) fans, thus lowering noise and power consumption considerably.

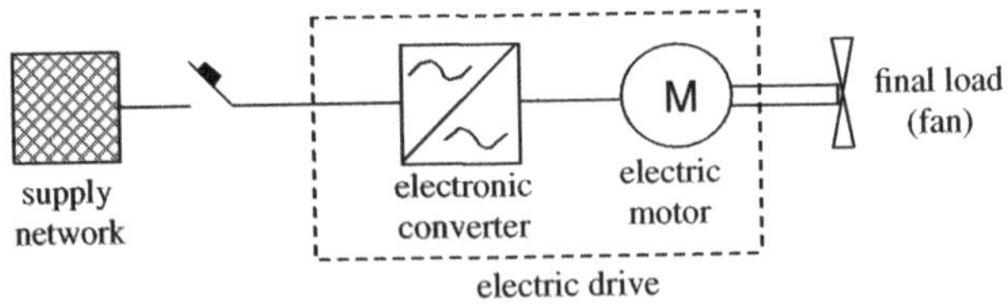

FIGURE 13.1. The electric drive as a small (partial) electric power system.

2. *In vehicle-testing laboratories, where cars are put on rollers.* When the car is asked to rotate its wheels, the rollers rotate and the torque on the rollers simulates the real-life torque generated on wheels by the wheel-road contact. In these labs, vehicles are tested at variable speeds (they are actually still, but their wheels rotate at variable speeds). Under these conditions, the evaluation of vehicle performance must take realistic cooling into account. Since in real life vehicles move against air, in these labs variable speed fans must be provided to produce an air flux against the car that has the same speed as the air hitting the vehicle when it is on the road, taking actual wheel speed into account.

Other very important examples of partial power systems are the *electric installations*. They are defined and dealt with in the following section.

13.2 ELECTRICAL INSTALLATIONS: A BASIC INTRODUCTION

Electrical installations are power systems aimed at final electricity consumption—that is, in homes, offices, and small companies. In electrical installation, electricity is generally taken from the public distribution network.

Recently the concept of electrical installation has slightly evolved, since the possibility of partial electricity generation inside electrical installations today is not uncommon. Indeed, today small companies or individual users can have in-house small photovoltaic electricity generators, whose energy adds up to that taken from the mains.

In small electrical installations such as those of homes or offices, the public distribution network delivers energy at a low-voltage level (LV). Larger installations—for example, for large companies or industries—may receive power at the medium voltage level or even at the high voltage level, but their analysis is far beyond the scope of this book. An example is shown in Figure 13.2, by means of an unifilar representation (compare with Figure 6.6).

Downstream of the point of connection (the so-called "point of common coupling" or PCC) with the public distribution network, a meter measures the energy consumed by the installation—that is, by the sum of all individual loads.

The energy (and power) entering the installation is conveyed, by means of a distribution board, to different sections of it. In a home, for instance, one section might be the kitchen, which is normally full of electrical appliances, and the other two or three could feed other zones of the home.

Normally it is a good rule to use an installation section for every home zone, so that when the circuit breaker situated at the beginning of a section is open (i.e., no current or voltage can be transferred downstream) the whole home zone is free of electricity hazards, while the other zones continue to function normally.

Each installation section has several potential loads, such as plugs (for sockets connection), lamps, or other loads. Except for sockets loads, these can be connected or disconnected from the main feeding line of the section through connection/disconnection switches ("sw" in the figure).

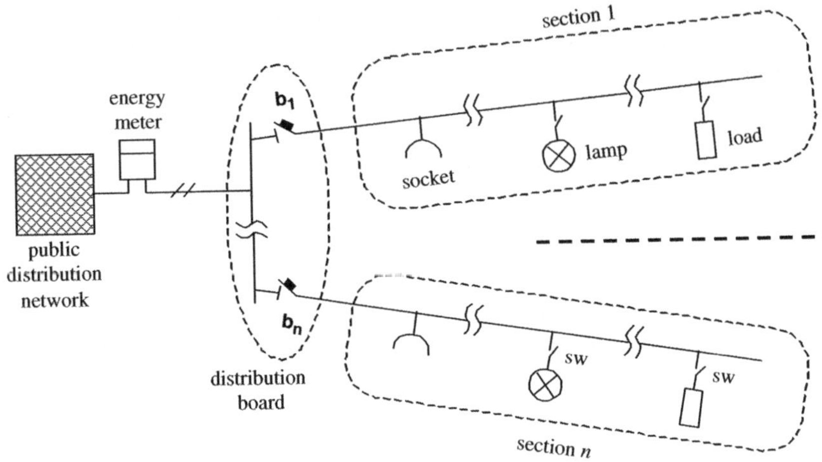

FIGURE 13.2. Diagram of a home or small office installation.

The different sections of the installation are fed though circuit breakers ($\mathbf{b_1}$ to $\mathbf{b_n}$ in Figure 13.2), which are able to interrupt currents either manually, because the installation user wants to disconnect electricity from a home zone, for instance, or automatically. An example of automatic operation is when the electricity supply must be interrupted because of a short-circuit to avoid damage to things or living bodies.

Focus on: *Terminology*

What we call *section* is instead called *circuit* in technical literature. It is very important to realize that in the electrical installation field of electrical engineering the word "circuit" has a totally different meaning from the way this word is used in Part II of this book and in all circuit theory books and scientific literature. This is why, since readers are not required to be electricity experts, a different word,—that is, *installation section*,—has been used to facilitate understanding.

The distribution scheme illustrated in Figure 13.2, which is typical for homes and small offices, is called "radial," since there is a central point (the distribution panel) from which different sections come and from which the distribution lines radiate outwards.

For more complex situations a multilevel radial scheme may be used, as illustrated in Figure 13.3. In this case, there are several distribution boards with two (or maybe more) levels of importance, representing a more hierarchical structure of the installation: a Main Distribution Board (MDB) and some Secondary Distribution Boards (SDB1 to SDBn). Also this case, the size of the area to be covered means that individual zones have to be grouped into larger installation areas, so as to reduce the number of lines from individual zones toward the main distribution board.

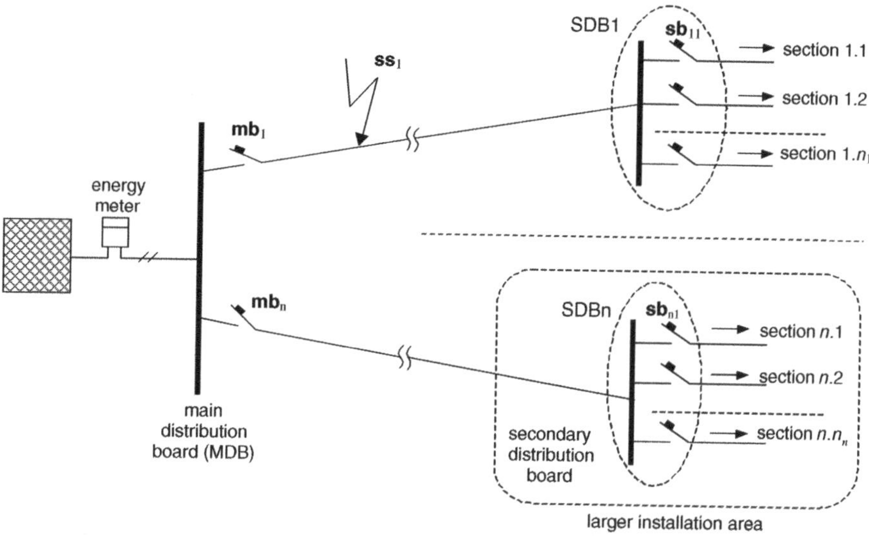

FIGURE 13.3. Distribution scheme with two levels of distribution boards.

When a secondary distribution board breaker such ad $\mathbf{sb}_{1.1}$ opens, it interrupts electricity supply only to Section 1.1, while if a main breaker such as $\mathbf{mb}_2$ opens, supply to a larger installation area is interrupted. Therefore, whenever possible, only the breakers immediately upstream of the fault must open. Breakers such as $\mathbf{mb}_1$ and $\mathbf{mb}_2$ are needed to protect the system in case of faults along the line connecting the main (or primary) to a secondary distribution board. If, for instance, the short circuit $\mathbf{ss}_1$ occurs, it is mandatory for $\mathbf{mb}_1$ to open, since the action of the downstream breakers $\mathbf{sb}_{11}$ to sb_{n1} will not interrupt the supply to the short circuit.

The purpose of these circuit breakers is to interrupt supply in three cases:

a. whenever a fault can cause a short circuit, which in turn can damage equipment and indirectly incur safety risks (because the short-circuit current is very large and causes cables to overheat which in turn might cause fire);
b. whenever damage to the insulation of cables could cause voltage to be applied to outside parts of electrical appliances that may be touched by humans or livestock, thus creating a high risk of injury or even death;
c. when the user wants to eliminate the supply of parts of the installation, to allow maintenance for example.

These three functions are commonly (but not mandatorily) combined into single devices, such as the circuit breakers shown in Figure 13.3.

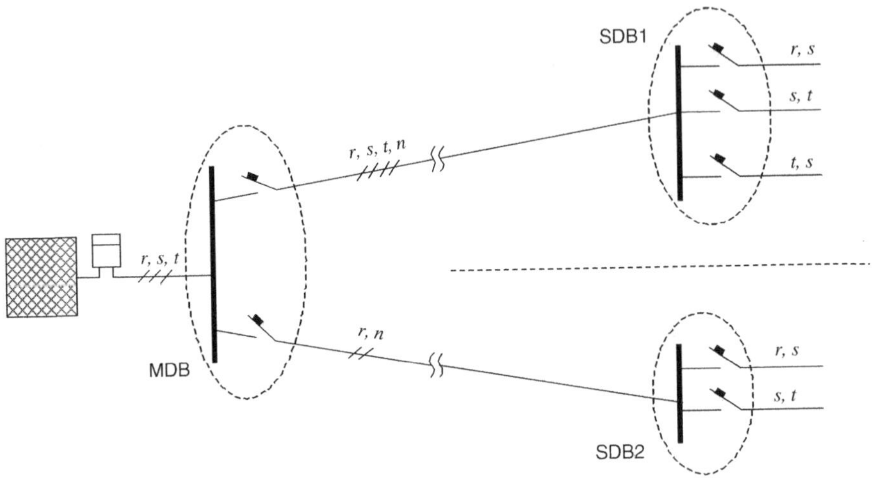

FIGURE 13.4. An example of a three-phase electrical installation, with some single-phase lines (the conductors between MDB and SDB2 might be *s,n* or *t,n*).

The number of levels of distribution boards can be greater than two. For instance, a multiple-storey house, containing a single family, might have three-level architecture:

- *a main distribution board*, located on the ground floor, immediately after the point of supply for the public distribution network;
- *floor subdistribution boards* on each floor;
- possibly some *final distribution boards* in different zones of the most loaded floors.

The examples shown in Figures 13.2 and 13.3 refer to smaller, single-phase installations, in which the conductors of each line connecting switchboards to loads are constituted by two wires. Only for more complex installations, especially in Europe, connection with the public distribution network is three-phase, and at least some of the lines connecting switchboards to loads are three-phase. An example of three-phase installation is shown in Figure 13.4. In this example the lines up to Secondary Distribution Boards SDB1 and SDB2 are three-phase four-wire, while the lines going out from these boards are single phase (phase-to-neutral supply). These lines are designed in such a way that the load is as uniform as possible between phases *r, s, t*. Thus, the three-phase lines are as uniformly loaded as possible, and the three-phase system as balanced as possible (remember that only *balanced* three-phase systems were examined in Chapter 6).

This can be achieved, for instance, when several phase-to-neutral lines are fed by MDB, connecting them from to line-neutral pairs: one can be *r, n*, another *s, n*, and another *t, n*.

More information about the way users' installations are connected to the public distribution network will be provided in Chapter 15.

More in Depth: *Series and shunt distribution*

In the previous section, even if not explicitly stated, it was taken for granted that the best way of distributing electricity to loads is by the so-called *shunt connection*. However, this was not the case at the beginning of electrification, in the two first decades of the twentieth century, when a lot of discussion took place on this.

Therefore, to satisfy the reader's curiosity, here are some comments about shunt and series connection.

In principle, a distribution system aims to deliver the electricity produced by a generator, which is usually very large, to a considerable number of users (= loads), each one usually very small. The autonomy of each load must be ensured, so that each device can modify its power request without changing the operating conditions of the others. Basically, the connection between a generator and a set of users can be carried out in two ways, illustrated in Figure 13.5. In the first case, loads are shunt-connected, and in the second they are series-connected.

In order to allow each user to change his or her own consumption—or even to be connected or disconnected without disturbing others—the generator must be regulated in such a way as to keep constant the common voltage in the first case and keep constant the common current in the second.

In a distribution system operated at constant voltage, each user is connected to the power line by a switch, which guarantees independence from the other loads. As the number of users grows, the voltage is kept constant, while the current provided by the generator increases. A short-circuit between the conductors of the power line causes a severe malfunction called "fault", because the regulating system, which is aimed at keeping the voltage constant, gives rise to a dangerous overcurrent.

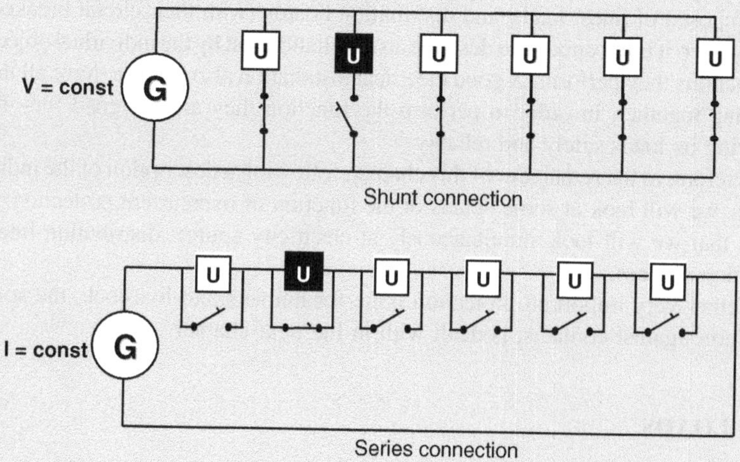

FIGURE 13.5. Shunt and series connection of loads in a distribution system.

Conversely, an interruption of the circuit has rather less severe consequences, since it simply shuts off the supply to downstream users, who are put "out of service".

In the series distribution system, each user is rigidly connected to the power line and is excluded from the service through a bypass switch; the higher the number of users fed at a constant current, the higher the voltage required at the generator terminals. In this case, a short circuit in the power line is not a fault, rather a disservice, because only downstream users are unsupplied. An interruption of the circuit is, instead, a proper fault: the current vanishes but the generator tends to keep it constant; thus very high voltages are generated and distributed to all users located upstream of the interruption.

The shunt distribution system has prevailed since the beginning of electrification, for many reasons. Among them is the possibility of connecting and disconnecting small devices with independent plug–socket couplings without requiring any other action in the system; in the series system, instead, the bypass switch must always be manoeuvred before connecting or disconnecting the user. The main advantage of the shunt distribution system derives from the fact that it maintains the voltage at a constant value everywhere, so that it is not dangerous for humans, regardless of the extension of the system and the presence of faults. It is known, in fact, that a voltage exceeding a few hundred volts is unsafe for residential use and small industrial applications—that is, in environments where no special precautions are adopted and no specialized personnel are employed.

For these reasons, electricity distribution systems, except in rare, special cases, were developed according to the *shunt connection of loads, operated at constant voltage.*

Looking at Figures 13.2 and 13.3, it can be seen that basically electrical installations are composed of lines, loads, and distribution boards (with their circuit breakers).

However, it is not enough to describe an installation just by the individual objects and the functions they perform. A good electrical installation also needs to have all its parts operating together, in order to perform the function they are designed for—that is, supplying its loads safely and reliably.

Therefore, in the remainder of this chapter, following a description of the individual objects, we will look at some details of the function of overcurrent protection, which means that we will look simultaneously at electricity source, distribution lines, and protection devices.

Another very important protection issue for humans and livestock, the so-called protection against contacts, is dealt with in the next chapter.

13.3 LOADS

Loads of a low-voltage installation are small devices that absorb power. More precisely, in general they absorb active and reactive powers (P and Q: see Chapter 5).

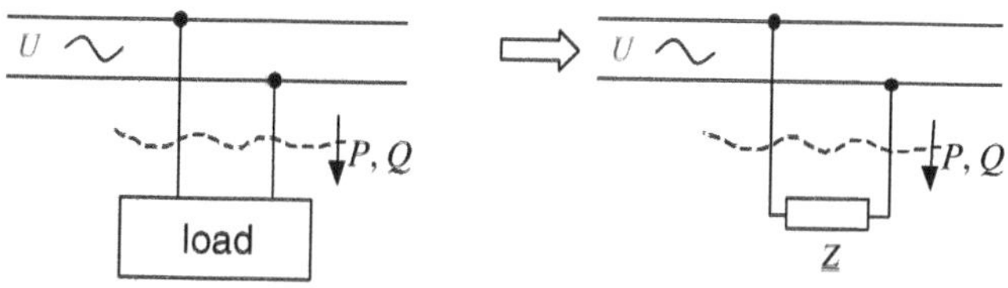

FIGURE 13.6. A single-phase load, modelled as a complex impedance.

Active power absorption means that some power drawn from the installation (coming from the public electricity distribution grid) is converted inside the load into some other form of energy, normally heat.

Heat is directly produced in heaters and ovens. It is *indirectly* produced in other cases. Consider a fan, for example. In this case electricity is first converted (after some power loss) into kinetic energy of air, but when air movement stops, this kinetic energy has been converted into heat.

The absorption of reactive power is related to the fast oscillation of instantaneous power around a null average (again, see Chapter 5), which occurs when inductors or capacitors are involved. The amplitude of these oscillations, with the addition of the appropriate sign, is what is called reactive power Q. By convention, positive reactive power is absorbed by inductors and generated by capacitors.

Examples of loads of a home installation are heaters (and ovens), fans, electronic appliances such as TV sets and PCs, and subsystems containing several elementary loads, such as dishwashers containing heaters and motors.

It is common practice to consider home loads as modelled by equivalent impedances, based on the absorption of P and Q. Consider Figure 13.6: the voltage U is shown without a polarity reference, because U is only the rms value and is a positive number without the need of a reference. This representation simply says that the rms value of the AC voltage between the two conductors of the line is equal to U. From equations

$$P = UI\cos\varphi \quad Q = UI\sin\varphi \quad S = \sqrt{P^2 + Q^2} = UI \quad U = ZI$$

it is

$$Z = \frac{U}{I} = \frac{U}{S/U} = \frac{U^2}{S} = \frac{U^2}{\sqrt{P^2 + Q^2}} \qquad (°)$$

and

$$\varphi = \tan^{-1}\frac{Q}{P} \qquad (°)$$

Equations (°) and (°°) give the amplitude and angle of the complex impedance $\underline{Z} = Ze^{j\varphi}$ as a function of line voltage and absorbed active and reactive power. Therefore, they allow our loads to be modelled as impedances.

Electric appliances normally bear a *nameplate* containing useful information such as

- nominal voltage U_n,
- nominal frequency f_n,
- nominal absorbed power P_n, and
- nominal current I_n (optional).

Nominal values, frequently referred to as *rated*, are those for which the appliance is designed. For the appliance to perform correctly, actual values should not differ too widely from these nominal values.

In particular, the actual (rms) voltage to which the load is connected must not differ from U_n by more than ±10%.

As far as frequency tolerance is concerned, loads not containing electric motors are highly tolerant of frequency variations. Thus frequency is really not a concern since in most countries it is kept under strict control to ensure the correct operation of the power system (some details in Chapter 15).

In the previous list, nominal power P_n and current I_n are absorbed by the load when it is fed by a source having as voltage the nominal voltage.[1]

From the nameplate data, it is easy to compute Q_n (the reactive power absorbed when the load itself is fed from a source with voltage U_n) and $\underline{Z}_n$, the impedance equivalent to the load, which constitutes a valid model of this load:

$$S_n = U_n I_n \Rightarrow Q_n = \sqrt{S_n^2 - P_n^2} \Rightarrow Z_n = \frac{U_n}{I_n}, \qquad \varphi_n = \tan^{-1}\frac{Q_n}{P_n}$$

and therefore

$$\underline{Z}_n = Z_n e^{j\varphi_n}$$

It is apparent that the equivalent impedance correctly models the load only when the voltage at its terminals is exactly equal to the nominal value U_n. As an approximation it can, however, be assumed that the computed equivalent impedance in nominal conditions $\underline{Z}_n$ stays sufficiently valid also when voltage is not exactly equal to U_n, but it is very near to it.

[1]And as frequency the nominal frequency, although this parameter, as mentioned above, is normally not critical.

Example 1. Evaluate the equivalent impedance of a fan heater, supposedly fed at the nominal voltage, with the following nameplate data:

$$U_n = 120\,\text{V}, \qquad I_n = 18\,\text{A}, \qquad f_n = 60\,\text{Hz}, \qquad P_n = 2\,\text{kW}$$

$$S_n = U_n I_n = 120 \cdot 80 = 2160\,\text{var}$$

$$Q_n = \sqrt{2160^2 - 2000^2} = 815.8\,\text{VA} \Rightarrow Z_n = \frac{120}{18} = 6.667\,\Omega$$

$$\varphi_n = \tan^{-1}\frac{815.8}{2000} = 22.19°$$

More in Depth: *Load modelling*

Modelling loads as impedances is common practice in electric power engineering, when loads are fed at voltages close to the nominal values (a common situation), and the values of P and Q are required, rather than their variation with voltage U.

If the variation of P and Q is to be evaluated as a function of voltage, the impedance model is often not adequate. In fact for an impedance it is

$$P = \frac{U^2}{Z}\cos\varphi \equiv U^2, \quad Q = \frac{U^2}{Z}\sin\varphi \equiv U^2 \qquad \text{(since } \cos\varphi \text{ does not usually vary with voltage)}$$

This quadratic dependence of P and Q on voltage is not the only possibility. Consider, for instance (here k means a quantity that remains constant when U varies):

1. *Asynchronous motors.* When the voltage is modified, the motor speed remains substantially constant and therefore also the steady-state load torque, which depends on this speed, is constant. Therefore $P \cong k$; moreover, it can be seen that $Q \cong k\,U^2$.
2. *Electronic converters feeding their loads (e.g., controlled rectifiers).* These converters are normally feedback-regulated so as to supply their loads at the set power. Therefore it is: $P \cong k$, $Q \cong k$.

When aggregate loads are involved (for instance, the whole load of a home or a building), their behaviour as a whole is a weighted average of the behaviour of its individual components.

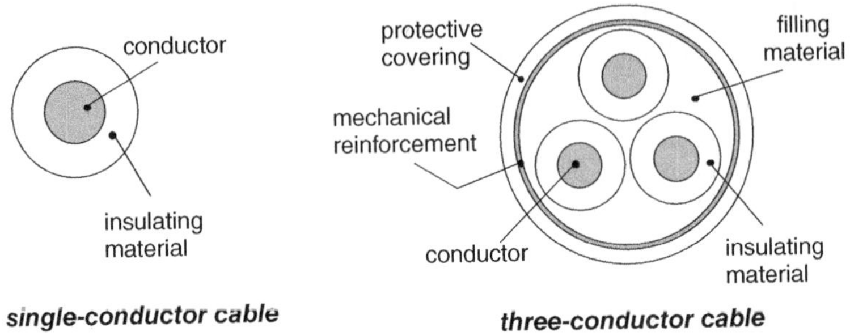

FIGURE 13.7. Typical construction of single-conductor or three-conductor cables.

13.4 CABLES

In electrical installations the lines that connect switchboards to loads or sockets are constituted by cables, which have a structure very different from that of overhead lines, pictures of which are shown at the very beginning of Chapter 6.

Cables are composed of conductors and their insulation, as shown in Figures 13.2 and 13.3. They can contain one or more conductors, usually three, which are adequate for three-phase systems. They are shaped like wires. From their cross section, shown in Figure 13.7, it can be seen that the conductor is a cylinder (whose length is greater than its diameter) which carries current; the conductor is covered with an external layer that insulates it from the outside; a three- or multiple-conductor cable needs additional filling material and protective covering.

13.4.1 Maximum Permissible Current and Choice of the Cross-Sectional Area

When current flows through a cable, it generates heat according to Joule's law. In fact, the cable has its own resistance R, and resistance per unit length R_l, and therefore the heat produced per unit of time and length is

$$P_{\mathrm{j}} = R_l I^2 \tag{13.1}$$

If current is in ampere and resistance in ohm/metre, P_{j} is in W/m or J/(m · s).

Consider a cable initially without current, in thermal equilibrium with its environment: it will have the same temperature as the air (or other substance) surrounding it.

When current starts to flow, let it, for simplicity's sake, be constant: heat is generated according to equation (13.1). This partly heats the cable and partly is dispersed to the outside.

To consolidate this idea, consider a free-standing cable surrounded only by air. Thermal analysis can be carried out according to the sketch shown in the left-hand part of Figure 13.8, in which $R_{\theta l1}$ and $R_{\theta l2}$ represent the thermal resistances, per unit length, between conductor and air: $R_{\theta l1}$, in particular, represents the resistance due to

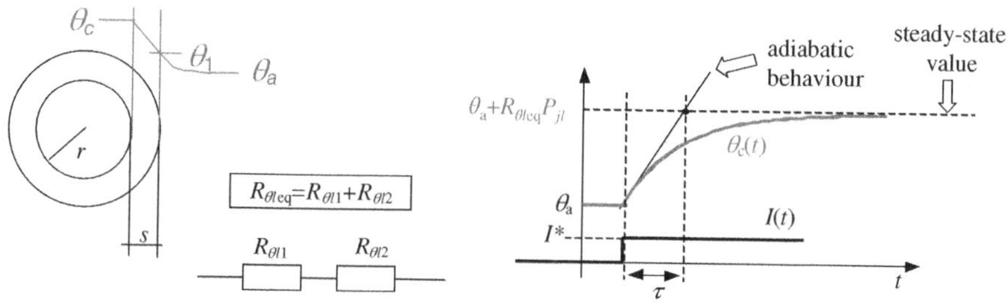

FIGURE 13.8. Thermal transient of a simple free-standing cable (conductor with insulation) surrounded only by air.

the insulation, and $R_{\theta l2}$ is related to the convection phenomenon of air around the cable.

> If the reader is not familiar with conduction and convection, it is sufficient to know that the thermal power (joule per second) that flows through a thermal resistance is proportional to the temperature difference at its two ends. For instance, for the insulator it is $P_{\theta f}=(\theta_c-\theta_1)^*R_{\theta l}1$, $P_{\theta f}$ being the heat flowing in J/s. The two thermal resistances can be combined as a series, exactly like electrical resistances, and therefore it can also be written that $P_{\theta f}=R_{\theta leq}(\theta_c-\theta_a)$, with $R_{\theta leq}=R_{\theta l1}+R_{\theta l2}$.

The behaviour of conductor temperatures over time can be obtained by solving the following simple differential equation which equates the heat produced, per unit length, to the sum of that absorbed by the conductor heat capacitance and that dispersed in the environment:

$$P_{jl}=C_{\theta l}\frac{d\theta_c}{dt}+(\theta_c-\theta_a)/R_{\theta leq} \tag{13.2}$$

where $C_{\theta l}$ is the thermal capacity (also known as heat capacity) of the conductor, per unit length. To find the solution of equation (13.2), it is easier if the following auxiliary variable is defined:

$$\theta=\theta_c-\theta_a$$

Consequently, equation (13.2) can be rewritten as

$$P_{jl}=C_{\theta l}\frac{d\theta}{dt}+\theta/R_{\theta leq}$$

or:

$$\theta_\infty=\tau\frac{d\theta}{dt}+\theta \qquad (^\circ)$$

$$\theta(0)=0$$

where $\theta_\infty=P_{jl}\,R_{\theta leq}$ and $\tau=R_{\theta leq}\,C_{\theta l}$.

TABLE 13.1. Maximum Operating Continuous Temperatures θ_c per Type of Insulation

Type of Insulation	Steady-State Limit θ_c (°C)
Thermoplastic (PVC)	70
Thermosetting (XLPE or EPR rubber)	90

It is easy to verify that equation (°) is solved by

$$\theta = \theta_{\infty} - \theta_{\infty} e^{-t/\tau}$$

The situation is described by the plot shown in the right-hand part of Figure 13.8.

Before the transient starts, the conductor has the same temperature as the surrounding air; when the current, rms value, suddenly reaches I^*, thus generating constant thermal power per unit length P_{jl}, the temperature begins to change. At first the heat produced is almost entirely accumulated in the conductor heat capacity. If no heat were dispersed in the environment through the thermal resistance $R_{\theta l\mathrm{eq}}$, the conductor temperature $\theta_c(t)$ would continue along the straight line indicated as "adiabatic behaviour" in Figure 13.8. In reality, as long as the conductor temperature rises above θ_a, the heat dispersion in the environment increases; therefore, more and more of the generated heat is transferred to the air surrounding the cable. The analysis and design of the initial (adiabatic) behaviour and final (steady-state) behaviour are of interest in electrical installation.

An important part of the plot of the right-hand part of Figure 13.8 is the horizontal straight line characterizing the steady-state thermal behaviour of the cable. It clearly shows that the steady-state temperature of the cable depends on room temperature θ_a and the product between the power dissipated in the cable per unit length and the thermal resistance per unit length.

On the other hand, all cables can be subjected to their maximum operating temperature for unlimited time.[2] The maximum operating temperatures for the most common types of insulation are shown in Table 13.1. The shown data are those used in International standards.

The data on the table give the amount of heat that can be afforded for a given room temperature:

$$P_{jl\max} = R_l I_z^2 = (\theta_c - \theta_a)/R_{\theta l\mathrm{eq}} \qquad (13.3)$$

in which the following definition is relevant:

[2]The maximum temperature allows the cable to have acceptable electrical qualities at its end of life, normally set at 30 years.

Definition: *Maximum permissible current (of an insulated conductor, of a cable)* I_z

The maximum permissible current is the maximum value of current that an insulated conductor or cable can carry indefinitely without reducing its normal life expectancy. It is normally indicated by the symbol I_z.

In some standards it is named *current-carrying capacity.*

For instance, at a room temperature of 30 °C, a PVC cable has a margin of over-temperature of 40 °C that in turn will determine the maximum current the cable can carry. If room temperature is 40 °C, the over-temperature margin will be reduced to 30 °C.

In the case of free-standing conductors in air, equation (13.3) allows us to compute the maximum continuous current for the cable I_z.

Indeed, for the sake of conciseness, the value of $R_{\theta leq}$ was not developed in equation (13.3). Readers who feel at ease with thermal computations can easily verify that it is $R_{\theta leq} = (\alpha \cdot s + k)/(2\pi rhk)$, where h is the heat transfer coefficient (for convection) and k and s are the thermal conductivity and thickness of the insulation layer, respectively.

Unfortunately, real-life situations are different from the free-standing reference, mainly because:

- The cable is placed in conduits or on trays (see Figure 13.9).
- In the vicinity of the cable there are other loaded cables that contribute to warming the surrounding air.

These two difficulties make computation more involved, and it is preferable to find the actual values on tables that are available also in International standards (as IEC 60364-5-52).

By way of some simple examples, Table 13.2 shows the values of I_z in two common situations. It is apparent that the higher the number of loaded cables, the

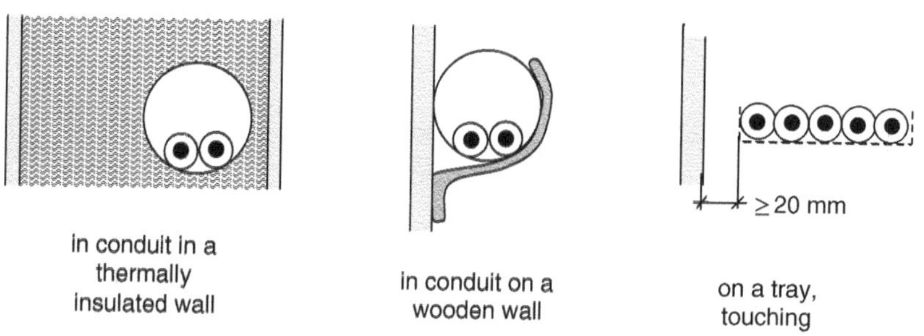

FIGURE 13.9. Some installation methods for cables.

TABLE 13.2. Some Examples of Current-Carrying Capabilities I_z in Amperes (PVC Insulated Cables and θ_a = 30 °C)

Nominal Cross-Sectional Area of the Conductor (mm^2)	Insulated Conductors in Conduit in a Thermally Insulated Wall		Insulated Conductors in Conduit on a Wooden Wall	
# of Conductors Loaded	2	3	2	3
1.5	14.5	13.5	17.5	15.5
2.5	19.5	18	24	21
4	26	24	32	28
6	36	31	41	36

lower the current capability (each cable warms the air surrounding the others). Moreover, installation in an insulated wall is a poorer means of dispersing heat than a conduit on a wooden wall.

It is fairly easy to find current-carrying capabilities at room temperatures which are different from the one to which a current-carrying capability table such as Table 13.2 refers.

Consider again the formula (13.3):

$$R_l I_z^2 = (\theta_{\text{lim}} - \theta_a)/R_{\theta leq}$$

If I_{z1} is at room temperature θ_{a1} and I_{z2} at θ_{a2}, it is

$$\frac{I_{z1}^2}{I_{z2}^2} = \frac{\theta_{\text{lim}} - \theta_{a1}}{\theta_{\text{lim}} - \theta_{a2}}$$

Consider now the reference capability at the reference temperature θ^* (found in the tables), I_z^*. The I_z at the generic temperature is then

$$I_{z1} = I_z^* \sqrt{\frac{\theta_{\text{lim}} - \theta_{a1}}{\theta_{\text{lim}} - \theta^*}}$$

For instance, to correct I_z values referring to 30 °C for use of a PVC cable at a room temperature of 40 °C, one should compute (see Table 13.1)

$$I_{z1} = I_z^* \sqrt{\frac{70 - 40}{70 - 30}} = 0.866 I_z^*$$

All cables of an installation must be designed to carry currents that are below their maximum permissible current.

Therefore the electrician or installation designer must find an estimate of the maximum current the cable is called to carry for long periods of time during the

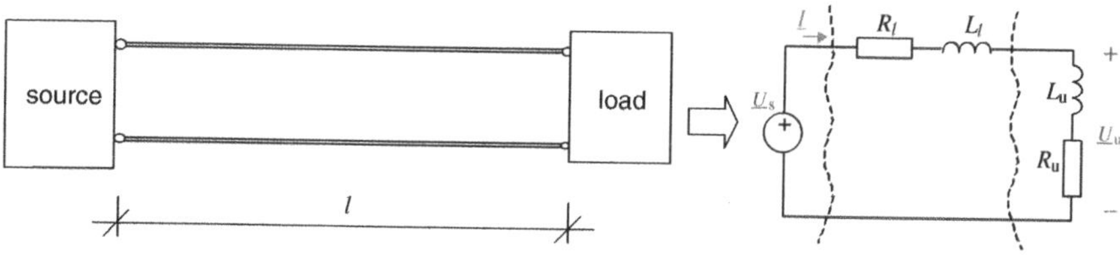

FIGURE 13.10. A cable line between source and load and its equivalent metacircuit.

installation life. This current is called *design current* I_B. Therefore, all cable maximum permissible currents must be coordinated with I_B, according to the following simple rule:

$$I_B \leq I_z \tag{13.4}$$

Moreover, some means must be provided to guarantee that currents much larger than I_z cannot flow through the cable, except for limited times. This is a topic called overload protection and will be dealt with in Section 13.6.

13.5 DETERMINING VOLTAGE DROP

A cable line between a source and a load can be modelled using the metacircuit introduced in Chapter 5, as shown in Figure 13.10. The resistance R_l contains the sum of the resistance of the upper and lower conductors of the line. The inductance L_l models the effects of self-induction in the rectangular surface of the transmission line, bound by the conductors and source and load.

When dealing with electrical installations, it is often necessary to evaluate the so-called *scalar voltage drop* (or simply voltage drop)

$$\Delta U = (|\underline{U}_s| - |\underline{U}_u|)/\sqrt{2}$$

that is, the difference in the rms values of the voltages one can measure near the source and near the load.[3]

A formula which gives a sufficiently close approximation of this scalar voltage drop can be easily found by referring to the phasor diagram of the circuit shown in Figure 13.11. Here, to ease analysis, think the phasors as having as amplitude the rms values of the corresponding sine waves.

Instead of evaluating the difference in the moduli of $\underline{U}_s$ and $\underline{U}_u$, we compute, as an approximation, the difference between the projection of $\underline{U}_s$ on the direction of $\underline{U}_u$ and

[3]The factor $\sqrt{2}$ is needed when phasors have as their amplitude the peak of the sinusoidal waves they represent. In other cases, when the amplitude of phasors is directly the rms value of voltages, this factor must be dropped.

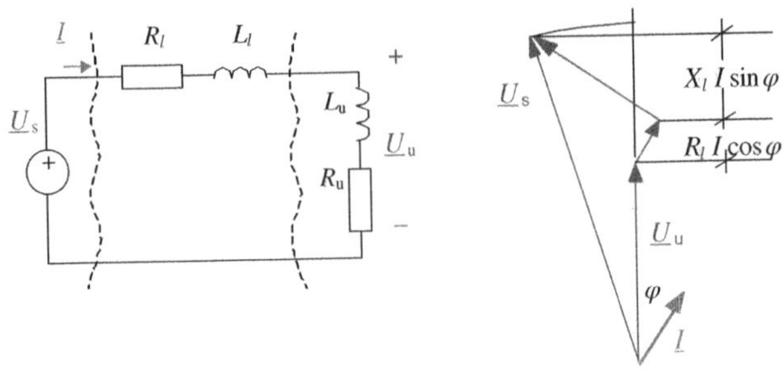

FIGURE 13.11. Diagram to compute the approximate formula of scalar voltage drop.

the modulus of $\underline{U}_u$ itself. Therefore, consulting the phasor diagram of Figure 13.11, we can write

$$\Delta U = U_s - U_u = I(R_l \cos\varphi + X_l \sin\varphi) \tag{13.5}$$

In low-voltage installations, except for cases of cables with cross-sectional areas greater than $95\,\text{mm}^2$, it easy to verify (by consulting the cable manufacturer's documentation) that it is $R_l \gg X_l$; and therefore

$$\Delta U \cong I R_l \cos\varphi, \quad R_l = 2\rho\frac{l}{A}, \qquad A > 95\,\text{mm}^2 \tag{13.6}$$

In large installations, much larger cross-sectional areas are used, and in these cases the full formula is adopted (13.5).

Example 2. A two-wire copper, PVC-insulated cable must be installed in a conduit on a wooden wall and must supply a 2-kW 500-var single-phase load at a distance of 30 m from a 120-V AC source. Determine the cross-sectional area of the cable, considering a maximum voltage drop of 4%.

First the current flowing in the line conductors I_B must be computed.

$$S = UI \quad \Rightarrow I_B = S/U = \sqrt{2000^2 + 500^2}/120 = 17.18\,\text{A}$$

From Table 13.2 it can be seen that a cross-sectional area $A = 1.5\,\text{mm}^2$ is adequate for this purpose (it will imply $I_z = 17.5$ A) from a thermal point of view. However, the voltage-drop criterion must also be satisfied. Since it is apparent that A will be much smaller than $95\,\text{mm}^2$, formula (13.6) can be used. So

$$\cos\varphi = P/S = 2000/\sqrt{2000^2 + 500^2} = 0.970$$
$$R_l = \Delta U/(I \cos\varphi) = 0.04 \cdot 120/(17.18 \cdot 0.970) = 0.288\,\Omega$$

To compute the area corresponding to R_l, it should be borne in mind that the copper resistivity is $\rho = 18 \cdot 10^{-9}\Omega\text{m} = 18\,\text{m}\Omega(\text{mm}^2/\text{m})$, thus:

$$A = \frac{2\rho l}{R_l} = \frac{2 \cdot 0.018 \cdot 30}{0.288} = 3.75\,\text{mm}^2$$

Thus the cross-sectional area is determined by the voltage drop requirement. The standardized value just above the computed value of $3.75\,\text{mm}^2$ is $4\,\text{mm}^2$ (Table 13.2). Using this conductor it will be $I_z = 32\,\text{A} \gg I_B$ and $\Delta U < 4\%$.

13.6 OVERCURRENTS AND OVERCURRENT PROTECTION

Any current exceeding the maximum permissible current is called an *overcurrent*. Measures are taken to protect electrical cables from the effects of overcurrents, that is, currents exceeding I_z. In practical installations two scenarios of overcurrents can occur:

- Overcurrents can occur in an electrically undamaged circuit, caused by excess load on a cable, for instance, when all the sockets in a room are loaded with heavy loads. In this case the current exceeds I_z by a limited amount and is called *overload current* or simply overload; overloads can be left on the system for a significant time without significant cable insulation damage.
- The overcurrent is caused by a "short circuit". A short circuit is a fault in which two conductors, intended as having different potentials in normal service, come into contact with each other. This can happen, for instance, when the cord connecting an iron to its plug is damaged and the two live conductors touch each other. In these cases the current flowing in the cable is much greater than I_z, and must be eliminated as fast as possible. Such currents are called *short-circuit currents*.

Overload and short-circuit currents must be dealt with in different ways, and this will be discussed in the next two sections; however, it is very common to combine protection against overcurrents and short-circuit currents in single devices—that is, the commonly used circuit breakers, which will be discussed in Section 13.6.3.

13.6.1 Overloads

Overloads having a small entity and time are tolerated in electrical installations. To facilitate the determination of protection, overloads are dealt with, by international agreement, in a conventional way: overloads up to 45% can persist, while overloads over 45% must be eliminated within a conventional tripping time (i.e., one or two hours depending on various factors, including local standards).

TABLE 13.3. Maximum Permissible Temperatures in Short-Circuit Conditions

	θ_c [°C]	θ_{lim} [°C]	k [A$\sqrt{s}$/mm^2]
Cu + PVC	70	160	115
Cu + EPR	90	250	143
Al + PVC	70	160	74

This is summarized by the following formula[4]:

$$I_2 \leq 1.45 I_z \tag{13.7}$$

This condition will be discussed again in Section 13.6.3.

where I_2 is the current that is certainly interrupted within the conventional tripping time.

All the breakers shown in Figures 13.2 and 13.3 must have the ability to protect the cables they feed from overload. For instance, $\mathbf{mb}_1$ will protect the cable between $\mathbf{mb}_1$ and SDB1, while $\mathbf{sb}_{11}$ will protect the cable feeding section 1.1.

13.6.2 Short Circuits

Short-circuit currents must be interrupted as fast as possible. What does "as fast as possible" mean in quantitative terms?

The immediate effect of very large currents flowing during a short circuit is that the conductor and its insulating material heat up considerably. We must ensure that the temporary high temperature reached by the conductor does not damage the insulator excessively.

This is obtained using the following formula:

$$I^2 t_i < k^2 A^2 \tag{13.8}$$

where:

- I is the rms value of the short-circuit current
- t_i is the time needed by the circuit breaker to interrupt the short-circuit current
- A is the cross-sectional area of the conductor in mm^2
- k is a function of the material of the conductor and insulation material and can be read from Table 13.3.

The maximum tolerable temperature in short-circuit conditions are those shown in Table 13.3, thus completing the information shown in Table 13.1, which show only temperatures θ_c that can persist continuously over time. In the following the subscript "i" of t will often be dropped.

[4]From IEC 60364-4-43.

More in Depth: *Evaluation of cable heating during short-circuit*

The reader might wonder where (13.8) comes from.

Consider the transient shown in Figure 13.8. Short circuits must be interrupted very fast. Therefore, in the thermal transient shown in the right part of the figure, the part to be taken into account is that immediately after the onset of current I^*, for a shorter duration than the thermal time constant τ. This means that the short circuit can be analysed under the assumption that the thermal transient is substantially adiabatic.

Equation (13.2) thus simplifies into

$$P_j = C_\theta \frac{d\theta_c}{dt}$$

that is,

$$RI^2 = \rho\frac{l}{A}I^2 = C_{\theta l}\frac{d\theta_c}{dt} \Rightarrow \rho\frac{l}{A}I^2\,dt = clA\,d\theta_c \Rightarrow I^2\,dt = A^2\frac{c}{\rho}\,d\theta_c$$

where c is the conductor's specific heat (thermal capacitance per volume.)

To correctly integrate this differential equation, the variability of the conductor's (copper or aluminium) resistivity to temperature must be taken into account. Therefore, if θ_{lim} is the maximum temperature permissible in a short circuit, it will be

$$\int_0^{t_i} I^2 \cdot dt = A^2 \cdot \int_{\theta_c}^{\theta_i} \frac{c}{\rho(\theta)}\,d\theta$$

where t_i and θ_i are breaking time and conductor temperature at breaking time. Finally,

$$\theta_i \le \theta_{\text{lim}} => \int_0^{t_i} I^2 \cdot dt \cong I^2 t_i \le A^2 \cdot \int_{\theta_c}^{\theta_{\text{lim}}} \frac{c}{\rho(\theta)}\,d\theta = k^2A^2$$

where k is the value used in (13.8), which, as stated, depends on the conductor (values of c and $\rho(\theta)$), and insulating material (value of θ_{lim}). The quantity $\int_0^{t_i} I^2 \cdot dt \cong I^2 t_i$ is called breaker *let-through energy*.

TABLE 13.4. Symbols of Different Switches Used in Electrical Installations

Disconnector (or Isolator)	Load-Break Switch	Circuit Breaker Form 1[a]	Circuit Breaker Form 2
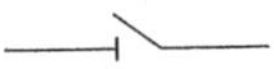			

[a]The two symbols shown as form 1 and form 2 of circuit breakers are considered equivalent in common practice, and they will be interchangeable in this book also. International standards make some distinction between the two, which is not, however, very clear.

13.6.3 Breaker Characteristics and Protection Against Overcurrents

There are different kinds of switching devices in electrical installations. The main symbols and names are shown in Table 13.4. The corresponding switches are:

- *Disconnector (or Isolator).* This has the purpose of isolating part of an installation from another, so that when they are open it is guaranteed that the voltage and current cannot be transferred between the two parts. If, for instance, in the drawing of Figure 13.2 switch $\mathbf{b}_1$ (which combines the functions of a disconnector and a circuit breaker) is open, personnel can operate safely on section 1 of the installation.
- *Load-Break Switch.* These are switches normally used in houses—for instance, to switch a lamp on and off. They are capable of opening the normal load current and are manually operated. They do not provide automatic protection of the circuit they control.
- *Circuit Breaker.* These are capable of tripping (i.e., switching from on to off position) automatically, when the current exceeds given thresholds. They are capable of cutting short-circuit currents. Circuit breakers are always capable of performing the functions of load-break switches. Although not always guaranteed, they usually also have the functions of a disconnector.

Circuit breakers have significant parameters that must be defined and understood, before we go on to explore how they must be chosen to guarantee adequate protection:

Definitions: *Characteristic circuit-breaker parameters*

- *Nominal Current I_n.* This is the reference current that breaker characteristics refer to. A breaker can be called to carry I_n indefinitely, without any damage to the breaker and without tripping.
- *Regulated Current I_r.* In some breakers it is possible to set the overload tripping threshold to a value lower than I_n, in order to comply with different installation conditions. This set value is the regulated current I_r, which can be carried indefinitely without tripping.

- *Conventional Overload Trip Current I_2.* This is the current that certainly causes the breaker to trip within a conventional tripping time t_2. Time t_2 may be 1 hour or 2 hours according to local standards.
- *Breaking Capacity.* This is the maximum current the breaker is capable of breaking (under given conditions).

Since any breaker can be called to carry I_n (or I_r) indefinitely, the following prescription must be set:

Rule: *Coordination of I_B and breaker nominal or regulated current*

$$I_B \leq I_n \qquad (\text{or } I_B \leq I_r \text{ for breakers that have } I_r \text{ instead of } I_n)$$

For a breaker to be able to protect a cable against overcurrents, the overload protection rule (13.7) must also be verified. However, standards require that circuit breakers comply with the following correspondence between I_n (or I_r) and I_2:

$$I_2 \leq 1.45 I_n \qquad (\text{or } I_2 \leq 1.45 I_r)$$

Therefore the condition (13.7) is verified simply by using the following rule:

Rule: *Circuit breaker overload protection condition*

When a cable is protected from overload by a circuit breaker, it must be

$$I_n \leq I_z \qquad (\text{or } I_r \leq I_z \text{ for breakers that have } I_r \text{ instead of } I_n)$$

Overload protection can be ensured also by other devices called fuses. For them the general rule (13.7) still applies, but $I_B \leq I_n \leq I_z$ will no longer be sufficient, since for fuses $I_2 > 1.45 I_n$. However, this is beyond the scope of this book.

Before issuing the third rule—that is, the short-circuit protection rule—we first consider how circuit breakers trip.

Circuit breakers normally have delayed-instantaneous (or thermal-magnetic) tripping curves. An example is shown in Figure 13.12a. Note the two curves, related to minimum and maximum tripping time, respectively. Indeed, the breaker is not able to guarantee a single time value for which a current can persist before tripping, but

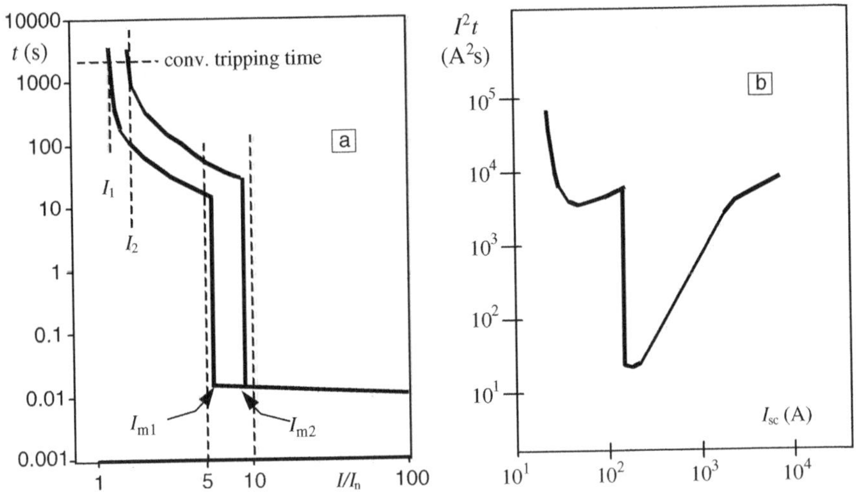

FIGURE 13.12. Sample tripping (*left*) and I^2t (*right*) curves of circuit breakers (solid lines).

only a range. Similarly, for a given time, an interval of currents that cause tripping in that time is given.

For instance, the minimum and maximum currents I_1 and I_2 are determined according to the conventional overload tripping time which, as we already said, is one or two hours, depending on national standards.

There is a zone (i.e., for currents below I_{m1}) for which the time to open the circuit is dependent on current: the larger the current, the shorter the time. This is normally achieved by sensors based on thermal effects. Since they require some time to heat up, they have a natural tendency to operate with times that depend on the value of the current. In another zone of the protective scheme—that is, for currents above I_{m2}—the intervention (tripping) is as fast as possible, and tripping time is quasi-independent of the amount of current flowing through the breaker. The zone above I_{m2} is called *range of instantaneous tripping*, since no intentional time delay has been added.

It is natural to combine the breaker and cable characteristics in such a way that the thermal part protects against overloads, thus avoiding tripping for only temporary overcurrents, while the instantaneous (magnetic) part operates against short circuits, since they require tripping which is as fast as possible.

The transition zone between the two regions of the curves is very important: low values of I_m guarantee better cable protection, but could more easily cause unwanted tripping for temporary overloads. The opposite occurs for large values of I_m.

By international convention, the breakers are classified according to where the transition zone lies between time-delayed and instantaneous tripping: this

TABLE 13.5. Standard Design of the Transition Between Delayed and Instantaneous Parts of Tripping Characteristic of Circuit Breakers[a]

Type	Range of Permissible I_{m1} and I_{m2} Values
B	$3I_n - 5I_n$
C	$5I_n - 10I_n$
D	$10I_n - 20I_n$

[a] I_n = circuit breaker nominal current.

classification[5] is shown in Table 13.5. For each type, both I_{m1} and I_{m2} must be within the indicated range.

To ensure that a cable is well protected against short circuits, equation (13.8) must be satisfied:

> **Rule:** *Short-circuit protection condition for circuit breaker*
>
> When a cable is protected from short circuit by a circuit breaker, equation (13.8) must be satisfied; that is, the I^2t allowed by the breaker to flow through the cable during tripping time t_i must not exceed the maximum permissible cable I^2t.

This can be graphically shown as indicated in Figure 13.12b. In practice, cable and breaker are correctly coordinated if all the possible short-circuit currents fall between I_a and I_b; if a circuit breaker is protecting a line such as that shown in Figure 13.13,

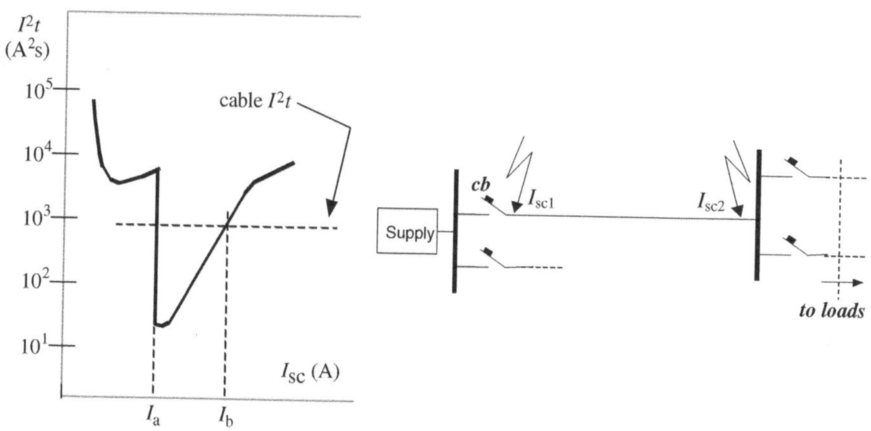

FIGURE 13.13. Coordination of the tolerable I^2t of a cable and the I^2t allowed to flow through the cable by the circuit breaker in the event of a short circuit.

[5] From International standard IEC 60898.

short-circuit currents at the beginning and at the end of the line must be evaluated. The farther a short-circuit current is from the power source, the smaller it is, so it will always be $I_{sc2} < I_{sc1}$. Therefore, the electrician will verify that: $I_{sc2} > I_a$ and $I_{sc1} < I_b$. If one of these cannot be verified, either the cable or the breaker must be changed so that both can be verified.

To summarize, a circuit breaker must protect a cable both against overload and short-circuit currents.

Overload protection is obtained by:

- First estimating the maximum design cable current I_B
- Choosing a cable which, in the given installation conditions, is such that $I_B < I_z$
- Choosing a breaker which is such that $I_B \leq I_n \leq I_z$

In addition to overload, short-circuit protection is achieved by obtaining the let-through energy of **cb** and the cable and then verifying this (see again Figure 13.13):

- $I_{sc2} > I_a$
- $I_{sc1} < I_b$

Example 3. Consider the cable analysed in Example 2. Imagine that the source can be modelled as a Thévenin equivalent, having as impedance a pure reactance $X_{Th} = 30\,m\Omega$. Find the short-circuit currents I_{sc1} and I_{sc2}.

Now determine a minimum value of I_2 and I^2t for the circuit breaker.

To determine I_{sc1} and I_{sc2} the circuit to be considered is the one shown below.

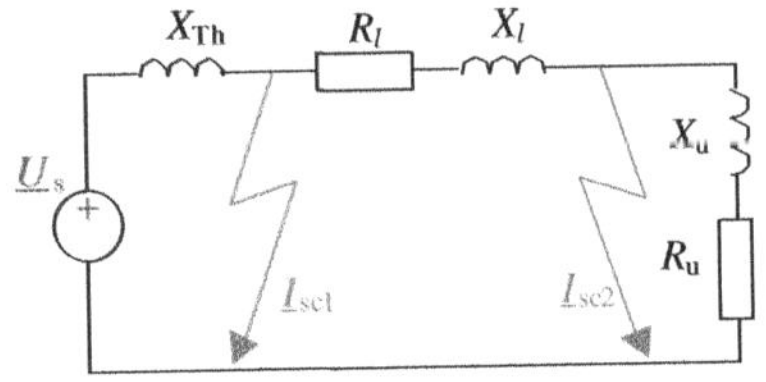

It is (remember, that $R_l = 288\,m\Omega$, X_l is negligible and $U_s = 120\,V$):

$$I_{sc1} = U_s / X_{Th} = 120/0.03 = 4000\ A$$

$$I_{sc2} = U_s / \sqrt{R_l^2 + X_{Th}^2} = 414\ A$$

and (taking k from Table 13.3)

$$I^2 t_{cable} = k^2 A^2 = (115 \cdot 4)^2 = 211.6\ A^2 s$$

The first requirement for the breaker is

$$I_n \geq I_B = 17.18\ A$$

Overload protection will be guaranteed by the further condition on I_n:

$$I_n \leq I_z \qquad \text{(that will imply } I_2 \leq 1.45\, I_z\text{)}$$

Finally, the computed values of I_{sc1}, I_{sc2}, and cable I^2t will have to be compared to the breaker I^2t in the manufacturer's documents.

In the large majority of cases (like this example), conditions $I_B \leq I_n \leq I_z$ will automatically single out a breaker that satisfies the short-circuit condition.

If this does not happen, a breaker with lower I^2t or a cable with a larger cross-sectional area A (and therefore larger I^2t) will have to be chosen.

Circuit-breaker manufacturers supply I^2t characteristics of their circuit breakers that facilitate the graphical verification of short-circuit conditions.

13.7 PROTECTION IN INSTALLATIONS: A LONG LIST

In Section 13.6, only thermal effects were taken into account in the coordination of circuit breakers and cables.

Electrical installations are very useful in our lives, but they must be well designed to be safe. This involves much more that protection against overcurrent. Just to give an idea, consider that International Standards[6] require the following kinds of protection:

- Protection against electric shock
- Protection against thermal effects
- Protection against overcurrents (overloads and short circuits)
- Protection against voltage disturbances and measures against electromagnetic influences
- Protection against power supply interruption

The aim of this book is not to train readers to become designers of electrical installations. However, it is important to understand how designers operate, in order to understand the results of what they do and to be able to communicate in a technically effective way with them if necessary.

Because of its great importance, protection against electric shock is the topic of the next chapter.

[6]International Electrotechnical Commission IEC 60364-1, Low-voltage electrical installations—Part 1: Fundamental principles, assessment of general characteristics, definitions.

14

ELECTRIC SHOCK AND PROTECTIVE MEASURES

> **For the Instructor**
>
> The topic of safety has an important place in a book on Electric Power Engineering Basics. We believe that any engineer with some knowledge of electricity should also have some basic knowledge of (a) the dangers associated with its use and (b) the main measures adopted by designers to minimize these risks.
>
> The aim of this chapter is not to teach students to design a safe electric system, but rather to build their awareness of the potential risks of electricity and to help them understand the purpose behind the measures they see in existing installations. This will also be useful to avoid practices that could lead to potential risks.

14.1 INTRODUCTION

This is a very special chapter. Design safety does not just involve technical knowledge and analysis, but also entails the designer's personal responsibility.

This is why the procedures for designing the safety aspects in any kind of engineering system are carefully defined in International Standards, thereby reducing individual discretion as much as possible. International Standards will thus be referred to frequently in this chapter.

Fundamentals of Electric Power Engineering: From Electromagnetics to Power Systems, First Edition. Massimo Ceraolo and Davide Poli.

Safety also involves complying with laws, which may vary according to the country and state in which the electrical installation is built and operates. An international book like this one can only provide some basic knowledge relating to international agreements on electricity safety. Engineers in charge of the actual designs must refer to national standards and law.

In this chapter, time-varying shapes are not used. All currents or voltages are represented by uppercase letters such as U or I. These symbols, as always in this book, indicate constant values when in DC and indicate rms values of sinusoids when in AC.

14.2 ELECTRICITY AND THE HUMAN BODY

14.2.1 Effects of Current on Human Beings

Current flowing in the body of a human being may have several types of effect, some of them harmful or even deadly. The greater the current or the exposition time, the more serious the effects. Frequency is also an important factor: the greatest dangers are caused by currents in the 15 to 100-Hz range, which includes the most commonly used frequencies of power systems all over the world, that is, 50 and 60 Hz.

DC currents are also dangerous, but in general they are significantly less dangerous than AC currents, when equal current (comparing DC current with rms AC) and exposition time are considered.

The following are the main effects of currents flowing through a human body. They are listed in increasing order of gravity, consequently also in order of intensity and/or duration of the current.

Involuntary Muscular Contraction. This is not harmful in itself if the current is kept under control (it is even used in gyms for stimulating muscular exercise). However, since it prevents voluntary muscle action, the victim may not be able to let go of the live part, so that the current continues to flow in the body, causing greater harm.

Respiratory Difficulties. Depending on its intensity, duration, and path, current flowing in the body can cause the involuntary contraction of respiratory muscles, as well as damage to (a) the neural activation pathways for these muscles and (b) the respiratory control mechanism in the brain. This respiratory block may prevent the victim from shouting for help. Normally, breathing difficulties disappear when the cause (i.e., the current) is removed; it may happen, however, that there is still some need for artificial respiration. If the respiratory block persists—for example, because the current is not interrupted—death will follow.

Ventricular Fibrillation. If a substantial part of the current entering the body traverses the heart, this current dominates over the physiological electrical impulses that coordinate the heart, causing *ventricular fibrillation*. This is a chaotic state in which the heart's fibres oscillate rapidly, preventing it from

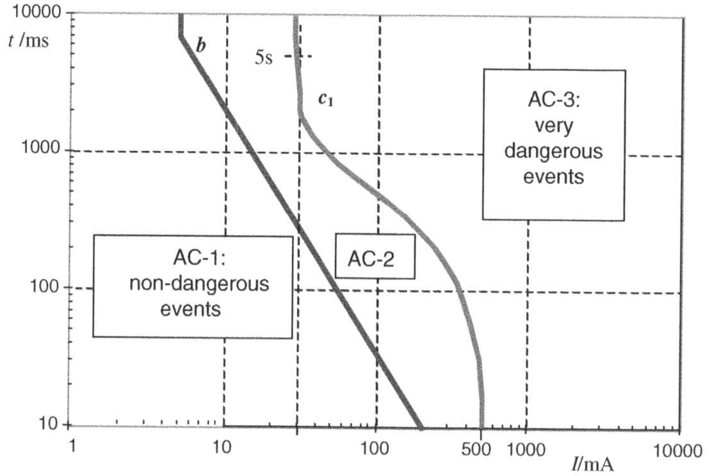

FIGURE 14.1. Conventional zones of effects of AC currents [1]. (left hand to feet; 15 to 100 Hz).

pumping blood. When ventricular fibrillation has started, it does not disappear naturally. Only fast medical treatment, usually with a defibrillator, can prevent sudden death.

From studies of the effects of current on animals, corpses, and other sources of information, a plot has been defined and internationally agreed upon. This contains the regions in which the various effects are more likely to occur. A simplified version of it is shown in Figure 14.1: the standard also defines a curve a and curves c_2 and c_3, not shown here for the sake of simplicity.

Naturally the zones have only statistical significance, since actual effects depend on the subject's age, sex, state of health, and so on.

The two curves in the plot define three zones with different levels of danger:

- **AC-1.** In this zone the current is perceived by the victim, but it is usually not enough to startle him or her.
- **AC-2.** In this zone perception is clear; it is also possible that involuntary muscular contractions occur, but usually there are no harmful effects.
- **AC-3:** This zone is characterized by extreme danger; the further away from curve c_1 (toward higher currents or longer duration), the likelier death becomes.

The reader might find it useful to draw a few facts from these plots and to always bear them in mind:

1. Currents below 5 mA can be considered entirely safe, regardless of the time they flow in the body.
2. Currents below 30 mA are fairly safe, since they are not likely to have very dangerous effects, whatever the duration.

3. Currents above 500 mA are always very dangerous, regardless of the time they flow in the body.
4. Curve c_1's dependence on time is negligible over 5 s; a current lasting 5 s is just as dangerous as one lasting much longer.

The zones defined in Figure 14.1 refer to left-hand-to-feet current paths; however, risks relating to other paths are also known and are obtained by the values shown in Figure 14.1 by multiplication by correction factors. For example:

- Right-hand-to-feet paths are 20% less dangerous: the plots in Figure 14.1 are thus often considered to be valid for any hand-to-feet (thus favouring safety); this is important since hand-to-feet is by and large the most common current path during accidents.
- A current flowing from both hands to both feet is as dangerous as from the left hand to both feet.
- A current flowing from hand to hand is 2.5 times less severe than from left hand to feet.

Example. A current of 225 mA, hand to hand, has the same likelihood of producing ventricular fibrillation as a current of 90 mA, left hand to both feet.

The plots shown in Figure 14.1, as mentioned, are valid for AC currents, at so-called "industrial frequencies," defined for the plot as the range 15–100 Hz. It was previously stated, however, that DC currents are less dangerous than AC ones, mainly because they interfere much less with the physiological currents that normally flow in the body. The latter have an impulse shape and are therefore much more similar to AC than DC. The danger limits of DC power systems are shown in Figure 14.2, which thus complements Figure 14.1.

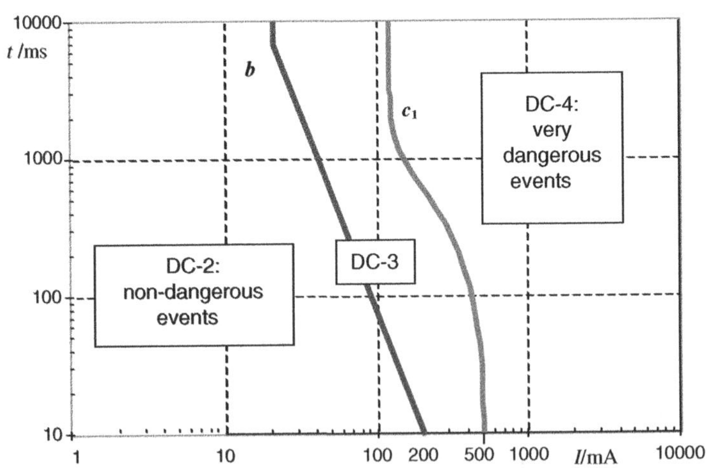

FIGURE 14.2. Conventional zones of DC current effects [1]. Path in the body: left hand to feet.

It is clear that DC currents are much less dangerous than AC ones for long durations while for very short durations the limits are the same; this is apparent considering that a current lasting 10 ms is seen by the body as a 10-ms impulse regardless of whether the electric system is DC or AC.

As we have seen, the dangers of electricity are related to the current flowing in the body, but it is more useful, for electrical installation design purposes, to derive from these curves other ones taking into account, instead of the current flowing in the body, the voltage between the conductive parts that can be touched by people and the ground. In other words, it would be useful to be able to express safety limits in terms of voltage-time couples instead of current–time couples.

To do so, an electrical model of the ground and of the human body is necessary, as shown in the next two sections.

14.2.2 The Mechanism of Current Dispersion in the Earth

When someone touches with a hand a part that has a potential different from that of the floor, and has his feet on the floor itself, he is traversed by a current (from hand to feet and vice versa) that depends on the voltage, the body, and the characteristics of the ground. The earth's behaviour is very important: clearly, for instance, no current flows in a floor made of perfectly insulating material, while in the case of wet earth in a garden, the current could be deadly.

In order to understand the earth's behaviour when traversed by a current, consider the system represented in Figure 14.3. The horizontal line represents the soil surface, that is, the ground.

In the ground are two fixed electrodes ***A***, and ***C***, which can be imagined as being vertical cylinders of about 1 metre in length. An additional electrode ***B*** can also be inserted into the earth at a distance *x* from ***A*** which can be anything from 0 to the distance between ***A*** and ***C***. A voltmeter connects the wires connected with ***A*** and ***C***.

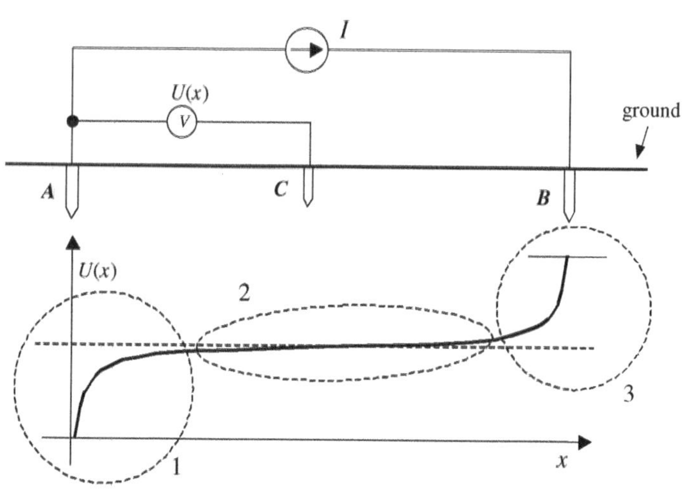

FIGURE 14.3. Potential differences of the earth's surface points when the earth is traversed by current I between electrodes ***A*** and ***C***.

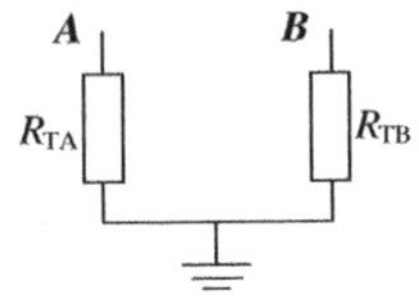

FIGURE 14.4. Lumped-component model of the earth between electrodes ***A*** and ***C***.

In the lower of the Figure 14.3 is a possible plot of the voltage $U(x)$ measured by voltmeter *V*.

The whole system can thus be imagined to be split into three zones:

- zone 1, around ***A***, in which the voltage initially rises fast, then gradually reduces its slope;
- central zone 2, in which the voltage can be assumed to be stable;
- zone 3, around ***C***, in which the voltage increases, with increasing slope, up to the voltage U_{AB}.

Furthermore, note that the voltage plot $U(x)$ is proportional to the current *I* circulating through the electrodes. Therefore, the earth below our feet has linear relation to current, although it is a distributed-parameter system. Therefore Ohm's law applies to it, and it can be effectively modelled by lumped resistors, choosing three important points: the two connections to electrodes ***A*** and ***B***, and the intermediate point between the two in which the voltage is most stable.

In Figure 14.4, note the earth symbol (⏚) used to represent the intermediate point with stable voltage.

The resistances R_{TA} and R_{TB} are the *earth resistances* of *earth electrodes* ***A*** and ***B***. Their actual values depend on soil resistivity and electrode shape and size. In particular, they are proportional to soil resistivity; they decrease, if shape is not modified, when the electrode size increases.

The actual computation of earth resistance might be complicated in the case of complex shapes; however, for the simple case of a linear vertical element inserted by a length *l* into soil with resistivity ρ_e, the following approximated formula can be used:

$$R_r = \rho_e / l$$

Unfortunately, soil resistivity varies dramatically with soil type. Values of 25–100 Ωm are typical for vegetal earth while it can reach 1000–5000 Ωm in rocks. It will soon be seen that smaller earth resistances enhance safety; therefore, it is much more difficult (requiring much larger electrodes) to make good earth electrodes in rocky ground than in ordinary soil.

14.2.3 A Circuital Model for the Human Body

In order to prevent electric shock, it is essential to correctly evaluate the behaviour of the human body when it is traversed by a current. Many studies have been carried out, mainly based on corpses and on animals.

One of the basic results is that the human body normally behaves *algebraically*. This means that if the body is seen from two points ***P*** and ***Q*** (e.g., the two hands), its constitutive equation is an algebraic equation, without derivatives or integrals:

$$U_{PQ} = f(I_{PQ}) \tag{14.1}$$

Equation (14.1) does not satisfy Ohm's law

$$U = RI$$

since the ratio U/I is not constant.

There are cases in which an algebraic model of the body is not sufficient, and a model also containing capacitors must be used; these are, however, beyond the scope of this book; for details see [1].

Even though the nonlinear nature of function f in (14.1) indicates that Ohm's law cannot be used, the fact that it is algebraic suggests that it can be described as a "resistance variable with the voltage applied":

$$U_{PQ} = R(U_{PQ})I_{PQ}$$

This can always be done when algebraic equations are used as constitutive equations, just defining

$$R_{PQ} = U_{PQ}/I_{PQ}$$

The reader might not feel comfortable about using nonlinear constitutive equations, since they are not among those studied in Part II of the book.

However, in Part II it was clearly stated that the validity of Kirchhoff's equations does not depend on the form of constitutive equations of the circuit elements. Therefore, the fact that they are nonlinear does not mean that circuits cannot be used to describe physical systems, including human bodies. It should simply be observed that non-linearity implies that superposition cannot be used.

Studies performed on the electrical behaviour of the human body have led to a model that considers it to be composed of subelements that represent the four arts and the trunk. However, they have also shown that it can be safely assumed that the trunk's resistance is negligible, and the four limbs share approximately the same resistance R_{ip} (the subscript standing for internal-partial).

The equivalent lumped component model of the body, therefore, can be assumed to be the one shown in the right-hand part of Figure 14.5. It allows equivalent resistances to be computed in different configurations:

- Hand-to-hand resistance $R_{h\text{-}h} = 2R_{ip}$
- Hand-to-both-feet resistance $R_{h\text{-}ff} = 1.5R_{ip}$
- Both-hands-to-both-feet resistance $R_{hh\text{-}ff} = R_{ip}$

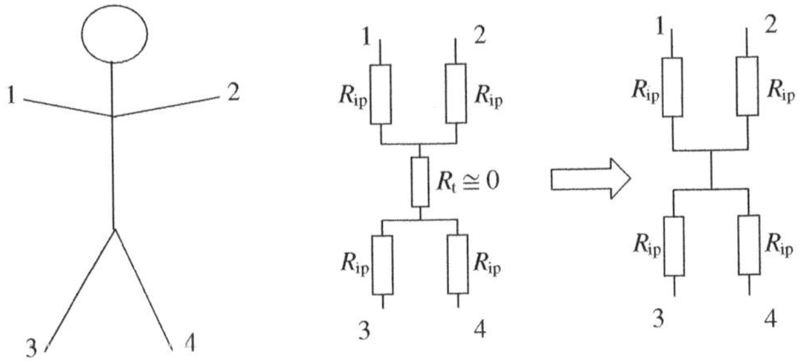

FIGURE 14.5. Lumped-parameter model of the human body.

TABLE 14.1. Hand-to-Hand Body Resistance with Dry Contact[a]

U/V:	25	50	75	100	125	225	700
R/Ω:	1750	1375	1125	990	900	800	620

[a]Values not exceeded for 5% of the population.

The value of R_{ip}, depends on the specific characteristics of the individual, on the applied voltage, and on the characteristic of the contact between body and live part.

International standards supply information about hand-to-hand resistance in various conditions. Table 14.1 only shows the impedance as a function of exceeded the voltage for 5% of the population in the case of contact in dry conditions.

This can be assumed to be the "statistical minimum resistance" of hand-to-hand resistance, that is, twice R_{ip}.

14.2.4 The Human Body in a Live Circuit

Now that we have drawn lumped parameter models (i.e., constitutive equations) of both earth and body, we can effectively evaluate what happens when the body comes into contact with a live part.

By and large, the most common scenario is when someone touches a live part with his hand while his feet are placed on the ground. This can be analysed with reference to Figure 14.6. The contents of the block "MV/LV transformer" change from country to country. The most common situations have already been discussed and presented in Figures 5.14 and 5.15, to which the reader is referred.

Point ***A*** is connected to, and thus (from a circuital point of view) coincides with, the LV transformer neutral ***N***. The earth electrode of the source MV/LV transformer is connected to $\boldsymbol{A} \equiv \boldsymbol{N}$.

Point ***C*** is the point at which the hand touches a live part, while point ***B*** represents the contact between the feet and the earth (ground).

This system can be transformed into a circuit (lumped-parameter model of the system), as shown in Figure 14.7a), in which:

- The whole power system feeding, through terminals ***A*** and ***C***, the right-hand part of the system displayed in Figure 14.6, has been modelled by means of its Thévenin equivalent ($\underline{E}_{CA}$, $\underline{Z}_{CA}$).

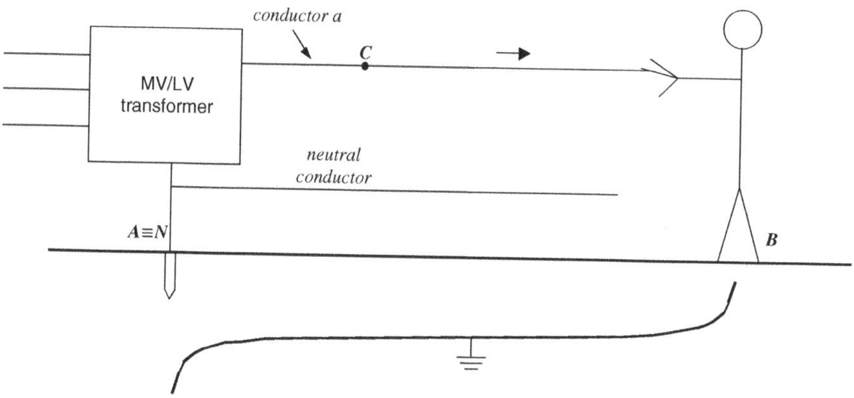

FIGURE 14.6. The circuital system created when a body touches a live part while standing on the ground.

- The earth's behaviour is modelled by means of its equivalent model of the type shown in Figure 14.4.
- The body's behaviour is modelled by means of a voltage-dependant resistance, as discussed in section 14.2.3.
- The person's feet on the ground work like an electrode: if the shoes' resistance is disregarded (i.e., is considered to be zero, this simplifies things and boosts safety), they are a conductor in Electrical contact with the earth; the corresponding earth resistance is named $R_{BT.}$
- The subscript "b" stands for "body," while "B" stands for "point ***B***."

In practical situations, it is safe to say that the neutral to earth resistance R_{NE} and the Thévenin impedance Z_{CA} are negligible compared to the body's resistance and to R_{BT}. Indeed, R_b and R_{BT} are on the order of hundreds of ohms, while R_{NT} and Z_{CA} are at most one or two. Therefore, it is normal practice to work on the simplified version of the circuit of Figure 14.7a, shown in Figure 14.7b.

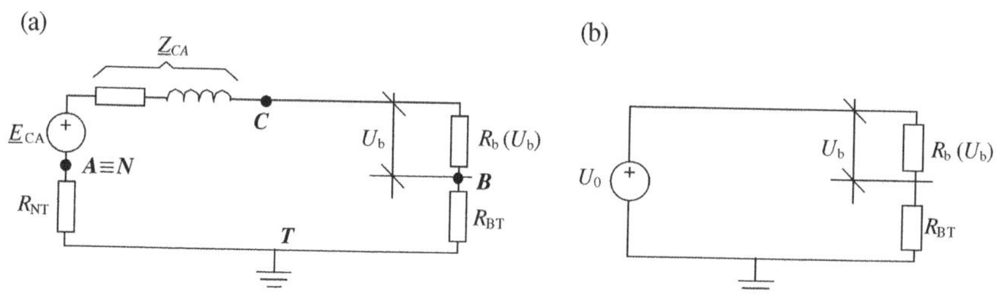

FIGURE 14.7. A circuit containing models for earth and body: (a) More detailed circuit. (b) Simplified circuit.

TABLE 14.2. Maximum Allowed Voltages U_{lim} (for unlimited time) Between Exposed Conductive Parts and Earth

	Normal Conditions	Increased Safety Required
AC	50 V	25 V
DC	120 V	60 V

As regards the numerical value of R_{BT}, it is fairly obvious that this is extremely variable and depends on the material below the person's feet and on any metallic piping just below the floor.

Following extensive research, it has been internationally agreed to assume two conventional values of R_{BT}:

- A value of 1000 Ω, when feet-ground contact and the body's sensitivity to current are normal;
- A lower value of 200 Ω when a general increase in safety is required.

This happens in medical locations (details in [11]) and, for some nations, such as Italy, in locations where the feet make contact with wet ground, such as agricultural and horticultural premises and construction and demolition sites.

When designing safety of installations, it is useful to take as a reference the total voltage between floor and the parts likely to be touched *before actual contact takes place*. In Figure 14.7b, this is indicated as U_0. When the person touches point ***C*** with his hand, the voltage between hand and feet will be lower, equal to U_C.

Now, the simple circuit shown in Figure 14.7b, along with what we've learned regarding resistance R_b and R_{BT}, allows us to determine the maximum value of U_0 which can persist safely for long periods in an electrical installation: this is the one that causes the current limit that can safely traverse the body for long periods (i.e., 30 mA for AC, 150 mA for DC).

With simple computations (not included here) and some approximations, we can draw the maximum values for voltages U_{lim} which can remain in an installation for long periods since they are not expected to cause an electric shock. These are shown in Table 14.2.

This table shows, once again, that, when voltage is equal, DC systems are much safer than AC ones.

14.2.5 System Earthing: TT, TN, and IT

The majority of risks come from contact between hands or other body parts and live elements of the electric circuit—that is, at a different potential from that in the ground where the user places his feet. For this reason, some parts of the system are usually *earthed* (i.e., connected to earth) in such a way as to reduce risk.

In order to understand electrical safety, it is also very important to understand *exposed conductive parts* (here also called ECPs).

Definition: *Exposed Conductive Parts (ECPs)*

These are conductive parts which can be readily touched and which are not normally live, but which may become live under faulty conditions.

Typical exposed conductive parts are the enclosure walls of electrical appliances, operating handles, and so on.

Conductive parts that can readily be touched that are enclosures containing live parts do not necessarily belong to the category of ECPs; indeed, some devices are built with such high levels of insulation between live parts and enclosures that the probability of external parts even (if they are conductive) becoming live as a consequence of a fault is negligible. Equipment with this characteristic is called *Class II equipment* and is identified by the symbol ⧈.

The basic system earthing structures are called TT, TN, and IT. The individual letters have the following meaning:

- T indicates direct connection of the point to earth (T stands for terra, the Latin word for earth).
- N indicates direct connection of the element to the neutral point (N stands for neutral).
- Iindicates that the point is insulated from earth.

Moreover:

- The first letter always indicates the connection of a point of the electrical system (usually the neutral point or when a neutral point is not available, a line conductor).
- The second letter indicates the connection of all the ECPs of an electrical installation.

These codes can be better understood by using the system drawings shown in Figures 14.8 to 14.10, which also show the meaning of the residual current I_r, as discussed and explained in Section 14.4.

Figure 14.8 refers to a TT system. The upper part of the figure shows a three-wire three-phase system, while in the lower part is a single-phase installation (such as the one usually found in homes), taking energy from one of the three conductors of a three-phase system. Both neutral point of the electricity source and exposed conductive parts are connected to the earth, thus producing the earth resistances R_N and R_T.

Figure 14.9 refers to two different versions of a TN system: the so-called TN-C ("C" stands for combined), in which the functions of protection (PE) and neutral (N) are combined in a single conductor called PEN, and the so-called TN-S ("S" stands for separated), in which the two conductors are distinct from each other.

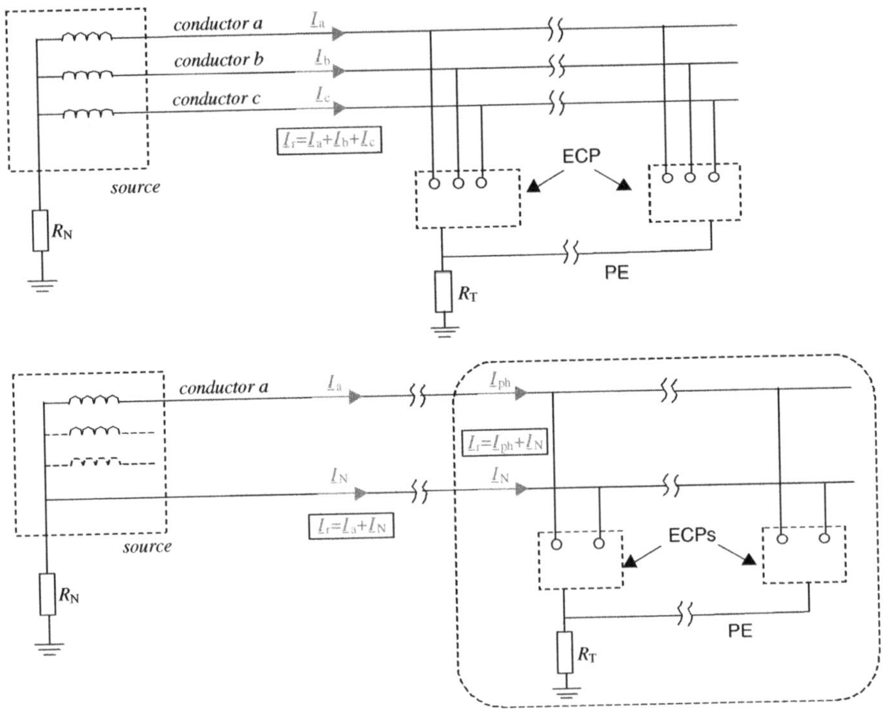

FIGURE 14.8. TT system earthing. *Top*: for a three-wire three-phase system. *Bottom*: for a single-phase (line-neutral) system.

In any case, the neutral point of the source (i.e., the star centre of the secondary winding of the MV/LV transformer) is connected directly to earth (i.e., to the earth electrode), while the ECPs are connected to the neutral point through the PE/PEN conductor: this justifies the code TN. Some information on how these systems operate to guarantee the user's safety will be given in Section 14.3.3, along with their advantages and disadvantages.

Finally, Figure 14.10 refers to an IT system: the neutral point of the source is isolated from the earth, while the ECPs are earthed by connecting them the earth by means of the PE conductor.

Again, information on how these systems operate to guarantee the user's safety will be shown in Section 14.3.3.

14.3 PROTECTION AGAINST ELECTRIC SHOCK

14.3.1 Direct and Indirect Contacts

Protection against electric shock is ensured differently for two types of risks: *direct contact* and *indirect contact*.

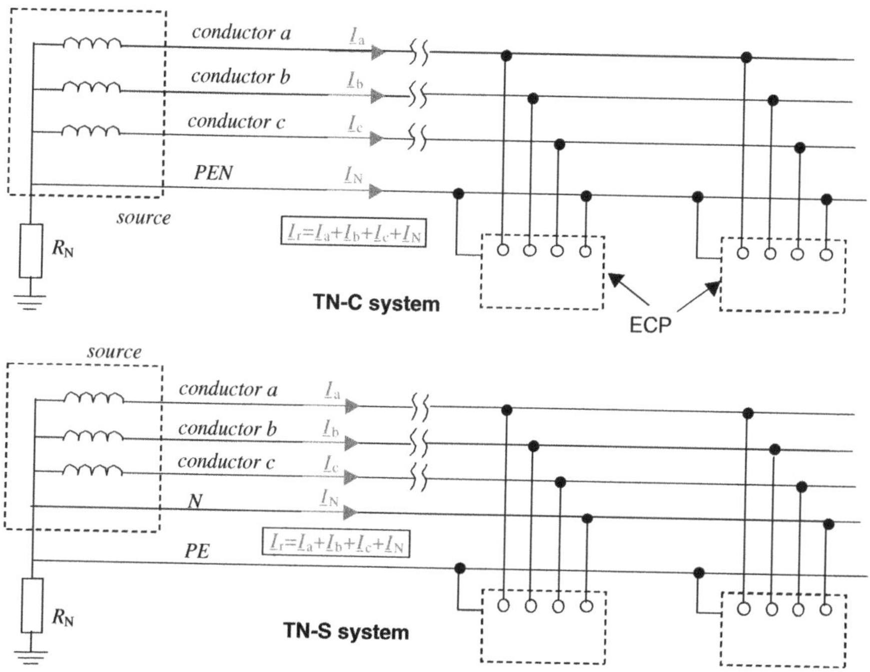

FIGURE 14.9. TN three-phase system earthing. *Top*: TN-C system. *Bottom*: TN-S system.

Direct contact, by definition,[1] is when people or animals come into electric contact with live parts. By contrast, an indirect contact is, by definition,[2] when people or animals come into electric contact with exposed conductive parts which have become live under faulty conditions.

These two modes of contact are very different from both a technical and behavioural point of view and are dealt with in very different ways both in the relevant standards and when designing electrical installations. The very definitions of direct and indirect contacts show that only indirect contacts involve a fault in the installation. Therefore, protection against direct contacts is also called basic protection, and protection against indirect contact is also called fault protection.

14.3.2 Basic Protection (Protection Against Direct Contact)

Basic protection is mainly achieved by surrounding the live parts with insulating material. In civil environments, except for very low voltage appliances, everything live (conductor wires, live parts inside appliances) is surrounded by insulating materials. The minimal requirement for live parts in environments which are

[1]IEV item number 195-06-03.
[2]IEV, item number 195-06-04.

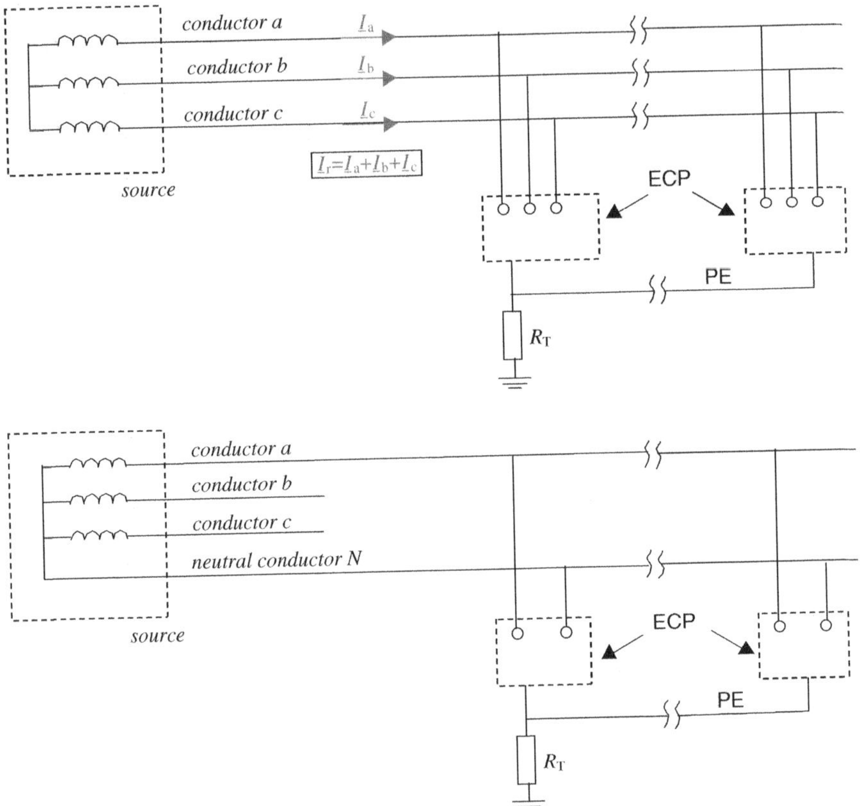

FIGURE 14.10. IT system earthing. *Top*: for a three-wire three-phase system. *Bottom*: for a single-phase system.

accessible to the public (i.e., excluding spaces where only people with specific training in electricity are allowed to enter) is that the so-called *jointed test finger* generally should not be able to enter. If able to enter partially, it must have adequate clearance from hazardous parts; that is, it must not be able to touch or get too near to any live part. The test finger is a device with articulation similar to that of a biological finger, with three joints and phalanges.

In some cases, requirements are more stringent. For instance, for holes on horizontal surfaces (onto which conducting wires could accidentally fall), connecting with something below that contains live parts, the *testing wire* must not be able to gain enter. The testing wire is a straight conducting wire with a diameter of one millimetre.

In spaces accessible to trained personnel, requirements are less rigid.

All components of electrical installations must have an IP code. IP codes express the ability of these components to resist access by solid or liquid materials. While the reader is invited to check standards [12] for complete information about IP codes, some basic information is given here.

IP codes, composed of the code "IP" and two digits, are easily understood as follows:

- The first digit indicates the degree of protection against the penetration of solid objects.
- The second digit indicates the degree of protection against the penetration of water.

Table 14.3 shows the meaning of some of the numerical digits employed.

If we consult Table 14.3, we can see that an enclosure with IP20 degree protection is protected against penetration by the back of a hand (simulated by a sphere) and a finger (simulated by the jointed test finger), but it has no protection against water. For outdoor installation (in gardens, etc.), it is common to require at least IP34.

When the value of one digit is not specified, it can be replaced by the letter X. This way, a IPX4 degree indicates level 4 protection against the penetration of water, as well as unspecified protection against the penetration of solid objects.

In some cases, IP codes also have a letter in addition to the two digits. Here it is only stated that code IPXXB is a slightly less strict version of IP2X, and IP XXD is slightly less strict than IP4X.

The qualitative prescriptions shown above regarding horizontal and vertical surfaces are thus translated into quantitative measures by the following prescriptions (IEC 60364):

- Live parts shall be inside enclosures or behind barriers providing at least a degree of protection of IPXXB or IP2X.
- Horizontal top surfaces of barriers or enclosures which are readily accessible shall provide a degree of protection of at least IPXXD or IP4X.

14.3.3 Fault Protection (Protection Against Indirect Contact)

Fault protection, except in the case of SELV systems which are discussed in the next section, are attained using *protective earthing, protective equipotential bonding, and automatic disconnection in the event of a fault.*

Protective Earthing. Exposed conductive parts (clearly defined in Section 14.2.5) must be connected by means of protective conductors to a *system earthing*. This will contribute to maintaining the ECP potential as close as possible to the potential of the earth on which humans place their feet. This minimizes the chance of being exposed to a dangerous difference of potential between hands and feet.

Protective Equipotential Bonding. This is constituted by conductors connecting all large metallic parts in a building to the building's system earthing, in such a way as to enhance even greater "equipotentiality." Otherwise, in some circumstances, non-negligible potential differences could appear between the ground and these large parts (e.g., the water piping), thus introducing dangers in the environments.

TABLE 14.3. Meaning of the Most Common IP Codes

Protection Against Penetration of Solid Objects			*Protection Against Water Penetration*		
Degree of Protection	*Meaning*	*Condition*	*Degree of Protection*	*Meaning*	*Condition*
1	Protected against access to hazardous parts with the back of a *hand*	The access probe, a sphere of 50 mm Ø, shall have adequate clearance from hazardous parts.	1	Protected against vertically falling water *drops*	Vertically falling drops shall have no harmful effects.
2	Protected against access to hazardous parts with a *finger*	The jointed test finger of 12-mm Ø, 80-mm length, shall have adequate clearance from hazardous parts.	2	Protected against vertically falling water *drops* when enclosure is *tilted* up to 15°	Vertically falling drops shall have no harmful effects when the enclosure is tilted at any angle of up to 15° on either side of the vertical.
3	Protected against access to hazardous parts with a *tool*	The access probe of 2.5 mm Ø shall not penetrate.	3	Protected against spraying water	Water sprayed at an angle of up to 60° on either side of the vertical shall have no harmful effects.
4	Protected against access to hazardous parts with a *wire*	The access probe of 1.0 mm Ø shall not penetrate.	4	Protected against *splashing water*	Water splashed against the enclosure from any direction shall have no harmful effects.

5	Protected against access to hazardous parts with a wire and *partially* against *dust*	In addition to level 4, ingress of dust is not totally prevented, but dust shall not penetrate in a quantity to interfere with the satisfactory operation of the apparatus or to impair safety.	5	Protected against water *jets*	Water projected in jets against the enclosure from any direction shall have no harmful effects.
6	Protected against access to hazardous parts with a wire and against *dust*	In addition to level 4, ingress of dust is totally prevented.	6	Protected against *powerful water jets*	Water projected in powerful jets against the enclosure from any direction shall have no harmful effects.

Automatic Disconnection in the Event of a Fault: General Case. The above two measures are combined with automatic disconnection in the event of a fault: a protective device automatically interrupts the electricity supply in the event of a fault of negligible impedance between the line conductor and an ECP; this interruption will remove any dangerous potential on ECPs. Subsequently, electricians can analyse the installation, find the fault, and remove it before activating the power supply again.

The protective device may be a circuit breaker, with the appropriate fault protection characteristics, already in the installation for overcurrent protection (Chapter 13). This can only happen for TN systems, in which, because of their nature, faults to ECPs create large currents which can cause a circuit breaker designed to protect against overcurrent (short-circuit) to trip.

It is instead impossible in TT systems, for which the currents created by faults to ECPs are much weaker. It will be seen in Section 14.4 that for these systems special protective devices exist, called RCDs, that are able to fulfil the purpose of disconnecting supply in the event of a fault to ECPs.

Automatic Disconnection in Case of a Fault; TN Systems. To ensure that the disconnection is fast enough, in the case of TN systems the following requirement must be fulfilled:

$$Z_a \cdot I_a \leq U_0 \tag{14.2}$$

where:

Z_a is the impedance of the fault loop comprising the source, the line conductor up to the point of the fault, and the protective conductor between the point of the fault and the source;

I_a is the current causing automatic tripping (i.e., opening) of the disconnecting device in a time $t_{i\text{-TN}}$ specified by the standards, which is in the most common case (i.e., nominal system voltage not higher than 230 V) equal to 0.4 s;

U_0 is the nominal line-to-earth voltage (e.g., in Europe 230 V, in the USA 120 V).

The rationale behind this rule can easily be understood by looking at the circuit which occurs when a fault to ECP occurs in a TN system (Figure 14.11, showing only the case of TN-S system, for the sake of brevity). During the fault it is obviously $Z_a \cdot I_f = U_0$, where I_f is the fault current (rms value). Equation (14.2) therefore prescribes

$$I_a \leq I_f$$

that is, the current I_a, which the device is able to interrupt in a very short time $t_{i\text{-TN}}$, must be lower than the actual fault current.

This, combined with the fact that all protection devices have an interruption curve that is an decreasing function of time (i.e., lower times correspond to larger currents), justifies what is asserted in the following rule:

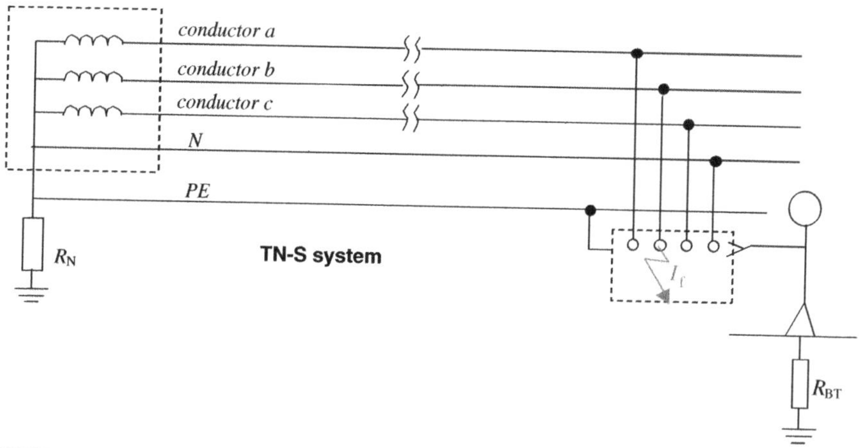

FIGURE 14.11. The circuit that determines when a fault to ECP occurs in a TN system.

> **Rule:** *TN systems fault protection*
>
> In TN systems equation (14.2) applies, which prescribes that every fault current is interrupted in a time lower than $t_{i\text{-}TN}$ (= 0.4 s for nominal line to earth voltages lower than 230 V).

Automatic Disconnection in the Event of a Fault: TT Systems. In the case of TT systems, fault-to-ECP currents are much lower than those of TN systems, and in order to be detected they require that the protective device be a residual current device (RCD, see Section 14.4), which is very sensitive even to small currents toward the system earthing.

The rule to ensure that the disconnection is fast enough in the case of TT systems is as follows:

$$R_a \cdot I_{\Delta n} \leq U_{lim} \tag{14.3}$$

where:

R_a is the sum $R_T + R_{PE} \cong R_T$ of the resistance of the earth electrode and the protective conductor for the exposed conductive parts (normally R_T dominates);

$I_{\Delta n}$ is the rated residual operating current of the RCD; that is, the residual current the RCD is able to interrupt in a time $t_{i\text{-}TT}$, specified by the standards, which in the most common case is equal to 0.2 s;

U_{lim} is the maximum voltage that can stay for unlimited time between the ECPs and the ground without disconnection; its values are those shown in Table 14.2.

The rationale behind this rule can easily be understood by looking at the circuit that we have when there is a fault to ECP in a TT system (Figure 14.12). The residual

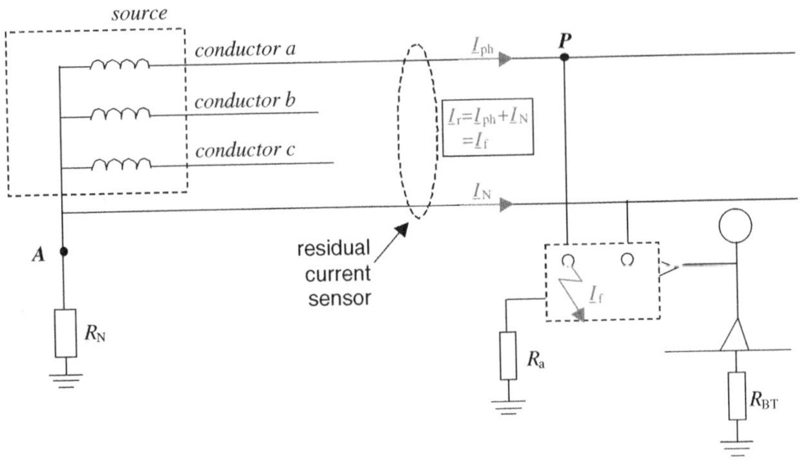

FIGURE 14.12. The circuit that determines when a fault to ECP occurs in a TT system.

current sensor will sense the residual current $\underline{I}_r$ which will be equal to the fault current I_f. The fault product $I_f \times R_a$ will be equal to the rms voltage on the ECPs—that is, the voltage that the body touches and that can cause electric shock.

It is thus clear that protection intervenes whenever the voltage between the ECP and the far earth (point in the circuit indicated by the earth symbol ⏚, overcomes the voltage limit U_{lim}—that is, the safety limit shown in Table 14.2 at the end of Section 14.2.4, which is normally, for AC systems, equal to 50 V.

This is summarized in the following rule:

> **Rule:** *TT systems fault protection.*
>
> In TT systems equation (14.3) applies, which prescribes that every fault current causing voltages on ECPs higher than U_{lim} is interrupted in a time lower than $t_{i\text{-}TT}$ (= 0.2 s for nominal line-to-earth voltages up to 230 V).

14.3.4 SELV Protection System

It is widely known that electricity at very low voltage is totally safe. For instance, children's and even babies' toys are operated with battery cells with voltages on the order of a few volts. Voltages that are considered to be intrinsically safe are called extra-low voltages and, in normal conditions, are up to 25 V if in AC and up to 60 V if in DC.

Very low voltages can be used to protect against direct and indirect contacts: they are officially called safety extra-low voltage (SELV) and protective extra-low voltage (PELV) systems.

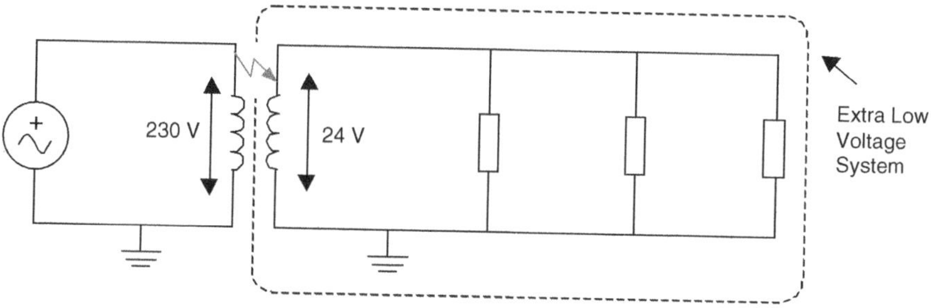

FIGURE 14.13. Need to provide a clear-cut separation between non-ELV and ELV systems for them to be SELV or PELV systems.

The difference between SELV and PELV is beyond the scope of this book; note only that the maximum safety level is reached in SELV systems.

SELV and PELV systems are designed so that voltage cannot exceed the value of extra-low voltage: under normal conditions and under single-fault conditions. This means that the source of energy must have extra-low voltage, and the electrical separation from systems having higher voltage must be very clear-cut.

The concept is clarified in Figure 14.13, in which a possible fault inside the transformer separating a 230-V system from a 24-V system can cause unsafe voltages in the installation connected to the secondary of the transformer.

14.4 THE RESIDUAL CURRENT DEVICE (RCD) PRINCIPLE OF OPERATION

In previous sections it was stated that the most common and effective device for automatic disconnection of the electric supply caused by a fault is the so-called residual current device.

Let us define, in compliance with international standards, the residual current:

> **Definition:** *Residual current*
>
> A residual current is the sum of instantaneous values of all currents flowing in a single-phase or in a three-phase system.

To better understand this concept see Figures 14.8, 14.9, and 14.10, where the value I_r of the residual current is shown as a function of the currents flowing in the wires, paying attention to correct directions.

In a healthy single-phase or three-phase system, the sum of all instantaneous currents delivered by the supply (i.e., the residual current) is virtually zero. This can be easily understood, considering that Kirchhoff's law applies to all loads and all the conductors.

Naturally, actual residual currents are only approximately null, since the insulator material of all the installation components is not *perfectly* insulating; and some drainage current, albeit very small, flows between the conductors and the ECPs. This results in a net residual current flow, which is again small. It should be stressed that in the circuits shown in Figures 14.8, 14.9, and 14.10, in the absence of fault, the residual current is exactly zero, since circuits are ideal models of reality; however, more detailed models dealing with phenomena related to the dispersion of current between conductors and ECPs, will be able to show the presence of those small residual currents also in healthy systems. The entity of residual currents in healthy systems depends on their physical extension, ranging from a few milliamperes in small systems and reaching up to several hundred milliamperes in the largest ones.

This situation of null residual current I_r changes dramatically when a fault occurs between one of the conductors and one exposed conductive part, as shown just as an example for a TT system in Figure 14.12. In this example, the single-phase residual current I_r, which could be measured by the shown *residual current sensor*, is far from zero, being

$$\underline{I}_r = \underline{I}_{ph} + \underline{I}_N = \underline{I}_f$$

A residual current sensor can therefore be effectively used to measure the fault to ECP currents.

Residual currents can be detected in several ways. One typical means is shown in Figure 14.14. This method is interesting because its operation is very easy to understand, being based on Faraday's and Ampère's laws. Here the residual current sensor is constituted by a rectangular torus of ferromagnetic material containing three windings. Windings one and two have exactly the same number of turns: $N_1 = N_2 = N_{12}$.

If we ignore the leakage flux, which is acceptable in these devices, the flux circulating in the magnetic circuit is

$$\underline{\Phi} = (N_1\underline{I}_1 - N_2\underline{I}_2)/\mathcal{R} = N(\underline{I}_1 - \underline{I}_2)/\mathcal{R} = N_{12}\underline{I}_r/\mathcal{R}$$

In which $\mathcal{R}$ is the magnetic circuit reluctance and the residual current is

$$\underline{I}_r = \underline{I}_1 - \underline{I}_2$$

The flux Φ, or any other quantity proportional to it, is therefore a measure of the residual current I_r. In particular the electromotive force at the terminals of coil 3 (with N_3 turns) is proportional to I_r and thus constitutes a good measure of it.

When this device feeds a healthy installation, we have $I_r = 0$, and also the voltage across coil 3 is zero.

If, on the contrary, a fault in the installation fed by the wires 1 and 2 (in which I_1 and I_2 flow) causes some residual current to be present, a significant residual flux is induced in the torus, which causes an electromotive force to be generated on winding

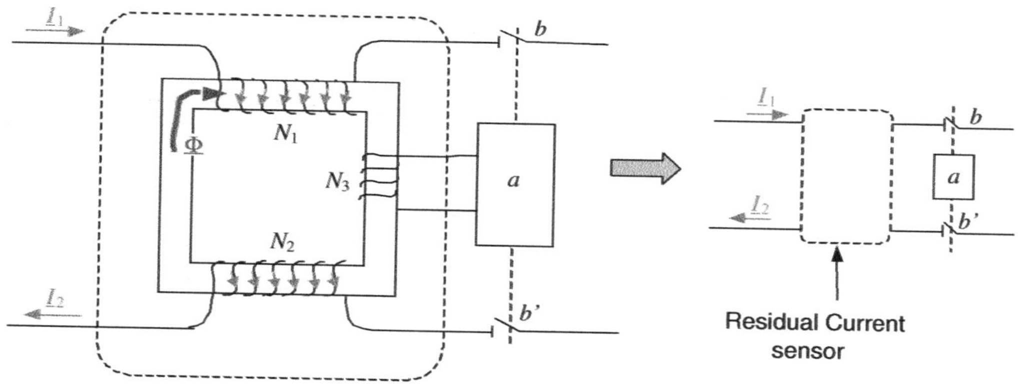

FIGURE 14.14. One possible version of a residual current device.

three. This then feeds the actuator *a*, which causes the breaker to trip (i.e., to open the circuit).

The *a* box absorbs a very low current $\underline{I}_a$, which does not significantly influence Φ.

14.5 WHAT ELSE?

This section simply aims to stress that we have only dealt with the basics of safety. Designing an electrical installation requires attention to many aspects which could not be included in this or the previous chapter, simply because of the introductory nature of this book.

Protective actions to be considered were listed at the end of Chapter 13. Here they are again, with an indication in brackets of the standards containing the corresponding measures to be taken, if available.

- Protection against electric shock [3]
- Protection against thermal effects [4]
- Protection against overcurrents [5]
- Protection against voltage disturbances and measures against electromagnetic influences [6]
- Protection against power supply interruption

REFERENCES

1. International Electrotechnical Commission IEC/TS 60479-1. Effects of current on human beings and livestock—Part 1: general aspects.
2. International Electrotechnical Commission IEC 60364-1. Low-voltage electrical installations—Part 1: Fundamental principles, assessment of general characteristics, definitions.
3. International Electrotechnical Commission IEC/TS 60364-4-41. Low-voltage electrical installations—Part 4-41: Protection for safety—Protection against electric shock.
4. International Electrotechnical Commission IEC 60364-4-42. Electrical installations of buildings—Part 4-42: Protection for safety—Protection against thermal effects.
5. International Electrotechnical Commission: IEC 60364-4-43. Electrical installations of buildings—Part 4-43: Protection for safety—Protection against overcurrent.
6. International Electrotechnical Commission: IEC 60364-4-43. Electrical installations of buildings—Part 4-44: Protection for safety—Protection against overvoltage disturbances and electromagnetic disturbances.
7. International Electrotechnical Commission: IEC 60364-5-51. Electrical installations of buildings—Part 5-51: Selection and erection of electrical equipment—Common rules.
8. International Electrotechnical Commission: IEC 60364-5-52. Electrical installations of buildings—Part 5-52: Selection and erection of electrical equipment—Wiring systems.

9. International Electrotechnical Commission: IEC 60364-5-53. Electrical installations of buildings—Part 5-53: Selection and erection of electrical equipment—Isolation, switching and control.
10. International Electrotechnical Commission: IEC 60364-5-54. Electrical installations of buildings—Part 5-54: Earthing arrangements, protective conductors and protective bonding conductors.
11. International Electrotechnical Commission: IEC 60364-7-710. Electrical installations of buildings—Part 7-710: requirements for special installations or location—Medical locations.
12. International Electrotechnical Commission: IEC 60529. Degrees of protection provided by enclosures (IP Code). This is also an ANSI standard with the number ANSI/IEC 605329-2004.

15

LARGE POWER SYSTEMS: STRUCTURE AND OPERATION

For the Instructor

This chapter is not concerned with turning readers into experts on large power systems; there are specific textbooks available which serve this purpose. Instead, its aim is to provide readers with a basic idea of these systems so they have sufficient background when dealing with the LV installation of an electric machine or drive.

A good introductory book in electrical power systems, covering the topics of this chapter and much more, is the [bp1].

15.1 AGGREGATION OF LOADS AND INSTALLATIONS: THE POWER SYSTEM

In the previous chapter, we provided information on the structure of LV installations.

LV installations, both single-phase and three-phase, are usually just the extreme terminal parts of a much larger system which comprises power plants, transformation power stations, and extra-high voltage, high-voltage, and medium-voltage lines, as well as, in many cases, trunks of low-voltage lines. The development and operation of these large power systems require special expertise.

Fundamentals of Electric Power Engineering: From Electromagnetics to Power Systems, First Edition.
Massimo Ceraolo and Davide Poli.

In this chapter we provide some basic characteristics of large power system structures and operations, starting with an explanation of why they are built this way.

15.2 TOWARD AC THREE-PHASE SYSTEMS

The number of users supplied by a single generator can be very high. In fact, the power of user devices rarely exceeds a few kilowatts and in many cases it is a fraction of a kW. Conventional generators, instead, have sizes ranging from 10 MW to 1000 MW. The basic scheme for feeding such a large number of users at constant voltage using a single generator is the tree structure ideally depicted in Figure 15.1, assuming, for instance, that a 100-MW generator is required to supply 100,000 1-kW users. In this example, the system voltage was set at a value which was not dangerous for humans, 100 V. Each peripheral line supplies 10 users; proceeding upstream, toward the generator, each node gathers 10 branches. The voltage being constant, at each node the currents are multiplied by 10. The practical unfeasibility of such a distribution system is evident, since it would require a generator rated for 1 MA and power lines able to transport 100 kA.

To reduce the current to reasonable values, it is essential to renounce the 100 V voltage in the parts of the system that are not directly in contact with users, which means increasing voltages in the upstream branches where the transmitted power is higher. This can be achieved in economic terms, but only in alternating current (AC), using transformers,[1] invented around 1885 (by Tesla and others).

This consideration is one of the main reasons why electric power systems, initially conceived and built in DC in the early 1880s,[2] rapidly converted to AC as soon as their extension grew and transformers were made available on an industrial

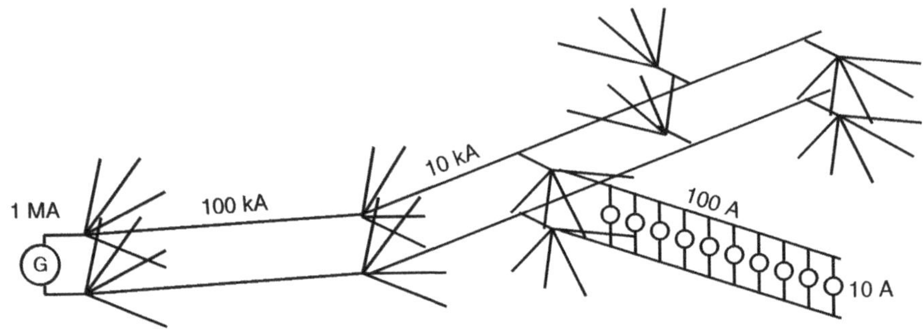

FIGURE 15.1. Tree structure of a distribution system operated at constant voltage.

[1]Nowadays, the introduction of DC/DC converters (see Chapter 8) extends the potential uses of DC distribution systems, at least for small-power applications. In large plants, for reasons of reliability, cost, and efficiency, transformers and AC distribution systems are used.

[2]In 1881 the Edison Electric Illuminating Company of New York inaugurated the very first electricity distribution system worldwide, consisting of short underground cables. The Pearl Street station was fed by four 250-hp boilers, supplying steam to six 110-V DC generators. Similar plants were installed in the next two years in London and Milan.

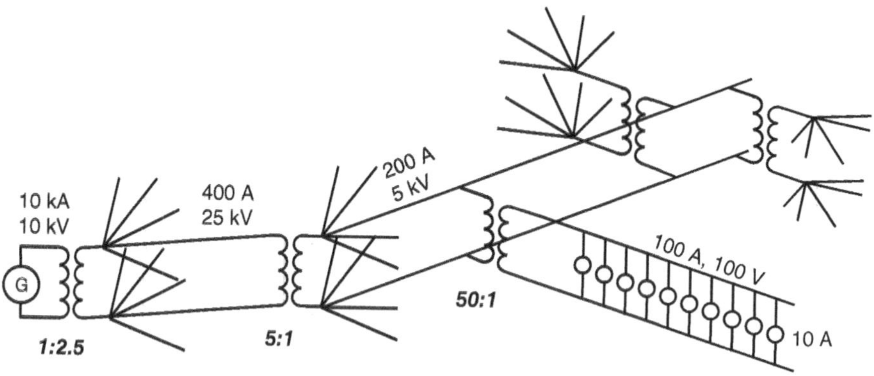

FIGURE 15.2. Tree structure of an AC distribution system with multiple voltage levels.

scale. The first commercial single-phase AC distribution system was installed in Oregon in 1889–1890.

It is worth noting that the invention and improvement of the incandescent light bulb in the 1870s made lighting the first publicly available energy application of electrical power. Previously, electricity was basically applied to telecommunications (telegraph, 1837).

Using transformers, it is possible to choose the most convenient voltage–current pair for each point of the system on the basis of safety, efficiency, and capability. This can be done, for instance, for the system presented in Figure 15.1, as described in Figure 15.2.

Another reason which fostered the development of the alternating current is that circuit breakers—that is, devices designed to interrupt short-circuit currents—can take advantage of it.

Indeed, breakers contain two mechanical contacts which separate from each other when the breaker is required to trip. If current is flowing when the opening action is required, an electric arc is generated through these contacts that must somehow be extinguished. This is much easier in AC than in DC, since in AC the breaker can exploit the current's periodical zero crossing.

Circuit breakers soon proved to be essential in order to eliminate the faulty part of the system and to limit the effects of short circuits on system reliability, safety, and continuity of supply. Lastly, AC alternators are definitely more reliable than DC generators.

The discovery of the rotating magnetic field by G. Ferraris in 1885 and the advent of the induction motor (Tesla, 1888) finally led to the use of the polyphase AC system, in particular the three-phase system. The polyphase system prevailed over the single-phase system because of the possibility of generating rotating magnetic fields using fixed windings (see Section 10.2.1); this opportunity simplified all rotating machines used to convert electrical energy into mechanical and vice versa,[3] and made them more reliable and efficient. Only the peripheral branches of the system, to which small-size users are connected, are single-phase for reasons of economy.

[3]It is also worth remembering that three-phase induction motors are able to provide a constant mechanical torque (see Section 11.2).

The first commercial, three-phase 2.3-kV system was installed by the Southern California Edison Company in 1893. Since then, electric power networks have always been *three-phase AC systems*, operating at constant[4] voltage as we saw in Section 12.2 (shunt connection).

15.3 ELECTRICITY DISTRIBUTION NETWORKS

In order to supply a huge number of small users by means of a single barycentric power plant, which was the usual task of the earliest power systems, the most basic distribution network has a tree structure. As we saw previously, this tree has branches which decrease in importance from the centre to the periphery, while links have a voltage proportional to their power flow. The final ones, to which the majority of users are connected, always operate at low voltage (LV); this makes it safe for people without any professional skills to use electricity for domestic purposes. Voltages of up to 1 kV (phase-to-phase, rms), in AC are normally considered to be LV.

The number of voltage levels, starting from the power plant to reach the smallest loads, is variable. Nowadays the most common is a three-level distribution system, which includes a high-voltage (HV) level, a medium-voltage (MV) level, and, finally, a LV distribution network.[5] For reasons we will see later, the three levels are commonly referred to as subtransmission grid and as primary and secondary distribution network. Normally, voltages larger than LV levels and up to 35 kV (line–line, rms) are considered as MV.[6] More information about the voltage levels used in different regions is given in Section 15.4.

Figure 15.3 shows the evolution from a simple subtransmission grid to a power system comprising downstream distribution networks. The corresponding single-line diagrams are also shown.

In single-line diagrams, readers can recognize symbols already shown in the previous chapters of this book for the generator and the transformer. In the same diagrams, long longitudinal links indicate power lines. A short transversal dash represents a "busbar"—that is, a conductor used to enable a common connection between several components. Oblique dashes emanating from this busbar indicate the beginning of other power lines, which correspond to other branches of the tree.

Voltages are chosen by distribution companies according to load density and for local historical reasons.

[4]In the case of AC systems, this terminology obviously means that voltage is constant in amplitude.
[5]Generally speaking, LV networks have nominal (phase-to-phase, rms) voltages up to 1 kV. The boundary between MV and HV depends on the country concerned and is normally between 30 kV and 100 kV. International standards tend to treat voltages of up to 35 kV differently from those beyond; this value is therefore a natural threshold between MV and HV systems and is internationally recognized as such.
[6]In many European countries the same levels are often called primary, secondary and tertiary (or LV) distribution networks, which can cause misunderstandings.

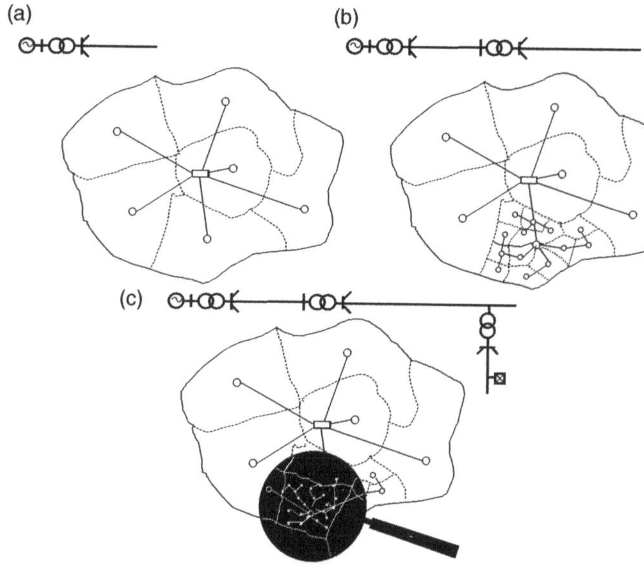

FIGURE 15.3. Radial-type schemes. (a) Subtransmission grid. (b) Adding a primary distribution network. (c) Adding a secondary distribution network.

In Europe, LV networks are all three-phase systems (see Chapter 6), whose structure is shown in Figure 6.13. LV distribution systems have a radius of action of a few hundred metres and all operate at 400 V (phase-to-phase, for supplying three-phase users) to 230 V (phase-to-neutral, for feeding single-phase users). With reference to Chapter 6, note that the ratio between phase-to-phase and phase-to-neutral voltages is close to $\sqrt{3}$. MV networks range from 6–10 kV to 20–35 kV (depending on the considered country and on the load density), covering a radius of a few kilometres, while HV networks can reach 150 kV, with a radius on the order of a few tens of kilometres.

In North America, power for a typical home is derived from a transformer that reduces the voltage to 240/120 V (both for single-phase users) distributing electricity with a three-wire line, as shown in Figure 6.14. The transformer is single-phase, with a secondary voltage of 240 V; a plug in the midpoint of its secondary winding allows a third wire to be used to supply 120-V users. Three-phase four-wire systems, similar to the one depicted in Figure 5.13, are also in use; in this case, voltages are usually 120/208 V or 277/480 V.[7]

The tree structure, while very simple and cheap, does not guarantee the best quality of service in terms of continuity of supply, because each node has only one feeding source. To provide backup power to the main nodes, power networks soon become *meshed*, especially at the level of HV distribution, where per-kilowatt costs required to increase the continuity of supply are lower. Here the "mesh" word takes a slightly

[7]When a slash is used with reference to three-phase systems, the first number represents the line-to-neutral voltage while the second number represents the phase-to-phase; both are rms values.

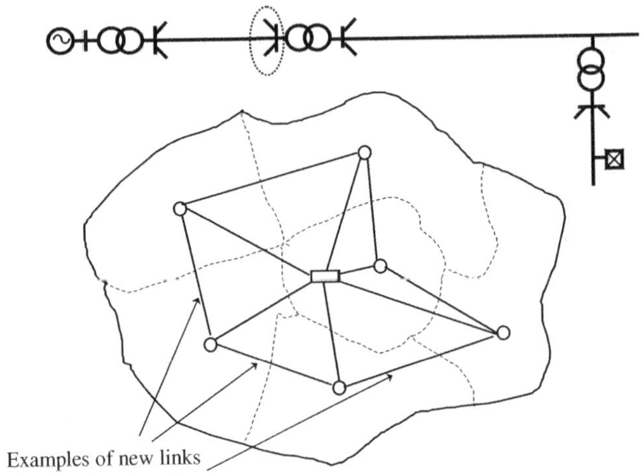

FIGURE 15.4. Meshed subtransmission grid.

different meaning from that used in circuit analysis: it indicates a loop of three-phase transmission lines and covers areas of square kilometres, where they are loops inside circuits (see Figure 15.4). Note how the single-line diagram is modified at the receiving subtransmission busbar (highlighted by a circle): the oblique dashes indicate incoming power lines, hence a meshed structure.

MV and LV distribution networks may be meshed (usually in urban districts) or radial (as in many rural areas), depending on their extension and on the load density. Some networks have a meshed structure but are radially operated, keeping open some links, basically to simplify automatic disconnection of faults.

15.4 TRANSMISSION AND INTERCONNECTION GRIDS

In the earliest power systems, during the two last decades of the nineteenth century, it was possible to install small power plants in a barycentric position with respect to the users, often in the city centre (Manhattan 1881, London 1882, Milan 1883). Afterwards, as power systems rapidly grew in size and number of users, for reasons of layout, environmental impact, and localization of the primary energy, power plants supplying large urban areas were required to be placed far away from the load centres.

In this case, a distribution network like the one shown in Figure 15.5a, where many HV nodes are supplied by the same power plant, is in general not economical.

It is, in fact, much cheaper and efficient to convey energy up to the centre of gravity of the loads by means of a single higher capacity power line (see Figure 15.5b). Such a line, usually operated at a higher voltage than the supplied HV distribution system, performs the so-called functions of energy "transmission." Please note how the single-line diagram is modified to incorporate a further voltage level (thicker line).

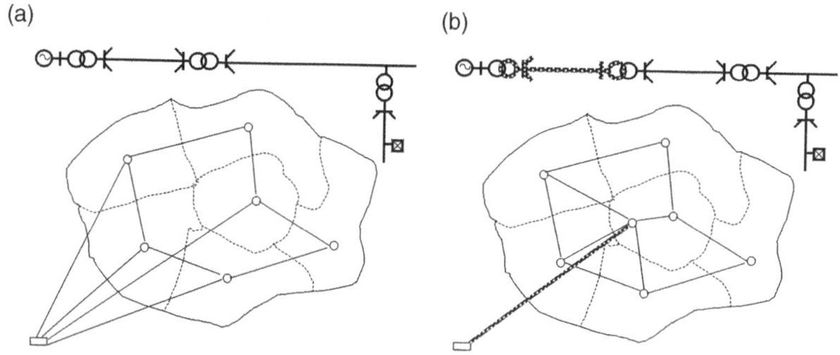

FIGURE 15.5. Transmission link between the power station and the center of gravity of loads.

For evident reasons of continuity of service, the main nodes of the HV distribution system require a second power supply. This usually comes from the power station intended to serve as an adjacent distribution system (see Figure 15.6). Such a function is generally referred to as "interconnection" and corresponds to the meshing of the transmission grid. Note how the single-line diagram is modified at the receiving transmission busbar to indicate a meshed structure, similar to what we have seen for the subtransmission grid.

This scheme soon prevailed over the idea (typical of the earliest power systems) of independent electric islands, each one supplied by its own power plant. The synchronous operation of contiguous islands, up to a single, large, interconnected system, brings many advantages, such as:

- the existence of a common reserve for the generation system, in order to increase the continuity of service to consumers, ensuring reliability in the event of contingencies such as unscheduled outages;

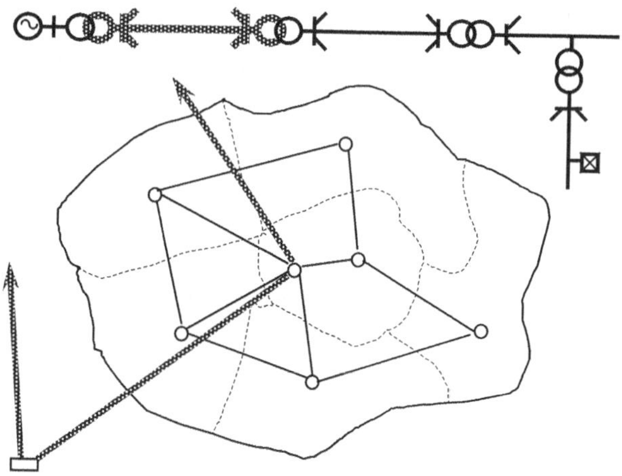

FIGURE 15.6. Interconnection between adjacent load areas.

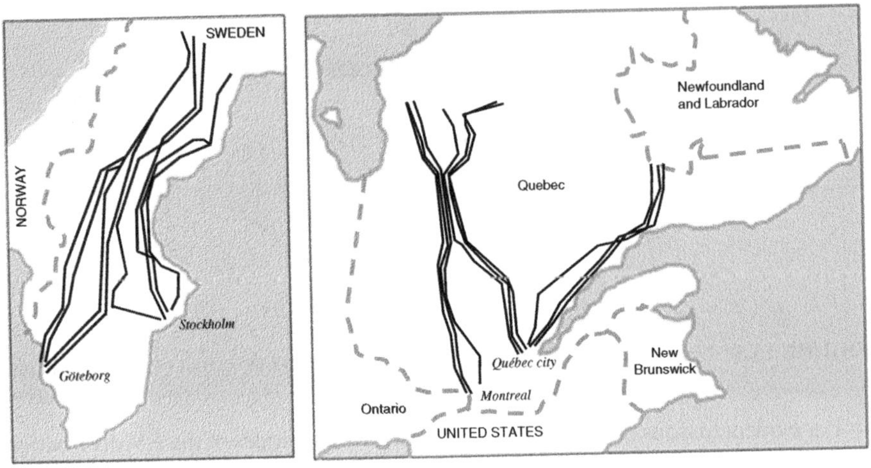

FIGURE 15.7. Swedish and Canadian transmission links.

- a peak value of the total load which is lower than the sum of the peak values of the single islands, which are generally not simultaneous;
- cheaper operation, thanks to a better integration between different forms of generation installed in adjacent areas, like thermoelectric units and hydro power plants;
- lower frequency deviation from nominal value, as will be discussed in Section 15.6.

In many countries, where loads and generators are distributed in a relatively uniform way, the link between all contiguous distribution systems leads to the creation of an extensive meshed grid with very high voltage, which simultaneously performs the functions of transmission and interconnection, as happens in many portions of the European and US grids. Within these grids, which collect the energy generated by the large-scale production plants to the main HV nodes, the functions of transmission and interconnection are intimately related and almost indistinguishable.

In some particular cases, however, it is possible to identify power lines which explicitly fulfil the typical function of transmission paths, usually for conveying energy toward regions with high levels of consumption by exploiting primary resources available in scarcely inhabited areas or even desert. This happens, for example, to transmit the hydropower produced in subarctic Sweden or in northern Québec to the more populated southern regions (Figure 15.7).

In Europe, the transmission–interconnection grid is operated at 400 kV and partially at 230 kV. In the United States and Canada, voltages of 230 kV, 345 kV, 500 kV, 735 kV, and 765 kV are commonplace. Transmission voltages above 230 kV are usually referred to as extra-high voltages (EHV). Since the early 1990s, also 1000-kV and 1100-kV lines for long-distance transmission purposes were energized. It is

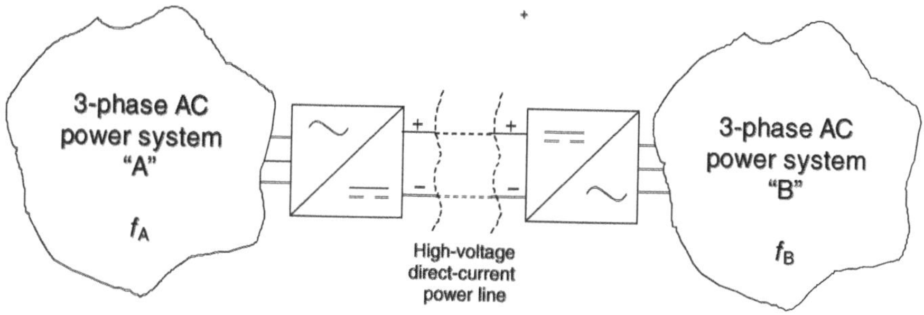

FIGURE 15.8. Structure of an HVDC link.

evident that a lot has happened since 1896, when an 11-kV three-phase line was transmitting 10 MW from Niagara Falls to Buffalo, over a distance of 20 miles.

Since the 1950s, interconnection between power systems can also be ensured for very long distances (overhead lines) or across the sea (submarine cables), using DC lines, the so-called high-voltage direct-current links (HVDC), currently operating at up to 800 kV. Two stations, used for converting AC to DC and DC to AC respectively, are placed at the extremes of the line. If two AC power systems are connected only through a DC link, they can be operated at different frequencies, as shown in Figure 15.8.

15.5 MODERN STRUCTURE OF POWER SYSTEMS AND DISTRIBUTED GENERATION

Based on what we have learned, Figure 15.9 summarizes the typical structure of a traditional power system. The depicted scheme includes:

- a production park composed of large-scale synchronous generators, connected to the transmission–interconnection grid by means of step-up transformers; the alternator's voltage is usually in the range 15–30 kV, depending on its power size: higher voltages are not allowed due to insulation requirements; the prime movers are usually hydraulic turbines, steam turbines (from coal, oil, nuclear fuel), or gas turbines;
- a meshed EHV transmission–interconnection grid (not less than 230 kV), whose nodes are the so-called "main stations," which in turn feed downstream grids;
- a meshed HV subtransmission grid (60–150 kV); for traditional reasons, medium-scale old generators (10–50 MW) are still connected to this grid through step-up transformers; also large industrial customers (the so-called "HV loads") are directly supplied by this grid, in the sense that they operate their own internal distribution network with cables and step-down transformers;

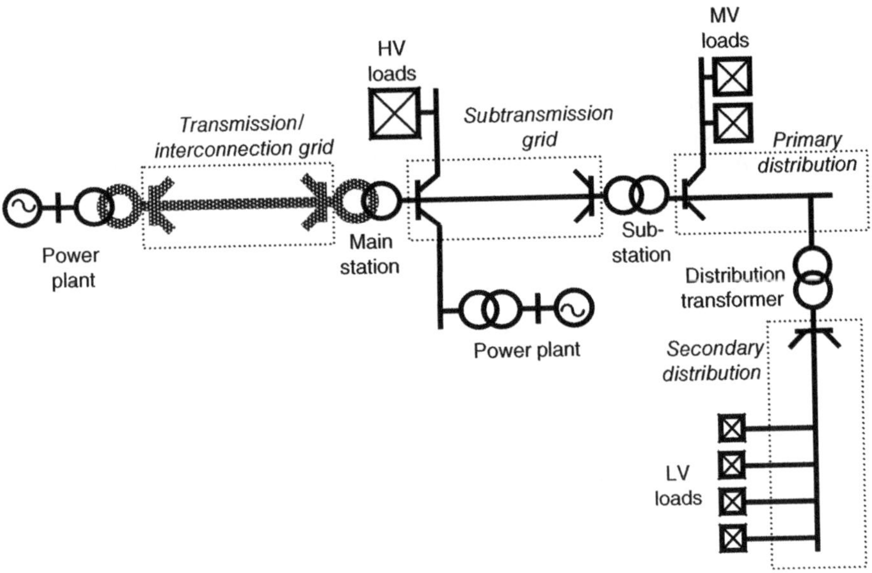

FIGURE 15.9. Traditional structure of a power system.

- an MV *primary* distribution network, fed by HV/MV "substations" and usually radial, except in areas of high load density; this network supplies MV loads and MV/LV step-down "distribution transformers";
- an LV *secondary*, radial, distribution network that supplies small industrial loads, as well as residential and commercial users.

Since the last decades of the twentieth century, the progressive advent of the so-called distributed generation[8] (DG) has introduced the installation of small-scale power plants directly connected to the MV and LV distribution networks (Figure 15.10). DG uses a wide range of technologies, including gas turbines, diesel engines, solar photovoltaic (PV), wind turbines, fuel cells, biomass, and small hydroelectric generators. Compared to traditional power systems, basically supplied by a few large-scale production plants located far away from load centres, this huge fleet of small-scale generators disseminated over the whole system constitutes a real novelty. Such a change of paradigm is full of opportunities but is not exempt from operational difficulties, as discussed in Section 15.8.

The recent success of DG generation is mainly due to the frequent use of well-subsidized renewable sources and to the improved efficiency performances of modern small-scale plants. Electricity production close to load centres not only reduces long-distance transmission losses, but also enables the exploitation of "waste heat"

[8]Also referred to as embedded generation. Distributed generation is distinct from "dispersed generation," which is typified by generators (usually based on diesel engines) that provide power to their loads as "electric islands"—that is, without connection to an external large grid.

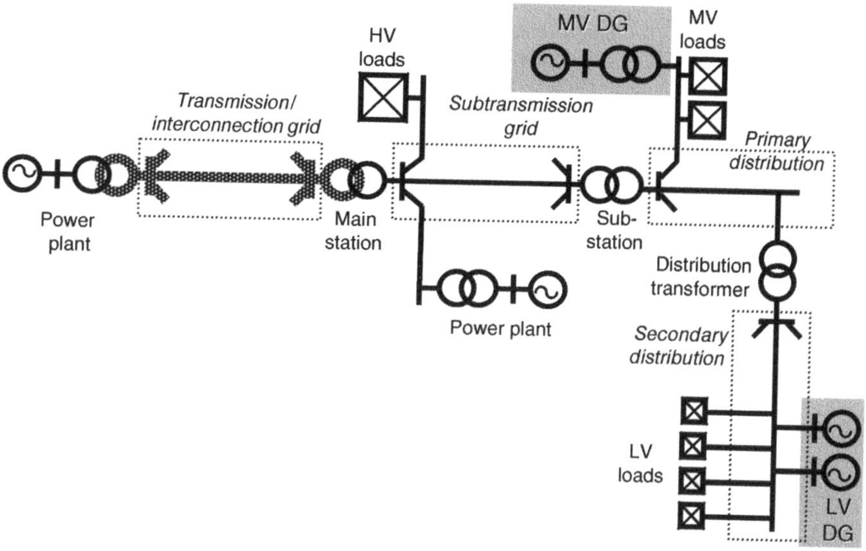

FIGURE 15.10. Structure of a power system with distributed generation (DG).

resulting from many forms of thermoelectric generation; long-distance heat transmission, in fact, would not be cost-effective. Some DG units that use conventional fuel-burning engines are then designed to operate as combined heat and power systems (CHP, or cogeneration), capable of providing heat to buildings or industrial processes. Combined cooling, heat, and power (CCHP, or trigeneration) is also possible, using absorption chillers that convert waste heat into cooling.

The installation of a DG unit by an electricity user, in combination with a preexistent load, is also referred to as "self-production". The perfect balance between production and consumption power profiles is not strictly required; in fact the self-producer can buy extra demand from the network and sell (at other times) production surplus to the distributor.

The impact of distributed generation can be so significant that the power produced at medium and low voltage can even exceed the load supplied by the corresponding distribution network; in this case, the power flow at the local HV/MV substation reverses and the primary distribution network supplies the upstream subtransmission grid. For the sake of simplicity, Figure 1.1 in Chapter 1 did not take this possibility into account. A more rigorous scheme is depicted in Figure 15.11, where double arrows indicate the possibility of reverse power flow at the HV/MV substation.

15.6 BASICS OF POWER SYSTEM OPERATION

A significant peculiarity marks a deep distinction between electric power systems and any other energy infrastructure: since electricity cannot be directly stored, generation and load must be continuously kept in perfect balance.

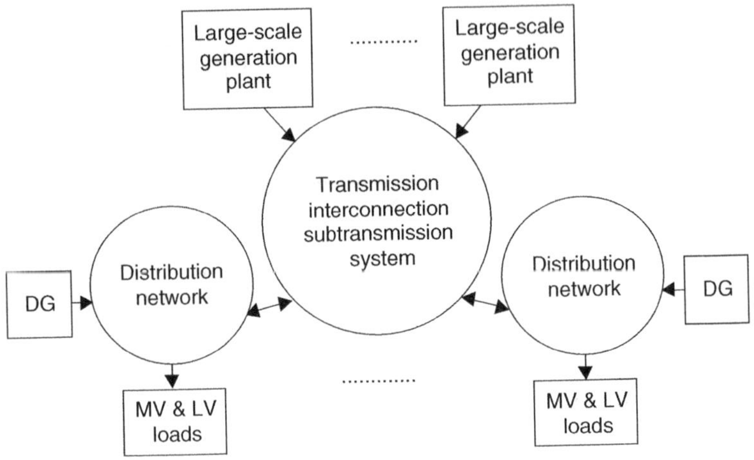

FIGURE 15.11. Possible energy flows with distributed generation (DG).

Direct storage of electricity might be possible inside capacitors or inductors whose stored energy is, as seen in Chapter 5, $(1/2)CU^2$ and $(1/2)LI^2$ respectively (U being the capacitor voltage and I the conductor current). However, the quantity of energy stored in an ordinary capacitor or an inductor is very low with respect to needs. Electricity can be indirectly stored by first converting it into other forms of energy. This can be done, for instance, in electrochemical (batteries), kinetic (flywheels), or gravitational form (water reservoirs). Even indirect storage is rather limited, due to economic and technical reasons.

Any mismatch between power supply and demand causes the loss of torque equilibrium at the rotors of all synchronous generators: their rotational speed, thus the frequency of generated electromotive force, increases or decreases depending on the sign of the imbalance. The need for keeping the rotor's speed within strict tolerances,[9] as well as contractual requirements in terms of power quality,[10] require that the power balance between generation and load be kept constant and, if needed, promptly restored.

The master rule that electricity has to be promptly available on demand permeates any consideration in the planning and operation of power systems. On the other hand, the strict relationship between system balance and grid frequency is another important reason (in addition to those already mentioned in Section 15.2) why AC systems soon surpassed the first DC grids: the frequency value can be used as a common signal to highlight and correct imbalances, without requiring any other communication infrastructure between generators and loads.

[9]Underspeed operation can be dangerous for the generator shaft due to possible mechanical resonances, while overspeed operation must be limited due to centrifugal forces.

[10]Frequency of voltage provided to the users must be very close to its rated value (usually ±1%), since it influences the performance of many appliances—for instance, motor speed.

Traditionally, single users are left free to modify their own power consumption, provided that it does not overcome a contractual limit, basically set to avoid network overload. The power balance used to be ensured by generators, required to meet the constantly changing load demand and to cope with unscheduled outages, by means of "frequency regulation" (see Section 15.6.1). Nowadays the increasing share of intermittent and nonprogrammable generations, like photovoltaic plants and wind farms, requires more complex balancing methods, where energy storage and controllable loads are directly involved in maintaining a constant equilibrium between supply and demand. This will be discussed in Section 15.8.

However, operating a real-time frequency regulation is not enough to guarantee the load demand to be met over time ("followed"). In fact, all generators in operation[11] must always keep an adequate power margin between their nominal rating and their present production level, for regulation and balancing purposes. Furthermore, the number of generators required to meet the load can differ widely between night and day due to the significant variation in demand; last but not least, the start-up of a cold thermoelectric unit can require up to 12–15 hours[12] and must be scheduled ahead of time. For all these reasons, a complex organizational machine is required, over different time periods, in order to ensure the load is met.

- **Planning:** With regard to the timing of installing generation, transmission, and distribution infrastructures, system adequacy must be guaranteed years or decades in advance to forecast future load peaks; strategic planning involves long-term choices like grid development and the amount, type, size, and location of power plants.
- **Mid-term operation:** The time span is usually one year; typical issues are scheduling the maintenance of power plants and allocating the use of hydro resources to each week of the year.
- **Short-term operation:** This is run 24 hours before the energy delivery, with 15 or 60 minute steps. Main tasks are load forecasting, unit commitment (which power plant must be spinning at each hour of the following day?) and Power Dispatching (what power will each spinning generator deliver?). The selected solution must not only meet the load without congesting the power links, but also guarantee security margins for real-time balancing operations (spinning and nonspinning generation reserve).
- **Real-time operation:** Rigorous compliance with unit commitment and power dispatching ensures that frequency regulation operated by generators constantly meets the load demand, which fluctuates continuously and randomly. Sudden contingencies like generator outages, unexpected load variations, and line faults are still possible; a quick protection system and constant supervision by a system control room are then required. In the most severe cases, defence plans are activated for emergency corrective actions like load shedding.

[11]Also called *spinning generators*.

[12]More than one day in the case of a nuclear power plant.

15.6.1 Frequency Regulation

To keep the power system in balance, most spinning generators are provided with an automatic control device, which regulates the mechanical power supplied by the local *prime mover* (i.e., the mechanical source of power) depending on the measured difference between the actual and the nominal rotor speed. This mechanism is called **load-frequency control** or, especially in Europe, **primary frequency regulation.**

Figure 15.12 illustrates how this control is effected by a single generator. When the frequency—hence the rotor speed—equals its nominal value f_0 (typically 50 or 60 Hz), the prime mover provides the mechanical power P_{ref}, corresponding to the desired power output of the generator. If rotors slow down and the frequency drops—for example, due to a sudden increase in load demand—the controller boosts the power output provided by the prime mover,[13] according to the characteristic shown in Figure 15.12. As a result, the active power supplied by this generator increases, thus helping to meet the additional load: the lower the slope of the load-frequency characteristic, the higher the share of the extra demand that is met by the generator.

In large power systems, load-frequency control is performed by hundreds or thousands of generators that run synchronously to supply the load, meeting its random variations and coping with unexpected outages of electrical components. Thanks to the simultaneous action of this huge number of generators, constantly ready to provide balance, the frequency fluctuations in highly interconnected systems are very small.

Primary regulation is able to quickly restore the power balance, usually within 20–30 seconds. However, the final frequency always differs from the initial one; otherwise the balancing action provided by primary regulators would be null: Figure 15.12 clearly shows that if $f = f_0$, P would be brought back to P_{ref}. Moreover, primary regulation is autonomously provided by generators without any kind of central supervision; power flows in the transmission grid, particularly in the links between adjacent countries ("tie-lines"), are consequently modified. To

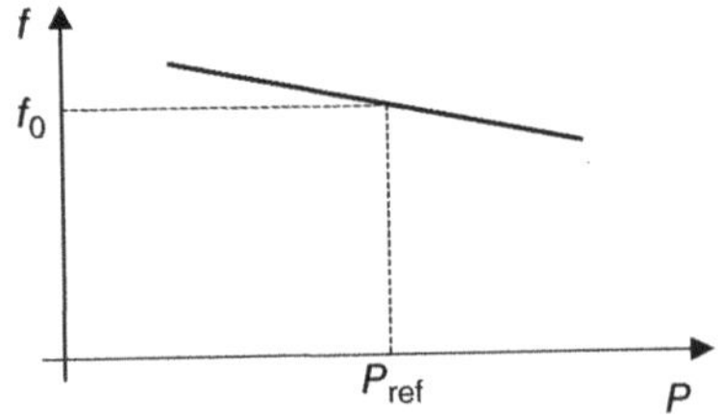

FIGURE 15.12. Load-frequency characteristic: primary regulation.

[13]This can be done by operating the valve which regulates the flow of steam or water to the turbine, in the case of thermoelectric or hydroelectric power plants respectively.

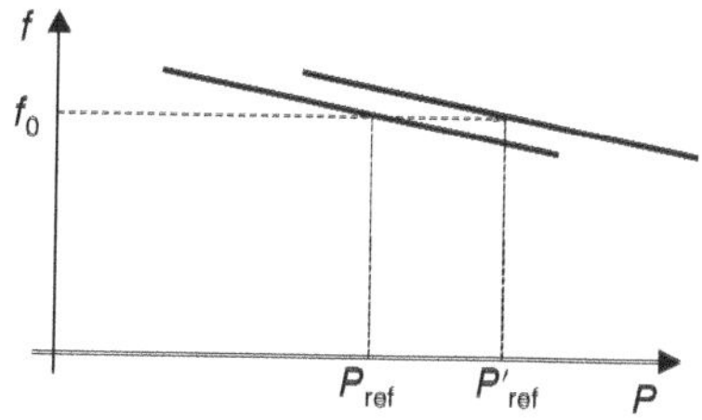

FIGURE 15.13. Load-frequency characteristic: secondary regulation.

restore nominal frequency and the commercial value of power flows at interconnections, a further frequency regulation is required.

This **secondary frequency regulation**, or **automatic generation control (AGC)**, is slower (its final steady state is reached in 10–15 minutes); it is activated by a central controller (one for each interconnected country) and is performed by a predefined subset of large generators. In the previous example, these would be committed to providing extra power in order to re-increase the frequency up to its nominal value, thus relieving primary regulators and finally meeting the entire additional load. They do this by shifting their load-frequency characteristic (Figure 15.13), so they can provide a higher power output P'_{ref} when the nominal frequency is restored. The sum of terms $P'_{ref} - P_{ref}$, extended to all generators performing AGC, must equal the extra load demand.

Scheduled power flows at interconnections can be restored simply by activating the secondary regulation exclusively in the country where the initial event occurred that caused the power mismatch. This area can be easily identified even when the reasons for the power imbalance are unknown; in fact, a parameter called Area Control Error (ACE), which is an opportune combination of measures of frequency and power flows across tie-lines, is constantly calculated and monitored in each country: secondary regulation must be carried out only in the area where ACE is non-null.

Other supplementary controls periodically reduce the integral of the frequency error to zero, so that appliances that depend on the grid frequency—like many clocks—can run correctly.

Primary and then secondary regulations modify the working point of the power system, thus altering the scheduled operation of the generators which perform AGC. The valuable power margin made available by these generators for balancing purposes must, however, be restored to be ready to cope with future contingencies; furthermore, power plants under AGC are not necessarily the cheapest ones. For this reason, a third, even slower, regulation is usually activated to restore their scheduled production within a few hours. This **tertiary regulation** is carried out on an economic basis by selecting the cheapest power plants available to relieve generators under AGC and meet the entire extra demand (economic dispatching).

Primary and secondary frequency regulations are always performed by automatic controllers, which act on spinning generators. Conversely, the times involved in tertiary regulation are long enough to also allow the use of nonspinning generators, committing the cheapest and quickest of them to start-up. In many countries, economic dispatching is manually operated.

The described balancing methods basically rely on conventional generators, like thermoelectric and hydroelectric power plants, which are relatively consistent and easily to control. On the contrary, new forms of production, like solar or wind power plants, inject a power profile into the grid which strictly depends on the real, current, and random availability of sun or wind. The increasing penetration of such an intermittent non-controllable generation is today pushing toward a deep revision of balancing methods. This will be discussed in Section 15.8.

15.6.2 Voltage Regulation

The operation of an electric system is not limited to maintaining the balance between production and load, which means dispatching the active power. In fact, the system operator must also keep voltages at grid nodes close to their nominal value, usually within a range of ±5%. This is because electric appliances do not tolerate large voltage variations, both for safety and performance reasons. Low voltages dim lights and slow down induction motors[14]; high voltages can cause the saturation of the magnetic cores of all electrical machines or, even worse, the flashover of insulation materials.

The primary way of keeping voltages in the desired range is by controlling the excitation of synchronous generators in power plants. As discussed in Chapter 11, the field current of an alternator directly affects the amplitude of its electromotive force (EMF). Unless the machine is connected to a prevailing network, changing EMF E_r results in the voltage U being modified at the machine terminals. This is evident from Figure 15.14, where the external grid is described by its Thévenin equivalent.

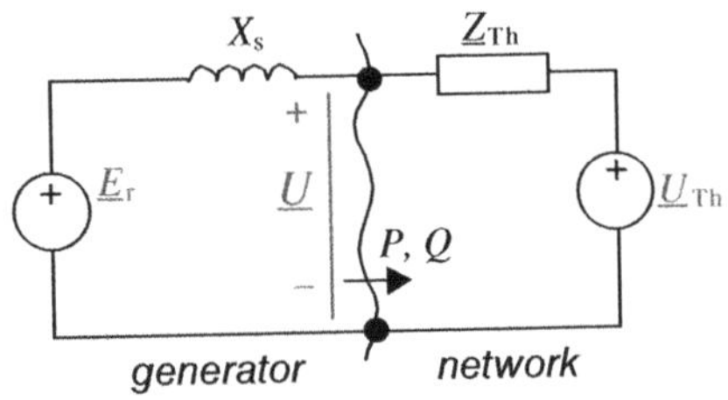

FIGURE 15.14. Connection of a synchronous generator to a nonprevailing network ($Z_{Th} \neq 0$): the higher E_r is, the higher U is.

[14]As we saw in Chapter 12, for a given value of slip, the mechanical torque provided by an induction motor is proportional to the square of its terminal voltage (see for instance eq. 12.10, considering that, except at the very beginning of the motor starting, $U_s \cong U_1$). As a consequence, if supply voltage drops, the torque-speed curve depicted in Figure 12.8 or 12.9 flattens, thus crossing the load curve at a lower speed.

Every synchronous generator is then provided with an **automatic voltage regulator** (AVR) that controls the field current in such a way that the voltage U at the machine terminals is kept at a desired value, or setpoint (**primary voltage regulation**). Moreover, system operators make opportune real-time calibrations of setpoints (**secondary voltage regulation**) in order to also keep the voltage of contiguous load nodes within a proper range. While frequency regulation is operated at system level (frequency is the same at all nodes[15]), voltage regulation has more local features, since an AVR has a significant influence only on nearby nodes.

As we have already seen [see equation (10.8)], the amplitude of the electromotive force directly affects the reactive power Q exchanged between the synchronous machine and the network it is connected to: overexcitation corresponds to injecting reactive power into the grid, and vice versa. For this reason:

- Secondary voltage regulation is also called "reactive power dispatching," since it defines how the reactive power required by the loads is generated by different power plants; as the load demand of reactive power changes throughout the day, also reactive power dispatching is time-varying.
- Any appliance supplying reactive power, like *R–C* loads, increases voltage at the local and nearby nodes; any device absorbing reactive power, like *R–L* loads, decreases voltage at the local and nearby nodes;
- Any device able to adjust its supply or absorption of reactive power can provide voltage regulation, exactly like a synchronous machine. Among shunt-connected passive devices, inductive reactors, capacitors, and their combination (Static Var Controllers, SVC) are usual. Power electronics is nowadays providing converters which generate AC electromotive forces which are controllable in amplitude and phase, and capable of exchanging P and Q in the four quadrants exactly like a synchronous machine (obviously P must be supported by an energy source or a storage). Among them, STATCOM are devices unable to exchange active power, so they appear similar to a synchronous machine without the prime mover: since they provide a voltage which is always in phase with U_{Th}, they can be used for the purposes of voltage regulation, by absorbing or providing reactive power.

Voltage regulation can be performed by installing special types of transformers called *tap-changer transformers* in the grid. These have an adjustable turns ratio N_2/N_1, and therefore the on-line changing of this ratio changes the ratio of its output voltage to input. If coil number one is connected to a constant voltage, for instance, an increase of N_2/N_1 implies a rise in the secondary voltage (i.e., connected with the coil with N_2 turns). Changing the turns ratio is called *tap-changing*.

[15] Transient phenomena involving the presence of different frequencies at grid nodes (rotor oscillations) are beyond the scope of this book.

On-load tap-changing, which means tap-changing without disconnecting loads, is usual for EHV/HV and for HV/MV transformers.

Tap-changing used to be the main tool for voltage regulation on the distribution networks. Nowadays the massive presence of distributed generation offers new opportunities to network operators, provided that DG is involved in frequency and voltage regulation.

15.7 VERTICALLY INTEGRATED UTILITIES AND DEREGULATED POWER SYSTEMS

The first power systems were installed and operated by private business companies. As the social importance of electricity grew, such companies became providers of a "public service," the so-called investor-owned monopolistic utilities. Free enterprise, natural monopoly over a given territory, and a relatively weak regulatory framework were the features of this first era.

In many countries, after World War II, mass electrification required common standards to be applied to the different regional contexts, especially as far as the connection of rural users was concerned. Business initiatives pursued by investor-owned companies could not always guarantee the same connection rights and uniform quality levels. For these reasons, in many countries the electric service was nationalized.

Both investor-owned and public utilities were usually organized as vertically integrated companies, in the sense that they were directly responsible for most of the generation capacity and for transmission and distribution facilities.

The rationale for combining all functions in a single firm, also including long-term planning, dispatching, R&D, and sales, was basically related to the physics of electricity delivery and to the need to keep generation and consumption always in balance. This challenging task requires a central authority to govern supply and demand of electricity along the power grid. There were also economic reasons for vertical integration. Low-cost production requires the simultaneous optimization of generator dispatching and transmission capacity; long-run profitability requires the coordination of investment decisions at all stages of the chain, from generators to the LV distribution network.

Vertically integrated utilities performed all the activities described in the previous paragraph at minimum cost, optimizing the use of their own assets under predefined quality standards. Tariffs applied to users were basically cost-driven.

Following the success of deregulation in many sectors of the economy, such as telecommunications, electricity systems have undergone a similar transition since the early 1990s. In many countries, vertically integrated utilities were progressively "unbundled"; that is, their functions of generating power, transmitting it over EHV and HV lines and distributing power to final users, were assigned to different companies, introducing competition both in generation and supply. Nowadays market structures and regulatory policies are extremely diversified around the world; various gradations of unbundling are in use, and there is no standard market model. However, some common features can be outlined.

- Transmission and distribution remain most efficiently organized as natural monopolies whose activities are strictly regulated in order to ensure non-discriminatory access to the transport infrastructure by all commercial operators (third-party access).
- On the wholesale market, generating companies (GENCOs) offer their energy output to retailers and, almost always, to large end-users; this can be performed through yearly agreements (bilateral contracts) or in more complex market-places managed by an independent market operator, like the day-ahead spot market. In the latter case, bid-based, security-constrained economic dispatching derives from a classic supply and demand equilibrium, usually setting prices on an hourly basis. Pay-as-bid or marginal price mechanisms can be adopted. In order to avoid congestions on grid bottlenecks, market clearing is performed by superimposing power flow constraints, which results in zonal or nodal pricing.
- In the retail market, retailers sell energy to end-use customers in the form of bilateral contracts; small clients can usually purchase electricity also from their local distributor at a regulated tariff. Consumers rarely face real-time pricing; in any case in the mid-term the retail market reflects the average prices set by the wholesale market.
- An independent system operator (ISO) is responsible for system reliability and security; for this reason, it must always ensure an adequate amount of generation reserve for real-time balancing; generators offer this availability in "ancillary service markets" operated by the ISO in the afternoon before the day of energy delivery, just after the end of the energy market sessions. The economic settlement can be *ex-ante* (generators are paid for their availability) or *ex-post* (only actual real-time balancing operations are remunerated). In the first case, the ISO usually charges all customers for power reserve, pro-quota; in the latter, balancing costs can be charged to the producers or consumers who caused the mismatch between demand and supply the ISO is also responsible for defence plans and for system restoration after a blackout. In many countries, the ISO directly operates the transmission system.

In many countries, competition has promoted efficiency and stimulated the refurbishment of old power plants. Short-term operation of deregulated power systems has frequently reduced energy prices, at least net of other factors like fuel costs. Nevertheless, there is no clear evidence regarding the effectiveness of the electricity market in terms of long-term system adequacy—that is, its structural capacity for meeting load. According to many economists, market-driven planning can cause dangerous fluctuations of available resources, namely the generating park, which means the uncontrolled alternation between long periods of lack and of excess generation capacity. These "boom–bust cycles" have a serious impact on system reliability and quality of supply. In some cases, regulators and policy makers are currently introducing "capacity payment" mechanisms to stimulate GENCOs to install new plants also during low-price periods, instead of postponing investments until the next price spikes.

15.8 RECENT CHALLENGES AND SMART GRIDS

The transition toward a low-carbon economy, while coping with an increasing energy demand,[16] is rapidly changing the way electricity is produced and consumed. The electric sector is today undergoing fundamental changes, mainly due to the opportunities, but also the challenges, relating to the increasing impact of distributed generation.

The interaction between this huge number of new producers/consumers ("prosumers"), randomly embedded into networks, requires a higher level of "intelligence" for the power systems of the future. This appears crucial for balancing the electric system, if we consider that the major part of DG is from intermittent, nonprogrammable, and nondispatchable energy sources like wind and sun.

Smart grids (SG) are an essential element in facilitating this transformation. The smart grid is an electricity network that can intelligently integrate the actions of all users connected to it—generators, consumers, and those that do both—in order to efficiently deliver sustainable, economic, and secure electricity supplies (Figure 15.15).

A smart grid employs innovative ICT tools for communication, control, and self-healing purposes, in order to:

- better facilitate the connection and operation of generators of all sizes and technologies;
- make wide use of energy storage technologies to keep the power system in balance even with increasing penetration of nonprogrammable renewable energy sources (Figure 15.16);

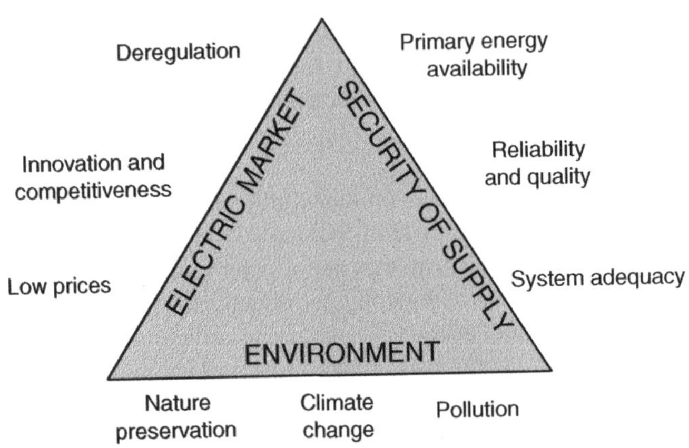

FIGURE 15.15. Smart grid main drivers.

[16] According to the International Energy Agency, world energy demand will increase by 40% in the period 2010–2030.

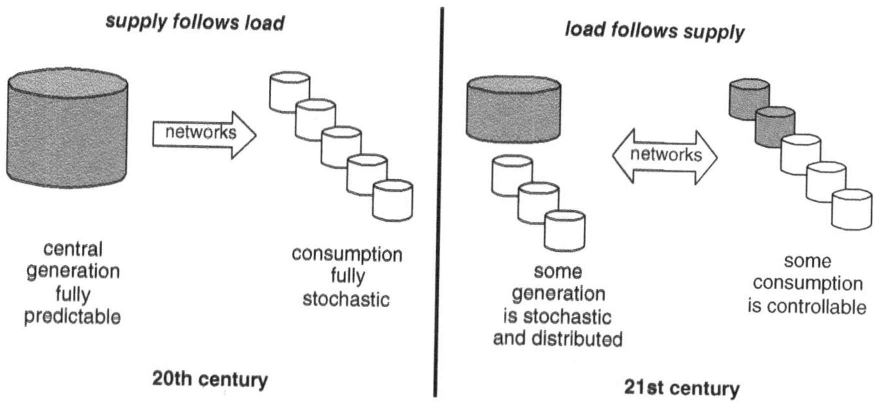

FIGURE 15.16. The new paradigm for balancing demand and supply.

- provide consumers with more information and options for choosing energy providers;
- allow consumers to play a part in optimizing the operation of the system—for example, reducing power spikes and shifting consumption to off-peak hours;
- significantly reduce the environmental impact of the whole electricity supply system;
- optimize the use of present assets, deferring new installations;
- maintain or improve the existing levels of system reliability, quality, and security of supply, while operating resiliently against (cyber) attacks and natural disasters;
- foster the development of integrated transnational markets and the interaction between different infrastructures (electricity, gas, mobility, water, etc.).

To cope with increasing energy demand, SGs use more brain (i.e., a better use of the existing assets), rather than more muscles (i.e., the installation of new generation and transmission facilities).

The power systems of the future aspire to be more reliable, secure, economic, environment-friendly and efficient. Large-scale central generators will be joined by a huge fleet of small prosumers, thereby substituting the top-down approach shown in Figure 15.9 (unidirectional power flows going from bulk generation toward users) with a more democratic network of resources, as depicted in Figure 15.17.

Smart grids will also enable active participation by small end-users (demand response and demand side management). The grids will give consumers information, control, and options that enable them to engage in new electricity and ancillary markets. Distribution system operators will treat willing consumers as resources in the day-to-day operation of the grid. Well-informed consumers will modify consumption as they balance their demands and resources with the electric system's capability of meeting these demands.

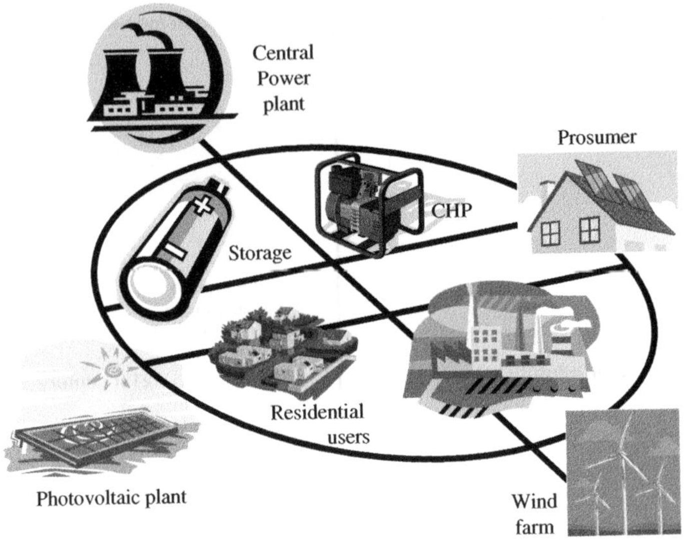

FIGURE 15.17. A smart grid.

15.9 RENEWABLE ENERGY SOURCES AND ENERGY STORAGE

It has been seen that new distribution grids and future smart grids take advantage of renewable energy sources, and of energy storage.

To better understand these networks, in this section some information about these components is provided.

15.9.1 Photovoltaic Plants

Photovoltaic (PV) cells directly convert the energy of solar radiation into DC electricity. A typical silicon PV cell is composed of a thin wafer, consisting of an ultra-thin layer of phosphorus-doped (N-type) silicon, on top of a thicker layer of boron-doped (P-type) silicon; this is what is called a P–N junction. When sunlight strikes the surface of the cell, the photons contained in the sun's rays cause extra electrons and holes to be generated above their normal equilibrium and an electrical field originates through the junction (photovoltaic effect). If the cell is connected to a load, this potential gives rise to a flow of current (electrons).

The performance of a solar cell is measured in terms of its efficiency in turning the power of sunlight (around $1\,kW/m^2$ at noon on a cloudless day at the equator) into electricity. There are many kinds of photovoltaic cells. Monocrystalline cells use very pure silicon, thus obtaining efficiencies of 17–20% (2013) at greater investment expense; they give their best with direct solar radiation and perfect alignment with solar rays. Polycrystalline silicon consists of multiple small crystals; efficiencies are usually around 13–16% (2013); these cells are particularly appreciated for locations

FIGURE 15.18. Photovoltaic panels installed on a roof.

with prevailing diffused solar radiation or when alignment with solar rays is not perfect due to installation constraints. Thin-film solar cells are obtained by depositing one or more thin layers of photovoltaic material onto a substrate.[17] These cells are the newest and cheapest, their strongpoints being their flexibility and their increasing efficiencies; at present (2013), commercial modules operate at around 9%, but future panels are expected to have performances similar to polycrystalline modules. The efficiency of the first solar cells, built in the 1950s, was lower than 4%.

Groups of 35–50 cells are electrically configured into modules (or "panels," usually ranging from 150 to 300 W) and these into arrays. Many arrays constitute a solar field (Figure 15.18). Using inverters, PV systems can produce AC electricity compatible with any conventional appliances (islanded or "stand-alone" systems) or inject their production into the distribution network with which they are in parallel (grid-connected systems).

Although the photovoltaic effect was discovered by the French physicist Edmond Becquerel as early as 1839, the first photovoltaic cells were produced in the late 1950s and throughout the 1960s were principally used to provide electrical power for satellites. During the 1970s, improvements in the performance and quality of PV modules helped to reduce costs and opened up a number of opportunities for powering remote islanded applications, including battery charging for navigational aids, signals, telecommunications equipment, and other critical, low power needs. Following the energy crises of the same years, there were significant efforts to increase the use of renewable energies to substitute fossil fuels, thus improving the efficiency of solar energy conversion and developing increasingly larger grid-connected PV systems. These produce electricity for residential, industrial, and commercial uses or simply inject electricity into the grid like a conventional power.

Nowadays, photovoltaic applications range from fractions of watts (consumer electronic devices, including calculators, watches, radios, and other small battery charging applications) to many megawatts (large grid-connected power plants).

[17]Depending on which material is deposited, we can obtain amorphous silicon (a-Si), cadmium telluride (CdTe), copper indium gallium selenide (CIS/CIGS), and organic photovoltaic cells (OPC).

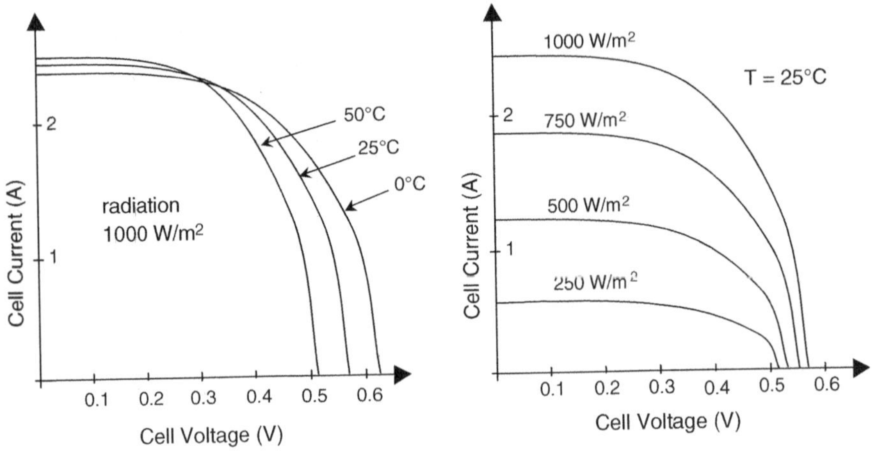

FIGURE 15.19. *I–U* characteristics of a photovoltaic cell. Dependence on temperature (*left*) and on solar radiation (*right*).

Figure 15.19 shows the typical *I–U* characteristics of a photovoltaic cell, which are strongly dependent on solar radiation and cells, in terms of both *short-circuit current* (the current in the diagrams corresponding to zero voltage) and *no-load voltage* (the voltage corresponding to zero current).

A PV array consists of several cells in series, to produce reasonably high DC voltage. Voltages between 50 and 100 V are common. To interface the arrays with an AC grid, it is thus necessary to first raise the electricity produced then convert it into AC.

The conventional structure of a grid-connected photovoltaic system is depicted in Figure 15.20. The inverter can be operated by controlling the amplitude and the displacement of the generated voltage with respect to the grid voltage, so as to handle the active and reactive power injected into the network (see Section 8.3.3.1). It is usual to operate at $\cos\varphi = 1$, that is, without exchange of reactive power with the external grid; in this case, the inverter is controlled so as to obtain an AC current that is in phase with the grid voltage; the amplitude of the AC current is instead controlled in order to keep U_{dc2} constant, which corresponds to injecting exactly the active power coming from the PV cells into the grid; the presence of the capacitor also reduces voltage oscillations at the DC circuit section.

Assuming U_{dc2} to be perfectly regulated by the inverter, from the point of view of the DC–DC converter it is a constant voltage source, while the panels can be considered as a current source. Readers can easily recognise the typical structure of a step-up DC–DC converter,[18] with active power flowing from left to right (see Section 8.3.2): since U_{dc2} is pretty much fixed, the converter's conduction ratio entails the voltage U_{dc1}, hence the working point of the cells on the actual *I–U* characteristic.

[18]This is shown in Figure 8.28, in which the "source" is the PV panel, whose E_{src} is, however, a function of the delivered current, according to the plots of Figure 15.19.

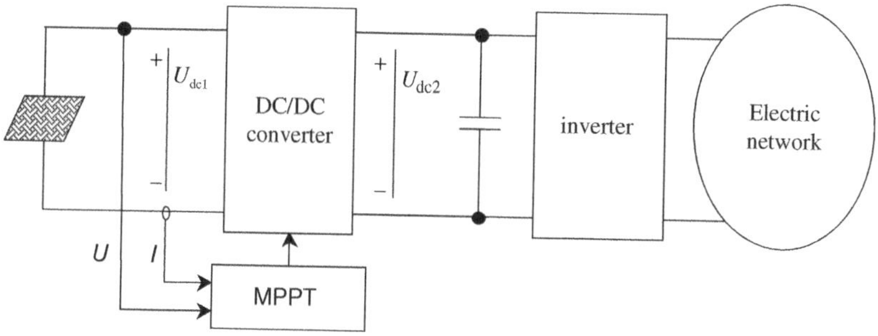

FIGURE 15.20. Typical structure of a grid-connected solar plant.

A maximum power point tracker (MPPT) algorithm is aimed at searching for the conduction ratio that maximizes the active power provided by the cells, as far as selecting the working point shown in the following figure. Since the actual *I–U* characteristic is not known beforehand, this search is usually heuristic ("perturb and observe" techniques), based on the maximization of the product of measured voltage and current (Figure 15.21).

Single-axis or dual-axis trackers, able to slowly move panels to follow the sun's trajectory, are particularly effective in regions that receive a large amount of direct radiation (production increased by 20–30%). In diffused light (i.e., under cloud or fog), tracking has little or no value.

With respect to conventional power plants, photovoltaic systems are penalized by a lower yearly kilowatt-hour production per installed kilowatt, which is usually in the range of 800–1700 kWh/kW depending on the latitude. To promote the installation of photovoltaic systems, state incentives are usually provided in the form of capital support or tariff bonus, in addition to the normal remuneration of electricity. At lower latitudes, the progressive decrease of installation costs per kilowatt makes the investment cost-effective also in the absence of incentives ("grid parity").

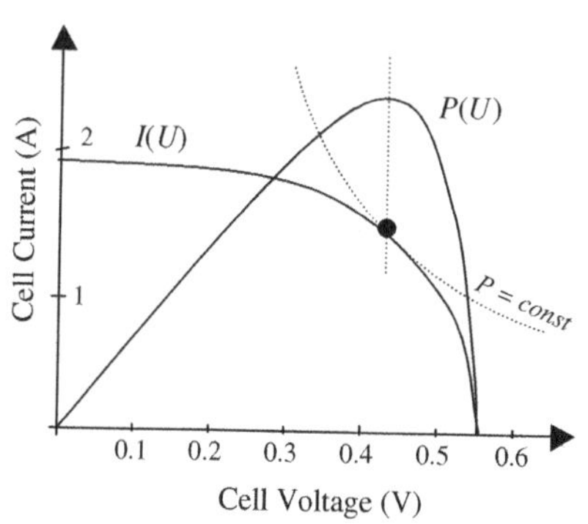

FIGURE 15.21. Point of maximum power on the *I–U* characteristic.

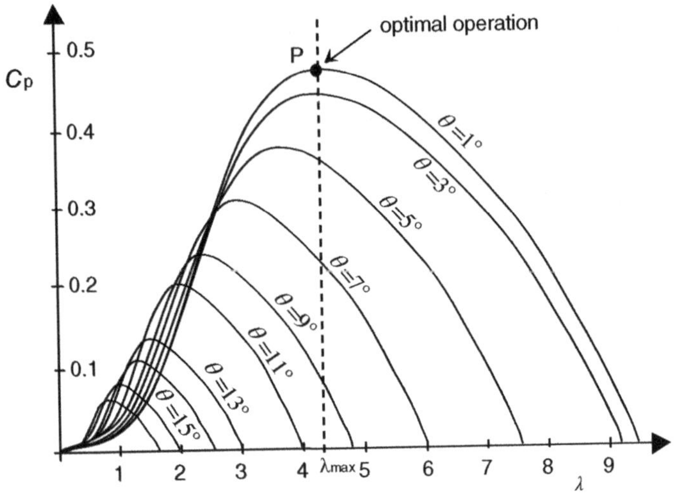

FIGURE 15.22. C_p as a function of the tip–speed ratio (example of a 3-blades turbine).

15.9.2 Wind Power Plants

The total wind power flowing through an imaginary area A, perpendicular to wind speed, is related to the kinetic energy of the air in motion:

$$P_{\text{tot}} = \tfrac{1}{2}\dot{m}v^2 = \tfrac{1}{2}(\rho A v)v^2 = \tfrac{1}{2}\rho A v^3$$

where $\dot{m}$ is the mass flow rate (kg/s), v is the wind speed, and ρ is the density of air. Wind power is then proportional to the third power of the wind speed. The power actually provided by a wind turbine is clearly lower:

$$P_{\text{w}} = C_{\text{p}}\tfrac{1}{2}\rho A v^3$$

where C_{p} is the so-called *coefficient of performance*. According to theory, C_{p} cannot be higher than 16/27 (0.5927, Betz limit); otherwise the wind speed behind the turbine would be null. As depicted in Figure 15.22, for a given turbine, C_{p} is a function of the pitch angle θ (which is null when the blades are vertical to the wind direction) and of the tip-speed ratio λ:

$$\lambda = \frac{\Omega r}{v}$$

where Ω is the turbine rotational speed and r is the radius of blades, hence Ωr equals the tangential speed at the blade tip.

Empirically, the maximum C_{p} is obtained for $\theta \cong 0°$ and $\lambda = \lambda_{\text{max}} \cong 4\pi/n$, where n is the number of blades; in the specific case depicted in the previous figure, $\lambda_{\text{max}} \approx 4.2$,

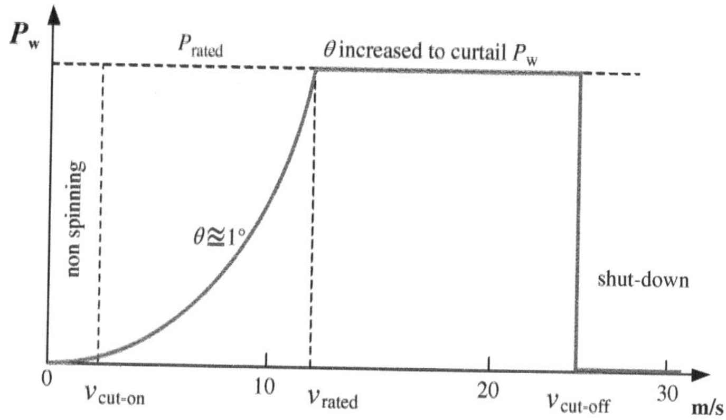

FIGURE 15.23. Available power P_w as a function of wind speed.

but the function $C_p(\lambda)$ is strongly dependent on the actual shape of blades. Optimal performance is then achieved by operating the turbine always at point P, for any wind speed: θ is kept constant at a very low value, while Ω is modified according to the wind speed in order to maintain $\lambda \approx \lambda_{max}$. If λ_{max} is not known beforehand or the wind speed cannot be measured, "perturb and observe" algorithms can be used in real time to perform a heuristic search for the optimal rotational speed, in analogy with MPPT procedures of photovoltaic plants.

If wind speed overtakes the so-called "rated speed" (12–15 m/s), the available power P_w would exceed the design power of the turbine, P_{rated}. The pitch-angle θ is then intentionally increased so as to reduce C_p and curtail P_w. In Figure 15.23, observe the cubic dependency of P_w on wind speed, below v_{rated}. When wind speed exceeds the "cut-off" value (usually around 25 m/s), the turbine is stopped for security reasons. Conversely, under the "cut-on" wind speed (3–4 m/s) the turbine is not able to start up, due to the very low value of P_w.

Three main schemes are currently in use:

Scheme A: Induction Machine, Directly Connected to the Grid. In this scheme, restricted to small-size turbines, the wind pushes a squirrel-cage induction machine above its synchronous speed. As discussed in Chapter 12, the slip is negative and the machine operates as a generator. According to Figure 12.8, the rotational speed is practically constant for any value of the torque provided by the turbine (for generator operation it is between Ω_0 and $\tilde{\Omega}$, but typically nearer to Ω_0); for this reason, Ω cannot be controlled to achieve the optimum C_p at all times and efficiency is jeopardized, especially at low and high wind speeds.

Another disadvantage of this scheme is that the electrical machine always draws reactive power from the grid, and shunt-connected capacitors are then required to overcome the lagging power factor.

In very small turbines, the pitch-angle regulation is not available. In this case, P_w is roughly limited to P_{rated}, taking advantage of the torsional rigidity of blades, designed to stall above v_{rated}.

Scheme B: Doubly-Fed Induction Generator (DFIG). As we saw in Chapter 12, in a normal ("singly fed") induction machine only the stator is fed, while the rotor is in short-circuit; the frequency f_r of electromagnetic forces and currents induced in the rotor depends on stator frequency and on slip: $f_r = sf$. The slip s, in turn, is defined by the balance between traction and load torques, as depicted in Figure 12.9. As discussed in Example 1 of Chapter 12, f_r determines the speed n_{rr} of the rotor magnetic field with respect to the rotor itself: $n_{rr} = 60f_r/p$, where p is the number of pole pairs and n_{rr} is measured in rpm. Whatever the mechanical load, the rotor and stator magnetic fields are synchronous; therefore $n_{rr} + n = n_0$, where n is the rotor speed and n_0 is the speed of the stator rotating field (synchronous speed, strictly fixed by the grid frequency and the number of pole pairs).

In a doubly fed wound-rotor induction machine, f_r and consequently n_{rr} are imposed by an external electrical supply. Since the two magnetic fields must be synchronous, the equation $n_{rr} + n = n_0$ remains valid: by setting f_r, hence n_{rr}, the operator can regulate the rotor speed n, regardless of load conditions and stator frequency.

Let us assume n_{rr}^* and n^* to be the values corresponding to the singly fed induction generator in the present wind conditions, where n_{rr}^* is slightly lower than zero. If $n_{rr} < n_{rr}^* < 0$, then n is higher than n^* and the doubly fed machine increases its speed. If $n_{rr}^* < n_{rr} < 0$, and even more so when $n_{rr} > 0$ (currents injected into the rotor change their sequence and the rotor field reverses its sense of rotation with respect to the rotor), then the machine slows down: $n < n^*$.

The rotor winding is supplied through the slip-rings/brushes coupling already described in Chapter 11 concerning additional rotor resistors and motor starting. Brushes are usually supplied by power electronics, which allows f_r to be set as desired (Figure 15.24). Both converters are PWM devices (see Section 8.3.3). The generator-side converter simply imposes voltages with desired f_r to the rotor; the other converter is operated in order to keep the DC-busbar at constant voltage.

When f_r is set so that $n > n^*$, the generator-side converter receives active power through the rotor (usually $\frac{1}{4} \div \frac{1}{3}$ of the turbine's mechanical power) and operates as a rectifier; the other converter must inject such power into the grid; hence it works as an inverter, generating voltages always ahead of the grid reference (like a synchronous generator; see Section 11.3.1).

Conversely, when $n < n^*$ the rotor winding draws active power from the generator-side converter, which operates as an inverter; the other converter draws such power from the grid, thus operating as a rectifier (its voltages are lagging to the grid reference).

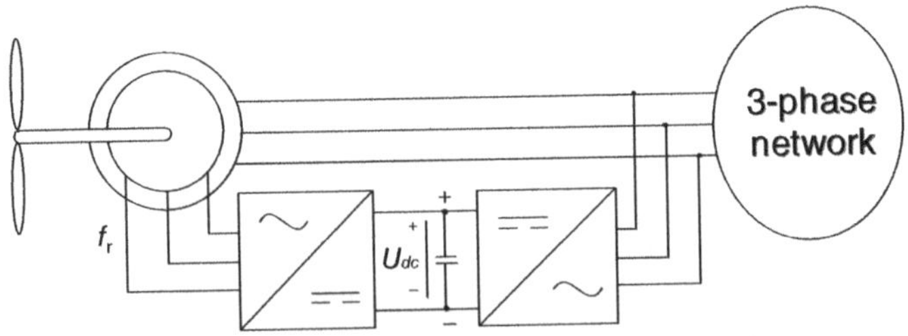

FIGURE 15.24. Doubly fed, wound-rotor induction generator.

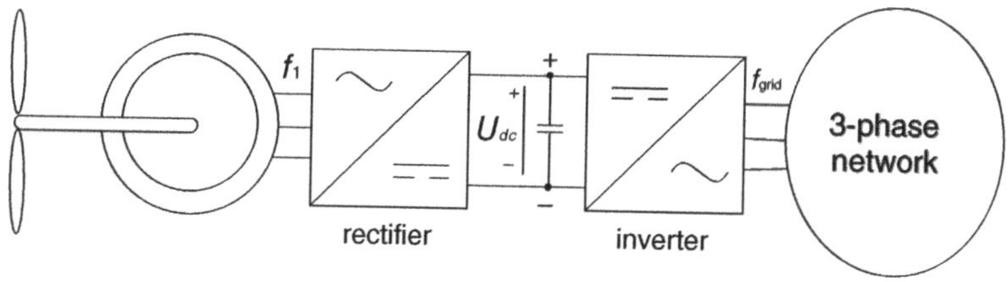

FIGURE 15.25. Permanent-magnet synchronous generator, connected to the network through power electronics.

The main advantage of this scheme is that turbine speed can be regulated at will in a typical range of ±30 percent around the synchronous speed, thus achieving the maximum C_p for many wind conditions.[19] Furthermore, by controlling the amplitude of the currents injected into the rotor windings, it is possible to regulate the reactive power absorbed or provided by the generator, exactly as in a synchronous machine; this should not surprise readers, since a synchronous machine can be thought of as a particular doubly fed induction machine, where the rotor is DC-supplied ($f_r = n_{rr} = 0$).

The only drawback of this scheme with respect to the squirrel-cage machine is that slip-rings and brushes require maintenance and can reduce machine reliability.

Scheme C: Power Electronics Connected Generator. In this case, a permanent-magnet synchronous generator is connected to the network through a power electronics interface (Figure 15.25).

The grid-side converter is a PWM inverter. The converter provides voltages whose amplitude and phase are both controllable at will; they determine the reactive and active power exchanged with the external network, exactly like a conventional synchronous generator. In particular, their amplitude can be varied at will, according to a desired power factor (leading or lagging). Their phase is instead regulated so as to maintain U_{dc} constant, thus injecting into the grid all the active power coming from the generator; this phase control also makes sure that the inverter follows the grid frequency, however much it varies.

Also the generator-side converter is a PWM device, but in this case it operates as a rectifier: the active power flows from the AC to the DC side. The frequency f_1 of the voltages imposed by the converter at the machine terminals determines the rotational speed of the synchronous generator: The electrical machine simply operates as if it were in parallel to a (frequency-varying) network, its electromotive forces being pushed by the prime mover always ahead of the converter's voltages. Since the turbine speed can be set at will by changing f_1, the maximum C_p is achieved in a very wide wind-speed range.

Schemes (a) and (b) require gearing mechanisms that adapt the slow turbine speed with the high synchronous speed of the generator; for practical reasons, in fact, the number of magnetic poles of the electrical machine cannot exceed 4–6. The gear box

[19]This range is the usual compromise between the cost of power electronics and the efficiency of wind energy conversion.

affects both the efficiency and reliability of the plant. In scheme C, f_1 can be low at will and the gear box is not required.

The only drawback of this scheme with respect to the DFIG is that the converter must be sized for the entire nominal power of the turbine, which in the past limited the maximum power rating of scheme (c), due to economic and technical reasons.

15.9.3 Energy Storage

It was seen in Section 15.6 that electricity cannot be directly stored, and therefore generation (meaning the sum of all the electric powers delivered by generators) and load (i.e., the sum of powers absorbed by all the individual loads) must be kept in perfect balance. The situation is depicted in Figure 15.26a.

This is a very simplified scheme of a large power system. The system involves some components that generate power (gen 1 to gen n) and others that absorb power (load 1 to load m).

In fact, the sum of the powers delivered by the generators must equal the power absorbed by the loads. The black lines between generators and loads represent ideal

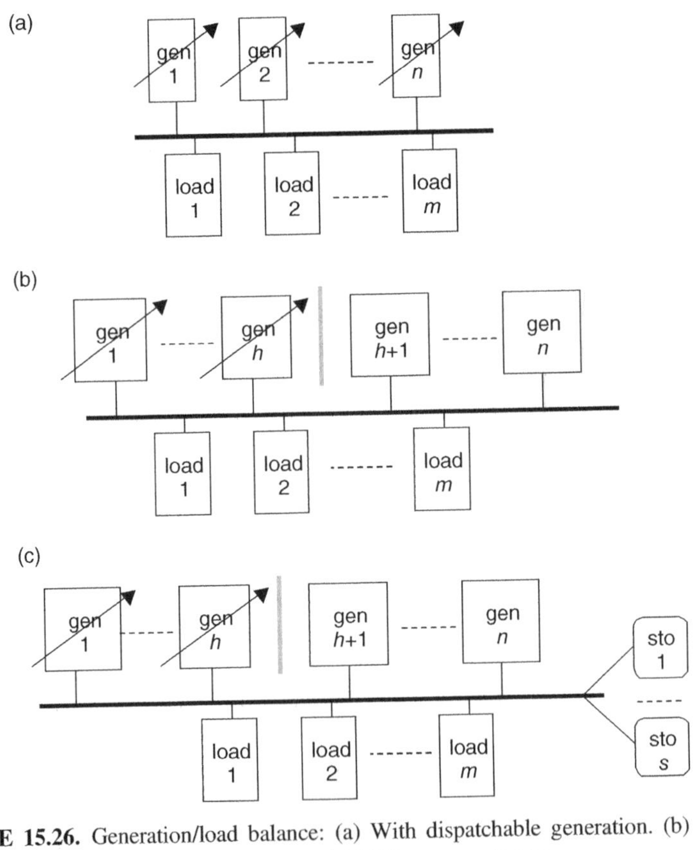

FIGURE 15.26. Generation/load balance: (a) With dispatchable generation. (b) With dispatchable and nondispatchable generators. (c) With energy storage.

lines (which therefore neither absorb nor deliver energy) linking all the components to a virtual common node, which summarizes and represents the transmission and distribution grid. This is what is called a "single-busbar model" of the power system.

The transverse arrow, which in diagrams in general indicates the adjustability, here shows that the power produced by the corresponding generators can be suitably varied to meet system needs. When this happens, the generators are said to be *dispatchable*.

In Figure 15.26a all the generators are dispatchable. However, it often happens that we are not able to, or do not want to, modify the power some of the generators produce. This is often the case of electric generators fed by renewable energy sources: whenever sun or wind is available, we want to convert the whole power they are able to produce into electricity; otherwise it is lost forever. These generators are called nondispatchable, and they are shown in the figure without transverse arrows. As discussed in Section 15.6, power dispatching involves both short-term generation scheduling and real-time operation of power plants.

On the other hand, we usually want to give loads the freedom to vary at will (within given technical or commercial limits). For instance, in households we can turn lights, heaters, and machines on or off at will, with the only limitation that the maximum allotted power must not be overcome.

In this framework, any load variation is promptly compensated for by an equal variation of generation only from dispatchable generators (Figure 15.26b). The others continue to produce all the power coming from their source. This might cause the dispatchable generators to be called to operate in low-efficiency regions. Indeed a power plant will have lower efficiency when it is called to operate loads which are too low or when the variation of the required power over time is too fast.

It might even happen that the variability of nondispatchable generation and/or loads is too much or too fast to be compensated for by dispatchable generators. In these cases, if the network arrangement is the one shown in Figure 15.26b, it might happen that we are forced to shed some loads or some generators from the grid, in order to keep the power system in balance.

The situation can be improved by resorting to the scheme depicted in Figure 15.26c, in which one or more energy storage systems are added to the network, like the ones indicated in the figure as sto 1 to sto s. These systems can be physically placed near generators or loads, or far from both. The effects of their position, indeed, are not very significant for the purposes of system balancing and can be neglected for our purposes.

A storage system is normally thought of as a system which is able to store electric energy. Indeed it does something different: it converts energy into a form that can be stored when it is absorbing power and then converts it back into electricity, when it is delivering power.

Typical storage systems are:

- *Hydraulic Reservoirs.* In this case, when electric power must be absorbed, water is moved by electric pumps from a lower location to a higher one, so electricity is converted into the gravitational potential energy of water (pumping operation). When the plant is required to deliver power, water is released by the upper location and sent to the lower one, and the potential energy is converted into kinetic energy by a turbine that in turn moves an alternator, thus generating electricity (turbine operation). The round-cycle efficiency of a hydraulic reservoir might be in the range 70–85%. In some cases, pumping and turbine operations are performed using the same, reversible hydraulic machine, and in others they are performed by employing a pump and a turbine.

- *Electrochemical Batteries.* Batteries used in electric systems are very similar to the batteries commonly used in everyday life, like the batteries of our cars, laptop computers, or mobile phones. When they are required to store energy, electrochemical batteries convert energy into potential chemical energy. This potential energy can be converted back into electricity when the grid needs power (and energy). The round-cycle efficiency of modern electrochemical batteries is around 90%.
- *Other Systems.* A lot of other possibilities are potentially usable, and sometimes actually used, as energy storage systems. Energy can be stored as kinetic energy into flywheels, into potential energy of compressed air, and even into the magnetic energy of very large inductors, operating at extremely low temperatures, so that superconductivity is exploited.[20] These other systems are, however, much rarer than hydraulic or electrochemical energy reservoirs and will therefore not be dealt with hereafter.

In order to operate, hydraulic reservoirs require a special land conformation: there must be a lower and higher basin, and space to allocate pumps, turbines and electric machines. Therefore these reservoirs are typically very large, allowing energy storage of at least several tens or hundreds of megawatt-hours, and can even be in the range of gigawatt-hours.

On the other hand, electrochemical batteries can be very small, and therefore they are more adequate for distributed, small storage systems. That's why they are becoming more and more important in modern distribution grids—that is, grids with distributed generation and smart grids. Battery energy storage systems used in power systems can range from 1 kWh to a few tens of megawatt-hours.

The building block of a battery energy storage is the *electrochemical reversible cell*. Cells can be produced using different technologies. Here we list the two most important ones.

- *Lead-Acid Cells.* These are based on chemical reactions that take advantage of the specific electrochemical properties of lead. Lead-acid cells are very widely

[20]Superconductivity is a special mode of operation of some conductors that occurs at very low temperatures. In these conditions the electric resistance of the material is practically null, and therefore current can virtually flow in a superconductive material for an indefinite time. For instance, mercury becomes superconductive at a temperature of around 6 K. Storage systems based on superconductivity are called superconductive magnetic energy storage systems (SMES).

used in engineering applications. For instance, they are the source of energy of nearly all cars in the world, when the internal-combustion engine is off. They are a mature and cheap technology. For distributed generation systems, however, they are not so attractive, since the number of charge/discharge cycles allowed (before the battery reaches its end-of-life) is relatively low: typically 500 cycles.

- *Lithium-Based Cells.* These are based on chemical reactions that take advantage of the specific electrochemical properties of lithium. Lithium cells are very widely used in engineering applications, specifically in consumer electronics: nearly all of today's mobile phones, tablets, and laptop computers are fed by lithium-based cells. Cells of consumer electronics devices contain very small amounts of energy compared to power system needs. Today there is a fast-growing offer of larger lithium-based cells, for stationary and vehicular applications. However, their technology is less mature than lead-acid batteries, and their cost per unit of stored energy is higher. Nevertheless, they are the most important technology for new applications, because of their fast-growing competitiveness and also because their cycle life (= life measured in cycles) is much larger than lead-acid's: typically 5000 cycles. Lithium cells are not of one single type. At the time of this writing (fall 2013), the most common types are NCA (nickel–cobalt–aluminium), NMC (nickel–manganese–cobalt), LMO (lithium–manganese oxide), LTO (lithium–titanate oxide), and LFP (lithium-iron phosphate), but the situation might change in the coming years.

Cells are characterized mainly by their *nominal capacity*, usually expressed in Ah. This represents the electric charge the considered cell can deliver under specific testing conditions. When discharged at currents which are different from the nominal current, the delivered charge varies.

For instance, when delivering a constant current of 5 A, a cell having $C_n = 50$ Ah (measured using a 10-A constant discharge current), will be totally discharged after more than 10 hours.

When discharged at currents which are higher than the nominal current, the charge the cell is able to deliver reduces. The amount of reduction depends on the type of cell.

A typical "power–oriented" lithium cell[21] that delivers 50 Ah when discharged at 5 A might deliver something around 45 Ah when discharged at 10 A. The delivered charge of lead-acid batteries is much more highly dependent on the discharge current.

Although batteries are designated by their nominal capacity (in Ah), the user is mainly interested in *the energy* the cell can deliver.

[21]When a cell is "power-oriented," it is designed for large discharge currents (and relatively low discharge times, say a few minutes or tens of minutes, even though longer discharge times are always possible). "Energy-oriented" cells do their best when discharged at smaller currents—that is, with discharge times of at least one hour.

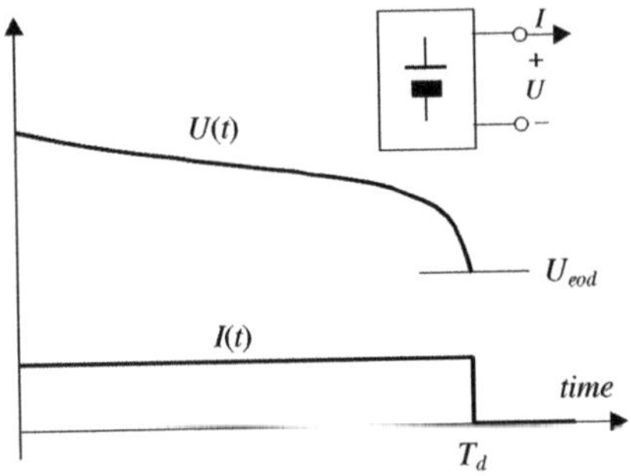

FIGURE 15.27. A typical constant-current cell discharge.

Conversion from capacity to energy is easy when constant current discharge is involved. When cells are discharged at constant current (Figure 15.27), their voltage continuously reduces, with a slope which is initially quite constant. Toward the end of discharge the voltage reaches a "knee," and then decreases rapidly. The cell is considered to be discharged when its end-of-discharge voltage U_{eod} is reached.

The energy the cell delivers during discharge can be computed as follows:

$$E_{\mathrm{disch}} = \int_{T_{\mathrm{d}}} U(t)I(t)\,\mathrm{d}t = I\int_{T_{\mathrm{d}}} U(t)\,\mathrm{d}t = QU_{\mathrm{avg}}$$

with

$$Q = IT_{\mathrm{d}} \text{ and } U_{\mathrm{avg}} = \frac{1}{T_{\mathrm{d}}}\int_{T_{\mathrm{d}}} U(t)\,\mathrm{d}t$$

Therefore E_{disch} is the delivered charge times that average voltage.

For instance, a typical lead-acid battery with $C = 50$ Ah, when discharged rather slowly, shows an average discharge voltage of 2 V per cell, thus delivering $50 \cdot 2 = 100$ Wh.

A typical lithium NMC cell, when discharged in approximately one hour, might have an average discharge voltage of 3.9 V and, therefore, a 50-Ah cell might deliver 195 Wh.

The energy and power available from a single cell are not adequate for power system applications. Normally, several cells in series are used. Depending on the case, the number of cells in series can be very large (even several hundred cells in series!).

In some cases, cells in parallel are also considered. A *battery* is a set of several cells, in either series or series/parallel combinations. Naturally, the energy of a battery composed of all equal cells is equal to the energy stored in a single cell, multiplied by the total number of cells.

Consider a battery composed of m parallel rows, each composed in turn of n cells in series. The whole battery has nm cells. The following formulas apply ("batt" stands for battery):

$$Q_{\text{row}} = Q_{\text{cell}}, \qquad E_{\text{row}} = nE_{\text{cell}} = n(U_{\text{avg,cell}} \cdot I) = U_{\text{avg,row}} \cdot I$$
$$Q_{\text{batt}} = mQ_{\text{row}}, \qquad E_{\text{batt}} = mE_{\text{row}} = mU_{\text{avg,row}} \cdot I = mU_{\text{batt,row}} \cdot I$$
$$E_{\text{batt}} = nmE_{\text{cell}}$$

In these formulas we disregarded the effects of connections between cells to create rows and batteries. Indeed, since the connection cables have a finite resistance, when the battery is discharged it converts a very small quantity of energy into loss (heat), which is not transferred to the load. Therefore, the battery energy is slightly smaller than nmE_{cell}.

APPENDIX: TRANSMISSION LINE MODELLING AND PORT-BASED CIRCUITS

A.1 MODELLING TRANSMISSION LINES THROUGH CIRCUITS

In Chapter 5 we introduced the so-called *basic rule* for modelling circuital systems through circuits, which required us to neglect (a) conduction and displacement currents through wires and (b) EMF induced by the effects of Faraday's law in the circuital systems space between wires.

There are cases, however, in which these cannot be neglected. The most important is the example of very long high-voltage power lines that transmit power over a distance of many kilometres, for which the cumulative effects of displacement currents between conductors and of the self-induced voltages can be significant in line operation.

In other cases, such as medium-voltage lines, it could be advisable to take into account the effects of self-induction while still neglecting displacement currents between conductors.

In this appendix we show that if self-induction cannot be disregarded, the very concept of circuit becomes loose, and special constraints must be introduced to continue to use the usual circuit analysis techniques, and in particular Kirchhoff's laws.

We do this by taking as reference a two-wire transmission line, the most important case for theoretical and practical reasons.

Fundamentals of Electric Power Engineering: From Electromagnetics to Power Systems, First Edition.
Massimo Ceraolo and Davide Poli.

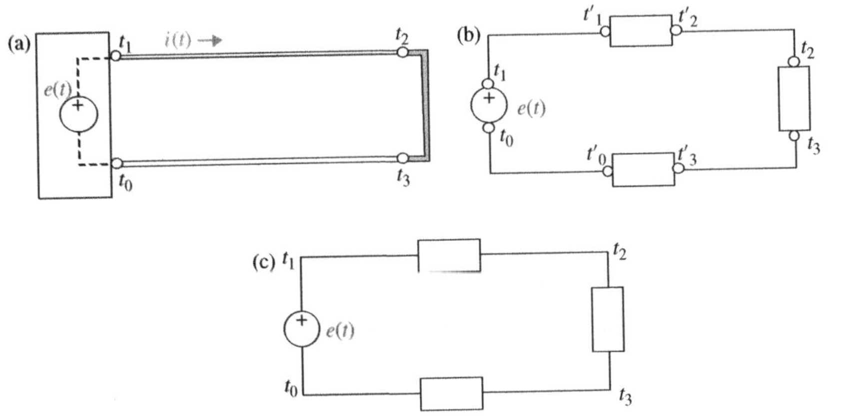

FIGURE A.1. A physical system containing a transmission line (a) and a lumped components model candidate (b and c).

A.1.1 Issues and Solutions When Displacement Currents are Neglected

Consider the physical system depicted in Figure A.1a, for which we postulate that it is possible to build a voltage source—that is, a device able to generate a time-varying voltage—on its left side. This is known to be true from common engineering practice, and it should be part of the student's previous knowledge.

The remainder of the circuit is constituted by a two-conductor line (let these conductors be two copper cylinders), connecting the voltage source with the load which is another cylindrical conductor possibly made of a different conducting material whose resistivity is higher than that of the line.

Although simple, this is a distributed-parameter physical system, governed by electromagnetics equations. Can this system be modelled by a *circuit* of the type shown in Figures A.1b and A.1c, where the four conductive circuit elements are substituted by branches?

If we look now at Figure A.1b, terminals are set exactly at the borders of lumped components and are connected to each other by ideal wires, as graphically shown. Once this is clear, the terminal representation is normally omitted, and the circuit is written as in Figure A.1c, where terminals are not shown and no distinction is made between the points of ideal wires, since all these points share the same current and the same potential.

The time variability of electrical quantities changes things radically with respect to the DC case:

- Faraday's law applies and must be taken into account.
- Current flows not only within wires, but also between the upper and lower wire. In fact, the two wires, along with the insulating material between them, behave

as a capacitor, the dielectric being the insulating material. Therefore there are displacement currents between them.

The practical experience of electrical engineers has highlighted the fact that it is possible to analyse transmission lines which are not very long (i.e., no longer than 50–80 km), operating at 50 or 60 Hz, while neglecting the effects of conduction and displacement currents between the wires.

For now, then, we will disregard the effects of these wire-to-wire currents; they will be considered at a later stage. As a consequence of this hypothesis, the current flowing inside the wires remains the same, from left-end terminals to right-end ones.

The first question to answer for the circuital system of Figure A.1a, is whether it is possible to define a potential function at the four terminals t_0, t_1, t_2, and t_3, if self-induction EMF cannot be neglected. If the answer is yes, KVL applies and a circuit can be used to effectively model the system. The existence of a potential function for the circuital system of Figure A.1a is equivalent to stating that the voltage across two of the terminals does not depend on the path used for measuring it. Therefore, we can imagine installing an *ideal voltmeter* in the circuital system and evaluating whether the voltmeter reading depends on its position.

A *voltmeter* is a device which is able to show the difference of potential at the two points of the circuital system to which it is connected. An *ideal voltmeter* is a device that draws no current (or an infinitesimal current) when it is inserted into the circuit and therefore its insertion does not alter the previous equilibrium of the system.

Consider the two ideal voltmeters V_1, and V_2, and the corresponding measuring wires depicted in Figure A.2. Because the voltmeters are ideal, the currents flowing in the wires connecting t_1 and t_2 with the voltmeters are negligible with respect to the

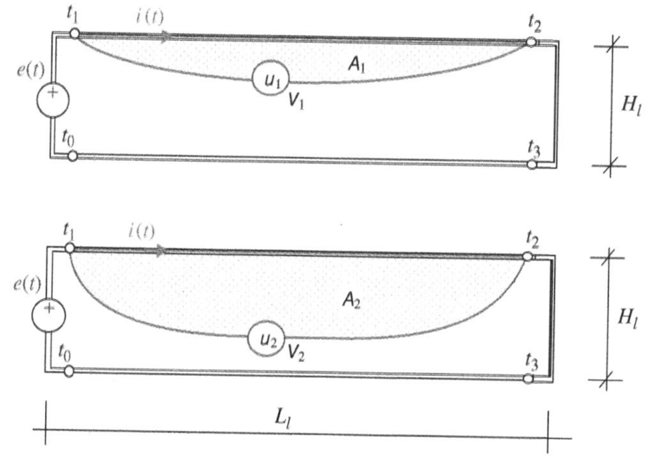

FIGURE A.2. Experimental evaluation of u_{12} of the system shown in Figure A.1.

main circuit current $i(t)$. The following equations can be written, combining Faraday's law and Ohm's law for the shaded areas A_1, and A_2:

$$
\begin{aligned}
A_1 &: u_1(t) = R_{12}i(t) + \frac{\mathrm{d}}{\mathrm{d}t}\psi_{A1}(t) \\
A_2 &: u_2(t) = R_{12}i(t) + \frac{\mathrm{d}}{\mathrm{d}t}\psi_{A2}(t)
\end{aligned}
\tag{A.1}
$$

where

- R_{12} is the resistance of the system's upper wire between t_1 and t_2.
- ψ_{A1}, ψ_{A2} fluxes are linked with surfaces of areas A_1 and A_2, respectively.

Undoubtedly, voltages u_1 and u_2 measured by voltmeters V_1 and V_2, respectively, are different from each other, because the two flux linkages ψ_{A1} and ψ_{A2} are different, being caused by the same electric field due to $i(t)$ but integrated in two different areas A_1 and $A_2 \gg A_1$.

Therefore, since the quantity measured as voltage between two points depends on the path in which we install the voltmeters, it is impossible to define the potentials of terminals t_1 and t_2. Since the existence of potential in all the connection points (element terminals) implies the applicability of the KVL and vice versa, KVL itself cannot be applied in the standard way, and therefore no standard circuit can be created for the system containing the transmission line. These results can be repeated whenever it is not deemed possible in a circuital system to neglect self-induction EMFs in the space between the wires that connect the components of the system component. The following can thus be stated:

> **Result:** *Circuit modelling of transmission lines*
>
> In a circuital system, when the self-induction electromotive force cannot be neglected in the meshes created by lines connecting devices, a potential function of all the points of the system cannot be defined, and therefore no standard circuit can be defined to model the system.

A system containing a long transmission line, however, in which the length is much greater than the distance between the conductors, can be analysed.

Consider Fig. A.3, in which $L_l \gg H_l$. In this case, the following equations can be written as

$$
\begin{aligned}
u_1(t) &= e(t) - R_c i(t) - \frac{\mathrm{d}}{\mathrm{d}t}\psi_{A_0+A_1}(t) \\
u_2(t) &= e(t) - R_c i(t) - \frac{\mathrm{d}}{\mathrm{d}t}\psi_{A_0+A_2}(t)
\end{aligned}
\tag{A.2}
$$

where R_c is the sum of upper and lower conductor resistance.

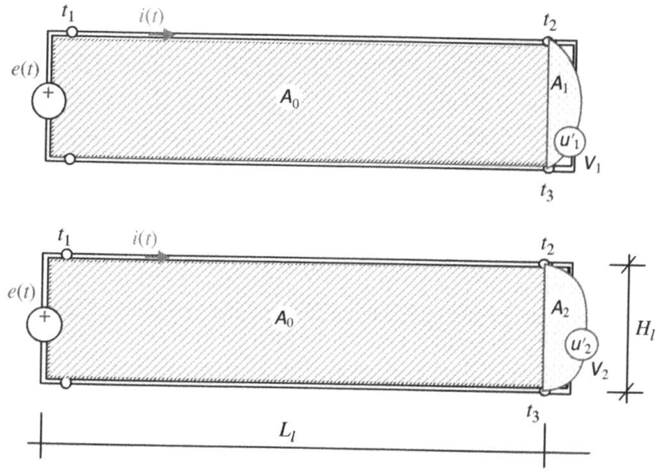

FIGURE A.3. Experimental evaluation of u_{23} of the system shown in Fig. A.1.

Since it is assumed $L \gg H$, it can also be said that $A_1 \ll A_0$ and $A_2 \ll A_0$, and therefore the flux $\psi_{A0\,+\,A1}$ linked with area A_0+A_1 and the flux $\psi_{A0\,+\,A1}$ linked with A_0+A_2 can be assumed to be equal to each other, as follows:

$$\frac{\mathrm{d}}{\mathrm{d}t}\psi_{A_0\,+\,A_1}(t) \cong \frac{\mathrm{d}}{\mathrm{d}t}\psi_{A_0\,+\,A_2}(t) \cong \frac{\mathrm{d}}{\mathrm{d}t}\psi_{A_0}(t) = \frac{\mathrm{d}}{\mathrm{d}t}(Li(t)) = L\frac{\mathrm{d}i(t)}{\mathrm{d}t}$$

where L is the self-inductance coefficient of the line and is related to the area A_0.

The two equations (A.2) can therefore be collapsed into a single equation:

$$u(t) = e(t) - R_c i(t) - L\frac{\mathrm{d}i(t)}{\mathrm{d}t} \tag{A.3}$$

Equation (A.3) can be the equation describing the electrical behaviour of the metacircuit[1] shown in Figure A.4.

The word “metacircuit” and the dashed lines indicate that this is not a true circuit since it cannot be used to compute voltage differences between a terminal existing at the sending end of the transmission line and one existing at the receiving end, but only voltage differences between terminals at the same end of the line. This metacircuit concept, though important, is usually ignored in circuit and electrical engineering textbooks.

Nevertheless, any electrical engineer knows that when long transmission lines are involved, it is not correct to evaluate voltage differences between terminals situated at two opposite ends of the line. To our knowledge, a rational analysis of why this

[1]The word “metacircuit” was invented here to make it clear that the drawing shown is not actually a circuit, since not all potential differences between terminals have physical meaning and can be computed.

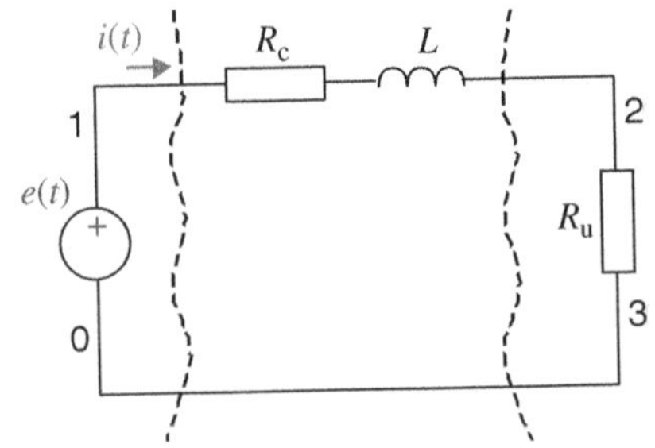

FIGURE A.4. A "metacircuit" representing the circuital system depicted in Figure A.1.

happens has not been previously published, at least in the more widespread literature; the contribution of this appendix is thus significant.

In the next section the analysis is repeated while also taking displacement currents into account.

A.1.2 Steady-State Analysis Considering Displacement Currents

Consider again a two-wire transmission line (Figure A.5), this time taking into account the displacement current. Consider in particular a single line segment of length dx (the shaded area in the figure).

This segment will have an infinitesimal inductance $\mathrm{d}L = L'\mathrm{d}x$ if L' indicates the inductance per unit length, and it will have an infinitesimal resistance $\mathrm{d}R = R'\mathrm{d}x$ if R' indicates the resistance per unit length.

The induction phenomenon along with the ohmic drop can be taken into account by writing the following equation:

$$u(x) - \mathrm{d}L\frac{\mathrm{d}i}{\mathrm{d}x} - \mathrm{d}R \cdot i = u(x + dx) = u(x) + \frac{\partial u}{\partial x}\,\mathrm{d}x$$

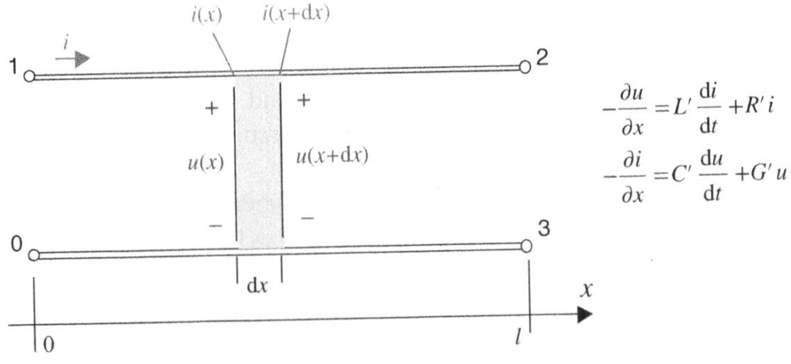

FIGURE A.5. Two-wire transmission line equations.

or:

$$u(x) - L^{'}\mathrm{d}x\frac{\mathrm{d}i}{\mathrm{d}x} - R^{'}\mathrm{d}x \cdot i = u(x) + \frac{\partial u}{\partial x}\,\mathrm{d}x$$

then

$$L^{'}\frac{\mathrm{d}i}{\mathrm{d}x} + R^{'} \cdot i = -\frac{\partial u}{\partial x}$$

In a similar way, conduction and displacement currents across the two wires in a segment of length dx can be found by resorting to conductance per unit length G' and capacitance per unit length[2] G':

$$i(x) - C^{'}\mathrm{d}x\frac{\mathrm{d}u}{\mathrm{d}x} - G^{'}\mathrm{d}x \cdot u = i(x + \mathrm{d}x) = i(x) + \frac{\partial i}{\partial x}\mathrm{d}x$$

then

$$C^{'}\frac{\mathrm{d}u}{\mathrm{d}x} + G^{'} \cdot u = -\frac{\partial i}{\partial x}$$

These developments justify the equations in Figure A.5.

If we consider the line operating in steady state at the angular frequency ω, the same equations can be written as follows (uppercase underlined quantities indicate the complex numbers that represent phasors):

$$\begin{aligned} -\frac{\partial \underline{U}(x)}{\partial x} &= j\omega L^{'} \cdot \underline{I}(x) + R^{'}\underline{I}(x) = \underline{Z}_{l}\underline{I}(x) \\ -\frac{\partial \underline{I}(x)}{\partial x} &= j\omega C^{'} \cdot \underline{U}(x) + G^{'}\underline{U}(x) = \underline{Y}_{l}\underline{U}(x) \end{aligned} \qquad \text{(A.4)}$$

Equation (A.4) can be easily solved by the steps described in all books dealing with transmission lines and omitted here for simplicity; the following are given instead:

$$\begin{aligned} \underline{U}(x) &= \underline{U}(0) \cdot \cosh \underline{K}x - \underline{Z}_{\mathrm{c}}\underline{I}(0) \cdot \sinh \underline{K}x \\ \underline{I}(x) &= -\frac{\underline{U}(0)}{\underline{Z}_{\mathrm{c}}} \cdot \sinh \underline{K}x + \underline{I}(0) \cdot \cosh \underline{K}x \end{aligned} \qquad \text{(A.5)}$$

where

$$\underline{Z}_{\mathrm{c}} = \sqrt{(R^{'} + j\omega L^{'})/(G^{'} + j\omega C^{'})} \quad \text{and} \quad \underline{K} = \sqrt{(R^{'} + j\omega L^{'}) \cdot (G^{'} + j\omega C^{'})}$$

[2]This is the capacitance per unit length of the capacitor, whose conductor plates, which are parallel to each other, are the two wires and the dielectric is the air between them.

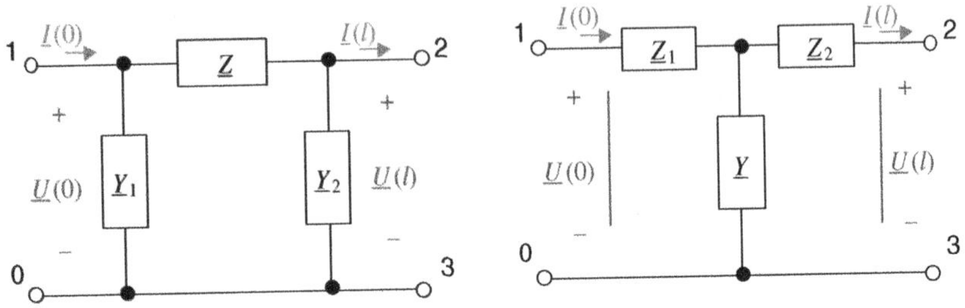

FIGURE A.6. Circuits candidates to model the system of Figure A.5 in steady state.

Suppose that now it is of interest to know only the behaviour of the line at the two ends: sending end (terminals **0–1**) and receiving end (terminals **2–3**). From equation (A.5) it follows that

$$\underline{U}(l) = \underline{D}\cdot\underline{U}(0) - \underline{B}\cdot\underline{I}(0), \qquad \underline{U}(0) = \underline{A}\cdot\underline{U}(l) + \underline{B}\cdot\underline{I}(l)$$
$$\underline{I}(l) = -\underline{C}\cdot\underline{U}(0) + \underline{A}\cdot\underline{I}(0), \qquad \underline{I}(0) = \underline{C}\cdot\underline{U}(l) + \underline{D}\cdot\underline{I}(l) \qquad \text{(A.6)}$$

where

$\underline{A} = \underline{D} = \cosh \underline{K}l$, $\underline{B} = \underline{Z}_c \sinh \underline{K}l$, $\underline{C} = \sinh \underline{K}l/\underline{Z}_c$ (note that $\underline{AD} - \underline{BC} = 1$)

As for the simpler case above, consider now the circuits depicted in Figure A.6.

$$\underline{Z} = \underline{B}, \quad \underline{Y}_1 = \underline{Y}_2 = (\underline{A} - 1)/\underline{B} \quad \underline{Z}_1 = \underline{Z}_2 = (\underline{A} - 1)/\underline{C}$$

It is easy to verify that equations (A.6) are valid for both circuits. However, it would be erroneous to state that the circuits in Figure A.6 are "steady-state equivalents" of the transmission line: those circuits, indeed, are branch-based circuits and, as such, allow the determination of quantities such as voltage $\underline{U}_{AC}$, something which is impossible (because it has no physical meaning) through the equations of the physical line.

Therefore, the obtained steady-state line model can again (see Section A.1.1) be better represented graphically using the metacircuit concept (Figure A.7)

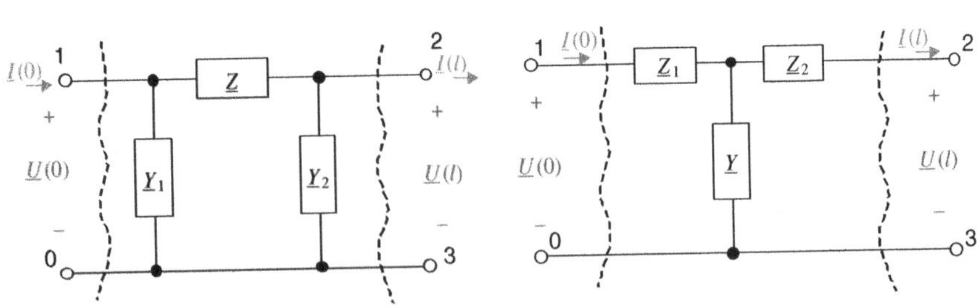

FIGURE A.7. "Metacircuits" representing a transmission line operating in steady state.

More in Depth: *Line models with circuits containing 1:1 transformers*

One could be tempted to avoid the introduction of the metacircuit concept for transmission lines operating in steady state and instead introduce a 1:1 ideal transformer at the beginning or the end of the models as shown in the case of the PI line model in the upper part of the diagram below.

This, however, would be tricky and is better avoided, since the use of this line equivalent would not prevent cross-impedances being added, such as the $\underline{Z}_{\text{new}}$ shown in the lower part of the diagram below. This is indeed erroneous because analysis of this network using Kirchhoff's equations would lead to results that do not correspond to the system physics.

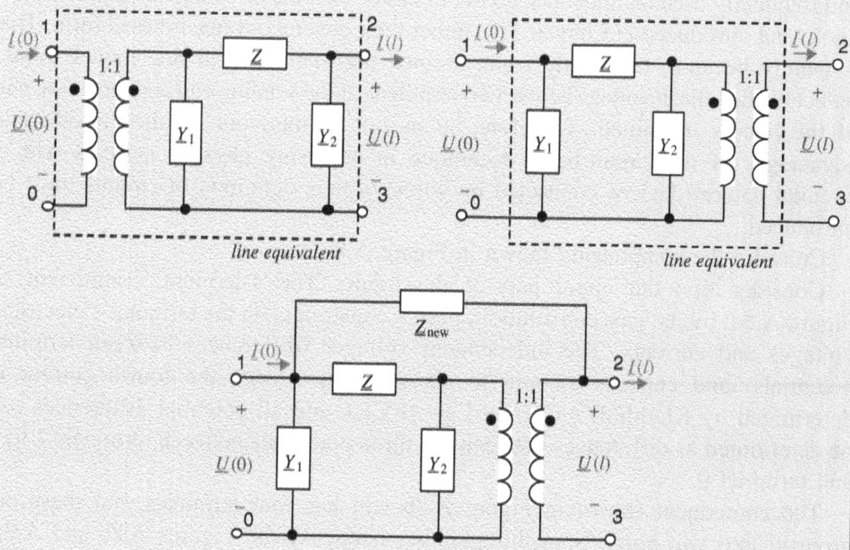

A.1.3 Practical Considerations

For low-voltage and medium-voltage lines (up to 35 kV nominal), considering the actual parameters (inductances per unit length and capacitances per unit length) and the limited lengths, the admittances $\underline{Y}$ can be neglected and the line can be represented by a metacircuit containing only the longitudinal impedance $\underline{Z}$ or $\underline{Z}_1 + \underline{Z}_2$.

Moreover, for low-voltage lines (1 kV nominal and below), for conductor sections up to 95 mm^2, the longitudinal reactance can also be neglected with respect to the resistance. For those lines, by and large the most common in domestic and civil environments, lines are modelled just by an equivalent resistor.

A more formal approach showing how to use modified circuits in steady state to analyse systems in which displacement currents between wires and self-induction voltages cannot be neglected is illustrated in the next section.

In high-voltage lines and very high voltage lines (e.g., in the range 132–800 kV) the Y component of the model can be neglected only for short line trunks, usually up to 100 km.

A.2 MODELLING LINES AS TWO-PORT COMPONENTS

A.2.1 Port-Based Circuits

In the previous section it was shown that transmission lines, which are very important and frequently used components, do not fit easily with the *circuit* concept used in this book (and introduced in Chapter 3) in either their general or branch-based form. This is mainly because, differently from circuits, the mathematical line model should reflect the fact that a voltage across two terminals at the sending end and receiving end of the line is undefined. Therefore, so as not to miss out on the considerable advantages of the circuit-based technique in analysing physical systems with a circuital nature (devices connected by wires), a new definition of circuits must be introduced.

Consider the components shown in Figure A.8.

Consider first the upper part of the figure. The 4-terminal component of Figure A.8a, by its very constitution, places constraints on the terminal's electrical voltages and currents. The independent voltages (differences between terminal potentials) and currents amount to six quantities, since the fourth current is determined by Kirchhoff's Current Law (KCL), and all potential differences can be determined as differences between the three potentials between terminals 1 to 3 and terminal 0.

The component shown in Figure A.8b still has four terminals, but these are grouped into two *ports*. Something similar happens with Figures A.8c and A.8a, where more terminals are involved, and each port contains m terminals instead of two.

The definition of the two-port$_m$ component outlined here is as follows[3]:

A two-port$_m$ component is an n-terminal component with the following special features:

1. The n terminals, which must be even in number, are grouped into two groups of $m = n/2$ terminals, called ports.
2. It is postulated that the sum of all currents entering any port is always zero.
3. When the two-port component is inserted into a circuit, the circuit must never be used to determine voltage differences between potentials of terminals belonging to different ports, since they are meaningless.

[3]This is a more structured definition of two-port components that can be found in textbooks (see Chapter 19 of [bc2]). The expression two-port$_m$ component is equivalent to "component with two m-terminal ports".

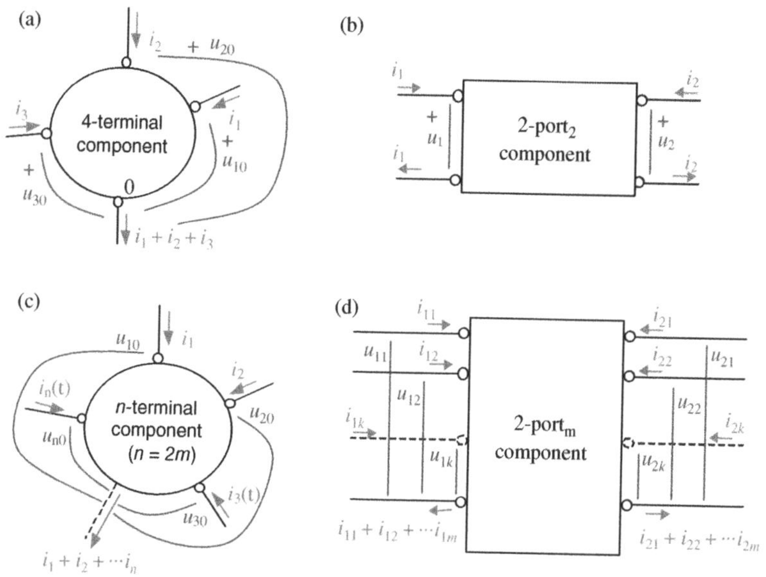

FIGURE A.8. From 4-terminal and n-terminal components to 2-port components.

Items 2 and 3 mean that KCL and Kirchhoff's Voltage Law (KVL) must be valid separately for each port (and each circuit section connected to that port) but not for the circuit as a whole.

A.2.2 Port-Based Circuit and Transmission Lines

The two-port$_2$ component is a suitable tool for modelling a circuit containing a transmission line such as the one depicted in Figure A.5.

When subsystems are connected to the left end and right end of the line, the whole system can be modelled by connecting the branch-based circuit modelling the left-hand subsystem on the left, the two-port$_2$ component modelling the line, and a branch-based circuit modelling the right-hand subsystem on the right.

The situation is summarized in Figure A.9. The circuital system containing the line is recalled in Figure A.9a for simplicity considering very simple subsystems at the left and right end of the line.

The line can be effectively modelled as a two-port$_2$ component as in Figure A.9b. Should conductive and displacement currents between the wires be disregarded, the two-port component behaviour is described by the single equation in Figure A.9c, while in the more general case in which these currents between wires are taken into consideration, in steady state, the line is described by a two-port component with the equations in Figure A.9d. The values of $\underline{A}, \underline{B}, \underline{C}, \underline{D}$ are those immediately given below equation (A.6).

Figures A.9e and A.9f show the metacircuit counterparts of the port-based representations of Figures A.9c and A.9d. The metacircuit approach is equivalent

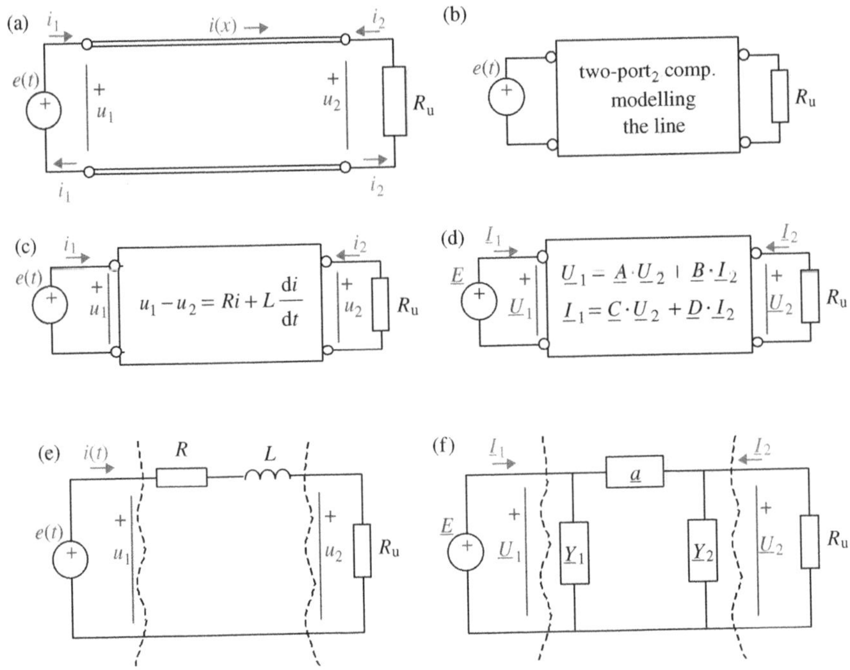

FIGURE A.9. The two-wire transmission line and related different circuital representations.

to the port-based one, but port-based components are more usually represented by their equations, rather than by circuit-like drawings.

In normal engineering usage, these curved–dashed lines are omitted. Normally this does not create problems, since the engineer retains a visual memory of them and will avoid the mistake of evaluating cross voltages between the ports, even when they are not explicitly shown.

A.2.3 A Sample Application

Transmission lines are used largely to transmit electric power at distances that range from some metres (household installations) to hundreds of kilometres (high-voltage three-phase transmission lines).

A significant example of a three-phase transmission system is shown in Figure A.10. It is composed of a *generator* (electric machine that generates electric power from mechanical power), some *transformers,* two *three-phase electric lines*, composed of three wires (e.g., the overhead conductors of high voltage lines) and the ground below them, and, finally, the *distribution network*, which acts as a load for the rest of the system shown.

In the figure, transformer 1 transfers electrical energy from the medium-voltage level to the high-voltage level and transformer 2 does the opposite.

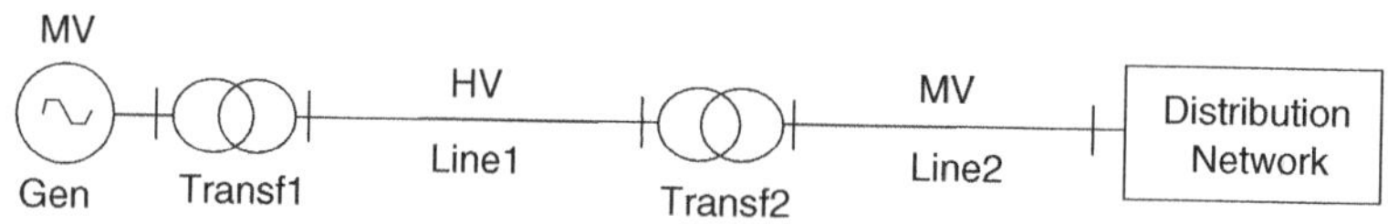

FIGURE A.10. Example of a power system.

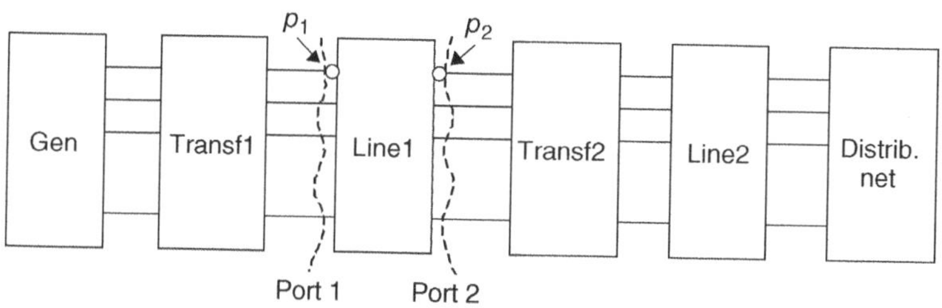

FIGURE A.11. Circuital representation of the power system of Figure A.10.

As usual in power system analysis, only a one-wire equivalent is shown. Indeed all apparatuses shown (generator, transformer, etc.) have four-terminal ports: the three phases plus an additional terminal which is normally connected to the ground. Consequently, the power system depicted in Figure A.10 can be represented as in Figure A.11.

The different apparatuses are either 4-terminal components (generator, distribution network), or two-port$_4$ components (lines, transformers).

The behaviour of each apparatus is described by its constitutive equation, similar to those given for the two-wire transmission line in Figure A.9, but more complex because of the higher number of terminals per port.

The connections between the port-based components shown in Figure A.10 indicate the equality of voltage of terminals connected to each other, as well as the equality of the current in the shown wires, paying attention to their signs.

This representation does not allow evaluation of the voltage difference between terminals belonging to two different ports of any two-port components, for example, $U_{12} = U_{p1} - U_{p2}$. This does not create limitations in the applications, because ports 1 and 2, and therefore the terminals p_1 and p_2, might be as much as several hundred kilometres from each other.

A.3 FINAL COMMENTS

This appendix has described a way to model transmission lines in circuit analysis even in cases in which it is not possible to neglect self-induced voltages (Faraday's law) and/or displacement currents between wires.

The approach outlined here is useful, and commonly adopted, in the analysis of very high voltage transmission networks.

The analysis presented in this appendix, however, gives only basic information geared to making the problem and the approach clear to students, but it is not sufficiently in-depth to give them adequate know-how for dealing with three-phase transmission systems containing long lines in a quantitative way.

SELECTED REFERENCES

Because of the introductory nature of this book, references have been reported only here and are carefully selected.

To ease understanding of the links to references, the first one or two digits indicate the nature of the reference itself. The first letter of any reference is "b" for books, "s" for standards, and "p" for scientific papers (the latter include just a couple of articles regarding specific topics that enlarge concepts explicitly discussed in the book).

BOOKS

Books about Electromagnetics

[be1] R. W. Chabay and B. A. Sherwood. *Matter & Interactions II—Electric & Magnetic Interactions*, John Wiley & Sons, Hoboken, NJ, 2007.

[be2] Fawwaz Ulaby. *Fundamentals of Applied Electromagnetics*, Prentice Hall, Upper Saddle River, NJ, 2007.

Circuit-Related Books

[bc1] M. E. Van Valkenburg. *Network Analysis*, Prentice Hall, Upper Saddle River, NJ, 2007 (First edition, 1974).

[bc2] C. Alexander and M. Sadiku. *Fundamentals of Electric Circuits*, third edition, McGraw-Hill Higher Education, New York, 2007.

Fundamentals of Electric Power Engineering: From Electromagnetics to Power Systems, First Edition.
Massimo Ceraolo and Davide Poli.

Books about Power Electronics, Machines, and Drives

[bm1] A. E. Fitzgerald, C. Kingsley, and S. Umans. *Electric Machinery*, McGraw-Hill, New York, 2002.

[bm2] N. Mohan, T. M. Undeland, and W. P. Robbins. *Power Electronics*, third edition, John Wiley & Sons, Hoboken, NJ, 2003.

[bm3] N. Mohan. *Power Electronics: A First Course*, John Wiley & Sons, Hoboken, NJ, 2011.

[bm4] N. Mohan. *Electric Machines and Drives: A First Course*, John Wiley & Sons, Hoboken, NJ, 2012.

[bm5] S. Sul. *Control of Electric Machine Drive Systems*, IEEE/Wiley, Hoboken, NJ, 2011.

[bm6] T. Crause, O. Wasynczuk, and S. D. Sudhoff. *Analysis of Electric Machinery*, IEEE Press, Piscataway, NJ, 2002.

Books about Power Systems Basics

[bp1] N. Mohan. *Electric Power Systems: A First Course*, John Wiley & Sons, Hoboken, NJ, 2012.

[bp2] Various authors. *Electrical Installation Guide 2010 According to IEC International Standards* (Schneider electric technical collection, available through the following Internet site: http://www2.schneider-electric.com/sites/corporate/en/customers/panel-builders/technical-resources/electrical-installation-guide.page.

General Books about Electrical Engineering

[bg1] Mulukutla S. Sarma. *Introduction to Electrical Engineering*, Oxford University Press, New York, 2001.

[bg2] Giorgio Rizzoni. *Principles and Applications of Electrical Engineering*, McGraw-Hill Higher Education, New York, 2004.

SCIENTIFIC PAPERS

[p1] R. F. Schiferl and T. A. Lipo. Power capability of salient pole permanent magnet synchronous motors in variable speed drive applications, *IEEE Transactions on Industry Applications*, Vol. 26, No. 1, January/February 1990, pp. 115–123.

[p2] G. Pellegrino, A. Vagati, P. Guglielmi, and B. Boazzo. Performance comparison between surface-mounted and interior PM motor drives for electric vehicle application, *IEEE Transactions on Industrial Electronics*, Vol, 59, No. 2, February 2012, pp. 803–811.

INTERNATIONAL STANDARDS

[s1] The Bureau International des Poids et measures home page: http://www.bipm.org/en/home/

[s2] The International System of Units (SI), National Institute of Standards and Technology (NIST), U.S. Department of Commerce, http://physics.nist.gov/cuu/Units/.

[s3] The International Electrotechnical Commission (IEC) about page: http://www.iec.ch/about/

[s4] The International Electrotechnical Vocabulary (IEV), accessible at the Electropedia Internet site, http://www.electropedia.org.

[s5] IEEE 519-1992: Recommended Practices and Requirements for Harmonic Control in Electrical Power Systems.

[s6] International standard IEC 60617: Graphical Symbols for Diagrams.

[s7] International standard IEC 60375: Conventions Concerning Electric and Magnetic Circuits.

[s8] International standard ISO 80000-4: Quantities and Units—Part 4: Mechanics.

[s9] International standard ISO 80000-6: Quantities and Units—Part 6: Electromagnetism.

[s10] IEEE Std 315: Graphic Symbols for Electrical and Electronics Diagrams.

ANSWERS TO THE PROPOSED EXERCISES

CHAPTER FOUR

4.1 **(a)** 20°C: $R = 0.0580\,\Omega$, $I = 206.9$ A, $P = 2483$ W
(b) 55°C: $R = 0.0661\,\Omega$, $I = 181.5$ A, $P = 2178$ W

4.2 **(a)** $9\,\Omega$
(b) $3\,\Omega$
(c) $15\,\Omega$
(d) $2\,\Omega$
(e) $10\,\Omega$
(f) $10\,\Omega$
(g) $6\,\Omega$

4.3 $I_1 = 7.5$ A, $I_2 = 2.5$ A

4.4 $I = 0.5$ A, $U_{AB} = 5$ V. U_{s1} charges U_{s2}

4.5 $R_G = 6\,\Omega$. $I_1 = 2$ A. $I_2 = I_3 = 1$ A

4.6 $U_{AB} = 8$ V

4.7 **(a)** $U_{Th} = 6$ V, $R_{Th} = 1.333\,\Omega$
(b) $U_{Th} = 5$ V, $R_{Th} = 5\,\Omega$
(c) $U_{Th} = 2$ V, $R_{Th} = 4\,\Omega$

Fundamentals of Electric Power Engineering: From Electromagnetics to Power Systems, First Edition.
Massimo Ceraolo and Davide Poli.

4.8 $U_{R4} = 6\,V$

4.9 $I_1 = 6\,A$, $I_2 = 1.8\,A$, $I_3 = -4.2\,A$

4.10 $I_1 = 0.2\,A$, $I_2 = 3.2\,A$

4.11 $U_I = 96\,V$. Power balance: $100 \cdot 0.2 + 96 \cdot 3 = 20 \cdot 0.2^2 + 30 \cdot 3.2^2$

4.12 $I_2 = 9.6\,A$, $I_3 = 2.4\,A$

4.13 $\Phi = 80.33$ lumen, $\mathscr{E} = 803.3\,Wh$

4.14 $I = 2.5\,A$

4.15 $U_{Th} = 3.2\,V$, $R_{Th} = 1.8\,\Omega$

4.16 $U_{Th} = 7\,V$, $R_{Th} = 4\,\Omega$

4.17 $P = 2.5\,W$

4.19 $R = R_{Th} = 2.9375\,\Omega$, $P = 14.66\,W$

4.20 $P = 20.22\,W$

4.21 $R = R_{Th} = 27.32\,\Omega$, $P = 0.141\,W$

CHAPTER FIVE

5.1 **(a)** $A = 10$, $\omega = 5$, $\theta = -60°$
(b) $A = 2$, $\omega = 3$, $\theta = 70°$
(c) $A = \sqrt{2}\,230$, $\omega = 2\pi 50$, $\theta = -80°$
(d) $A = \sqrt{2}\,127$, $\omega = 2\pi 400$, $\theta = 160°$

5.2 **(a)** $A = 2\sin 30° = 1$, $B = 2\cos(30°) = \sqrt{3}$
(b) $5\cos(t - 15°) = 5\sin(t + 85°)$, $A = 5\sin 75° = 4.830$, $B = 5\cos 75° = 1.294$
(c) $M = \sqrt{3^2 + 4^2} = 5$, $\theta = \tan^{-1}(3/4) = 36.87°$
(d) $M = \sqrt{5^2 + 1^2} = 5.1$, $\theta = \tan^{-1}(5/1) - 90° = -11.31°$

5.3 **(a)** $\underline{U} = 100\,V\angle 20°$
(b) $\underline{U} = 5\,V\angle 30°$
(c) $\underline{I} = 3\,A$
(d) $\underline{I} = 4\,A\angle 75°$

5.4 **(a)** $\underline{Z} = 5 + j20\,\Omega$
(b) $\underline{Z} = 10 - j100\,\Omega$
(c) $\underline{Z} = 10 + j90\,\Omega$
(d) $\underline{Z} = 1 + j0.95\,\Omega$
(e) $\underline{Z} = 30 - j500\,\Omega$

5.5 **(a)** $11 + 1j$
(b) $1 + 0.6j$

(c) $3.949 + 1.109j$
(d) $2 + 3j$
(e) $8 + 4j$

5.6 $\underline{I}_1 = 9 + 2j$ A

$\underline{I}_2 = 1 + 3j$ A

5.7 $\underline{U}_1 = 2.647 + 4.412j$ V

$\underline{U}_2 = 3.353 - 1.412j$ V

5.8 (a) $\underline{U}_{th} = 8.666 + 2.000j$ V, $\underline{Z}_{th} = 0.8461 + 0.8974j\ \Omega$
(b) $\underline{U}_{th} = -5 + 5j$ V, $\underline{Z}_{th} = 5j\ \Omega$
(c) $\underline{U}_{th} = -4 + 22j$ V, $\underline{Z}_{th} = 3 + j\ \Omega$

5.9 $i_1(t) = 3.993 \sin(2\pi 50t - 18.14°)$ A

$i_2(t) = 1.989 \sin(2\pi 50t - 69.63°)$ A

$i_3(t) = 3.164 \sin(2\pi 50t + 11.33°)$ A

5.11 $i_1(t) = 4.472 \sin(100t - 63.43°) + 4.938 \sin(314t - 140.9°)$ A

$i_2(t) = 4.472 \sin(100t - 63.43°) + 0.7863 \sin(314t - 50.97°)$ A

$u_{AB}(t) = 8.944 \sin(100t + 26.56°) + 4.938 \sin(314t + 39.03°)$ V

5.12 $i(t) = 0.5511 \sin(10t - 19.09°)$ A

5.13 $i_C(t) = 0.3204 \cos(4t + 88.1°)$ A $= 0.3204 \sin(4t + 178.1°)$ A, flowing from left to right

5.14 Gen.: $P = 189.7$ W, $Q = 62.16$ var

R_1: $P = 79.74$ W, $Q = 0$ var

R_2: $P = 9.891$ W, $Q = 0$ var

R_3: $P = 100.1$ W, $Q = 0$ var

L_2: $P = 0$ W, $Q = 62.16$ var

5.15 $P_{Us} = 10$ W, $P_{Is} = 12.19$ W, $P_R = 22.19$ W

5.16 $Q_I = -32.02$ var

5.17 $p_{avg} = P = UI \cos \varphi = 10 \cdot 3 \cdot \cos(65° - 5°) = 15$ W

$p_{max} = P + UI = 15 + 30 = 45$ W

5.19 $P = 1.127$ W, $Q = -0.389$ var

5.20 $u(t) = 4$ V when $0 < t < 1$, $u(t) = -6$ V when $2 < t < 3$, $u(t) = 2$ V when $3 < t < 4$, otherwise $u(t) = 0$

5.21 $C_1 = 2$ mF, $C_{0.8} = 1.25$ mF

CHAPTER SIX

6.1 $\underline{I}_a = 4.094\ \mathrm{A}\angle -2.109°$, $\underline{I}_b = 4.094\ \mathrm{A}\angle -122.1°$, $\underline{I}_c = 4.094\ \mathrm{A}\angle 117.9°$

6.2 $\underline{U}_{sab\lambda} = \sqrt{2}\cdot 230.9\ \mathrm{V}\angle 0° = 326.6\ \mathrm{V}\angle 0$

$\underline{I}_a = 11.66\ \mathrm{A}\angle -34.82°$

$\underline{I}_b = 11.66\ \mathrm{A}\ \angle -154.8°$

$\underline{I}_c = 11.66\ \mathrm{A}\ \angle 85.1°$

$\underline{I}_{ab} = 6.730\ \mathrm{A}\ \angle -4.824°$

6.3 $\underline{U}_{sab\lambda} = 326.6\ \mathrm{V}\angle -30°$, $\underline{Z}_\lambda = 4 + j\ \Omega$

$\underline{I}_a = 41.82\ \mathrm{A}\angle -69.81°$

$\underline{I}_b = 41.82\ \mathrm{A}\angle 170.2°$

$\underline{I}_c = 41.82\ \mathrm{A}\angle 50.19°$

$\underline{I}_{ab} = 24.14\ \mathrm{A}\angle -39.81°$

6.4 $\underline{I}_a = 58.07\ \mathrm{A}\angle -31.59°$

$\underline{I}_1 = 24.51\ \mathrm{A}\angle -44.72°$

$\underline{I}_2 = 34.66\ \mathrm{A}\angle -22.34°$

$\underline{U}_{ab} = 306.1\ \mathrm{V}\angle 18.97° = \sqrt{2}\cdot 216.4\ \mathrm{V}\angle 18.97°$

6.5 $\underline{I}_1 = 33.69\ \mathrm{A}\angle -83.37°$

$\underline{I}_2 = 38.73\ \mathrm{A}\angle 22.44°$

$\underline{I}_3 = 43.86\ \mathrm{A}\angle -25.20°$

$\underline{U}_{ab} = 443.0\ \mathrm{V}\angle 35.76°$

6.6 $\underline{I}_a = 17.33\ \mathrm{A}\angle -21.92°$

$\underline{I}_b = 12.56\ \mathrm{A}\angle 175.5°$

$\underline{I}_c = 6.538\ \mathrm{A}\angle 122.9°$

$\underline{I}_{NO} = 4.851\ \mathrm{A}\angle 105.8°$

6.7 $\underline{I}_a = 15.20\ \mathrm{A}\angle -10.43°$

$\underline{I}_b = 22.27\ \mathrm{A}\angle -166.4°$

$\underline{I}_c = 11.19\ \mathrm{A}\angle 63.44°$

$\underline{I}' = 2.642\ \mathrm{A}\angle 129.9°$

6.8 $S_{load} = 754.4\ \text{W} + j251.5\ \text{var}$

$S_{line} = 50.29\ \text{W} + j75.44\ \text{var}$

$S_{gen} = 804.7\ \text{W} + j326.9\ \text{var} = S_{load} + S_{line}$

6.9 $S_{load} = 3669\ \text{W} + j1223\ \text{var}$

$S_{line} = 1019\ \text{W} + j2038\ \text{var}$

$S_{gen} = 4688\ \text{W} + j3261\ \text{var} = S_{load} + S_{line}$

6.10 $S_{load} = 10492\ \text{W} + j2623\ \text{var}$

$S_{line} = 5246\ \text{W} + j10492\ \text{var}$

$S_{gen} = 15738\ \text{W} + j13115\ \text{var} = S_{load} + S_{line}$

6.11 $S_{gen1} = 1898\ \text{W} + j16326\ \text{var}$

$S_{gen2} = 18732\ \text{W} + j2485\ \text{var}$

Line power losses = 6202 W

6.12 $S_{gen} = 1634\ \text{W} + j3740\ \text{var} = S_{load} + S_{line} + S_{neutral}$

6.13 $pf = 0.936$

$C_{\lambda} = 367.5\ \mu\text{F}$

$C_{\Delta} = 122.5\ \mu\text{F}$

CHAPTER SEVEN

7.1 $N_2 = 25$ turns, $\phi = U_1/(\omega N_1) = U_2/(\omega N_2) = 19.1\ \text{mWb}$, $\hat{B} = \sqrt{2} \cdot \phi / S = 1.08\ \text{T}$

7.2 $I_1 = 2.4\ \text{A}$, $P_1 = P_2 = 496.8\ \text{W}$, $Q_1 = Q_2 = 240.6\ \text{var}$

7.3 $I_1 = 848.1\ \text{mA}$

7.4 $I_1 = 45\ \text{A}$, $I_2 = 7.5\ \text{A}$

S = 2.25 kVA, thus the transformer is overloaded by 12.5%

7.5 **(a)** $R_{add2} = 1\ \Omega$
(b) $R_{add1} = 100\ \Omega$

7.6 $N_1 = 245$ turns, $I_m = \dfrac{\hat{H}}{\sqrt{2} \cdot l \cdot N_1} = 0.182\ \text{A}$

7.7 **(a)** $3\ \Omega$
(b) $I_1 = 76.67\ \text{A}$, $I_2 = 19.17\ \text{A}$
(c) 11.76 kW
(d) 66.67%
(e) $16\ \Omega$

7.8

$$\eta = \frac{P_2}{P_2 + P_0 + P_{sc} \cdot (S/S_{rated})^2}$$

(a) $\eta_{100\%} = 0.9804$, $\eta_{75\%} = 0.9812$, $\eta_{50\%} = 0.9796$, $\eta_{25\%} = 0.9690$
(b) $\eta_{100\%} = 0.9756$, $\eta_{75\%} = 0.9766$, $\eta_{50\%} = 0.9747$, $\eta_{25\%} = 0.9615$

7.9 $P_2 = 180$ kW, losses $= 2927$ W

$P_0 = P_{copper} =$ losses/2 $= 1463.4$ W, $P_{sc} = P_{copper}/0.9^2 - 1806.7$ W

7.10 **(a)** $pf = 0.261$
(b) $R_m = 352.7\,\Omega$, $I_a = 0.652$ A
(c) $X_m = 95.3\,\Omega$, $I_\mu = 2.413$ A

7.11 $P_1 = 440$ W, $Q_1 = 392$ var, $U_1 = 256$ V

7.12 $\underline{Z}''_{sc} = 16.23 + j35.63$ mΩ, $I_{2sc} = 5874$ A

7.13 $U_2 = 230.6$ V, $I_1 = 20.28$ A, $\eta = 0.946$

7.14 **(a)** $U_2 = 225.9$ V (phase-to-phase), 130.4 V (phase)
(b) $I_1 = 3.608$ A (line), 2.083 A (phase)
(c) $I_2 = 191.7$ A (both line and phase)

7.15 $N_1 = 4619$ turns, $N_2 = 2656$ turns

$I_{1\ line} = 360.8$ A, $I_{2\ line} = 627.6$ A

7.16 **(a)** $I_{1line} = \dfrac{\sqrt{P_1^2 + Q_1^2}}{\sqrt{3}U_1} = 4.841$ A, where $P_1 = P_2$ and $Q_1 = Q_2$ (ideal transformer)

(b) $I_{1phase} = I_{1line}/\sqrt{3} = 2.795$ A

(c) $\hat{B} = \sqrt{2}\dfrac{U_1/\sqrt{3}}{\omega N_1 S} = 0.866$ T

(d) $N_1/N_2 = E_{1phase}/E_{2phase} = \dfrac{U_1/\sqrt{3}}{U_2} = 1.818$, thus $N_2 = 22$ turns

7.17 **(a)** 100 MVA, $(230/\sqrt{3}) : (34.5/\sqrt{3})$kV
(b) 100 MVA, $(230/\sqrt{3}) : 34.5$ kV
(c) 100 MVA, $230 : (34.5/\sqrt{3})$kV
(d) 100 MVA, 230:34.5 kV

CHAPTER TEN

10.1 $E = 565.5$ V

10.2 **(a)** $I_a = 14.00$ A
(b) $U = 559.9$ V
(c) $P_e = 7836$ W
(d) $T = 50.39$ Nm, $P_m = 7915$ W

10.3 **(a)** $E = 267.0$ V
(b) $U = U_{load} + \Delta U_{brushes} = 242$ V; $I_a = P_{cu}/(E - U) = 23.97$ A
(c) $R_a = P_{cu}/I_a^2 = 1.044\ \Omega$
(d) $P_e = 5754$ W
(e) $T = 40.75$ Nm

10.4 **(a)** $E = 207.2$ V
(b) $E = 192.8$ V

10.5 $I_a = 21.25$ A, $I_f = 1.25$ A, $E = 260.6$ V, $T = 44.07$ Nm

10.6 **(a)** $I_a = 22.86$ A
(b) $E = 193.1$ V
(c) $n = 1054$ rpm
(d) $P_m = 4415$ W
(e) $\eta = 96.57\%$

10.7 $T = kN_a\Phi_f\, U/R_a - (kN_a\Phi_f)^2\Omega/R_a$, $T = T_{load} = 48.80$ Nm
(a) $I_a = 27.89$ A
(b) $E = 191.6$ V
(c) $n = 1046$ rpm
(d) $P_m = 5344$ W
(e) $\eta = 95.81\%$

10.8 **(a)** $I = 100$ A, $I_a = 96$ A, $I_f = 4$ A
(b) $E = 390.4$ V
(c) $\eta = 93.70\%$

10.9 $T = 258.9$ Nm, $\eta = 93.19\%$

10.10 $n_0 = 315.1$ rpm

10.11 $T = 182.4$ Nm, $\eta = 95.5\%$

10.12 $n_2 = 817.0$ rpm, $T_1 = 135.0$ Nm, $T_2 = 75.92$ Nm

$T_1/T_2 = (I_1/I_2)^2$, $\Omega_1/\Omega_2 \approx I_2/I_1$

10.13 $T_{800\ rpm} = 52.52$ Nm, $n = 963.3$ rpm

CHAPTER ELEVEN

11.1

$p=1$	$p=2$	$p=3$	$p=4$
3000 rpm	1500 rpm	1000 rpm	750 rpm

11.2 $E=5693$ V (9861 V line-to-line), $\beta=23.83°$

$P=2.4$ MW, $Q=1.8$ Mvar

11.3 In the wye equivalent circuit: $R=0.033\ \Omega$ and $X=3\ \Omega$, then $E=392.2$ V and $\beta=27.77°$

Per phase, $E=679.3$ V

$P=42.5$ kW, $Q=26.34$ kvar

11.4 $E=3421.4$ V (5926 V line-to-line), $\beta=49.57°$

$P=2.7$ MW, $Q=-1.31$ Mvar

11.5 **(a)** $E=2020$ V (3500 V line-to-line), $\beta=0$, $I=0$, $Q=0$
(b) $E=2020$ V (3500 V line-to-line), $\beta=14.9°$, $I=249.5$ A, $Q=-196.2$ kvar, $pf=0.992$, $\varphi=7.45°$

Since $R=0$ and $E=U$, $\varphi=\beta/2$

11.6 $E=2233$ V (3868 V line-to-line), $\beta=10.73°$, $I=214.4$ A

11.7 $X=25\ \Omega$, $E=3025$ V (5239 V line-to-line), $\beta=18.55°$, $I=45.28$ A

11.8 **(a)** $E=5936$ V (10.28 kV line-to-line)
(b) $U=10.47$ kV (line-to-line), $P=1243$ kW, $Q=-408.6$ kvar

11.9 9 MW, 1109 A

11.10 $E=204.6$ V (354.3 V line-to-line), $\beta=20.04°$, $P=11.78$ kW, $Q=-7.30$ kvar

11.11 $E=284.2$ V (492 V line-to-line), $\beta=13.4°$, $P=11.78$ kW, $Q=7.30$ kvar

CHAPTER TWELVE

12.1 $n_0=750$ rpm, $n=712.5$ rpm, $f_r=2.5$ Hz, $n_{rr}=37.5$ rpm, $n_r=n+n_{rr}=n_0=750$ rpm

12.2 $n_0=1800$ rpm, $s=2.778\%$, $f_r=1.667$ Hz

12.3 **(a)** $n_0=1000$ rpm
(b) $n=970$ rpm
(c) $f_r=50$ Hz ($s=1$)
(d) $f_r=1.5$ Hz ($s=0.03$)

12.4 4 poles ($p = 2$), $n_0 = 750$ rpm, $s = 4\%$

12.5 $n_0 = 750$ rpm, $s = 5.333\%$, $pf = 0.883$, $T = 297$ Nm, $\eta = 85.91\%$

12.6 $T = 562.5$ Nm

12.7 $s = 1.958$, $f_r = 117.5$ Hz

12.8 $s = 2.87\%$, $n = 1457$ rpm

12.9 $s = 6.94\%$, $T = 56.16$ Nm

12.10 $\hat{s} = 0.4$, then from equation (12.11) $X = 1.25\,\Omega$

$\hat{T} = 612$ Nm, then from equation (12.11) $U_1 = 231.1$ V and $U_{\text{line-to-line}} = 400.3$ V

$s = 0.25$, then from equation (12.10) $T = 550.1$ Nm

12.11 Imposing $\hat{T} = 2T_{\text{starting}}$ with equation (12.11) and equation (12.10), written with $s = 1$: $X = (2 + \sqrt{3})R_r'$

From equation (12.10) (written with $s = 1$) and the same equation (written at full-load slip), by imposing $T_{\text{starting}} = T_{\text{full-load}}$ and $X = (2 + \sqrt{3})R_r'$ we obtain $s_{full\text{-}load} = 7.18\%$

12.12 From equation (12.12), we have

$$\frac{T_{starting}}{T_{full-load}} = \left(\frac{I_{starting}}{I_{full-load}}\right)^2 \frac{s_{full-load}}{s_{starting}} = 8^2 \cdot \frac{0.04}{1} = 2.56$$

With reduced voltage, we obtain

$$\frac{T_{starting}}{T_{full-load}} = \left(\frac{I_{starting}}{I_{full-load}}\right)^2 \frac{s_{full-load}}{s_{starting}} = 4^2 \cdot \frac{0.04}{1} = 0.64$$

The half starting current means half supply voltage, which reduces 4 times the starting torque.

12.13 From equation (12.12), we have

$$\frac{T_{starting}}{T_{full-load}} = \left(\frac{I_{starting}}{I_{full-load}}\right)^2 \frac{s_{full-load}}{s_{starting}} = 6^2 \cdot \frac{0.05}{1} = 1.8$$

Having reduced $\sqrt{3}$ times the voltage and then the starting current, we obtain

$$\frac{T_{starting}}{T_{full-load}} = \left(\frac{I_{starting}}{I_{full-load}}\right)^2 \frac{s_{full-load}}{s_{starting}} = (6/\sqrt{3})^2 \cdot \frac{0.05}{1} = 0.6$$

The result is correct: The starting torque is reduced 3 times, being proportional to U^2.

12.14 **(a)** $R_{phase} = 0.21\,\Omega$, because $R_{seen} = 2/3R_{phase}$. $R_{wye} = R_{phase}/3 = 0.07\,\Omega$
(b) 85%
(c) $P_0 = 956\,W$
(d) $P_0 - 3R_{wye}I_0^2 = 872\,W$, where $3R_{wye}I_0^2$ are copper losses in no-load condition.
(e) $P_{cu} = P_{el} - P_{mech} - 872 = 1768\,W$, whose $3R_{wye}I^2 = 612\,W$ in stator and the remaining 1156 W in rotor.
(f) $P_{ag} = P_{mech} + P_{cur} = 16.15\,kW$; $s = P_{cur}/P_{ag} = 0.071$

12.15 $(R_r + R_{add})/s_{start} = R_r/s_{full\text{-}load}$, then $R_{add} = 2.88\,\Omega$

12.16 **(a)** $R_{phase} = 0.243\,\Omega$, because $R_{seen} = 2/3R_{phase}$. $R_{wye} = R_{phase}/3 = 0.081\,\Omega$
(b) $P_0 - 3R_{wye}I_0^2 = 781.7\,W$
(c) $P_{cus} = 3R_{wye}I^2 = 875\,W$
(d) 17.94 kW
(e) 17.31 kW, 272 Nm
(f) 88.34%

12.17 35.25 A, 196.9 Nm

12.18 **(a)** 227.5 W
(b) 19.81 kW, 195.9 Nm
(c) 90%
(d) 476 μF

12.19 **(a)** 71.4%
(b) 87.6%
(c) 81.5%

12.20 **(a)** 509.3 Nm, 163.3 A, $s = 0.15$
(b) 149.4 Nm, 228.4 A

12.21 **(a)** 2.23%
(b) 34 A
(c) 1467 rpm
(d) 22.75 kW, 148.2 Nm

12.22 **(a)** 1455 rpm
(b) 5630 W, 37.0 Nm
(c) 393.3 V
(d) 5.08 A
(e) 0.139 A

INDEX

Fundamentals of Electric Power Engineering: From Electromagnetics to Power Systems, First Edition.
Massimo Ceraolo and Davide Poli.

Printed and bound by CPI Group (UK) Ltd, Croydon, CR0 4YY
16/09/2026
14963423-0005